CLYMER®

EVINRUDE/JOHNSON
OUTBOARD SHOP MANUAL
48-235 HP • 1973-1990

The world's finest publisher of mechanical how-to manuals

INTERTEC PUBLISHING

P.O. Box 12901, Overland Park, Kansas 66282-2901

Copyright ©1991 Intertec Publishing

FIRST EDITION
First Printing February, 1985

SECOND EDITION
Updated to include 1985 models
First Printing February, 1986

THIRD EDITION
Updated to include 1986 models
First Printing December, 1986

FOURTH EDITION
Updated to include 1987 models
First Printing June, 1987
Second Printing December, 1987
Third Printing June, 1988

FIFTH EDITION
First Printing February, 1989
Second Printing June, 1989
Third Printing November, 1989

SIXTH EDITION
Updated to include 1988 and 1989 models
First Printing March, 1990
Second Printing October, 1990

SEVENTH EDITION
Updated to include 1990 models
First Printing July, 1991
Second Printing January, 1992
Third Printing October, 1992
Fourth Printing July, 1993
Fifth Printing February, 1994
Sixth Printing November, 1994
Seventh Printing July, 1995
Eighth Printing August, 1996
Ninth Printing July, 1997
Tenth Printing, June, 1998
Eleventh Printing June, 1999
Twelfth Printing June, 2000

Printed in U.S.A.

CLYMER and colophon are registered trademarks of Intertec Publishing.

ISBN: 0-89287-555-0

Library of Congress: 91-55572

Technical illustrations by Steve Amos, Diana Kirkland, Mitzi McCarthy and Carl Rohkar.

Thanks to Marine Specialties, Sun Valley, California, and Ken's Boat Center, Burbank, California.

Tools shown in Chapter Two courtesy of Thorsen Tool, Dallas, Texas. Test equipment shown in Chapter Two courtesy of Dixson, Inc., Grand Junction, Colorado.

COVER: Photographed by Michael Brown Photographic Productions, Los Angeles, California. Assisted by Bill Masho.
- *Boat driven by Dubie.*
- *Photo boat courtesy of Cypress Gardens, Florida, and driven by Mike Monts de Oca.*
- *Thanks to Jerry Imber for letting us have the Ranger first.*

All rights reserved. Reproduction or use, without express permission, of editorial or pictorial content, in any manner, is prohibited. No patent liability is assumed with respect to the use of the information contained herein. While every precaution has been taken in the preparation of this book, the publisher assumes no responsibility for errors or omissions. Neither is any liability assumed for damages resulting from use of the information contained herein. Publication of the servicing information in this manual does not imply approval of the manufacturers of the products covered.

All instructions and diagrams have been checked for accuracy and ease of application; however, success and safety in working with tools depend to a great extent upon individual accuracy, skill and caution. For this reason, the publishers are not able to guarantee the result of any procedure contained herein. Nor can they assume responsibility for any damage to property or injury to persons occasioned from the procedures. Persons engaging in the procedure do so entirely at their own risk.

General Information	1
Tools and Techniques	2
Troubleshooting	3
Lubrication, Maintenance and Tune-up	4
Engine Synchronization and Linkage Adjustments	5
Fuel System	6
Ignition and Electrical Systems	7
Power Head	8
Gearcase	9
Pwer Trim and Tilt Systems	10
Oil Injection Systems	11
Sea Drives	12
Index	13
Wiring Diagrams	14

Intertec Book Division

President Cameron Bishop
Executive Vice President of Operations/CFO Dan Altman
Senior Vice President, Book Division Ted Marcus

EDITORIAL

Director of Price Guides
Tom Fournier

Senior Editor
Mark Jacobs

Editors
Mike Hall
Frank Craven
Paul Wyatt

Associate Editors
Robert Sokol
Carl Janssens
James Grooms

Technical Writers
Ron Wright
Ed Scott
George Parise
Mark Rolling
Michael Morlan
Jay Bogart
Ronney Broach

Inventory and Production Manager
Shirley Renicker

Editorial Production Supervisor
Dylan Goodwin

Editorial Production Coordinator
Sandy Kreps

Editorial Production Assistants
Greg Araujo
Dennis Conrow
Shara Meyer
Susan Hartington

Technical Illustrators
Steve Amos
Robert Caldwell
Mitzi McCarthy
Michael St. Clair
Mike Rose

MARKETING/SALES AND ADMINISTRATION

General Manager, Technical and Specialty Books
Michael Yim

General Manager, AC-U-KWIK
Randy Stephens

Advertising Production Coordinator
Kim Sawalich

Advertising Coordinator
Jodi Donohoe

Advertising/Editorial Assistant
Janet Rogers

Advertising & Promotions Manager
Elda Starke

Senior Art Director
Andrew Brown

Marketing Assistant
Melissa Abbott

Associate Art Director
Chris Paxton

Sales Manager/Marine
Dutch Sadler

Sales Manager/Manuals
Ted Metzger

Sales Manager/Motorcycles
Matt Tusken

Sales Coordinator
Paul Cormaci

Telephone Sales Supervisor
Joelle Stephens

Telemarketing Sales Representative
Susan Kay

Customer Service/Fulfillment Manager
Caryn Bair

Fulfillment Coordinator
Susan Kohlmeyer

Customer Service Supervisor
Terri Cannon

Customer Service Representatives
Ardelia Chapman
Donna Schemmel
Dana Morrison
April LeBlond

The following books and guides are published by Intertec Publishing.

CLYMER SHOP MANUALS
Boat Motors and Drives
Motorcycles and ATVs
Snowmobiles
Personal Watercraft

ABOS/INTERTEC/CLYMER BLUE BOOKS AND TRADE-IN GUIDES
Recreational Vehicles
Outdoor Power Equipment
Agricultural Tractors
Lawn and Garden Tractors
Motorcycles and ATVs
Snowmobiles and Personal Watercraft
Boats and Motors

AIRCRAFT BLUEBOOK-PRICE DIGEST
Airplanes
Helicopters

AC-U-KWIK DIRECTORIES
The Corporate Pilot's Airport/FBO Directory
International Manager's Edition
Jet Book

I&T SHOP SERVICE MANUALS
Tractors

INTERTEC SERVICE MANUALS
Snowmobiles
Outdoor Power Equipment
Personal Watercraft
Gasoline and Diesel Engines
Recreational Vehicles
Boat Motors and Drives
Motorcycles
Lawn and Garden Tractors

Contents

QUICK REFERENCE DATA IX

CHAPTER ONE
GENERAL INFORMATION 1
- Manual organization 1
- Notes, cautions and warnings 1
- Torque specifications 2
- Engine operation 2
- Fasteners 2
- Lubricants 8
- Gasket sealant 10
- Galvanic corrosion 11
- Protection from galvanic corrosion 13
- Propellers 14

CHAPTER TWO
TOOLS AND TECHNIQUES 21
- Safety first 21
- Basic hand tools 21
- Test equipment 26
- Service hints 28
- Special tips 30
- Mechanic's techniques 31

CHAPTER THREE
TROUBLESHOOTING 33
- Operating requirements 34
- Starting system 34
- Charging system 39
- Ignition system 45
- 1973-1977 CD 2 ignition troubleshooting 46
- 1978-1984 CD 2 ignition troubleshooting 55
- 1985-1988 CD 2 ignition troubleshooting 60
- 1989-1990 CD2USL ignition troubleshooting 65
- 1973-1978 CD 3 ignition troubleshooting 69
- 1976-1978 CD 6 ignition troubleshooting 75
- 1979-1984 CD 3 and 1979-1984 CD 6 ignition troubleshooting 82
- 1985-1988 CD 3 ignition troubleshooting 87
- 1989-1990 CD 3 ignition troubleshooting 94
- 1973-1977 CD 4 ignition troubleshooting 100
- 1978-1984 CD 4 ignition troubleshooting 109
- 1987 88 hp, 1985-1987 90 hp, 1985 115 hp and 1986-1987 110 hp CD 4 ignition troubleshooting 115
- 1988 88 hp, 90 hp and 110 hp CD 4 ignition troubleshooting 121
- 1989-1990 88 hp, 90 hp, 110 hp and 115 hp CD 4 ignition troubleshooting 126
- 1985-1987 120 and 140 hp CD 4 ignition troubleshooting 130
- 1988-1990 120 and 140 hp CD 4 ignition troubleshooting 135
- 1985 V6 and 1986-1988 150-175 hp CD 6 ignition troubleshooting 143
- 1989-1990 150 and 175 hp CD 6 ignition troubleshooting 148
- 1986-1987 200-225 hp CD ignition troubleshooting 161
- 1988-1990 200-225 CD ignition troubleshooting 167
- Ignition and neutral start switch 176
- Neutral start switch 177
- Fuel system 177
- Engine temperature and overheating 180
- Engine 182

CHAPTER FOUR
LUBRICATION, MAINTENANCE AND TUNE-UP 193
- Lubrication 193
- Storage 200
- Complete submersion 202
- Anti-corrosion maintenance 203
- Engine flushing 203
- Tune-up 204

CHAPTER FIVE
ENGINE SYNCHRONIZATION AND LINKAGE ADJUSTMENTS 217
- Engine timing 217
- Synchronizing 217
- Required equipment 217
- 1973-1988 50 hp (tiller models) 219
- 1973-1988 48, 50, 55 and 60 hp (2-cylinder remote control models) 223
- 1989-1990 48 and 50 hp models 227
- 60, 65, 70 and 75 hp (3-cylinder models) 230
- 65 hp (3-cylinder tiller models) 234
- V4 engines (except 1985-on 120-140 hp and 1986-on 88-115 hp) 238
- V4 engines (1986-on 88-115 hp) 243
- V4 engines (1985-on 120-140 hp) 246
- V6 engines (1975-1985) 252
- V6 engines (1986-on 150-175 hp) 257
- V6 engines (1986-on 200-225 hp) 259

CHAPTER SIX
FUEL SYSTEM 265
- Fuel pump 265
- Carburetors 268
- Carburetor (48-75 hp) 271
- V4 and V6 carburetor (except 1985-on 120-140 hp and 1986-on 200-225 hp) 275
- V4 and V6 carburetor (1985-on 120-140 hp and 1986-on 200-225 hp) 281
- Choke and primer solenoid service 284
- Anti-siphon devices 286
- Fuel tank 286
- Fuel line and primer bulb 286

CHAPTER SEVEN
IGNITION AND ELECTRICAL SYSTEMS 292
- Battery 292
- Battery charging system 299
- Electric starting system 302
- Starter motor 302
- Ignition system 306

CHAPTER EIGHT
POWER HEAD 317
- Engine serial number 318
- Fasteners and torque 319
- Gaskets and sealants 319
- Flywheel 319
- Power head 320
- Reed block service 364
- Thermostat service 365

CHAPTER NINE
GEARCASE 375
- Propeller 375
- Water pump 376
- Gearcase removal/installation (all models) 379
- Hydro-mechanical gearcase 381
- Mechanical gearcase (except 1989-1990 48 and 50 hp) 397
- Pressure and vacuum test 416

CHAPTER TEN
POWER TRIM AND TILT SYSTEMS 420
External power trim and tilt system............421
Internal power trim and tilt system 427
Trim and tilt motor testing 445
System removal/installation....................445

CHAPTER ELEVEN
OIL INJECTION SYSTEMS 448
OMC Economixer..........................448
OMC VRO system 455

CHAPTER TWELVE
SEA DRIVES 461
SelecTrim/Tilt ™ (1983-1990 1.6L) 461
SelecTrim/Tilt ™ (1987-1988 1.8L, 1989-1990 2.0L, 1984-1985 2.5L [S-type], 1984-1985 2.6L [S-type], 1986-1988 2.6L, 1987-1988 2.7L and 1989-1990 3.0L) 469

INDEX 475

WIRING DIAGRAMS 478

Quick Reference Data

TIMING SPECIFICATIONS

	Full Throttle Spark Advance degrees BTDC	Throttle Pickup degrees BTDC
48 hp		
1987-1988	19 ±1	2-4
1989-1990	19 ±1	—
50 hp		
1973-1975	19 ±1	3
1980-1981	19	2-4
1982-1988	19 ±1	2-4
1989-1990	19 ±1	—
55 hp		
1975-1976	19	4
1977-1979	19	3
1980-1981	19	2-4
1982-1983	19 ±1	2-4
60 hp		
1980-1981	21	2-4
1982-1985	21 ±1	2-4
1986-on	19 ±1	1 ±1 ATDC
65 hp		
1973	22	8
70 hp		
1974	20	TDC
1975-1979	17	TDC
1980-1981	19	TDC
1982	19 ±1	TDC
1983-1985	See note 1	TDC ±1
1986-1987	19 ±1	4 ±1 ATDC
1988	19 ±1	1 ±1 ATDC
1989-1990	19 ±1	1 ±1 BTDC
75 hp		
1975-1979	16	TDC
1980-1981	19	TDC
1982	19 ±1	TDC
1983-1985	See note 1	TDC ±1
1986-1987	19 ±1	1 ±1 ATDC
85 hp		
1973	28	5
1974	28	4
1975-1976	26	4
1977-1979	26	3-5
1980	28	4-6
88 hp	28 ±1	4 ±1
90 hp		
1981	28	4-6
1982	28 ±1	4-6
1983-1985	See note 1	3-5
1986-on	28 ±1	4 ±1
100 hp	28	4-6
110 hp	28 ±1	4 ±1
115 hp		
1973	26	5
1974	26	4

(continued)

TIMING SPECIFICATIONS (continued)

	Full Throttle Spark Advance degrees BTDC	Throttle Pick-up degrees BTDC
115 hp (continued)		
1975-1976	24	4
1977-1979	28	0-3
1980-1981	28	4-6
1982	28 ±1	4-6
1983-1984	See note 1	3-5
1990	28 ±1	4 ±1
120 hp		
1985	22	—
1986-on	18 ±1	—
135 hp		
1973	22	5
1974	22	4
1975-1976	20	4
140 hp		
1977-1979	28	0-3
1980-1981	28	4-6
1982	28 ±1	4-6
1983-1984	See note 1	3-5
1985	22	—
1986-on	18 ±1	—
150 hp		
1977-1979	26	6-8
1980-1981	28[2]	6-8
1982	28 ±1	6-8
1983-1985	See note 1	6-8
1986-on	32 ±1[3]	7 ±1
175 hp		
1977-1979	28	5
1980-1981	28[2]	6-8
1982	28 ±1	6-8
1983-1985	See note 1	6-8
1986-on	28 ±1	7 ±1
185 hp	See note 1	6-8
200 hp		
1975-1979	28	5
1980-1981	26[4]	6-8
1982	28 ±1	6-8
1983-1985	See note 1	6-8
1986-on	18 ±1	—
225 hp		
1986-on	18 ±1	—
235 hp		
1977-1979	28	6-8
1980-1981	30	6-8
1982	30 ±1	6-8
1983-1985	See note 1	6-8

1. Refer to timing decal on engine.
2. If timer base has red sleeves, 32° BTDC.
3. 1986-1987 150STL, 30 ±1° BTDC; 1988-1989 150STL, 28 ±1° BTDC.
4. 1981 CH model: timer base with red sleeves, 30° BTDC; 1981 CIA, CIB and CIM model: timer base with black sleeves, 28° BTDC.

RECOMMENDED SPARK PLUGS

Engine	Champion plug type	Gap (in.)
48 hp		
1987-1988	See note 1	See note 1
1989-1990	See note 2	See note 2
50 hp		
1973-1974	UL77V	—
1975-1984	L77J4	0.040
1985-1988	See note 1	See note 1
1989-1990	See note 2	See note 2
55 hp	L77J4	0.040
60 hp		
1980-1984	L77J4	0.040
1985-on	See note 1	See note 1
65 hp	UL74V	—
70 & 75 hp		
1974	UL77V	—
1975-1984	L77J4	0.040
1985-on	See note 1	See note 1
85 hp		
1973-1974	UL77V	—
1975-1980	L77J4	0.040
88 hp	See note 3	See note 3
90 hp		
1981-1984	L77J4	0.040
1985-1986	See note 1	See note 1
1987-1990	See note 3	See note 3
100 hp		
1979-1980	L77J4	0.040
110 hp		
1986	See note 1	See note 1
1987-1990	See note 3	See note 3
115 hp		
1973-1974	UL77V	—
1975-1984	L77J4	0.040
1990	See note 3	See note 3
120 hp	See note 1	See note 1
135 hp	UL77V	—
140 hp		
1977-1984	UL77V	—
1985-on	See note 1	See note 1
150 hp		
1977-1986	QUL77V or UL77V	—
1987-1990	See note 3	See note 3
175 hp		
1977-1986	QUL77V or UL77V	—
1987-1990	See note 3	See note 3
185 hp	QUL77V or UL77V	—
200 hp		
1975-1986	QUL77V or UL77V	—
1987-1990	See note 3	See note 3

(continued)

RECOMMENDED SPARK PLUGS (continued)

Engine	Champion plug type	Gap (in.)
225 hp	See note 3	See note 3
235 hp	QUL77V or UL77V	—

1. OMC recommends the use of Champion QL77JC4 or L77JC4 plugs gapped to 0.040 in. for sustained low-speed operation. For sustained high-speed use, OMC recommends Champion QL78V or L78V (non-adjustable gap).
2. OMC recommends the use of Champion QL78C plugs gapped to 0.030 in. for sustained low-speed operation. For sustained high-speed use, OMC recommends Champion QL16V (non-adjustable gap).
3. OMC recommends the use of Champion QL77JC4 or L77JC4 plugs gapped to 0.040 in. on models prior to 1989 and 0.030 in. on 1989-1990 models for sustained low-speed operation. For sustained high-speed use, OMC recommends Champion QUL77V or UL77V (non-adjustable gap).

BATTERY CAPACITY (HOURS)

Accessory draw	80 Amp-hour battery provides continuous power for:	Approximate recharge time
5 amps	13.5 hours	16 hours
15 amps	3.5 hours	13 hours
25 amps	1.8 hours	12 hours

Accessory draw	105 Amp-hour battery provides continuous power for:	Approximate recharge time
5 amps	15.8 hours	16 hours
15 amps	4.2 hours	13 hours
25 amps	2.4 hours	12 hours

IGNITION TROUBLESHOOTING

Symptom	Probable cause
Engine won't start, but fuel and spark are good	Defective or dirty spark plugs. Spark plug gap set too wide. Improper spark timing. Shorted stop button. Air leaks into fuel pump. Broken piston ring(s). Cylinder head, crankcase or cylinder sealing faulty. Worn crankcase oil seal.
Engine misfires @ idle	Incorrect spark plug gap. Defective, dirty or loose spark plugs. Spark plugs of incorrect heat range. Leaking or broken high tension wires. Weak armature magnets. Defective coil or condenser. Defective ignition switch. Spark timing out of adjustment.

(continued)

IGNITION TROUBLESHOOTING (continued)

Symptom	Probable cause
Engine misfires @ high speed	See "Engine misfires @ idle." Coil breaks down. Coil shorts through insulation. Spark plug gap too wide. Wrong type spark plugs. Too much spark advance.
Engine backfires	Cracked spark plug insulator. Improper timing. Crossed spark plug wires. Improper ignition timing.
Engine preignition	Spark advanced too far. Incorrect type spark plug. Burned spark electrodes.
Engine noises (knocking at power head)	Spark advanced too far.
Ignition coil fails	Extremely high voltage. Moisture formation. Excessive heat from engine.
Spark plugs burn and foul	Incorrect type plug. Fuel mixture too rich. Inferior grade of gasoline. Overheated engine. Excessive carbon in combustion chambers.
Ignition causing high fuel consumption	Incorrect spark timing. Leaking high tension wires. Incorrect spark plug gap. Fouled spark plugs. Incorrect spark advance. Weak ignition coil. Preignition.

APPROXIMATE STATE OF CHARGE

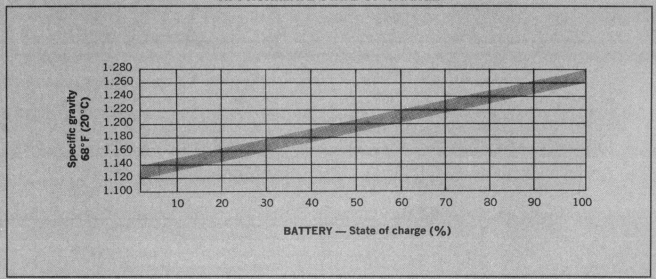

Introduction

This Clymer shop manual covers service and repair of all Evinrude/Johnson 2- and 3-cylinder inline engines (48 hp and above) and all V4 and V6 (85-235 hp) engines designed for recreational use from 1973-1990. Specific Sea Drive information is covered in Chapter 12. It does not cover similar displacement engines designed expressly for racing, sailing or commercial use. Step-by-step instructions and hundreds of illustrations guide you through jobs ranging from simple maintenance to complete overhaul.

This manual can be used by anyone from a first time owner/amateur to a professional mechanic. Easy to read type, detailed drawings and clear photographs give you all the information you need to do the work right.

Having a well-maintained engine will increase your enjoyment of your boat as well as assuring your safety offshore. Keep this shop manual handy and use it often. It can save you hundreds of dollars in maintenance and repair bills and make yours a reliable, top-performing boat.

Chapter One

General Information

This detailed, comprehensive manual contains complete information on maintenance, tune-up, repair and overhaul. Hundreds of photos and drawings guide you through every step-by-step procedure.

Troubleshooting, tune-up, maintenance and repair are not difficult if you know what tools and equipment to use and what to do. Anyone not afraid to get their hands dirty, of average intelligence and with some mechanical ability, can perform most of the procedures in this book. See Chapter Two for more information on tools and techniques.

A shop manual is a reference. You want to be able to find information fast. Clymer books are designed with you in mind. All chapters are thumb tabbed and important items are indexed at the end of the book. All procedures, tables, photos, etc., in this manual assume that the reader may be working on the machine or using this manual for the first time.

Keep this book handy in your tool box. It will help you to better understand how your machine runs, lower repair and maintenance costs and generally increase your enjoyment of your marine equipment.

MANUAL ORGANIZATION

This chapter provides general information useful to marine owners and mechanics.

Chapter Two discusses the tools and techniques for preventive maintenance, troubleshooting and repair.

Chapter Three describes typical equipment problems and provides logical troubleshooting procedures.

Following chapters describe specific systems, providing disassembly, repair, assembly and adjustment procedures in simple step-by-step form. Specifications concerning a specific system are included at the end of the appropriate chapter.

NOTES, CAUTIONS AND WARNINGS

The terms NOTE, CAUTION and WARNING have specific meanings in this manual. A NOTE provides additional information to make a step or procedure easier or clearer. Disregarding a NOTE could cause inconvenience, but would not cause damage or personal injury.

A CAUTION emphasizes areas where equipment damage could result. Disregarding a CAUTION could cause permanent mechanical damage; however, personal injury is unlikely.

A WARNING emphasizes areas where personal injury or even death could result from negligence. Mechanical damage may also occur. WARNINGS *are to be taken seriously.* In some cases, serious injury or death has resulted from disregarding similar warnings.

TORQUE SPECIFICATIONS

Torque specifications throughout this manual are given in foot-pounds (ft.-lb.) and either Newton meters (N·m) or meter-kilograms (mkg). Newton meters are being adopted in place of meter-kilograms in accordance with the International Modernized Metric System. Existing torque wrenches calibrated in meter-kilograms can be used by performing a simple conversion: move the decimal point one place to the right. For example, 4.7 mkg = 47 N·m. This conversion is accurate enough for mechanics' use even though the exact mathematical conversion is 3.5 mkg = 34.3 N·m.

ENGINE OPERATION

All marine engines, whether 2- or 4-stroke, gasoline or diesel, operate on the Otto cycle of intake, compression, power and exhaust phases.

4-stroke Cycle

A 4-stroke engine requires two crankshaft revolutions (4 strokes of the piston) to complete the Otto cycle. **Figure 1** shows gasoline 4-stroke engine operation. **Figure 2** shows diesel 4-stroke engine operation.

2-stroke Cycle

A 2-stroke engine requires only 1 crankshaft revolution (2 strokes of the piston) to complete the Otto cycle. **Figure 3** shows gasoline 2-stroke engine operation. Although diesel 2-strokes exist, they are not commonly used in light marine applications.

FASTENERS

The material and design of the various fasteners used on marine equipment are not arrived at by chance or accident. Fastener design determines the type of tool required to work with the fastener. Fastener material is carefully selected to decrease the possibility of physical failure or corrosion. See *Galvanic Corrosion* in this chapter for more information on marine materials.

Threads

Nuts, bolts and screws are manufactured in a wide range of thread patterns. To join a nut and bolt, the diameter of the bolt and the diameter of the hole in the nut must be the same. It is just as important that the threads on both be properly matched.

The best way to determine if the threads on two fasteners are matched is to turn the nut on the bolt (or the bolt into the threaded hole in a piece of equipment) with fingers only. Be sure both pieces are clean. If much force is required, check the thread condition on each fastener. If the thread condition is good but the fasteners jam, the threads are not compatible.

Four important specifications describe every thread:
   a. Diameter.
   b. Threads per inch.
   c. Thread pattern.
   d. Thread direction.

Figure 4 shows the first two specifications. Thread pattern is more subtle. Italian and British

GENERAL INFORMATION

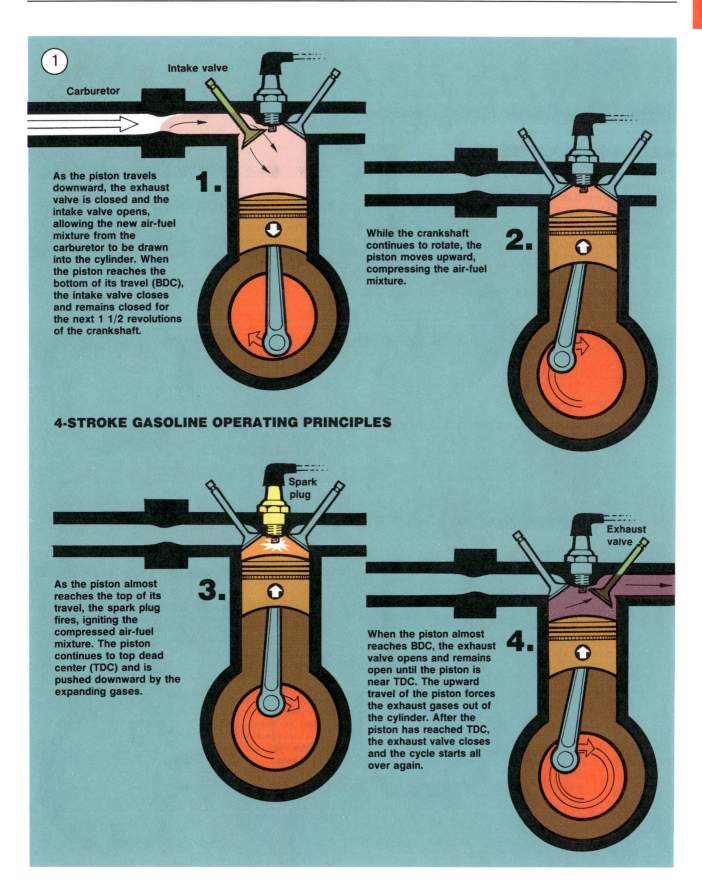

4-STROKE GASOLINE OPERATING PRINCIPLES

1. As the piston travels downward, the exhaust valve is closed and the intake valve opens, allowing the new air-fuel mixture from the carburetor to be drawn into the cylinder. When the piston reaches the bottom of its travel (BDC), the intake valve closes and remains closed for the next 1 1/2 revolutions of the crankshaft.

2. While the crankshaft continues to rotate, the piston moves upward, compressing the air-fuel mixture.

3. As the piston almost reaches the top of its travel, the spark plug fires, igniting the compressed air-fuel mixture. The piston continues to top dead center (TDC) and is pushed downward by the expanding gases.

4. When the piston almost reaches BDC, the exhaust valve opens and remains open until the piston is near TDC. The upward travel of the piston forces the exhaust gases out of the cylinder. After the piston has reached TDC, the exhaust valve closes and the cycle starts all over again.

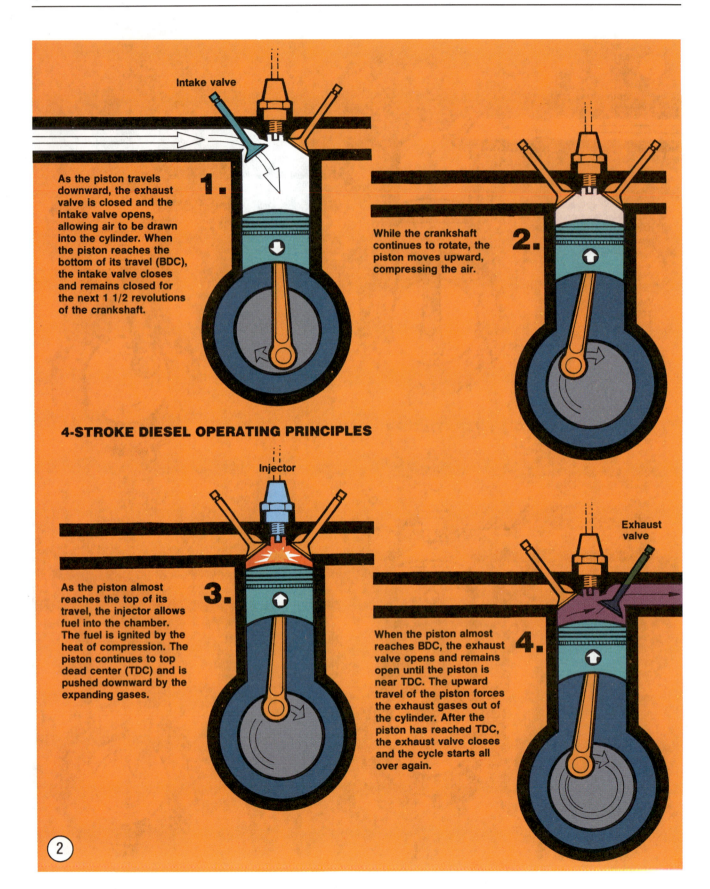

GENERAL INFORMATION

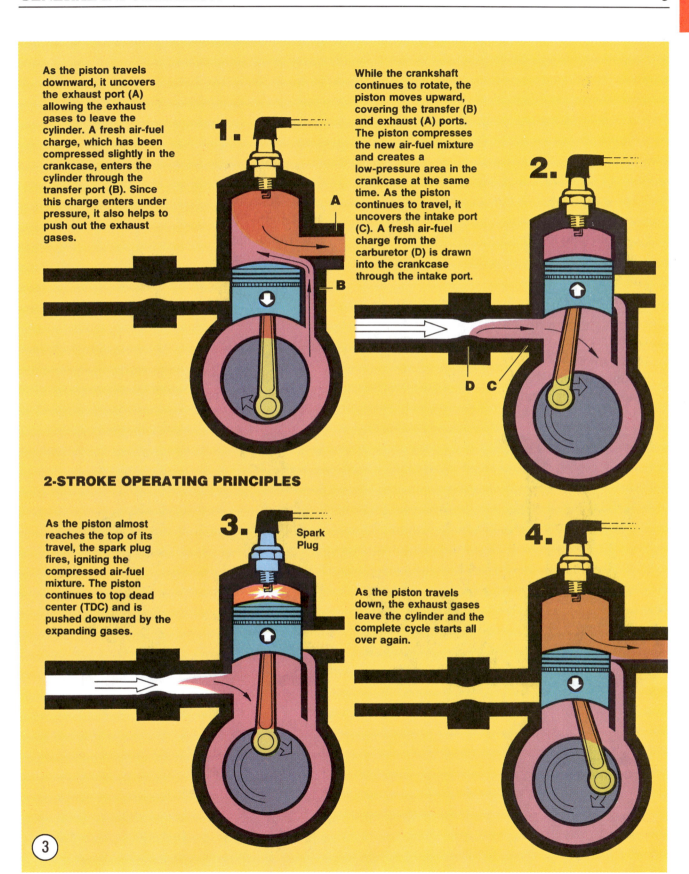

2-STROKE OPERATING PRINCIPLES

1. As the piston travels downward, it uncovers the exhaust port (A) allowing the exhaust gases to leave the cylinder. A fresh air-fuel charge, which has been compressed slightly in the crankcase, enters the cylinder through the transfer port (B). Since this charge enters under pressure, it also helps to push out the exhaust gases.

2. While the crankshaft continues to rotate, the piston moves upward, covering the transfer (B) and exhaust (A) ports. The piston compresses the new air-fuel mixture and creates a low-pressure area in the crankcase at the same time. As the piston continues to travel, it uncovers the intake port (C). A fresh air-fuel charge from the carburetor (D) is drawn into the crankcase through the intake port.

3. As the piston almost reaches the top of its travel, the spark plug fires, igniting the compressed air-fuel mixture. The piston continues to top dead center (TDC) and is pushed downward by the expanding gases.

4. As the piston travels down, the exhaust gases leave the cylinder and the complete cycle starts all over again.

standards exist, but the most commonly used by marine equipment manufacturers are American standard and metric standard. The threads are cut differently as shown in **Figure 5**.

Most threads are cut so that the fastener must be turned clockwise to tighten it. These are called right-hand threads. Some fasteners have left-hand threads; they must be turned counterclockwise to be tightened. Left-hand threads are used in locations where normal rotation of the equipment would tend to loosen a right-hand threaded fastener.

Machine Screws

There are many different types of machine screws. **Figure 6** shows a number of screw heads requiring different types of turning tools (see Chapter Two for detailed information). Heads are also designed to protrude above the metal (round) or to be slightly recessed in the metal (flat) (**Figure 7**).

Bolts

Commonly called bolts, the technical name for these fasteners is cap screw. They are normally described by diameter, threads per inch and length. For example, 1/4-20 × 1 indicates a bolt 1/4 in. in diameter with 20 threads per inch, 1 in. long. The measurement across two flats on the head of the bolt indicates the proper wrench size to be used.

Nuts

Nuts are manufactured in a variety of types and sizes. Most are hexagonal (6-sided) and fit

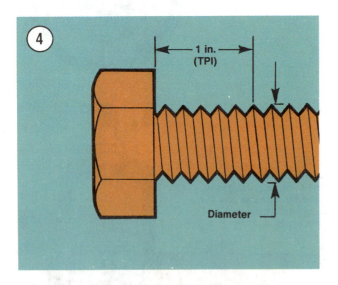

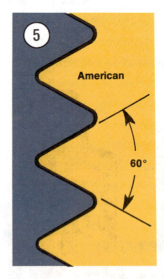

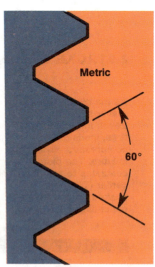

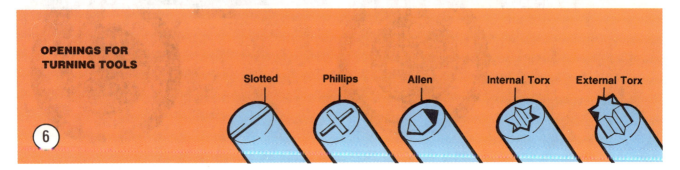

GENERAL INFORMATION

on bolts, screws and studs with the same diameter and threads per inch.

Figure 8 shows several types of nuts. The common nut is usually used with a lockwasher. Self-locking nuts have a nylon insert that prevents the nut from loosening; no lockwasher is required. Wing nuts are designed for fast removal by hand. Wing nuts are used for convenience in non-critical locations.

To indicate the size of a nut, manufacturers specify the diameter of the opening and the threads per inch. This is similar to bolt specification, but without the length dimension. The measurement across two flats on the nut indicates the proper wrench size to be used.

Washers

There are two basic types of washers: flat washers and lockwashers. Flat washers are simple discs with a hole to fit a screw or bolt. Lockwashers are designed to prevent a fastener from working loose due to vibration, expansion and contraction. **Figure 9** shows several types of lockwashers. Note that flat washers are often used between a lockwasher and a fastener to provide a smooth bearing surface. This allows the fastener to be turned easily with a tool.

Cotter Pins

Cotter pins (**Figure 10**) are used to secure special kinds of fasteners. The threaded stud

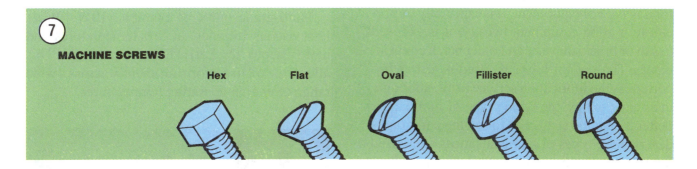

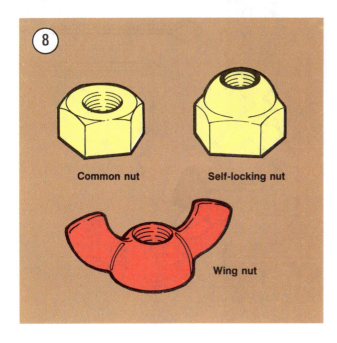

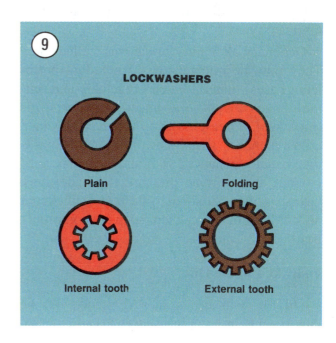

must have a hole in it; the nut or nut lock piece has projections that the cotter pin fits between. This type of nut is called a "Castellated nut." Cotter pins should not be reused after removal.

Snap Rings

Snap rings can be of an internal or external design. They are used to retain items on shafts (external type) or within tubes (internal type). Snap rings can be reused if they are not distorted during removal. In some applications, snap rings of varying thickness can be selected to control the end play of parts assemblies.

LUBRICANTS

Periodic lubrication ensures long service life for any type of equipment. It is especially important to marine equipment because it is exposed to salt or brackish water and other harsh environments. The *type* of lubricant used is just as important as the lubrication service itself; although, in an emergency, the wrong type of lubricant is better than none at all. The following paragraphs describe the types of lubricants most often used on marine equipment. Be sure to follow the equipment manufacturer's recommendations for lubricant types.

Generally, all liquid lubricants are called "oil." They may be mineral-based (including petroleum bases), natural-based (vegetable and animal bases), synthetic-based or emulsions (mixtures). "Grease" is an oil which is thickened with a metallic "soap." The resulting material is then usually enhanced with anticorrosion, antioxidant and extreme pressure (EP) additives. Grease is often classified by the type of thickener added; lithium and calcium soap are commonly used.

4-stroke Engine Oil

Oil for 4-stroke engines is graded by the American Petroleum Institute (API) and the Society of Automotive Engineers (SAE) in several categories. Oil containers display these ratings on the top or label (**Figure 11**).

API oil grade is indicated by letters, oils for gasoline engines are identified by an "S" and oils for diesel engines are identified by a "C." Most modern gasoline engines require SF or SG graded oil. Automotive and marine diesel engines use CC or CD graded oil.

Viscosity is an indication of the oil's thickness, or resistance to flow. The SAE uses numbers to indicate viscosity; thin oils have low numbers and thick oils have high numbers. A "W" after the number indicates that the viscosity testing was done at low temperature to simulate cold weather operation. Engine oils fall into the 5W-20W and 20-50 range.

Multi-grade oils (for example, 10W-40) are less viscous (thinner) at low temperatures and more viscous (thicker) at high temperatures. This allows the oil to perform efficiently across a wide range of engine operating temperatures.

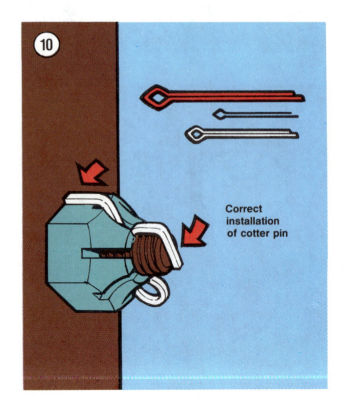

Correct installation of cotter pin

GENERAL INFORMATION

2-stroke Engine Oil

Lubrication for a 2-stroke engine is provided by oil mixed with the incoming fuel-air mixture. Some of the oil mist settles out in the crankcase, lubricating the crankshaft and lower end of the connecting rods. The rest of the oil enters the combustion chamber to lubricate the piston, rings and cylinder wall. This oil is then burned along with the fuel-air mixture during the combustion process.

Engine oil must have several special qualities to work well in a 2-stroke engine. It must mix easily and stay in suspension in gasoline. When burned, it can't leave behind excessive deposits. It must also be able to withstand the high temperatures associated with 2-stroke engines.

The National Marine Manufacturer's Association (NMMA) has set standards for oil used in 2-stroke, water-cooled engines. This is the NMMA TC-W (two-cycle, water-cooled) grade (**Figure 12**). The oil's performance in the following areas is evaluated:

a. Lubrication (prevention of wear and scuffing).
b. Spark plug fouling.
c. Preignition.
d. Piston ring sticking.
e. Piston varnish.
f. General engine condition (including deposits).
g. Exhaust port blockage.
h. Rust prevention.
i. Mixing ability with gasoline.

In addition to oil grade, manufacturers specify the ratio of gasoline to oil required during break-in and normal engine operation.

Gear Oil

Gear lubricants are assigned SAE viscosity numbers under the same system as 4-stroke engine oil. Gear lubricant falls into the SAE 72-250 range (**Figure 13**). Some gear lubricants are multi-grade; for example, SAE 85W-90.

Three types of marine gear lubricant are generally available: SAE 90 hypoid gear lubricant is designed for older manual-shift units; Type C gear lubricant contains additives designed for electric shift mechanisms; High viscosity gear lubricant is a heavier oil designed to withstand the shock loading of high-performance engines or units subjected to severe duty use. Always use a gear lubricant of the type specified by the unit's manufacturer.

Grease

Greases are graded by the National Lubricating Grease Institute (NLGI). Greases are graded by number according to the consistency of the grease; these ratings range from No. 000 to No. 6, with No. 6 being the most solid. A typical multipurpose grease is NLGI No. 2 (**Figure 14**). For specific applications, equipment manufacturers may require grease with an additive such as molybdenum disulfide (MOS^2).

GASKET SEALANT

Gasket sealant is used instead of pre-formed gaskets on some applications, or as a gasket dressing on others. Two types of gasket sealant are commonly used: room temperature vulcanizing (RTV) and anaerobic. Because these two materials have different sealing properties, they cannot be used interchangeably.

RTV Sealant

This is a silicone gel supplied in tubes (**Figure 15**). Moisture in the air causes RTV to cure. Always place the cap on the tube as soon as possible when using RTV. RTV has a shelf life of one year and will not cure properly when the shelf life has expired. Check the expiration date

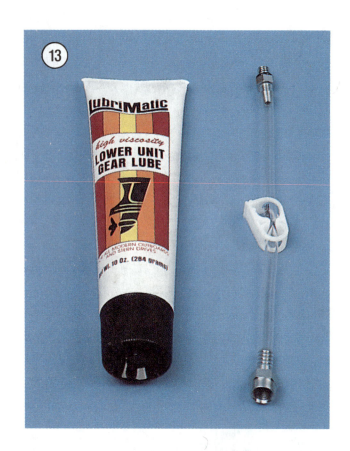

GENERAL INFORMATION

on RTV tubes before using and keep partially used tubes tightly sealed. RTV sealant can generally fill gaps up to 1/4 in. (6.3 mm) and works well on slightly flexible surfaces.

Applying RTV Sealant

Clean all gasket residue from mating surfaces. Surfaces should be clean and free of oil and dirt. Remove all RTV gasket material from blind attaching holes because it can create a "hydraulic" effect and affect bolt torque.

Apply RTV sealant in a continuous bead 2-3 mm (0.08-0.12 in.) thick. Circle all mounting holes unless otherwise specified. Torque mating parts within 10 minutes after application.

Anaerobic Sealant

This is a gel supplied in tubes (**Figure 16**). It cures only in the absence of air, as when squeezed tightly between two machined mating surfaces. For this reason, it will not spoil if the cap is left off the tube. It should not be used if one mating surface is flexible. Anaerobic sealant is able to fill gaps up to 0.030 in. (0.8 mm) and generally works best on rigid, machined flanges or surfaces.

Applying Anaerobic Sealant

Clean all gasket residue from mating surfaces. Surfaces must be clean and free of oil and dirt. Remove all gasket material from blind attaching holes, as it can cause a "hydraulic" effect and affect bolt torque.

Apply anaerobic sealant in a 1 mm or less (0.04 in.) bead to one sealing surface. Circle all mounting holes. Torque mating parts within 15 minutes after application.

GALVANIC CORROSION

A chemical reaction occurs whenever two different types of metal are joined by an electrical conductor and immersed in an electrolyte. Electrons transfer from one metal to the other through the electrolyte and return through the conductor.

The hardware on a boat is made of many different types of metal. The boat hull acts as a conductor between the metals. Even if the hull is wooden or fiberglass, the slightest film of water (electrolyte) within the hull provides conductivity. This combination creates a good environment for electron flow (**Figure 17**). Unfortunately, this electron flow results in galvanic corrosion of the metal involved, causing one of the metals to be corroded or eaten away

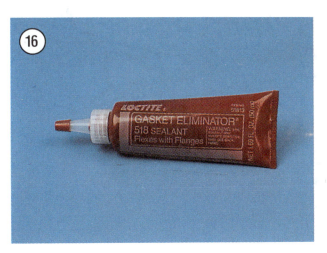

by the process. The amount of electron flow (and, therefore, the amount of corrosion) depends on several factors:

a. The types of metal involved.
b. The efficiency of the conductor.
c. The strength of the electrolyte.

Metals

The chemical composition of the metals used in marine equipment has a significant effect on the amount and speed of galvanic corrosion. Certain metals are more resistant to corrosion than others. These electrically negative metals are commonly called "noble;" they act as the cathode in any reaction. Metals that are more subject to corrosion are electrically positive; they act as the anode in a reaction. The more noble metals include titanium, 18-8 stainless steel and nickel. Less noble metals include zinc, aluminum and magnesium. Galvanic corrosion becomes more severe as the difference in electrical potential between the two metals increases.

In some cases, galvanic corrosion can occur within a single piece of metal. Common brass is a mixture of zinc and copper, and, when immersed in an electrolyte, the zinc portion of the mixture will corrode away as reaction occurs between the zinc and the copper particles.

Conductors

The hull of the boat often acts as the conductor between different types of metal. Marine equipment, such as an outboard motor or stern drive unit, can also act as the conductor. Large masses of metal, firmly connected together, are more efficient conductors than water. Rubber mountings and vinyl-based paint can act as insulators between pieces of metal.

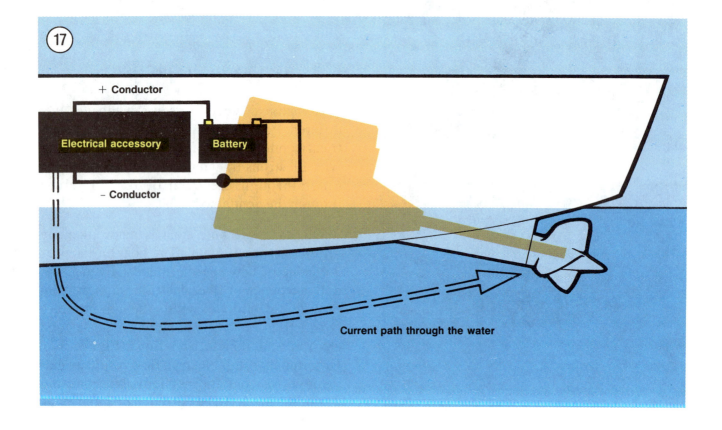

GENERAL INFORMATION

Electrolyte

The water in which a boat operates acts as the electrolyte for the galvanic corrosion process. The better a conductor the electrolyte is, the more severe and rapid the corrosion.

Cold, clean freshwater is the poorest electrolyte. As water temperature increases, its conductivity increases. Pollutants will increase conductivity; brackish or saltwater is also an efficient electrolyte. This is one of the reasons that most manufacturers recommend a freshwater flush for marine equipment after operation in saltwater, polluted or brackish water.

PROTECTION FROM GALVANIC CORROSION

Because of the environment in which marine equipment must operate, it is practically impossible to totally prevent galvanic corrosion. There are several ways by which the process can be slowed. After taking these precautions, the next step is to "fool" the process into occurring only where *you* want it to occur. This is the role of sacrificial anodes and impressed current systems.

Slowing Corrosion

Some simple precautions can help reduce the amount of corrosion taking place outside the hull. These are *not* a substitute for the corrosion protection methods discussed under *Sacrificial Anodes* and *Impressed Current Systems* in this chapter, but they can help these protection methods do their job.

Use fasteners of a metal more noble than the part they are fastening. If corrosion occurs, the larger equipment will suffer but the fastener will be protected. Because fasteners are usually very small in comparison to the equipment being fastened, the equipment can survive the loss of material. If the fastener were to corrode instead of the equipment, major problems could arise.

Keep all painted surfaces in good condition. If paint is scraped off and bare metal exposed, corrosion will rapidly increase. Use a vinyl- or plastic-based paint, which acts as an electrical insulator.

Be careful when using metal-based antifouling paints. These should not be applied to metal parts of the boat, outboard motor or stern drive unit or they will actually react with the equipment, causing corrosion between the equipment and the layer of paint. Organic-based paints are available for use on metal surfaces.

Where a corrosion protection device is used, remember that it must be immersed in the electrolyte along with the rest of the boat to have any effect. If you raise the power unit out of the water when the boat is docked, any anodes on the power unit will be removed from the corrosion cycle and will not protect the rest of the equipment that is still immersed. Also, such corrosion protection devices must not be painted because this would insulate them from the corrosion process.

Any change in the boat's equipment, such as the installation of a new stainless steel propeller, will change the electrical potential and could cause increased corrosion. Keep in mind that when you add new equipment or change materials, you should review your corrosion protection system to be sure it is up to the job.

Sacrificial Anodes

Anodes are usually made of zinc, a far from noble metal. Sacrificial anodes are specially designed to do nothing but corrode. Properly fastening such pieces to the boat will cause them to act as the anode in *any* galvanic reaction that occurs; any other metal present will act as the cathode and will not be damaged.

Anodes must be used properly to be effective. Simply fastening pieces of zinc to your boat in random locations won't do the job.

You must determine how much anode surface area is required to adequately protect the equipment's surface area. A good starting point is provided by Military Specification MIL-A-818001, which states that one square inch of new anode will protect either:

a. 800 square inches of freshly painted steel.
b. 250 square inches of bare steel or bare aluminum alloy.
c. 100 square inches of copper or copper alloy.

This rule is for a boat at rest. When underway, more anode area is required to protect the same equipment surface area.

The anode must be fastened so that it has good electrical contact with the metal to be protected. If possible, the anode can be attached directly to the other metal. If that is not possible, the entire network of metal parts in the boat should be electrically bonded together so that all pieces are protected.

Good quality anodes have inserts of some other metal around the fastener holes. Otherwise, the anode could erode away around the fastener. The anode can then become loose or even fall off, removing all protection.

Another Military Specification (MIL-A-18001) defines the type of alloy preferred that will corrode at a uniform rate without forming a crust that could reduce its efficiency after a time.

Impressed Current Systems

An impressed current system can be installed on any boat that has a battery. The system consists of an anode, a control box and a sensor. The anode in this system is coated with a very noble metal, such as platinum, so that it is almost corrosion-free and will last indefinitely. The sensor, under the boat's waterline, monitors the potential for corrosion. When it senses that corrosion could be occurring, it transmits this information to the control box.

The control box connects the boat's battery to the anode. When the sensor signals the need, the control box applies positive battery voltage to the anode. Current from the battery flows from the anode to all other metal parts of the boat, no matter how noble or non-noble these parts may be. This battery current takes the place of any galvanic current flow.

Only a very small amount of battery current is needed to counteract galvanic corrosion. Manufacturers estimate that it would take two or three months of constant use to drain a typical marine battery, assuming the battery is never recharged.

An impressed current system is more expensive to install than simple anodes but, considering its low maintenance requirements and the excellent protection it provides, the long-term cost may actually be lower.

PROPELLERS

The propeller is the final link between the boat's drive system and the water. A perfectly

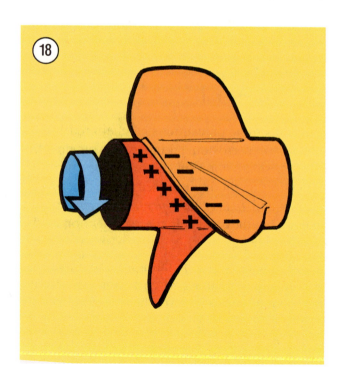

GENERAL INFORMATION

maintained engine and hull are useless if the propeller is the wrong type or has been allowed to deteriorate. Although propeller selection for a specific situation is beyond the scope of this book, the following information on propeller construction and design will allow you to discuss the subject intelligently with your marine dealer.

How a Propeller Works

As the curved blades of a propeller rotate through the water, a high-pressure area is created on one side of the blade and a low-pressure area exists on the other side of the blade (**Figure 18**). The propeller moves toward the low-pressure area, carrying the boat with it.

Propeller Parts

Although a propeller may be a one-piece unit, it is made up of several different parts (**Figure 19**). Variations in the design of these parts make different propellers suitable for different jobs.

The blade tip is the point on the blade farthest from the center of the propeller hub. The blade tip separates the leading edge from the trailing edge.

The leading edge is the edge of the blade nearest to the boat. During normal rotation, this is the area of the blade that first cuts through the water.

The trailing edge is the edge of the blade farthest from the boat.

The blade face is the surface of the blade that faces away from the boat. During normal rotation, high pressure exists on this side of the blade.

The blade back is the surface of the blade that faces toward the boat. During normal rotation, low pressure exists on this side of the blade.

The cup is a small curve or lip on the trailing edge of the blade.

The hub is the central portion of the propeller. It connects the blades to the propeller shaft (part of the boat's drive system). On some drive systems, engine exhaust is routed through the hub; in this case, the hub is made up of an outer and an inner portion, connected by ribs.

The diffuser ring is used on through-hub exhaust models to prevent exhaust gases from entering the blade area.

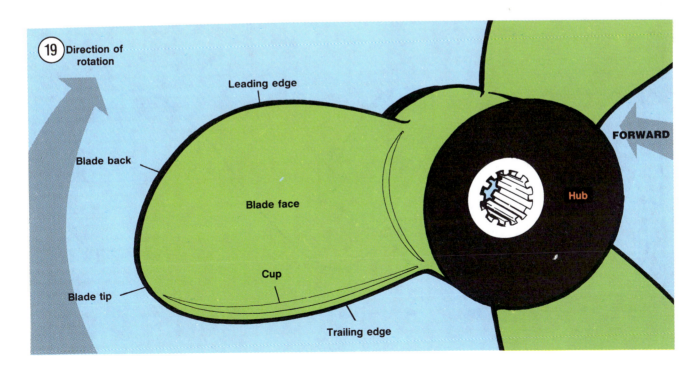

Propeller Design

Changes in length, angle, thickness and material of propeller parts make different propellers suitable for different situations.

Diameter

Propeller diameter is the distance from the center of the hub to the blade tip, multiplied by 2. That is, it is the diameter of the circle formed by the blade tips during propeller rotation (**Figure 20**).

Pitch and rake

Propeller pitch and rake describe the placement of the blade in relation to the hub (**Figure 21**).

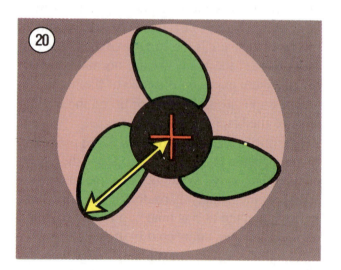

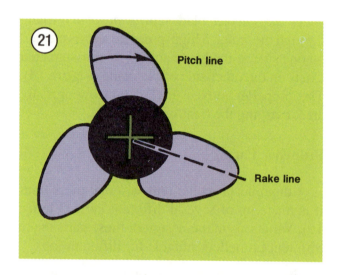

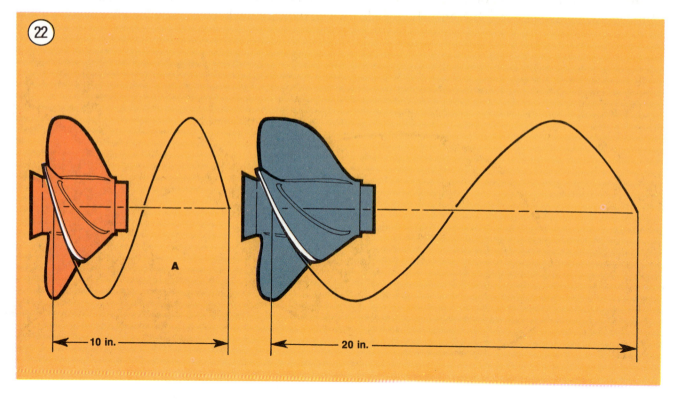

GENERAL INFORMATION

Pitch is expressed by the theoretical distance that the propeller would travel in one revolution. In A, **Figure 22**, the propeller would travel 10 inches in one revolution. In B, **Figure 22**, the propeller would travel 20 inches in one revolution. This distance is only theoretical; during actual operation, the propeller achieves about 80% of its rated travel.

Propeller blades can be constructed with constant pitch (**Figure 23**) or progressive pitch (**Figure 24**). Progressive pitch starts low at the leading edge and increases toward to trailing edge. The propeller pitch specification is the average of the pitch across the entire blade.

Blade rake is specified in degrees and is measured along a line from the center of the hub to the blade tip. A blade that is perpendicular to the hub (A, **Figure 25**) has 0° of rake. A blade that is angled from perpendicular (B, **Figure 25**) has a rake expressed by its difference from perpen-

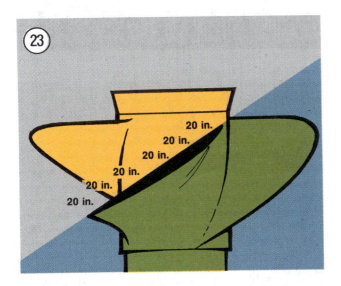

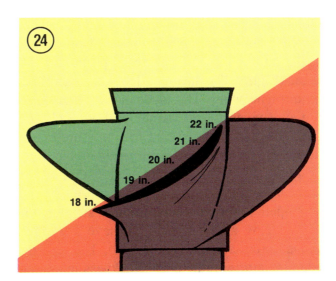

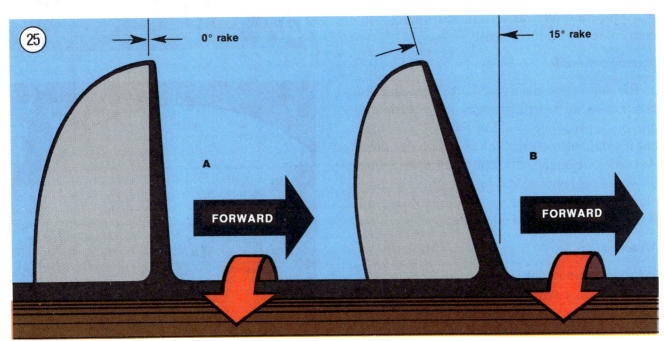

dicular. Most propellers have rakes ranging from 0-20°.

Blade thickness

Blade thickness is not uniform at all points along the blade. For efficiency, blades should be as thin as possible at all points while retaining enough strength to move the boat. Blades tend to be thicker where they meet the hub and thinner at the blade tip (**Figure 26**). This is to support the heavier loads at the hub section of the blade. This thickness is dependent on the strength of the material used.

When cut along a line from the leading edge to the trailing edge in the central portion of the blade (**Figure 27**), the propeller blade resembles an airplane wing. The blade face, where high pressure exists during normal rotation, is almost flat. The blade back, where low pressure exists during normal rotation, is curved, with the thinnest portions at the edges and the thickest portion at the center.

Propellers that run only partially submerged, as in racing applications, may have a wedge-shaped cross-section (**Figure 28**). The leading edge is very thin; the blade thickness increases toward the trailing edge, where it is the thickest. If a propeller such as this is run totally submerged, it is very inefficient.

Number of blades

The number of blades used on a propeller is a compromise between efficiency and vibration. A one-blade propeller would be the most efficient, but it would also create high levels of vibration. As blades are added, efficiency decreases, but so do vibration levels. Most propellers have three blades, representing the most practical trade-off between efficiency and vibration.

Material

Propeller materials are chosen for strength, corrosion resistance and economy. Stainless steel, aluminum and bronze are the most commonly used materials. Bronze is quite strong but

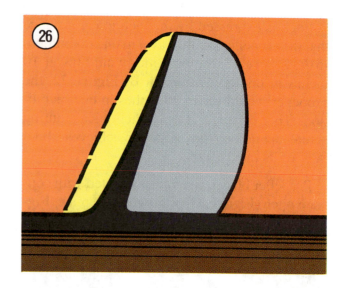

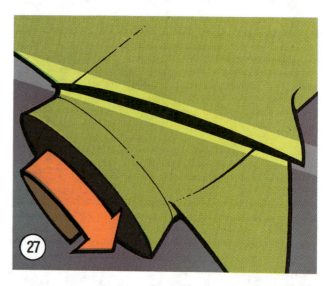

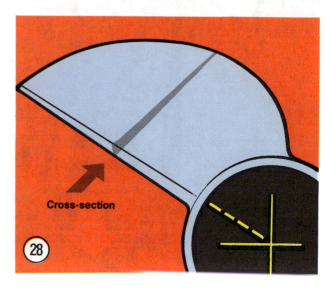

GENERAL INFORMATION

rather expensive. Stainless steel is more common than bronze because of its combination of strength and lower cost. Aluminum alloys are the least expensive but usually lack the strength of steel. Plastic propellers may be used in some low horsepower applications.

Direction of rotation

Propellers are made for both right-hand and left-hand rotation although right-hand is the most commonly used. When seen from behind the boat in forward motion, a right-hand propeller turns clockwise and a left-hand propeller turns counterclockwise. Off the boat, you can tell the difference by observing the angle of the blades (**Figure 29**). A right-hand propeller's blades slant from the upper left to the lower right; a left-hand propeller's blades are the opposite.

Cavitation and Ventilation

Cavitation and ventilation are *not* interchangeable terms; they refer to two distinct problems encountered during propeller operation.

To understand cavitation, you must first understand the relationship between pressure and the boiling point of water. At sea level, water will boil at 212° F. As pressure increases, such as within an engine's closed cooling system, the boiling point of water increases—it will boil at some temperature higher than 212° F. The opposite is also true. As pressure decreases, water will boil at a temperature lower than 212° F. If pressure drops low enough, water will boil at typical ambient temperatures of 50-60° F.

We have said that, during normal propeller operation, low-pressure exists on the blade back. Normally, the pressure does not drop low enough for boiling to occur. However, poor blade design

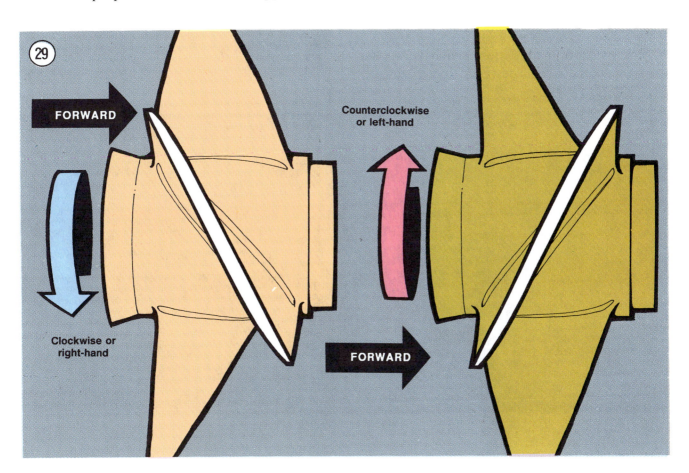

or selection, or blade damage can cause an unusual pressure drop on a small area of the blade (**Figure 30**). Boiling can occur in this small area. As the water boils, air bubbles form. As the boiling water passes to a higher pressure area of the blade, the boiling stops and the bubbles collapse. The collapsing bubbles release enough energy to erode the surface of the blade.

This entire process of pressure drop, boiling and bubble collapse is called "cavitation." The damage caused by the collapsing bubbles is called a "cavitation burn." It is important to remember that cavitation is caused by a decrease in pressure, *not* an increase in temperature.

Ventilation is not as complex a process as cavitation. Ventilation refers to air entering the blade area, either from above the surface of the water or from a through-hub exhaust system. As the blades meet the air, the propeller momentarily over-revs, losing most of its thrust. An added complication is that as the propeller over-revs, pressure on the blade back decreases and massive cavitation can occur.

Most pieces of marine equipment have a plate above the propeller area designed to keep surface air from entering the blade area (**Figure 31**). This plate is correctly called an "antiventilation plate," although you will often *see* it called an "anticavitation plate." Through hub exhaust systems also have specially designed hubs to keep exhaust gases from entering the blade area.

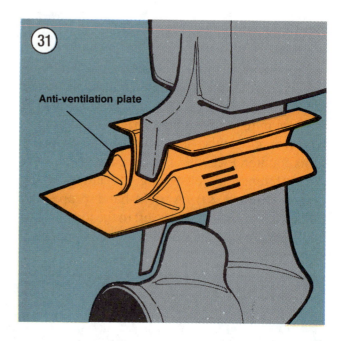

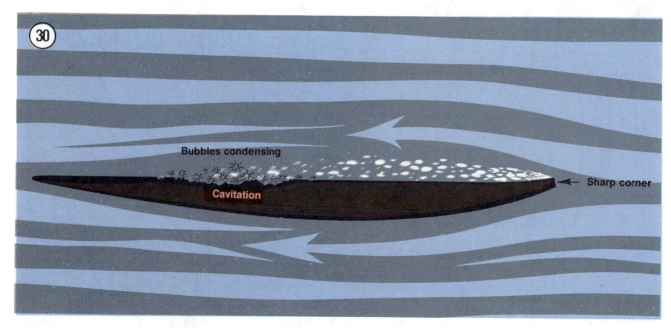

Chapter Two

Tools and Techniques

This chapter describes the common tools required for marine equipment repairs and troubleshooting. Techniques that will make your work easier and more effective are also described. Some of the procedures in this book require special skills or expertise; in some cases, you are better off entrusting the job to a dealer or qualified specialist.

SAFETY FIRST

Professional mechanics can work for years and never suffer a serious injury. If you follow a few rules of common sense and safety, you too can enjoy many safe hours servicing your marine equipment. If you ignore these rules, you can hurt yourself or damage the equipment.

1. Never use gasoline as a cleaning solvent.
2. Never smoke or use a torch near flammable liquids, such as cleaning solvent. If you are working in your home garage, remember that your home gas appliances have pilot lights.
3. Never smoke or use a torch in an area where batteries are being charged. Highly explosive hydrogen gas is formed during the charging process.
4. Use the proper size wrenches to avoid damage to fasteners and injury to yourself.
5. When loosening a tight or stuck fastener, think of what would happen if the wrench should slip. Protect yourself accordingly.
6. Keep your work area clean, uncluttered and well lighted.
7. Wear safety goggles during all operations involving drilling, grinding or the use of a cold chisel.
8. Never use worn tools.
9. Keep a Coast Guard approved fire extinguisher handy. Be sure it is rated for gasoline (Class B) and electrical (Class C) fires.

BASIC HAND TOOLS

A number of tools are required to maintain marine equipment. You may already have some of these tools for home or car repairs. There are also tools made especially for marine equipment repairs; these you will have to purchase. In any case, a wide variety of quality tools will make repairs easier and more effective.

Keep your tools clean and in a tool box. Keep them organized with the sockets and related drives together, the open end and box wrenches together, etc. After using a tool, wipe off dirt and grease with a clean cloth and place the tool in its correct place.

The following tools are required to perform virtually any repair job. Each tool is described and the recommended size given for starting a tool collection. Additional tools and some duplications may be added as you become more familiar with the equipment. You may need all standard U.S. size tools, all metric size tools or a mixture of both.

Screwdrivers

The screwdriver is a very basic tool, but if used improperly, it will do more damage than good. The slot on a screw has a definite dimension and shape. A screwdriver must be selected to conform with that shape. Use a small screwdriver for small screws and a large one for large screws or the screw head will be damaged.

Two types of screwdriver are commonly required: a common (flat-blade) screwdriver (**Figure 1**) and Phillips screwdrivers (**Figure 2**).

Screwdrivers are available in sets, which often include an assortment of common and Phillips blades. If you buy them individually, buy at least the following:

a. Common screwdriver—5/16 × 6 in. blade.
b. Common screwdriver—3/8 × 12 in. blade.
c. Phillips screwdriver—size 2 tip, 6 in. blade.

Use screwdrivers only for driving screws. Never use a screwdriver for prying or chiseling. Do not try to remove a Phillips or Allen head screw with a common screwdriver; you can damage the head so that the proper tool will be unable to remove it.

Keep screwdrivers in the proper condition and they will last longer and perform better. Always keep the tip of a common screwdriver in good condition. **Figure 3** shows how to grind the tip to the proper shape if it becomes damaged. Note the parallel sides of the tip.

Pliers

Pliers come in a wide range of types and sizes. Pliers are useful for cutting, bending and crimping. They should never be used to cut hardened objects or to turn bolts or nuts. **Figure 4** shows several types of pliers.

Each type of pliers has a specialized function. General purpose pliers are used mainly for holding things and for bending. Locking pliers are used as pliers or to hold objects very tightly, like a vise. Needlenose pliers are used to hold or bend small objects. Adjustable or slip-joint pliers can

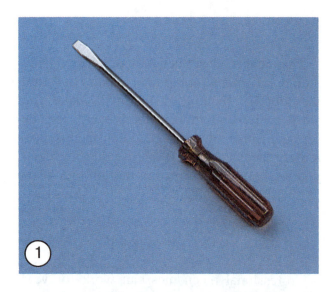

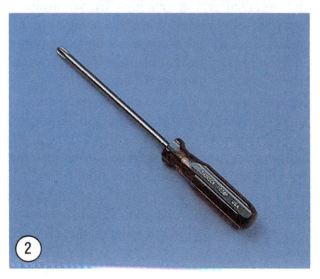

TOOLS AND TECHNIQUES   23

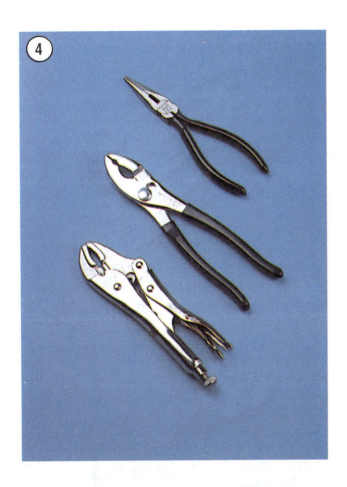

be adjusted to hold various sizes of objects; the jaws remain parallel to grip around objects such as pipe or tubing. There are many more types of pliers. The ones described here are the most commonly used.

Box and Open-end Wrenches

Box and open-end wrenches are available in sets or separately in a variety of sizes. See **Figure 5** and **Figure 6**. The number stamped near the end refers to the distance between two parallel flats on the hex head bolt or nut.

Box wrenches are usually superior to open-end wrenches. An open-end wrench grips the nut on only two flats. Unless it fits well, it may slip and round off the points on the nut. The box wrench grips all 6 flats. Both 6-point and 12-point openings on box wrenches are available. The 6-point gives superior holding power; the 12-point allows a shorter swing.

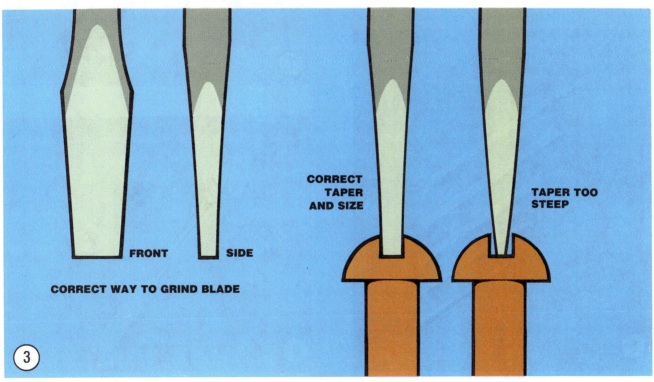

Combination wrenches, which are open on one side and boxed on the other, are also available. Both ends are the same size.

Adjustable Wrenches

An adjustable wrench can be adjusted to fit nearly any nut or bolt head. See **Figure 7**. However, it can loosen and slip, causing damage to the nut and maybe to your knuckles. Use an adjustable wrench only when other wrenches are not available.

Adjustable wrenches come in sizes ranging from 4-18 in. overall. A 6 or 8 in. wrench is recommended as an all-purpose wrench.

Socket Wrenches

This type is undoubtedly the fastest, safest and most convenient to use. See **Figure 8**. Sockets, which attach to a suitable handle, are available with 6-point or 12-point openings and use 1/4, 3/8 and 3/4 inch drives. The drive size indicates

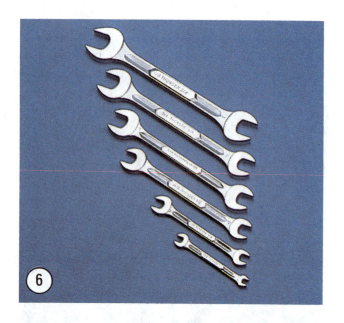

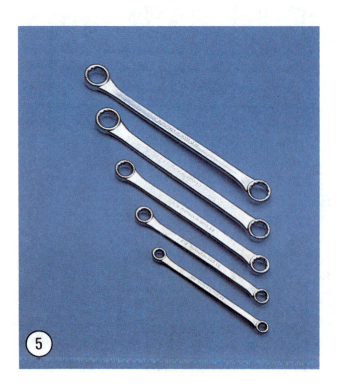

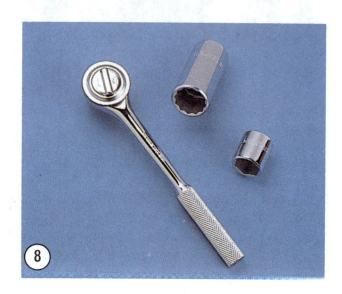

TOOLS AND TECHNIQUES

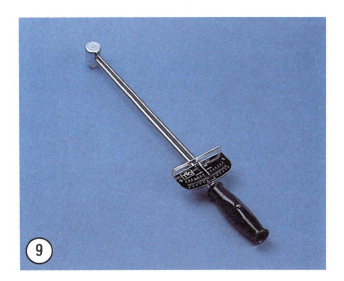

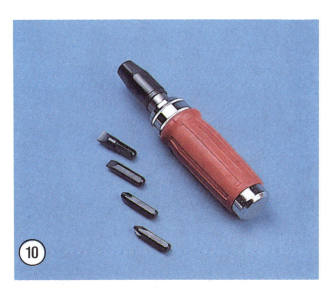

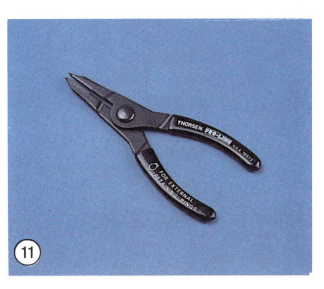

the size of the square hole that mates with the ratchet or flex handle.

Torque Wrench

A torque wrench (**Figure 9**) is used with a socket to measure how tight a nut or bolt is installed. They come in a wide price range and with either 3/8 or 1/2 in. square drive. The drive size indicates the size of the square drive that mates with the socket. Purchase one that measures up to 150 ft.-lb. (203 N•m).

Impact Driver

This tool (**Figure 10**) makes removal of tight fasteners easy and eliminates damage to bolts and screw slots. Impact drivers and interchangeable bits are available at most large hardware and auto parts stores.

Circlip Pliers

Circlip pliers (sometimes referred to as snap-ring pliers) are necessary to remove circlips. See **Figure 11**. Circlip pliers usually come with several different size tips; many designs can be switched from internal type to external type.

Hammers

The correct hammer is necessary for repairs. Use only a hammer with a face (or head) of rubber or plastic or the soft-faced type that is filled with buckshot (**Figure 12**). These are sometimes necessary in engine tear-downs. *Never* use a metal-faced hammer as severe damage will result in most cases. You can always produce the same amount of force with a soft-faced hammer.

CHAPTER TWO

Feeler Gauge

This tool has either flat or wire measuring gauges (**Figure 13**). Wire gauges are used to measure spark plug gap; flat gauges are used for all other measurements. A non-magnetic (brass) gauge may be specified when working around magnetized parts.

Other Special Tools

Some procedures require special tools; these are identified in the appropriate chapter. Unless otherwise specified, the part number used in this book to identify a special tool is the marine equipment manufacturer's part number.

Special tools can usually be purchased through your marine equipment dealer. Some can be made locally by a machinist, often at a much lower price. You may find certain special tools at tool rental dealers. Don't use makeshift tools if you can't locate the correct special tool; you will probably cause more damage than good.

TEST EQUIPMENT

Multimeter

This instrument (**Figure 14**) is invaluable for electrical system troubleshooting and service. It combines a voltmeter, an ohmmeter and an ammeter into one unit, so it is often called a VOM.

Two types of multimeter are available, analog and digital. Analog meters have a moving needle with marked bands indicating the volt, ohm and amperage scales. The digital meter (DVOM) is ideally suited for troubleshooting because it is easy to read, more accurate than analog, contains internal overload protection, is auto-ranging (analog meters must be recalibrated each time the scale is changed) and has automatic polarity compensation.

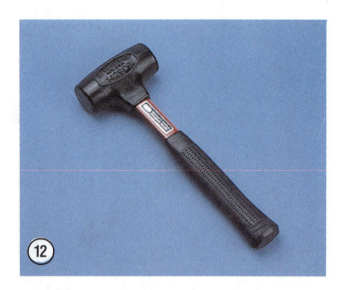

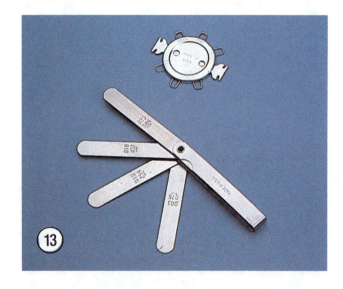

TOOLS AND TECHNIQUES

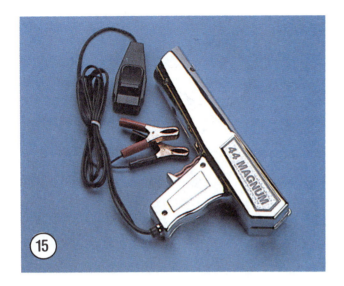

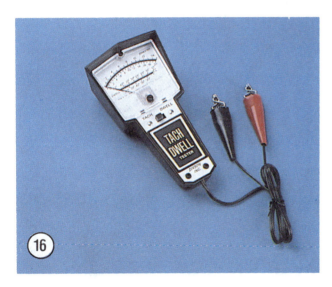

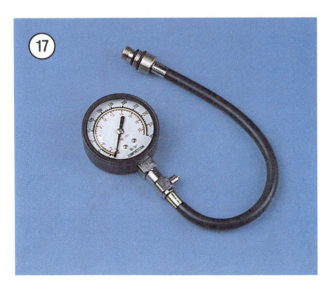

Strobe Timing Light

This instrument is necessary for dynamic tuning (setting ignition timing while the engine is running). By flashing a light at the precise instant the spark plug fires, the position of the timing mark can be seen. The flashing light makes a moving mark appear to stand still opposite a stationary mark.

Suitable lights range from inexpensive neon bulb types to powerful xenon strobe lights. See **Figure 15**. A light with an inductive pickup is best because it eliminates any possible damage to ignition wiring.

Tachometer/Dwell Meter

A portable tachometer is necessary for tuning. See **Figure 16**. Ignition timing and carburetor adjustments must be performed at the specified idle speed. The best instrument for this purpose is one with a low range of 0-1000 or 0-2000 rpm and a high range of 0-6000 rpm. Extended range (0-6000 or 0-8000 rpm) instruments lack accuracy at lower speeds. The instrument should be capable of detecting changes of 25 rpm on the low range.

A dwell meter is often combined with a tachometer. Dwell meters are used with breaker point ignition systems to measure the amount of time the points remain closed during engine operation.

Compression Gauge

This tool (**Figure 17**) measures the amount of pressure present in the engine's combustion chamber during the compression stroke. This indicates general engine condition. Compression readings can be interpreted along with vacuum gauge readings to pinpoint specific engine mechanical problems.

The easiest type to use has screw-in adapters that fit into the spark plug holes. Press-in rubber-tipped types are also available.

Vacuum Gauge

The vacuum gauge (**Figure 18**) measures the intake manifold vacuum created by the engine's intake stroke. Manifold and valve problems (on 4-stroke engines) can be identified by interpreting the readings. When combined with compression gauge readings, other engine problems can be diagnosed.

Some vacuum gauges can also be used as fuel pressure gauges to trace fuel system problems.

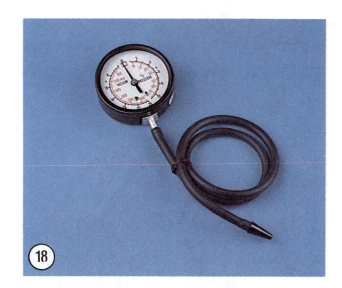

Hydrometer

Battery electrolyte specific gravity is measured with a hydrometer (**Figure 19**). The specific gravity of the electrolyte indicates the battery's state of charge. The best type has automatic temperature compensation; otherwise, you must calculate the compensation yourself.

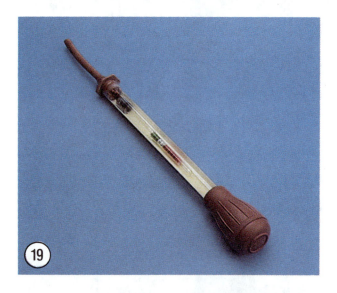

Precision Measuring Tools

Various tools are needed to make precision measurements. A dial indicator (**Figure 20**), for example, is used to determine run-out of rotating parts and end play of parts assemblies. A dial indicator can also be used to precisely measure piston position in relation to top dead center; some engines require this measurement for ignition timing adjustment.

Vernier calipers (**Figure 21**) and micrometers (**Figure 22**) are other precision measuring tools used to determine the size of parts (such as piston diameter).

Precision measuring equipment must be stored, handled and used carefully or it will not remain accurate.

SERVICE HINTS

Most of the service procedures covered in this manual are straightforward and can be performed by anyone reasonably handy with tools.

TOOLS AND TECHNIQUES

It is suggested, however, that you consider your own skills and toolbox carefully before attempting any operation involving major disassembly of the engine or gearcase.

Some operations, for example, require the use of a press. It would be wiser to have these performed by a shop equipped for such work, rather than trying to do the job yourself with makeshift equipment. Other procedures require precise measurements. Unless you have the skills and equipment required, it would be better to have a qualified repair shop make the measurements for you.

Preparation for Disassembly

Repairs go much faster and easier if the equipment is clean before you begin work. There are special cleaners, such as Gunk or Bel-Ray Degreaser, for washing the engine and related parts. Just spray or brush on the cleaning solution, let it stand, then rinse away with a garden hose. Clean all oily or greasy parts with cleaning solvent as you remove them.

> *WARNING*
> *Never use gasoline as a cleaning agent. It presents an extreme fire hazard. Be sure to work in a well-ventilated area when using cleaning solvent. Keep a Coast Guard approved fire extinguisher, rated for gasoline fires, handy in any case.*

Much of the labor charged for repairs made by dealers is for the removal and disassembly of other parts to reach the defective unit. It is frequently possible to perform the preliminary operations yourself and then take the defective unit in to the dealer for repair.

If you decide to tackle the job yourself, read the entire section in this manual that pertains to it, making sure you have identified the proper one. Study the illustrations and text until you have a good idea of what is involved in completing the job satisfactorily. If special tools or replacement parts are required, make arrangements to get them before you start. It is frustrating and time-consuming to get partly into a job and then be unable to complete it.

Disassembly Precautions

During disassembly of parts, keep a few general precautions in mind. Force is rarely needed to get things apart. If parts are a tight fit, such as

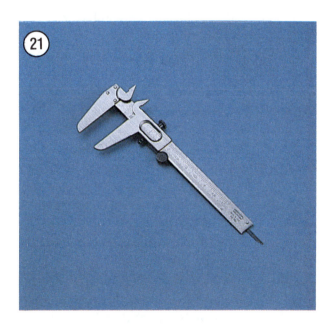

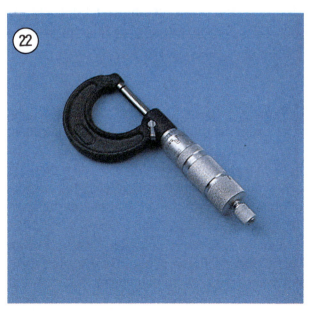

a bearing in a case, there is usually a tool designed to separate them. Never use a screwdriver to pry apart parts with machined surfaces (such as cylinder heads and crankcases). You will mar the surfaces and end up with leaks.

Make diagrams (or take an instant picture) wherever similar-appearing parts are found. For example, head and crankcase bolts are often not the same length. You may think you can remember where everything came from, but mistakes are costly. There is also the possibility you may be sidetracked and not return to work for days or even weeks. In the interval, carefully laid out parts may have been disturbed.

Cover all openings after removing parts to keep small parts, dirt or other contamination from entering.

Tag all similar internal parts for location and direction. All internal components should be reinstalled in the same location and direction from which removed. Record the number and thickness of any shims as they are removed. Small parts, such as bolts, can be identified by placing them in plastic sandwich bags. Seal and label them with masking tape.

Wiring should be tagged with masking tape and marked as each wire is removed. Again, do not rely on memory alone.

Protect finished surfaces from physical damage or corrosion. Keep gasoline off painted surfaces.

Assembly Precautions

No parts, except those assembled with a press fit, require unusual force during assembly. If a part is hard to remove or install, find out why before proceeding.

When assembling two parts, start all fasteners, then tighten evenly in an alternating or crossing pattern if no specific tightening sequence is given.

When assembling parts, be sure all shims and washers are installed exactly as they came out.

Whenever a rotating part butts against a stationary part, look for a shim or washer. Use new gaskets if there is any doubt about the condition of the old ones. Unless otherwise specified, a thin coat of oil on gaskets may help them seal effectively.

Heavy grease can be used to hold small parts in place if they tend to fall out during assembly. However, keep grease and oil away from electrical components.

High spots may be sanded off a piston with sandpaper, but fine emery cloth and oil will do a much more professional job.

Carbon can be removed from the cylinder head, the piston crown and the exhaust port with a dull screwdriver. *Do not* scratch either surface. Wipe off the surface with a clean cloth when finished.

The carburetor is best cleaned by disassembling it and soaking the parts in a commercial carburetor cleaner. Never soak gaskets and rubber parts in these cleaners. Never use wire to clean out jets and air passages; they are easily damaged. Use compressed air to blow out the carburetor *after* the float has been removed.

Take your time and do the job right. Do not forget that the break-in procedure on a newly rebuilt engine is the same as that of a new one. Use the break-in oil recommendations and follow other instructions given in your owner's manual.

SPECIAL TIPS

Because of the extreme demands placed on marine equipment, several points should be kept in mind when performing service and repair. The following items are general suggestions that may improve the overall life of the machine and help avoid costly failures.

1. Unless otherwise specified, use a locking compound, such as Loctite Threadlocker, on all bolts and nuts, even if they are secured with lockwashers. Be sure to use the specified grade

TOOLS AND TECHNIQUES

of thread locking compound. A screw or bolt lost from an engine cover or bearing retainer could easily cause serious and expensive damage before its loss is noticed.

When applying thread locking compound, use a small amount. If too much is used, it can work its way down the threads and stick parts together that were not meant to be stuck together.

Keep a tube of thread locking compound in your tool box; when used properly, it is cheap insurance.

2. Use a hammer-driven impact tool to remove and install screws and bolts. These tools help prevent the rounding off of bolt heads and screw slots and ensure a tight installation.

3. When straightening the fold-over type lockwasher, use a wide-blade chisel, such as an old and dull wood chisel. Such a tool provides a better purchase on the folded tab, making straightening easier.

4. When installing the fold-over type lockwasher, always use a new washer if possible. If a new washer is not available, always fold over a part of the washer that has not been previously folded. Reusing the same fold may cause the washer to break, resulting in the loss of its locking ability and a loose piece of metal adrift in the engine.

When folding the washer, start the fold with a screwdriver and finish it with a pair of pliers. If a punch is used to make the fold, the fold may be too sharp, thereby increasing the chances of the washer breaking under stress.

These washers are relatively inexpensive and it is suggested that you keep several of each size in your tool box for repairs.

5. When replacing missing or broken fasteners (bolts, nuts and screws), always use authorized replacement parts. They are specially hardened for each application. The wrong 50-cent bolt could easily cause serious and expensive damage.

6. When installing gaskets, always use authorized replacement gaskets *without* sealer, unless designated. Many gaskets are designed to swell when they come in contact with oil. Gasket sealer will prevent the gaskets from swelling as intended and can result in oil leaks. Authorized replacement gaskets are cut from material of the precise thickness needed. Installation of a too thick or too thin gasket in a critical area could cause equipment damage.

MECHANIC'S TECHNIQUES

Removing Frozen Fasteners

When a fastener rusts and cannot be removed, several methods may be used to loosen it. First, apply penetrating oil, such as Liquid Wrench or WD-40 (available at any hardware or auto supply store). Apply it liberally and allow it penetrate for 10-15 minutes. Tap the fastener several times with a small hammer; do not hit it hard enough to cause damage. Reapply the penetrating oil if necessary.

For frozen screws, apply penetrating oil as described, then insert a screwdriver in the slot and tap the top of the screwdriver with a hammer. This loosens the rust so the screw can be removed in the normal way. If the screw head is too chewed up to use a screwdriver, grip the head with locking pliers and twist the screw out.

Avoid applying heat unless specifically instructed because it may melt, warp or remove the temper from parts.

Remedying Stripped Threads

Occasionally, threads are stripped through carelessness or impact damage. Often the threads can be cleaned up by running a tap (for internal threads on nuts) or die (for external threads on bolts) through threads. See **Figure 23**.

Removing Broken Screws or Bolts

When the head breaks off a screw or bolt, several methods are available for removing the remaining portion.

If a large portion of the remainder projects out, try gripping it with vise-grip pliers. If the projecting portion is too small, file it to fit a wrench or cut a slot in it to fit a screwdriver. See **Figure 24**.

If the head breaks off flush, use a screw extractor. To do this, centerpunch the remaining portion of the screw or bolt. Drill a small hole in the screw and tap the extractor into the hole. Back the screw out with a wrench on the extractor. See **Figure 25**.

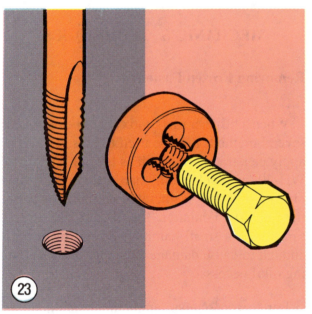

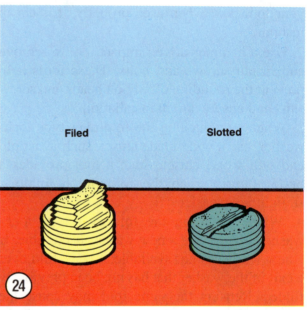

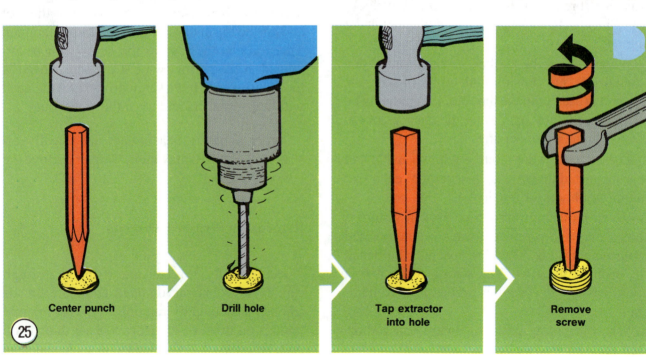

Chapter Three

Troubleshooting

Troubleshooting is a relatively simple matter when it is done logically. The first step in any troubleshooting procedure is to define the symptoms as fully as possible and then localize the problem. Subsequent steps involve testing and analyzing those areas which could cause the symptoms. A haphazard approach may eventually solve the problem, but it can be very costly in terms of wasted time and unnecessary parts replacement.

Never assume anything. Don't overlook the obvious. If the engine suddenly quits when running, check the easiest and most accessible spots first. Make sure there is gasoline in the tank, the fuel petcock is in the ON position, the spark plug wires are properly connected and the wiring harnesses are properly connected.

If a quick visual check of the obvious does not turn up the cause of the problem, look a little further. Learning to recognize and describe symptoms accurately will make repairs easier for you or a mechanic at the shop. Saying that "it won't run" isn't the same as saying "it quit at high speed and wouldn't start."

Gather as many symptoms together as possible to aid in diagnosis. Note whether the engine lost power gradually or all at once, what color smoke (if any) came from the exhaust and so on. Remember—the more complicated an engine is, the easier it is to troubleshoot because symptoms point to specific problems.

After the symptoms are defined, areas which could cause the problems should be tested and analyzed. You don't need fancy or complicated test equipment to determine whether repairs can be attempted at home. A few simple checks can save a large repair bill and time lost while the engine sits in a shop's service department.

On the other hand, be realistic and don't attempt repairs beyond your abilities. Service departments tend to charge heavily for putting together a disassembled engine that may have been abused. Some won't even take on such a job—so use common sense and don't get in over your head.

Proper lubrication, maintenance and periodic tune-ups as described in Chapter Four will reduce the necessity for troubleshooting. Even with the best of care, however, an outboard motor is prone to problems which will eventually require troubleshooting.

This chapter contains brief descriptions of each operating system and troubleshooting procedures to be used. Tables 1-3 present typical starting, ignition and fuel system problems with their probable causes and solutions. Tables 1-4 are at the end of the chapter.

OPERATING REQUIREMENTS

Every outboard motor requires 3 basic things to run properly: an uninterrupted supply of fuel and air in the correct proportions, proper ignition at the right time and adequate compression. If any of these are lacking, the motor will not run.

The electrical system is the weakest link in the chain. More problems result from electrical malfunctions than from any other source. Keep this in mind before you blame the fuel system and start making unnecessary carburetor adjustments.

If a motor has been sitting for any length of time and refuses to start, check the condition of the battery first to make sure it has an adequate charge, then look to the fuel delivery system. This includes the gas tank, fuel pump, fuel lines and carburetor(s). Rust may have formed in the tank, obstructing fuel flow. Gasoline deposits may have gummed up carburetor jets and air passages. Gasoline tends to lose its potency after standing for long periods. Condensation may contaminate it with water. Drain the old gas and try starting with a fresh tankful. If the carburetor is getting a satisfactory supply of good fuel, turn to the starting system.

STARTING SYSTEM

Description

Johnson and Evinrude outboard motors covered in this manual (except 1973 50 hp and 1980-1983 55 hp rope-start models) are equipped with an electric starter motor (**Figure 1**). The motor is mounted vertically on the engine. When battery current is supplied to the starter motor, its pinion gear is thrust upward to engage the teeth on the engine flywheel. Once the engine starts, the pinion gear disengages from the flywheel. This is similar to the method used in cranking an automotive engine.

The starting system requires a fully charged battery to provide the large amount of electrical current required to operate the starter motor. The battery may be charged externally or by an alternator stator and rectifier system which keeps the battery charged while the engine is running.

Starting Circuit

The starting circuit consists of the battery, starter motor, starter and choke switches, starter and choke solenoids, ignition switch, neutral start switch, a fuse (on some models) and connecting wiring.

Turning the ignition switch to START allows current to flow through the solenoid coil. The solenoid contacts close and allow current to flow from the battery through the solenoid to the starter motor.

A neutral start switch prevents current flow through the solenoid coil whenever the throttle lever is set beyond the START position. On models with remote control, the neutral start switch is located in the remote control box. On models without remote control, a plunger-operated switch rides on a throttle lever cam.

The choke solenoid electrically moves the choke valve linkage to close the choke for starting. A primer solenoid introduced on some 1980 models replaces the choke solenoid. When the ignition key is depressed while in the START mode, the fuel pump sends fuel to the primer solenoid, which introduces it into the bypass cover (1980-1981) or directly to the carburetor (1982-on). The primer solenoid system is used on all 1983 and later models covered in this manual.

General troubleshooting procedures are provided in **Table 1**.

TROUBLESHOOTING

CAUTION
Do not operate the starter motor continuously for more than 30 seconds. Allow the motor to cool for at least 2 minutes between attempts to start the engine.

Troubleshooting Preparation (All Models)

Before troubleshooting the starting circuit, make sure:
a. The battery is fully charged.
b. The control lever is in NEUTRAL.
c. All electrical connections are clean and tight.
d. The wiring harness is in good condition, with no worn or frayed insulation.
e. Battery cables are the proper size and length. Replace undersize cables or relocate the battery to shorten the distance between the battery and starter solenoid.
f. The fuse installed in the red lead between ignition switch and solenoid is good, if so equipped.
g. The fuel system is filled with an adequate supply of fresh gasoline that has been properly mixed with Johnson or Evinrude Outboard Lubricant. See Chapter Four.

Starting Difficulties With Older Engines

Many older 2-stroke engines are plagued by hard starting and generally poor running for which there seems to be no good cause. Carburetion and ignition are satisfactory and a compression test shows all is well in the engine's upper end.

What a compression test does not show is a lack of primary compression. The crankcase in a 2-stroke engine must be alternately under pressure and vacuum. After the piston closes the intake port, further downward movement of the piston causes the trapped mixture to be pressurized so it can rush quickly into the cylinder when the scavenging ports are opened. Upward piston movement creates a vacuum in the crankcase, enabling air-fuel mixture to be drawn in from the carburetor.

If the crankshaft seals or case gaskets leak, the crankcase cannot hold pressure or vacuum and proper engine operation becomes impossible. Any other source of leakage, such as defective cylinder base gaskets or porous or cracked crankcase castings, will result in the same conditions.

Older engines suffering from hard starting should be checked for pressure leaks with a small brush and soap suds solution. The following is a list of possible leakage points in the engine:
a. Crankshaft seals.
b. Spark plug threads.
c. Cylinder head joint.
d. Cylinder base joint.
e. Carburetor mounting flange(s).
f. Crankcase joint.

Troubleshooting (50 and 60 hp Tiller Electric With Push Button Start)

Refer to **Figure 2** for this procedure.
1. Disconnect the starter-to-solenoid cable at point 5. Disconnect the black ground lead at point

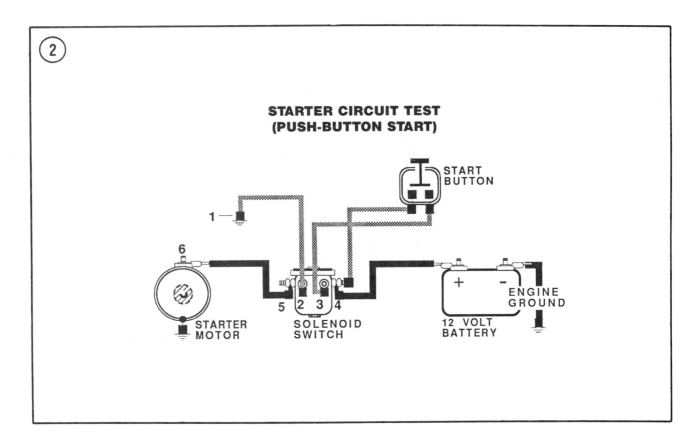

1. Connect a 12-volt test light between the lead and a good engine ground.
2. Turn the ignition switch to START. If the light comes on, reconnect the black lead and proceed to Step 8. If the light does not come on, proceed with Step 3.
3. Connect the test light between ground and point 2. If the light comes on, there is an open in the wiring between point 1 and point 2.
4. If the light does not come on, connect the test light between ground and point 3. Turn the ignition switch to START. If the light comes on, the solenoid is defective.
5. If the light does not come on, connect the test light between ground and point 4. If the light comes on, check the start switch.
6. If the light does not come on, check for an open in the wiring between point 4 and the positive battery terminal.
7. Connect the test light between ground and point 5. Depress the start switch. If the light does not come on, the solenoid is defective.
8. If the test light comes on in Step 7 and the solenoid clicks, connect the test light between ground and point 6. Reconnect the starter cable at point 6. Depress the start switch. If the test light comes on and the starter motor does not turn over, replace the starter motor. If the test light does not come on, check for a broken cable or poor connection in the wiring.

Troubleshooting (All Others)

CAUTION
Disconnect starter-to-solenoid cable to prevent starter engagement during Steps 1-8.

Refer to **Figure 3** for this procedure.
1A. With safety switch—Locate the safety switch at point 1 and remove the white wire. Connect a 12-volt test light between a good engine ground and the white wire.

TROUBLESHOOTING

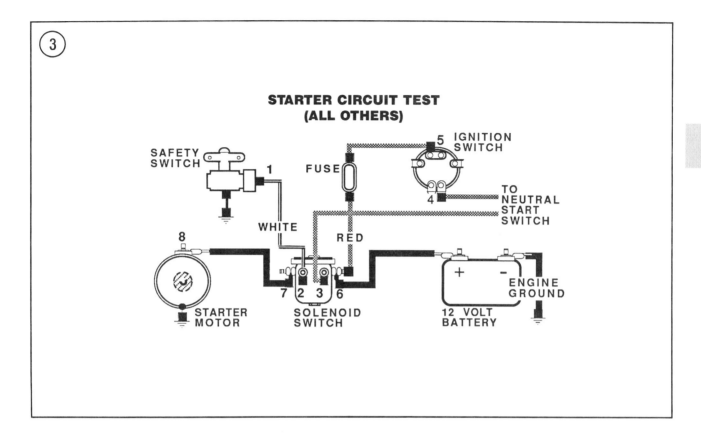

1B. Without safety switch—Disconnect the black solenoid lead at point 2. Connect a 12-volt test light between the solenoid terminal and a good engine ground.

2. Turn the ignition switch to START. If the light comes on, proceed with Step 3. If the light does not come on in Step 1A or 1B, proceed with Step 9.

3. With safety switch—Connect the white wire to the switch with the test lamp still connected. Turn the switch to the START position. If the light does not come on, proceed with Step 9. If the light comes on, the switch is not properly connected or it is defective. The throttle may be advanced too far for the switch to function properly.

NOTE
Turn the key to OFF before connecting or disconnecting the test light in the following steps. This will prevent the possibility of a shock.

4. Connect the test light between ground and point 2 and turn the key back to START. If the light comes on, there is an open in the wiring between point 1 and point 2.

5. If the light does not come on, connect the test light between ground and point 3. Turn the ignition switch to START. If the light comes on, the solenoid is defective.

6. If the light does not come on, connect the test light between ground and point 4. Turn the ignition switch to START. If the light comes on, the lead between point 3 and point 4 is loose, corroded or disconnected or the neutral start switch is open.

7. If the light does not come on, connect the test light between ground and point 5. Leave the ignition switch OFF. If the light comes on, the switch is defective.

8. If the light does not come on, check for an open or burned fuse between point 5 and 6 and correct as required. Connect the test light between ground and point 6. Leave the ignition

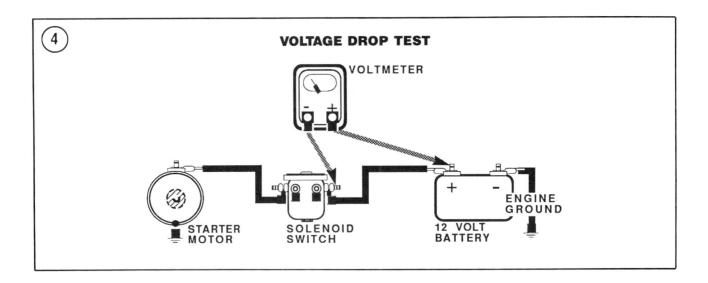

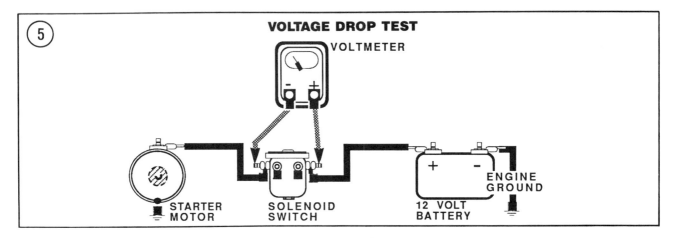

switch OFF. The light should light. If it does not, check for an open circuit between point 6 and the battery.

9. Reconnect the starter cable at the solenoid. Connect the test light between ground and point 7. Turn the ignition switch to START. The test light should come on and the solenoid should click. If the light does not come on, the solenoid is defective.

10. Connect the test light between ground and point 8. Turn the ignition switch to START. If the light comes on but the starter motor will not turn over, replace the starter motor. If the light does not come on, check for a broken cable or a poor connection in the wiring.

Voltage Drop Test

A systematic test from the positive battery terminal through the starting system and back to the negative battery terminal will locate any component or connection with enough resistance to cause hard starting. Holding the voltmeter leads against the terminals (instead of connecting them) will test the component and connection for high resistance. Refer to **Figures 4-7** for this procedure.

1. Disconnect the armature plate-to-power pack leads or 4-wire connector to allow cranking of the engine without starting.

2. Connect the red voltmeter lead to the positive battery terminal and the black voltmeter lead to the starter solenoid positive terminal. See **Figure**

TROUBLESHOOTING

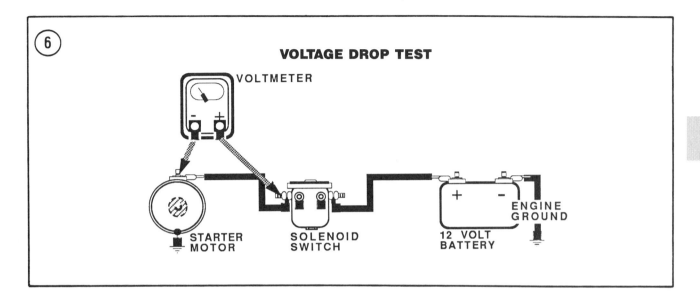

VOLTAGE DROP TEST

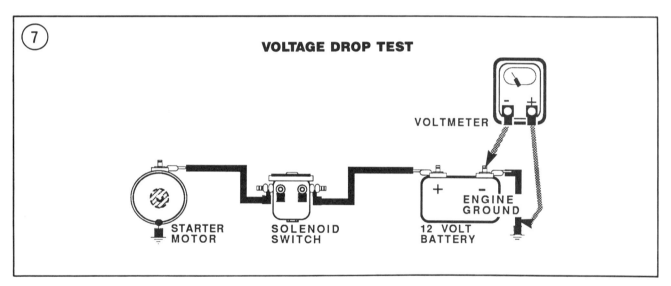

VOLTAGE DROP TEST

4. Crank the engine with the ignition switch. The voltmeter reading should be 0.3 volts or less.

3. Connect the voltmeter as shown in **Figure 5**. With the engine cranking, the voltmeter reading should be 0.2 volts or less.

4. Connect the voltmeter as shown in **Figure 6**. With the engine cranking, the voltmeter reading should be 0.2 volts or less.

5. Connect the red voltmeter lead to the common engine ground terminal as shown in **Figure 7**. Connect the black voltmeter lead to the negative battery terminal. With the engine cranking, the voltmeter reading should be 0.3 volts or less.

6. Clean, retighten or replace any connection, wiring or component that exceeds the specified voltage drop.

7. Reconnect the armature plate-to-power pack leads or 4-wire connector.

CHARGING SYSTEM

Description

The Johnson and Evinrude outboards covered in this manual through 1985 may be equipped with a 4, 5 or 6 amp unregulated or a 6, 10, 15 or 35 amp regulated alternator charging system. The 35 amp system is used with 185-235 hp V6 motors; the 6, 10 or 15 amp systems are used with most V4 and V6 models; 50-65 hp motors may have a 4, 5 or 6 amp system; 70 and 75 hp engines all use a 6 amp system. Check your owner's manual to determine the exact system rating on your motor.

For 1986-on, a 4, 3/6 or 6 amp unregulated or a 3/9 or 35 amp regulated alternator charging system is used. The 3/6 amp system is used on manual trim models while the 3/9 amp system is used with power trim models. The dual system design delivers 3 amps at idle or trolling and 6 or 9 amps at cruising speed.

The charging system on all models consists of permanent magnets cast in the flywheel (**Figure 8**), a stator assembly containing coils wound on a laminated iron core (**Figure 9**), a rectifier (**Figure 10**), the battery and connecting wiring. Regulated charging systems also have a voltage regulator and fuse. Flywheel rotation past the stator coils produces alternating current (AC), which is sent to the rectifier for conversion to direct current (DC). Starting with 1984 regulated charging systems, the rectifier and voltage regulator are combined in a single unit.

A malfunction in the charging system generally causes the battery to remain undercharged. Since the stator (**Figure 9**) is protected by its location underneath the flywheel (**Figure 8**), it is more likely that the battery, rectifier or connecting wiring will cause problems. The following conditions will cause rectifier damage:
 a. Battery leads reversed.
 b. Running the engine with the battery leads disconnected.
 c. A broken wire or loose connection resulting in an open circuit.

Troubleshooting Preparation

Before troubleshooting the charging circuit, visually check the following:

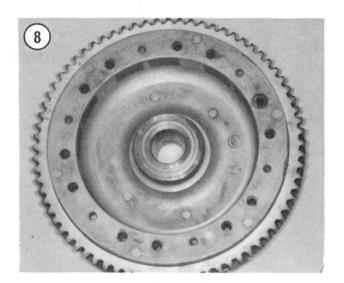

TROUBLESHOOTING

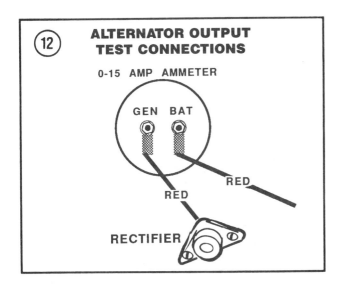

1. Make sure the red cable is connected to the positive battery terminal. If polarity is reversed, check for a damaged rectifier.

NOTE
A damaged rectifier will generally be discolored or have a burned appearance.

2. Check for corroded or loose connections. Clean, tighten and insulate with OMC Black Neoprene Dip as required.
3. Check battery condition. Clean and recharge as required.

4. Check wiring harness between the armature plate and battery for damaged or deteriorated insulation and corroded, loose or faulty connections. Repair or replace as required.
5. Check the fuse on regulated charging systems.

Alternator Output Quick Check

A quick check of the alternator output can be made with an induction ammeter (**Figure 11**). Fit the ammeter over the positive battery cable and run the motor at fully throttle in a test tank or in the water. The induction ammeter will show the alternator output. The total electrical load on the system from the engine and accessories cannot exceed the system specification.

Alternator Output Test

Perform this test for a more accurate reading of alternator output. Refer to **Figure 12** (typical 5 amp system) for this procedure.
1A. Unregulated system—Disconnect the red rectifier lead at the terminal board.
1B. Regulated system—Disconnect the red rectifier/regulator lead at the battery side of the starter solenoid.
2. Connect the disconnected rectifier or rectifier/regulator lead to the negative terminal of a 0-40 amp ammeter. Connect the wiring harness red lead (unregulated) or battery side of the starter solenoid (regulated) to the positive ammeter terminal.
3. With the engine in a test tank or on the boat in the water, start and run at full throttle. The ammeter should read approximately the same as the system output capacity.
4. If little or no charge is shown, test the stator and rectifier or rectifier/regulator as described in this chapter.

Cranking Discharge Test
(All 1984-on Models with Rectifier/Regulator)

1. With the engine in a test tank or on the boat in the water, connect a voltmeter across the

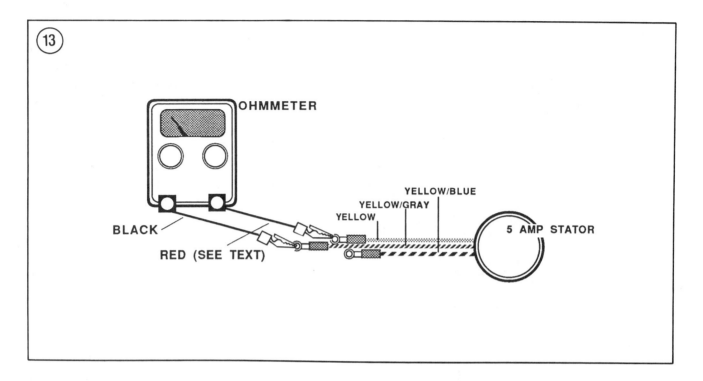

battery terminals and note the reading. If voltage is 12.5 volts or more, continue with Step 2. If voltage is less than 12.5 volts, proceed with Step 3.
2. Disconnect the ignition charge coil connector(s). Crank the engine in short bursts until the voltmeter reading with the key in the OFF position is less than 12.5 volts.
3. Disconnect the negative battery cable, then the positive battery cable.
4. Connect a 0-40 ammeter in series between the battery side of the starter solenoid and the rectifier/regulator red lead.
5. Reconnect the positive battery cable, then the negative battery cable. Reconnect the ignition charge coil connector(s), if disconnected.
6. Start the engine and run at approximately 4,500 rpm while watching the engine ammeter. The ammeter should read nearly full output (6, 9, 10, 15 or 35 amps, depending upon the system). As the engine continues to run, the voltage should stabilize at about 14.5 volts and the alternator output should start to decrease.
7. If the ammeter does not read as specified in Step 6, test the stator as described in this chapter.

If the stator is good, replace the rectifier/regulator.

Stator Resistance Test (1978-on 2-Cylinder)

Refer to **Figure 13** for this procedure.
1. Disconnect the yellow, yellow/gray and yellow/blue stator leads at the terminal block.
2. Set the ohmmeter on the low ohm scale.
3. Connect the black ohmmeter test lead to the yellow stator lead.
4. Connect the red test lead to the yellow/gray stator lead. Note the ohmmeter reading and compare to **Table 4**.
5. Move the red test lead to the yellow/blue stator lead. Note the ohmmeter reading and compare to **Table 4**.
6. Connect the ohmmeter test leads between a good engine ground and the yellow/blue stator lead. Note the meter reading, move the test lead from the yellow/blue to the yellow/gray stator lead and note that reading. Both connections should read infinity, indicating an open circuit.

TROUBLESHOOTING

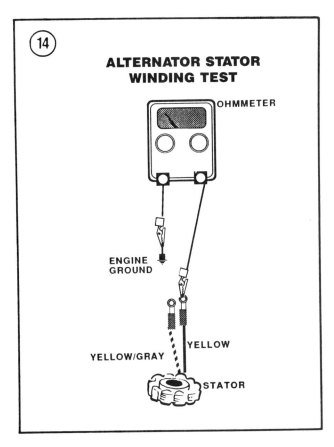

Figure 14 ALTERNATOR STATOR WINDING TEST

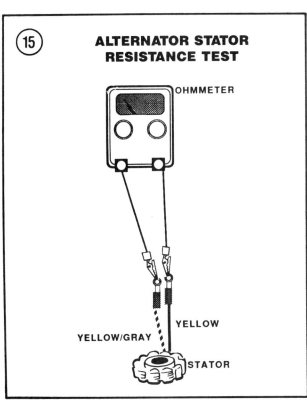

Figure 15 ALTERNATOR STATOR RESISTANCE TEST

If any other reading is shown, the stator is shorted to ground.

7. Replace the stator if any readings are other than specified.

Stator Resistance Test (All Others)

Refer to **Figure 14** and **Figure 15** for this procedure.

1. Disconnect the negative battery cable.
2. Set the ohmmeter on the high ohm scale.
3. Disconnect the yellow and yellow/gray stator leads at the terminal block.
4. Connect one ohmmeter test lead to ground (**Figure 14**). Alternately connect the other test lead to each stator lead. The meter should read infinity, indicating an open circuit. If any other reading is shown, the stator is shorted to ground.
5. Set the ohmmeter on the low ohm scale.
6. Connect the test leads between the disconnected stator leads (**Figure 15**). Note the reading and refer to **Table 4**. If the meter reads infinity, the stator windings are open.
7. Replace the stator if readings are other than specified.

Rectifier Test (2-cylinder Engine)

Refer to **Figure 16** and **Figure 17** for this procedure.

1. Ground one ohmmeter test lead at the rectifier case or a good engine ground. Connect the other test lead to the yellow/gray rectifier lead. See **Figure 16**. Note the ohmmeter reading.
2. Reverse the test leads and note the ohmmeter reading. The ohmmeter should read zero in one direction and infinity in the other. If the reading is the same in Step 1 and Step 2, the diode is defective and the rectifier should be replaced. High resistance indicates an open diode; low resistance indicates a shorted diode.
3. Repeat Step 1 and Step 2 to test the yellow and yellow/blue rectifer leads.

4. Connect one ohmmeter test lead to the red rectifier lead. Connect the other test lead to the yellow/gray rectifier lead. See **Figure 17**. Note the ohmmeter reading.
5. Reverse the test leads and note the ohmmeter reading. The ohmmeter should read zero in one direction and infinity in the other. If the reading is the same in both directions, the rectifier is defective.
6. Repeat Step 4 and Step 5 to test the yellow rectifier lead, then the yellow/blue lead.
7. Replace the rectifier if the readings are not as specified.

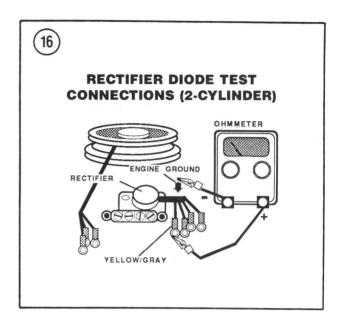

Rectifier Test (All 3-6 cylinder Except 1984-on 10-35 Amp Systems)

Refer to **Figure 18** for typical location and test connections.
1. Disconnect the rectifier leads at the terminal board.
2. With the ohmmeter on the high scale, connect one test lead to the yellow rectifier lead and the other test lead to ground. Note the meter reading.
3. Reverse the test leads and note the reading. The ohmmeter should read zero in one direction and infinity in the other. If the reading is the same in both directions, the rectifier is defective.
4. Repeat Step 2 and Step 3 to test the yellow/gray lead.
5. Replace the rectifier if the readings are not as specified.

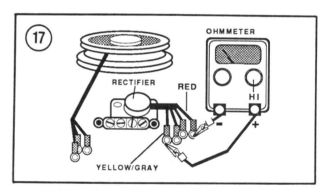

Regulator Test (All 1973-1983 V4 and 1976-1983 V6 with 10-35 Amp System)

1. Undercharged battery:
   a. Disconnect the regulator leads at the terminal block and connect an ammeter to the terminal block in their place.
   b. Start the engine and run at 4,300-4,600 rpm with all accessories off.
   c. If the ammeter reading indicates a full charge, replace the regulator. If there is no reading, check the stator and rectifier as described in this chapter.
2. Overcharged battery:
   a. Start the engine and run at 4,300-4,600 rpm with all accessories off for a minimum of 20 minutes.
   b. Check battery voltage with a voltmeter. If greater than 15 volts, replace the regulator.

Rectifier and Regulator Assembly Test (1984-on 10-35 Amp Systems)

See *Cranking Discharge Test* in this chapter.

TROUBLESHOOTING

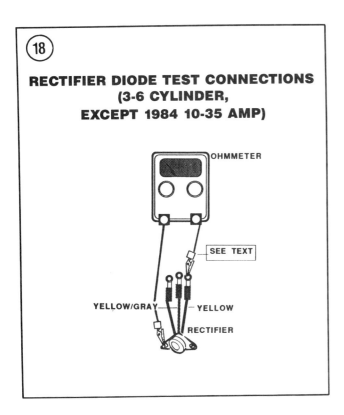

IGNITION SYSTEM

The wiring harness used between the ignition switch and engine is adequate to handle the electrical needs of the outboard. It *will not* handle the electrical needs of accessories. Whenever an accessory is added, run new wiring between the battery and accessory, installing a separate fuse panel on the instrument panel.

If the ignition switch requires replacement, *never* install an automotive-type switch. A marine-type switch must always be used.

Description

Variations of a flywheel magneto capacitor discharge (CD2, CD 3, CD 4 and CD 6) ignition system are used on all Johnson and Evinrude outboards covered in this manual. **Figure 19** is a schematic of the CD 3 ignition system showing the major components common to all variations. See Chapter Seven for a full description. General troubleshooting procedures are provided in **Table 2**.

Troubleshooting Precautions

Several precautions should be strictly observed to avoid damage to the ignition system.
1. Do not reverse the battery connections. This reverses polarity and can damage the rectifier or power pack unit.
2. Do not "spark" the battery terminals with the battery cable connections to check polarity.
3. Do not disconnect the battery cables with the engine running.

Troubleshooting Preparation (All Ignition Systems)

NOTE
To test the wiring harness for poor solder connections in Step 1, bend the molded rubber connector while checking each wire for resistance.

1. Check the wiring harness and all plug-in connections to make sure all terminals are free of corrosion, all connectors are tight and the wiring insulation is in good condition.
2. Check all electrical components grounded to the engine for a good connection.
3. Make sure that all ground wires are properly connected and the connections are clean and tight.
4. Check remainder of the wiring for disconnected wires and shorts or open circuits.
5. Make sure there is an adequate supply of fresh and properly mixed fuel available to the engine.
6. Check the battery condition on electric start models. Clean terminals and recharge the battery, if necessary.
7. Check spark plug cable routing. Make sure the cables are properly connected to their spark plugs.
8. Remove all spark plugs, keeping them in order. Check the condition of each plug. See Chapter Four.
9. Install a spark tester (**Figure 20**) between the plug wires and a good ground to check for spark

46 CHAPTER THREE

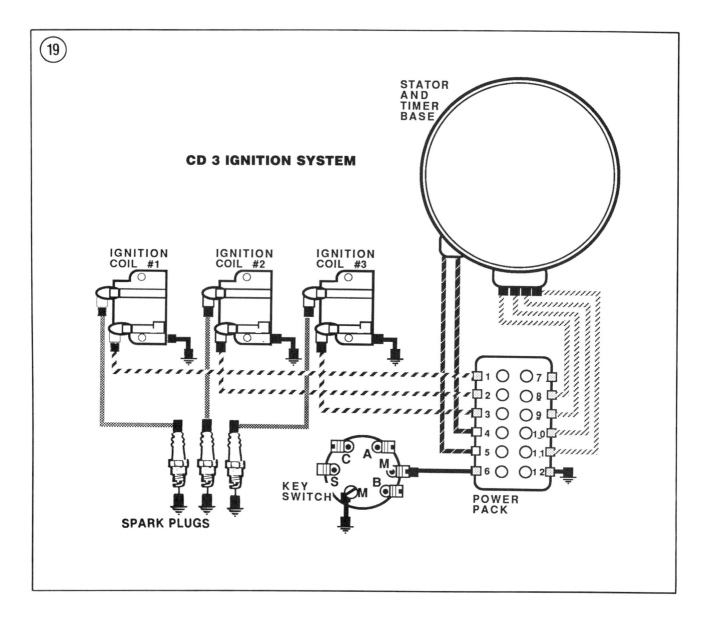

at each cylinder. See **Figure 21**. Set the spark tester air gap to 1/2 in. Crank the engine over while watching the spark tester. If a spark jumps at each plug gap, the ignition system is good.

10. If a spark tester is not available, remove each spark plug and reconnect the proper plug cable to one plug. Lay the plug against the cylinder head so its base makes a good connection and turn the engine over. If there is no spark or only a weak one, check for loose connections at the coil and battery. Repeat the check with each remaining plug. If the connections are good, the problem is most likely in the ignition system.

1973-1977 CD 2 IGNITION TROUBLESHOOTING

The CD 2 ignition is used on 2-cylinder engines. **Figure 22** shows the terminal board connections. Although many terminal boards have the wire colors embossed either on the board itself or on the cover (if used), not all models will use all of the wires and connections shown. Double check your connections by referring to **Figure 22**.

Figure 23 shows the power pack terminals. Refer to the diagram when reconnecting power

TROUBLESHOOTING

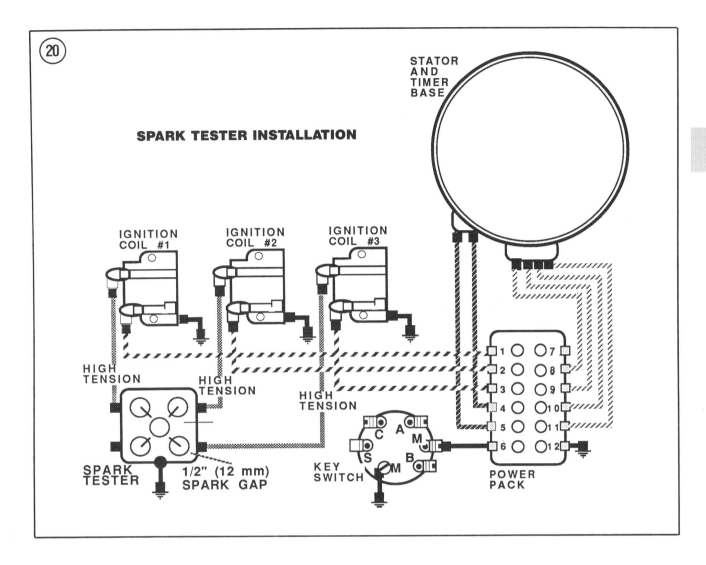

pack leads to avoid any misconnections that can cause damage to the power pack or CD 2 ignition.

The sensor coil (**Figure 24**) and timer base on electric start models are replaced as an assembly. The charge coils are part of the potted stator assembly. **Figure 25** shows the charge coils and stator without the potting material. If defective, replace the charge coils and stator as an assembly.

On manual start models, charge and sensor coils can be replaced individually. See Chapter Seven.

Coat all electrical terminal connections with OMC Black Neoprene Dip after reconnecting the leads to the terminal board or power pack.

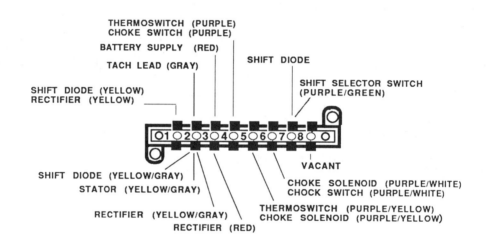

TROUBLESHOOTING

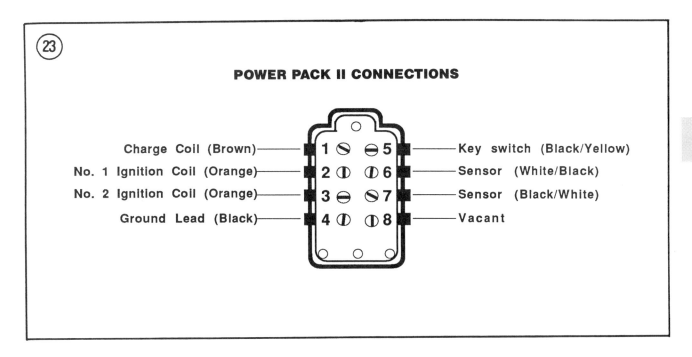

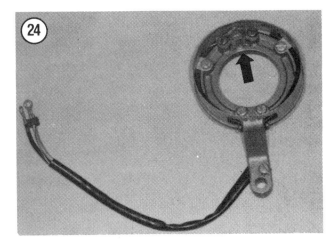

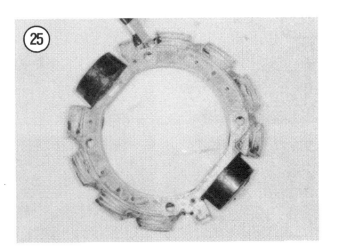

An S-80 or M80 neon test light (available from your Johnson or Evinrude dealer) is required for this procedure.

Ignition Output Test

Refer to **Figure 26** for this procedure.

1. Connect a spark tester as shown in **Figure 26**. Set the tester air gap to 1/2 in.
2. Crank the engine with the ignition switch while watching the spark tester. If strong, steady and alternate sparks appear at the spark tester, the ignition coil output is good.
3. If the sparks at one tester gap are weak or erratic or if there is no spark at one gap, perform the *Sensor Coil Input Test* in this chapter.
4. If the sparks are weak or erratic at both tester gaps, perform the *Charge Coil Output Test* in this chapter.

Sensor Coil Input Test

Refer to **Figure 27** for this procedure.

1. With the spark tester installed as shown in **Figure 27**, disconnect the sensor leads at terminal 6 and terminal 7 of the power pack.

CHAPTER THREE

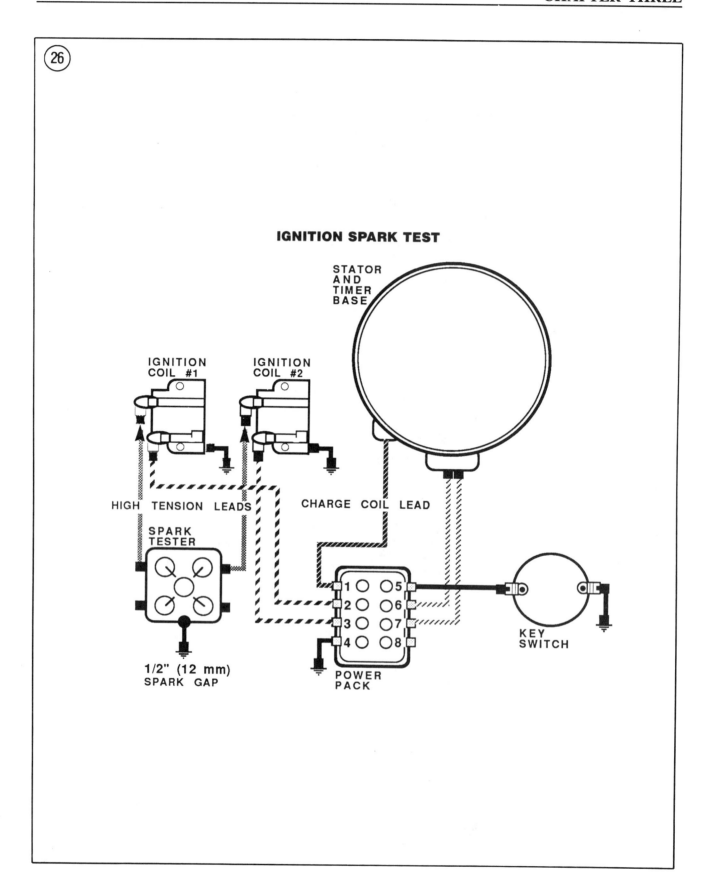

TROUBLESHOOTING

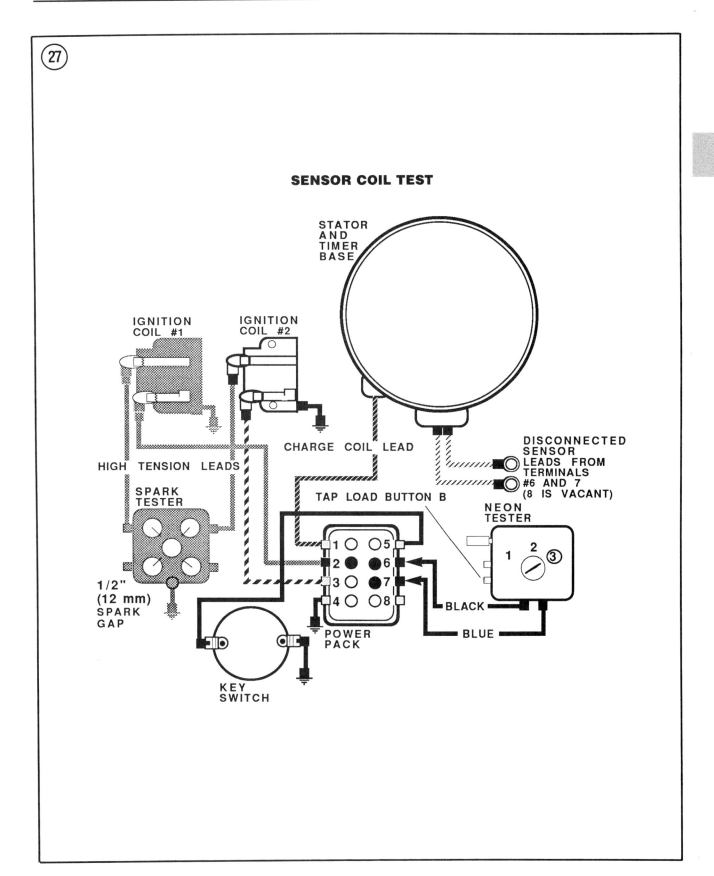

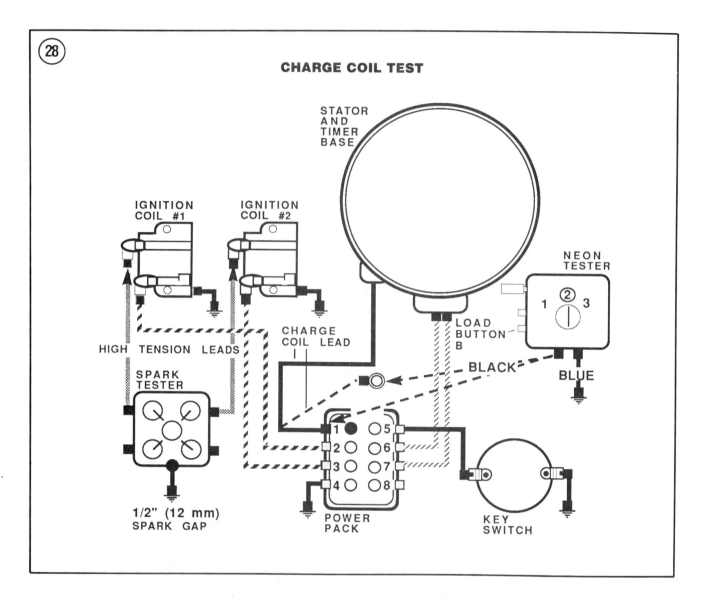

2. Connect the black test lead to terminal 6 and the blue test lead to terminal 7. Set the neon light switch on position 3.

3. Crank the engine with the ignition switch while watching the spark tester and rapidly tapping tester load button B. Reverse the test leads and repeat this step. The coils should fire alternately:
   a. If both coils fire at the same time, replace the sensor coil and timer base assembly.
   b. If there is no spark at either coil, perform the *Charge Coil Output Test* in this chapter.
   c. If there is no spark at one coil, perform the *Power Pack Output Test* in this chapter.

4. Reconnect the sensor leads at the power pack.

Charge Coil Output Test

Refer to **Figure 28** for this procedure.

1. Connect the black test lead at terminal 1 of the power pack and the blue test lead to a good engine ground.

2. Set the neon light switch on position 2. Crank the engine with the ignition switch while watching the tester neon light.

3. If the light glows steadily, perform the *Power Pack Output Test* in this chapter.

TROUBLESHOOTING

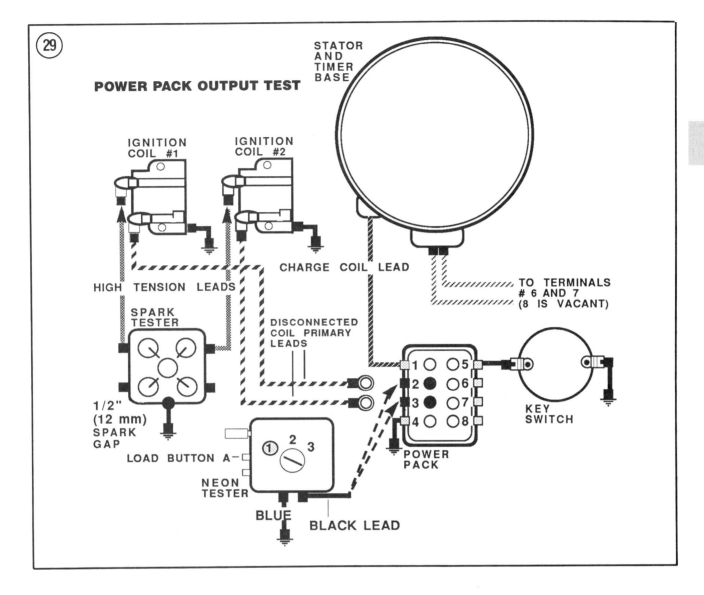

4. If the light glows intermittently or not at all, check the charge coil output lead for an open or a short to ground and make sure the power pack ground lead (terminal 4) is good.

5. Crank the engine after checking the leads in Step 4. If the light remains the same, disconnect the charge coil output lead from power pack terminal 1 and connect it to the black test lead. See **Figure 28**.

6. Depress tester load button B while cranking the engine:
   a. If there is a steady glow from the light, perform the *Power Pack Output Test* in this chapter.
   b. If the tester light does not glow or glows intermittently, replace the charge coil and stator assembly.

7. Reconnect the charge coil lead at the power pack.

Power Pack Output Test

Refer to **Figure 29** for this procedure.

1. Disconnect the ignition coil primary leads at power pack terminals 2 and 3.
2. Connect the black test lead to terminal 2 and the blue test lead to a good engine ground.

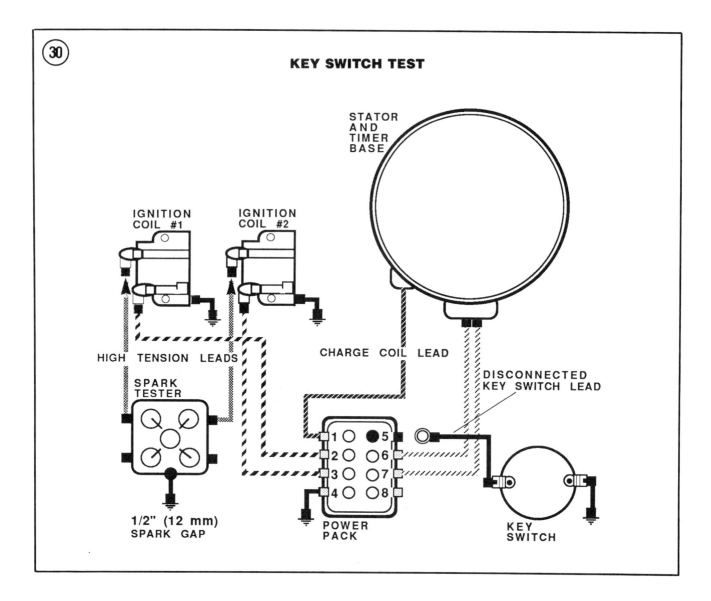

Figure 30 KEY SWITCH TEST

3. Set neon light switch on position 1 and depress load button A. Crank the engine with the ignition switch while watching the tester neon light.

4. Move the black test lead to terminal 3 and repeat Step 3:
   a. If the test light emits a steady glow on both outputs, replace the ignition coils.
   b. If there is not light on either output, perform the *Key Switch Test* as described in this chapter.
   c. If there is no light at one output or if the light is dim or intermittent on one or both outputs, replace the power pack.

5. Reconnect the coil leads at the power pack.

Key Switch Test

Refer to **Figure 30** for this procedure.

1. With the spark tester intalled as shown in **Figure 30**, disconnect the key switch lead at power pack terminal 5.

2. Crank the engine and watch the spark tester. If there is no spark or a spark at only one test gap, replace the power pack. If the spark is good, check the key switch lead for defects. If the lead is good, replace the key switch.

TROUBLESHOOTING

Sensor Coil Resistance Check

1. Disconnect the sensor leads from power pack terminals 6 and 7.
2. Connect an ohmmeter between the disconnected sensor leads and note the reading. If it is not 10-20 ohms, replace the sensor coil and timer base assembly.
3. Set the ohmmeter on the high scale. Connect the red test lead to either one of the disconnected sensor leads. Connect the black test lead to a good engine ground. If the ohmmeter needle moves, the sensor coil is grounded. Replace the sensor coil and timer base assembly.

Charge Coil Resistance Check

1. Disconnect the charge coil lead at power pack terminal 1.
2. Set the ohmmeter on the high scale and connect the test leads between the disconnected charge coil lead and a good engine ground.
3. If the ohmmeter reading is not 800-950 ohms (1973-1975) or 675-825 ohms (1976-1977) at room temperature (70° F), replace the charge coil and stator assembly.

1978-1984 CD 2 IGNITION TROUBLESHOOTING

The CD 2 ignition is used on 2-cylinder engines. The 4-wire connector plugs connect the charge and sensor coil leads to the power pack. The 3-wire connector plugs connect the power pack and ignition coils. No timing adjustments are required with this ignition. Correct timing will be maintained as long as the wires are properly positioned in the connectors.

If the ignition system produces a satisfactory spark and the engine backfires but will not start, the ignition timing may be 180 degrees off. Check to make sure that the black/white wire in the 4-wire connector is positioned in the connector terminals marked "B." Also check to make sure that the No. 1 coil (orange) wire is in the 3-wire connector B terminal and that it connects with the power pack (orange/blue) wire in the other connector B terminal.

Jumper leads are required for troubleshooting. Fabricate 4 leads using 8 inch lengths of 16-gauge wire. Connect a pin (OMC part No. 511469) at one end and a socket (OMC part No. 581656) with one inch of tubing (OMC part No. 519628) at the other. Ohmmeter readings should be made when the engine is cold. Readings taken on a hot engine will show increased resistance caused by engine heat and result in unnecessary parts replacement without solving the basic problem.

Output tests should be made with a Stevens or Electro-Specialties CD voltmeter and CD adapter.

Stop Button Elimination Test

Refer to **Figure 31** for this procedure.
1. Connect a spark tester as shown in **Figure 20**. Set the tester air gap to 1/2 in.
2. Separate the power pack-to-ignition coil 3-wire connector. Insert a jumper wire between the connector B terminals. Insert a jumper wire between the connector C terminals.
3. Crank the engine while watching the spark tester. If there is no spark or a spark at only one gap, remove the jumper wires and reconnect the connector plugs. Continue testing to locate the problem.
4. If a spark jumps both gaps alternately, the problem is in the stop button circuit or the emergency ignition cutoff switch.

Sensor Coil Resistance Test

1. Separate the power pack-to-armature plate 4-wire connector. Insert jumper wires in terminals B and C of the armature plate end of the connector.

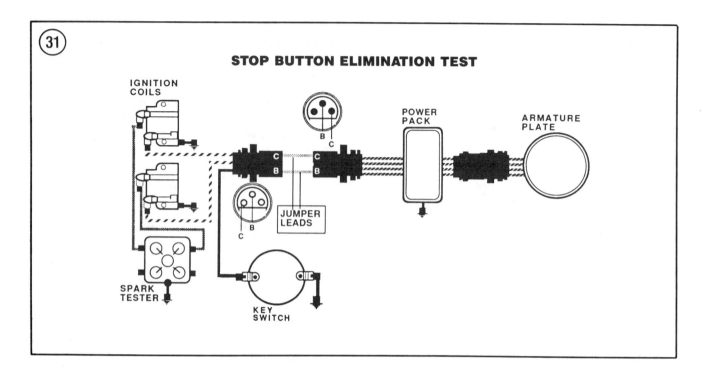

2. Connect an ohmmeter between the 2 jumper wires (**Figure 32**) and note the reading. If it is not 30-50 ohms, replace the sensor coil.

3. Set the ohmmeter on the high scale. Ground the black test lead at the armature plate and connect the red test lead to the C terminal jumper wire (**Figure 33**). The ohmmeter needle should not move. If it does, the sensor coil is grounded. Check for a grounded sensor coil lead before replacing the coil.

4. Remove the jumper wires and perform the *Charge Coil Resistance Test*.

Charge Coil Resistance Test

1. With the 4-terminal connector plug disconnected, insert jumper wires in terminals A and D of the armature plate end of the connector.

2. Connect an ohmmeter between the jumper wires (**Figure 34**) and note the reading. If it is not 400-550 ohms (electric start) or 500-650 ohms (manual start), replace the charge coil.

3. Set the ohmmeter on the high scale. Ground the black test lead at the armature plate and connect the red test lead to the A terminal jumper

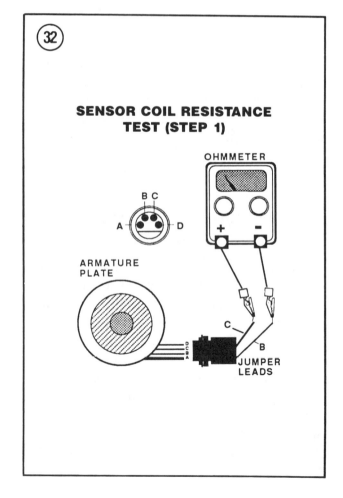

TROUBLESHOOTING

wire (**Figure 35**). The ohmmeter needle should not move. If it does, the charge coil is grounded. Check for a grounded charge coil lead before replacing the coil.

Charge Coil Output Test

Refer to **Figure 36** for this procedure.

1. Disconnect the 4-wire connector. Set the CD voltmeter switches to NEGATIVE and 500. Insert the red test lead in cavity A of the armature plate end of the connector. Ground the black test lead at the armature plate.
2. Crank the engine and note the meter reading.
3. Move the red test lead to cavity D of the connector and crank the engine again. Note the meter reading.
4. There should be no meter reading in Step 2 or Step 3. If there is, check for a grounded charge coil lead before replacing the coil.
5. Leave the red test lead in cavity D of the connector. Remove the black test lead from the armature plate and insert it in cavity A of the connector.
6. Crank the engine and note the meter reading. If it is less than 230 volts, replace the charge coil.

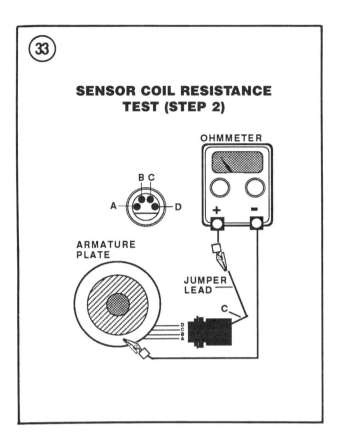

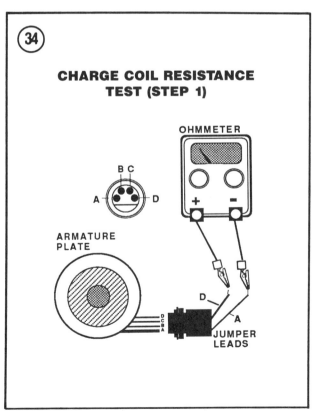

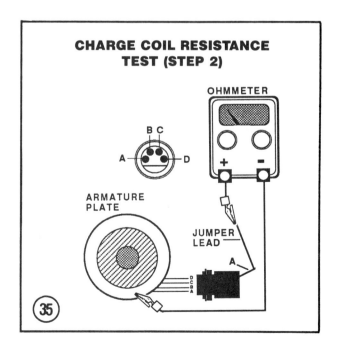

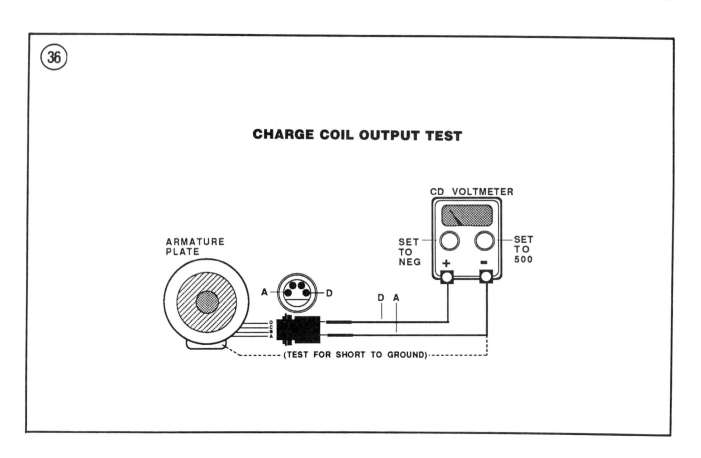

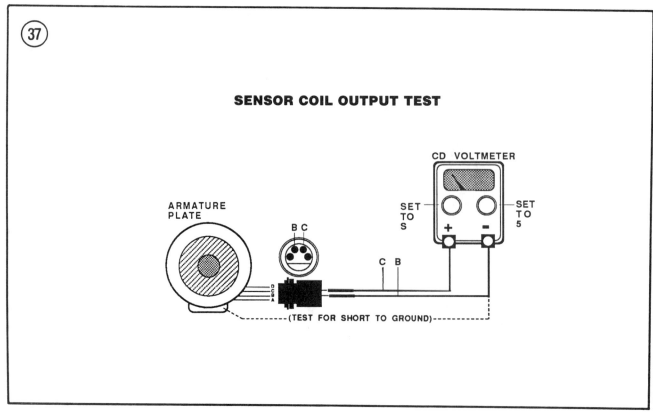

TROUBLESHOOTING

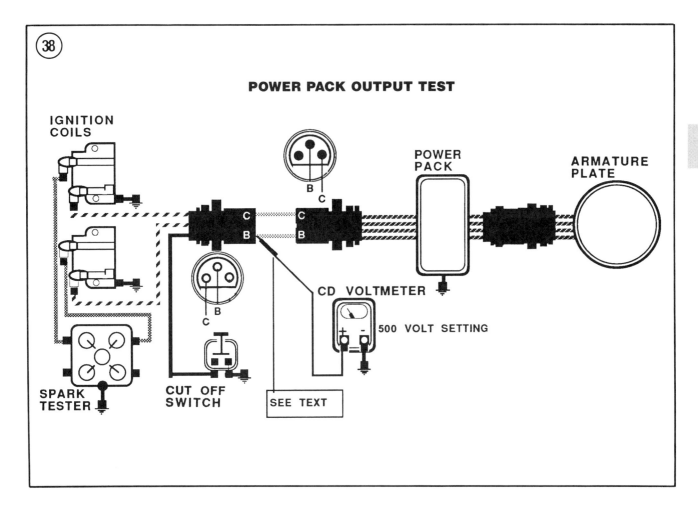

Sensor Coil Output Test

Refer to **Figure 37** for this procedure.

1. Disconnect the 4-wire connector. Set the CD voltmeter switches to S and 5. Insert the red test lead in cavity C of the armature plate end of the connector. Ground the black test lead at the armature plate.
2. Crank the engine and note the meter reading.
3. Move the red test lead to cavity B of the connector and crank the engine again. Note the meter reading.
4. There should be no meter reading in Step 2 or Step 3. If there is, check for a grounded sensor coil lead before replacing the coil.
5. Leave the red test lead in cavity B of the connector. Remove the black test lead from the armature plate and insert it in cavity C of the connector.
6. Crank the engine and note the meter reading. If it is less than 0.3 volts, replace the sensor coil.

Power Pack Output Test

Refer to **Figure 38** for this procedure.

1. Disconnect the 3-wire connector. Set the CD voltmeter switches to NEGATIVE and 500. Insert jumper wires between terminals B and C of the connector.
2. Connect the red test lead to the jumper lead at terminal B and ground the black test lead.
3. Crank the engine and note the meter reading.
4. Move the red test lead to the jumper lead at terminal C. Crank the engine and note the meter reading.
5. If the meter reading is 180 volts or more in Step 3 or Step 4, check the ignition coil(s). If

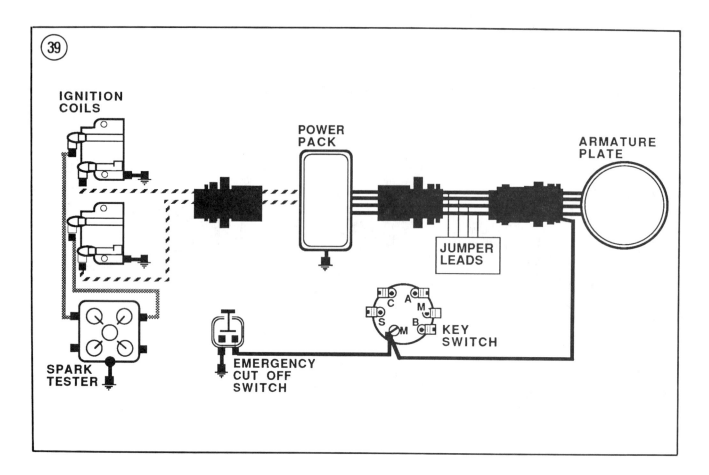

there is no reading, the power pack is probably defective. Substitute a known-good power pack and repeat the procedure.

Ignition Coil Resistance Test

1. Disconnect the high tension lead at the ignition coil.
2. Disconnect the 3-wire connector. Insert a jumper lead in terminal B of the ignition coil end of the connector.
3. Connect the ohmmeter red test lead to the B terminal jumper lead. Connect the black test lead to a good engine ground. The meter should read 0.1 ± 0.05 ohms.
4. Set the ohmmeter on the high scale. Move the black test lead to the ignition coil high tension terminal. The meter should read 225-325 ohms.
5. If the readings are not as specified in Step 3 or Step 4, replace the No. 1 ignition coil.

6. Move the jumper lead from terminal B to terminal C and repeat the procedure to test the No. 2 ignition coil.

1985-1988 CD 2 IGNITION TROUBLESHOOTING

The CD 2 ignition is used on 2-cylinder engines. A 5-wire connector plug connects the charge and sensor coil leads to the power pack. No timing adjustments are required with this ignition. Correct timing will be maintained as long as the wires are properly positioned in the connectors.

Jumper leads are required for troubleshooting. Fabricate 5 leads using 8 inch lengths of 16-gauge wire. Connect a pin (OMC part No. 511469) at one end and a socket (OMC part No. 581656) with one inch of tubing (OMC part No. 519628) at the other. Ohmmeter readings should be made when the engine is cold. Readings taken on a hot

TROUBLESHOOTING

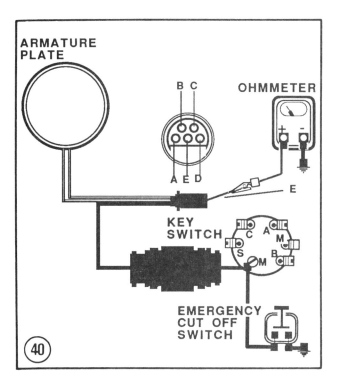

40

engine will show increased resistance caused by engine heat and result in unnecessary parts replacement without solving the basic problem.

Output tests should be made with a Stevens or Electro-Specialties CD voltmeter and CD adapter.

Stop Button/Key Switch Elimination Test

Refer to **Figure 39** for this procedure.
1. Connect a spark tester as shown in **Figure 20**. Set the tester air gap to 1/2 in.
2. Separate the power pack-to-armature plate 5-wire connector. Insert jumper wires between the A, B, C and D terminals of the connector.
3. Crank the engine while watching the spark tester. If there is no spark or a spark at only one gap, remove the jumper wires and reconnect the connector plugs. Continue testing to locate the problem.
4. If a spark jumps both gaps alternately, the problem is in the stop button/key switch circuit or the emergency ignition cutoff switch.

Stop Button Circuit Test

Refer to **Figure 40** for this procedure.
1. Separate the power pack-to-armature plate 5-wire connector.
2. Install the cap/lanyard assembly on the emergency ignition cutoff switch, if so equipped.
3. Insert a jumper wire in the connector E terminal. Connect the red ohmmeter test lead to the jumper wire and the black test lead to a good ground. The ohmmeter should read infinity.
4. If the meter reading is not as specified in Step 3, remove the emergency ignition cutoff switch from the stop button circuit (if so equipped).
   a. If the ohmmeter reads infinity, replace the cutoff switch.
   b. If the ohmmeter reads zero, replace the stop button.
5. Depress the stop button. The meter should read zero. If not, replace the stop button.
6. Remove the cap/lanyard assembly from the cutoff switch, if so equipped. The meter should read zero. If not, replace the cutoff switch.

Key Switch Stop Circuit Test

Refer to **Figure 40** for this procedure.
1. Separate the power pack-to-armature plate 5-wire connector.
2. Install the cap/lanyard assembly on the emergency ignition cutoff switch, if so equipped.
3. Insert a jumper wire in the connector E terminal. Connect the red ohmmeter test lead to the jumper wire and the black test lead to a good ground. The ohmmeter should read infinity.
4. If there is meter needle movement in Step 3, disconnect the black/yellow wire at the key switch M terminal with the ohmmeter still connected.
5. If the meter reads infinity with the wire disconnected in Step 4, test the key switch. If the meter reads zero, remove the emergency ignition cutoff switch from the circuit.
   a. If the ohmmeter reads infinity, replace the cutoff switch.

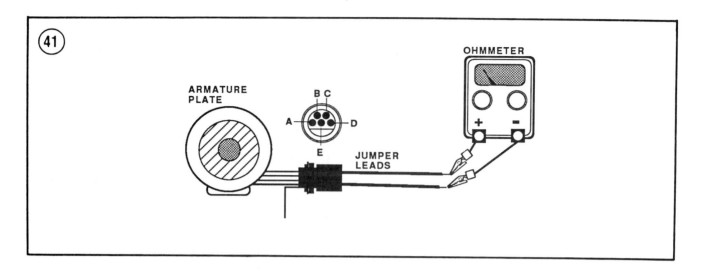

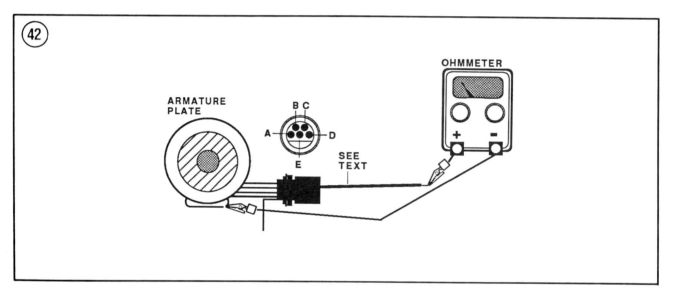

b. If the ohmmeter reads zero, check for a defect in the black/yellow wire between the key switch and 5-wire connector. Repair or replace as required.

Sensor Coil Resistance Test

1. Separate the power pack-to-armature plate 5-wire connector. Insert jumper wires in terminals B and C of the armature plate end of the connector.
2. Connect an ohmmeter between the 2 jumper wires (**Figure 41**) and note the reading. If it is not 40 ± 10 ohms, replace the sensor coil.
3. Set the ohmmeter on the high scale. Ground the black test lead at the armature plate and connect the red test lead alternately to the B and C terminal jumper wires (**Figure 42**). The ohmmeter needle should not move. If it does, the sensor coil is grounded. Check for a grounded sensor coil lead before replacing the coil.
4. Remove the jumper wires and perform the *Charge Coil Resistance Test*.

Charge Coil Resistance Test

1. With the 5-terminal connector plug disconnected, insert jumper wires in terminals

TROUBLESHOOTING

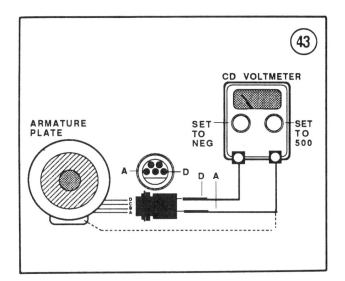

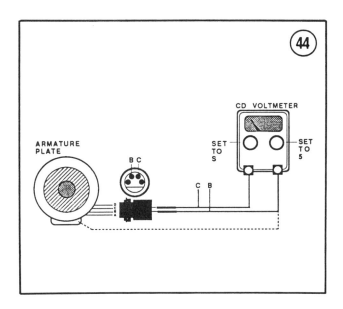

Charge Coil Output Test

Refer to **Figure 43** for this procedure.
1. Disconnect the 5-wire connector. Set the CD voltmeter switches to NEGATIVE and 500. Insert the red test lead in cavity A of the armature plate end of the connector. Ground the black test lead at the armature plate.
2. Crank the engine and note the meter reading.
3. Move the red test lead to cavity D of the connector and crank the engine again. Note the meter reading.
4. There should be no meter reading in Step 2 or Step 3. If there is, check for a grounded charge coil lead before replacing the coil.
5. Leave the red test lead in cavity D of the connector. Remove the black test lead from the armature plate and insert it in cavity A of the connector.
6. Crank the engine and note the meter reading. If it is less than 230 volts, check for a problem in component connectors and/or wiring. If no problem is found, replace the charge coil.

Sensor Coil Output Test

Refer to **Figure 44** for this procedure.
1. Disconnect the 5-wire connector. Set the CD voltmeter switches to S and 5; if using a Merc-O-Tronic Model 781 voltmeter, set the switches to POSITIVE and 5.
2. Insert the red test lead in cavity C of the armature plate end of the connector. Ground the black test lead at the armature plate.
3. Crank the engine and note the meter reading.
4. Move the red test lead to cavity B of the connector and crank the engine again. Note the meter reading.
5. There should be no meter reading in Step 3 or Step 4. If there is, check for a grounded sensor coil lead before replacing the coil.
6. Leave the red test lead in cavity B of the connector. Remove the black test lead from the armature plate and insert it in cavity C of the connector.

A and D of the armature plate end of the connector.
2. Connect an ohmmeter between the jumper wires (**Figure 41**) and note the reading. If it is not 575 ± 25 ohms, replace the charge coil.
3. Set the ohmmeter on the high scale. Ground the black test lead at the armature plate and connect the red test lead alternately to the A and D terminal jumper wires (**Figure 42**). The ohmmeter needle should not move. If it does, the charge coil is grounded. Check for a grounded charge coil lead before replacing the coil.

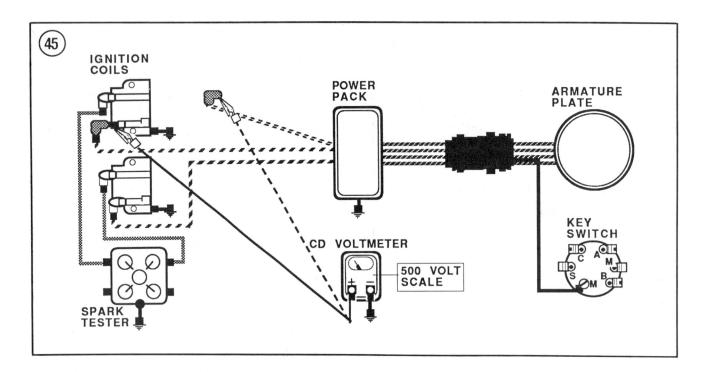

7. Crank the engine and note the meter reading. If it is less than 2 volts, check for a problem in the component connectors and/or wiring. If no problem is found, replace the sensor coil.

Power Pack Output Test

Refer to **Figure 45** for this procedure.
1. Disconnect the primary lead from each ignition coil. Install a terminal extender on the primary terminal of each coil, then reconnect the primary lead to the extender terminal.
2. Set the CD voltmeter switches to NEGATIVE and 500.
3. Connect the red test lead to the No. 1 coil terminal extender and ground the black test lead.
4. Crank the engine and note the meter. It should read 200 volts or more.
5. If the meter reading is less than 200 volts in Step 4, disconnect the primary lead from the No. 1 coil terminal extender. Connect the red test lead directly to the spring clip in the primary lead boot and repeat Step 4.
   a. If the meter now reads 200 volts or more, test the ignition coil as described in this chapter.
   b. If the meter still reads less than 200 volts, check the spring clip and primary lead wire for defects. If none are found, replace the power pack.
6. Repeat Steps 3-5 to test the No. 2 ignition coil.
7. Remove the terminal extenders from the coil terminals with a clockwise pulling motion. Reconnect the primary leads to the coil terminals. Make sure the orange/blue lead connects to the No. 1 coil.

Ignition Coil Resistance Test

1. Disconnect the high tension and primary leads at each ignition coil.
2. Connect the ohmmeter red test lead to the primary terminal of the No. 1 coil. Connect the black test lead to a good engine ground. The meter should read 0.1 ± 0.05 ohms.
3. Set the ohmmeter on the high scale. Move the black test lead to the ignition coil high tension terminal. The meter should read 225-325 ohms.
4. If the readings are not as specified in Step 2 or Step 3, replace the No. 1 ignition coil.
5. Repeat Steps 2-4 to test the No. 2 ignition coil.

TROUBLESHOOTING

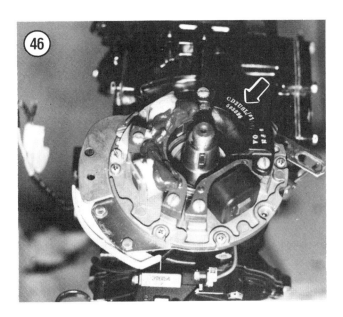

6. With the ohmmeter on the low scale, connect the red test lead to one end of the high tension lead and the black test lead to the other end to check wire resistance. Resistance should be near zero; if not, replace the high tension lead. Repeat this step to test the other high tension lead.

1989-1990 CD2USL IGNITION TROUBLESHOOTING

The CD2USL ignition system is used on 1989 and 1990 48 and 50 hp models. The ignition system is completely contained under the flywheel with the exception of the ignition coils. The breakdown of the CD2USL model number is as follows: CD-capacitor discharge; 2-two cylinders; U-under the flywheel; S-contains S.L.O.W. (speed limiting overheat warning) mode; L-contains rpm rev. limiter. Ignition system model number is printed on top of ignition module located beneath engine flywheel. The ignition module incorporates the power pack and the sensor coil into one assembly instead of two assemblies as on CD 2 ignition systems. See **Figure 46**.

S.L.O.W. Operation And Testing

The S.L.O.W. (speed limiting overheat warning) circuit is activated when the engine temperature exceeds 203° F. When activated, the S.L.O.W. mode will limit the engine speed to approximately 2000 rpm. To be deactivated, the engine must be cooled to 162° F and the engine speed must be slowed to an idle.

On models equipped with remote electric start, a blocking diode located in the engine wiring harness is used to isolate the S.L.O.W. warning system from other warning horn signals.

1. Install the engine in a test tank with the proper test wheel.
2. Connect a tachometer according to manufacturer's instructions.
3. Disconnect temperature sensor tan lead at connector (**Figure 47**).
4. Start engine and run at approximately 3500 rpm.
5. Connect engine wiring tan lead to a clean engine ground.
6. Note engine rpm. Engine rpm should reduce to approximately 2000 rpm. If not, check engine wiring for an open and if not found, replace ignition module. Refer to Chapter Seven.
7. Remote Electric Start Models—Calibrate an ohmmeter to R×100 scale. Connect one ohmmeter lead to engine wiring tan lead for

temperature sensor and other ohmmeter lead to tan wire connector in red engine harness connector. Note ohmmeter reading, then reverse leads or push "polarity" button on meter if so equipped and note reading. Continuity (near zero resistance) should be noted in one direction and no continuity or infinite resistance should be noted in other direction. If not, replace blocking diode in engine wiring harness.

Ignition System Output Test

1. Remove spark plug leads from spark plugs.
2. Mount a spark tester on the engine and connect a tester lead to each spark plug lead. Set the spark tester air gap to 1/2 in. See **Figure 48**.
3. Connect ignition system emergency cutoff clip and lanyard if so equipped.
4. Crank the engine over while watching the spark tester. If a spark jumps at each air gap, test the ignition system as outlined under *Running Output Test* in this chapter.
5. If a spark jumps at *only* one air gap, first make sure all connections are good, then retest. If only one spark is noted, test ignition system as outlined under *Ignition Plate Output Test* in this chapter.
6. If *no* spark jumps at either spark tester air gap, first make sure all connections are good, then retest. If no spark is noted, test ignition system as outlined under *Stop Circuit Test* in this chapter.

Stop Circuit Test

1. Connect spark tester as outlined under *Ignition System Output Test*.
2. Disconnect stop circuit connector. See **Figure 49**.
3. Make sure ignition system emergency cutoff clip and lanyard are installed if so equipped.
4. Crank the engine over while watching the spark tester. If a spark jumps at each air gap, test the stop circuit using the following ohmmeter tests. If *no* spark jumps at either spark tester air gap, first make sure all connections are good,

TROUBLESHOOTING

then retest. If no spark is noted, test ignition system as outlined under *Ignition Plate Output Test* in this chapter.

5. Calibrate an ohmmeter to Rx100 scale.

6A. Stop Button Models—Connect one ohmmeter lead to the terminal end of stop circuit connector (**Figure 49**) which leads to tiller handle stop button.

6B. Key Switch Models—Connect one ohmmeter lead to the terminal end of stop circuit connector (**Figure 49**) which leads to key switch.

7. Connect remaining ohmmeter lead to a good engine ground.

8A. Stop Button Models—Note ohmmeter. No continuity or infinite resistance should be noted. If continuity is noted, look for a short in the wiring and if not found, replace stop button assembly.

8B. Key Switch Models—Place key switch in "ON" position. Note ohmmeter. No continuity or infinite resistance should be noted. If continuity is noted, look for a short in the wiring and if not found, replace key switch assembly.

9A. Stop Button Models—Depress stop button or remove stop button clip and note ohmmeter. Continuity (near zero resistance) should be noted. If continuity is not noted, look for an open in the wiring and if not found, replace stop button assembly.

9B. Key Switch Models—Place key switch in "OFF" position. Note ohmmeter. Continuity (near zero resistance) should be noted. If continuity is not noted, look for an open in the wiring and if not found, replace key switch assembly.

Ignition Plate Output Test

WARNING
Disconnect spark plug leads to prevent accidental starting.

1. Remove primary leads from ignition coils.
2. Connect No. 1 cylinder ignition coil primary lead to the red lead of Stevens load adapter No. PL-88 and the black lead of load adapter to a good engine ground. See **Figure 50**.
3. Connect the red lead of a peak-reading voltmeter to the connector end of the red lead on Stevens load adapter No. PL-88. Connect the black lead of the peak-reading voltmeter to a good engine ground.
4. Set the peak-reading voltmeter on "POS" and "500".
5. Crank the engine over while noting the peak-reading voltmeter. The meter should show at least 175 volts.
6. Repeat Steps 2-5 for No. 2 cylinder ignition coil primary lead.
7. If both primary leads show at least 175 volts, then refer to *Ignition Coil Resistance Test* in this chapter.
8. If only one primary lead shows at least 175 volts, the ignition module must be replaced. Refer to Chapter Seven.
9. If neither primary lead shows sufficient output, then refer to *Charge Coil Resistance Test* in this chapter.

Charge Coil Resistance Test

WARNING
Disconnect spark plug leads to prevent accidental starting.

1. Remove automatic rewind starter, if so equipped, and engine flywheel. Refer to Chapter Eight.
2. Remove the two ignition module mounting screws.
3. Raise ignition module to expose the two bullet connectors. See **Figure 51**.
4. Disconnect the two bullet connectors.
5. Calibrate an ohmmeter to Rx100 scale.
6. Connect one ohmmeter lead to the terminal of the brown wire leading to the charge coil and connect the remaining ohmmeter lead to the terminal of the brown/yellow wire leading to the charge coil.
7. The ohmmeter should show between 535-585 ohms.
8. To test charge coil for shorts-to-ground, connect the black ohmmeter lead to a good engine ground and alternately connect the ohmmeter red lead to the terminal ends of the brown and brown/yellow wires leading from the charge coil.
9. No continuity or infinite resistance should be noted on both wires.
10. Replace charge coil if test results are not within specifications.
11. If the charge coil test good, then the ignition module must be replace.

Running Output Test

1. Remove primary leads from ignition coils and connect Stevens terminal extenders No. TS-77 to ignition coil, then reconnect primary leads.
2. Connect the red lead of a peak-reading voltmeter to the No. 1 cylinder ignition coil terminal extender. Connect the black lead of the peak-reading voltmeter to a good engine ground.
3. Set the peak-reading voltmeter on "POS" and "500".
4. Make sure the outboard motor is mounted in a suitable test tank or testing fixture with a

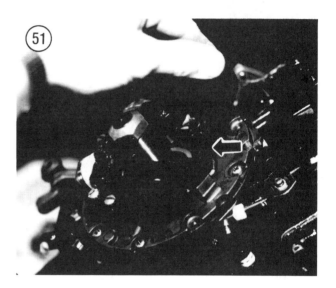

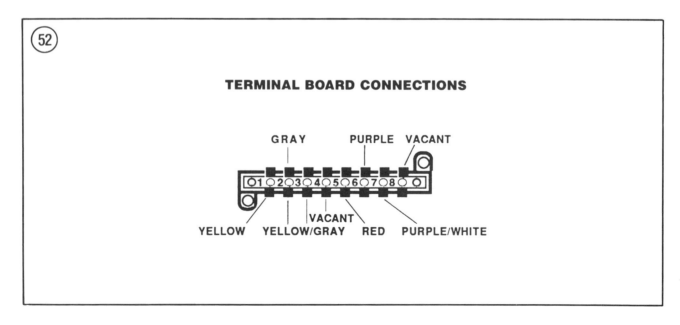

TROUBLESHOOTING

suitable loading device (test wheel or dynometer).

5. Start the engine and operate at the rpm ignition trouble is suspected.
6. The peak-reading voltmeter should show at least 200 volts.
7. Repeat Steps 2-6 to test No. 2 cylinder.
8. If either cylinder shows less than 200 volts, test charge coil as outlined under *Charge Coil Resistance Test* in this chapter.
9. If charge coil test good, then replace ignition module. Refer to Chapter Seven.

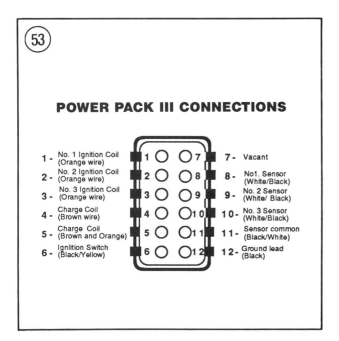

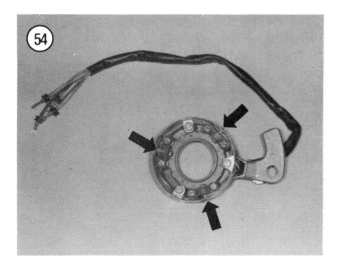

10. Remove ignition coil terminal extenders, then reconnect primary leads.

Ignition Coil Resistance Test

1. Disconnect the primary and high tension (secondary) coil leads at the ignition coil to be tested.
2. With an ohmmeter set on the low scale, connect the red test lead to the coil primary terminal. Connect the black test lead to a good engine ground (if the coil is mounted) or the coil ground tab (if the coil is unmounted). The meter should read 0.1 ± 0.05 ohms.
3. Set the ohmmeter on the high scale. Move the black test lead to the ignition coil high tension terminal. The meter should read 225-325 ohms.
4. If the readings are not as specified in Step 2 or Step 3, replace the ignition coil. See Chapter Seven.

1973-1978 CD 3 IGNITION TROUBLESHOOTING

The CD 3 ignition is used on 3-cylinder engines covered in this manual.

Figure 52 shows the terminal board connections. Although many terminal boards have the wire colors embossed either on the board itself or on the cover (if used), not all models will use all of the wires and connections shown. Double check your connections by referring to **Figure 52**.

Figure 53 shows the power pack terminals. Refer to the diagram when reconnecting power pack leads to avoid any misconnections that can cause damage to the power pack or CD 3 ignition.

The 3 sensor coils are potted in the timer base (**Figure 54**) and are replaced as an assembly. The charge coils are located in the potted stator assembly. **Figure 25** shows the charge coils and stator without the potting material. If defective, replace the charge coils and stator as an assembly.

Coat all electrical terminal connections (except ground stud, nut and terminal motor cable-to-bypass cover) with OMC Black Neoprene Dip after reconnecting the leads to the terminal board or power pack.

An S-80 or M80 neon test light (available from your Johnson or Evinrude dealer) is required for this procedure.

Ignition Output Test

Refer to **Figure 55** for this procedure.
1. Connect a spark tester as shown in **Figure 55**. Set the tester air gap to 1/2 in.
2. Crank the engine with the ignition switch while watching the spark tester. If strong, steady and alternate sparks appear at the spark tester, the ignition coil output is good.
3. If the sparks at 1 or 2 tester gaps are weak or erratic or if there is no spark at 1 or 2 gaps, perform the *Sensor Coil Output Test* in this chapter.
4. If the sparks are weak or erratic at all 3 tester gaps, perform the *Charge Coil Output Test* in this chapter.
5. If there is no spark from any of the coils, perform the *Key Switch Test* in this chapter.

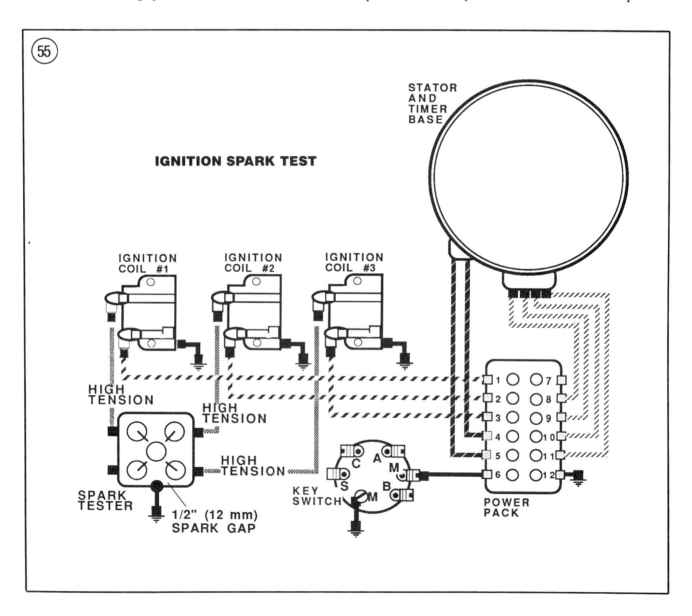

TROUBLESHOOTING

Sensor Coil Output Test

Refer to **Figure 56** for this procedure.

1. With the spark tester installed as shown in **Figure 56**, disconnect the sensor leads at terminals 8, 9, 10 and 11 of the power pack.
2. Mark cylinder number on each ignition coil lead and disconnect at terminals 2 and 3.
3. Connect the black test lead to terminal 8 and the blue test lead to terminal 11. Set neon light switch on position 3.
4. Crank the engine with the ignition switch while watching the spark tester and rapidly tapping tester load button B:

   a. If there is a spark on the ignition coil output, check the sensor coil resistance as described in this chapter.

   b. If there is no spark, perform the *Power Pack Output Test* in this chapter.

5. Reconnect the coil lead at terminal 2 and disconnect the coil lead at terminal 1. Move the black test lead to terminal 9 and repeat Step 4.
6. Reconnect the coil lead at terminal 3 and disconnect the coil lead at terminal 1. Move the black test lead to terminal 10 and repeat Step 4.

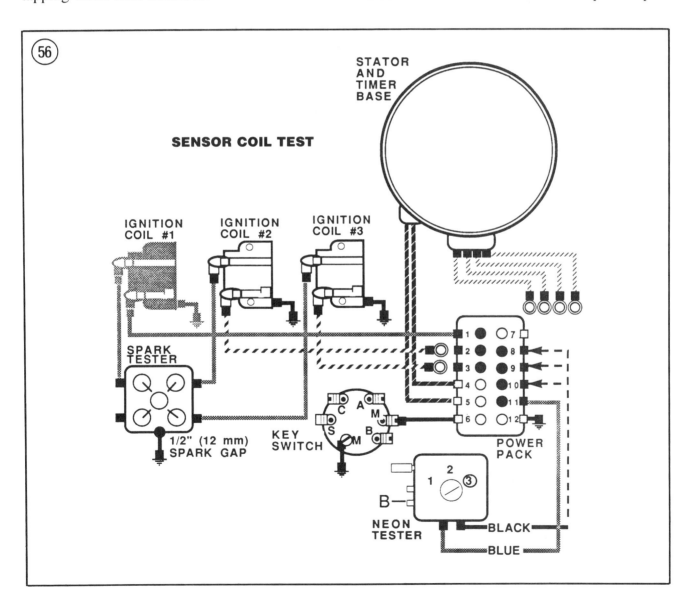

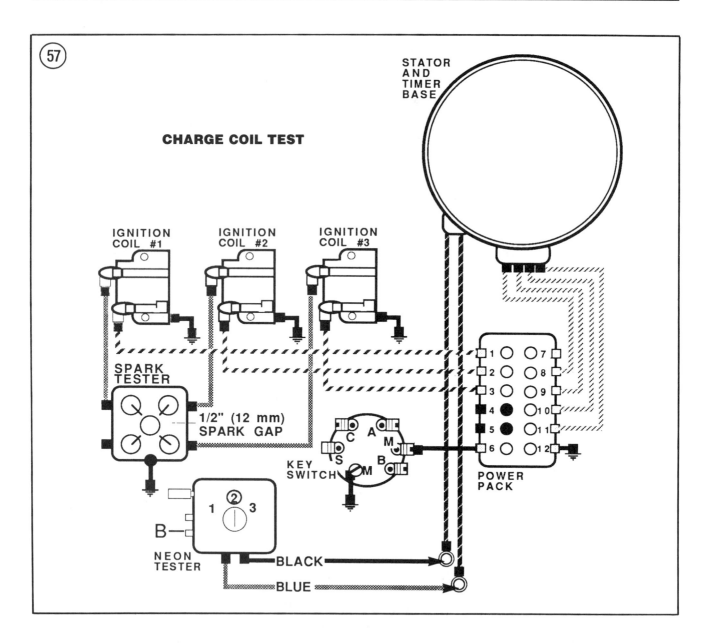

Charge Coil Output Test

Refer to **Figure 57** for this procedure.

1. Disconnect the charge coil leads at power pack terminals 4 and 5.

2. Connect the neon tester leads between the 2 disconnected charge coil leads.

3. Set neon light switch on position 2. Depress load button B and crank the engine with the ignition switch while watching the tester neon light:

   a. If the light glows steadily, perform the *Sensor Coil Output Test* in this chapter.
   b. If the light glows intermittently, check the charge coil output leads for an open or a short to ground and make sure the power pack ground lead (terminal 12) is good. If leads are good, replace the charge coil and stator assembly.
   c. If the light does not glow, replace the charge coil and stator assembly.

4. Reconnect the charge coil leads at the power pack.

TROUBLESHOOTING

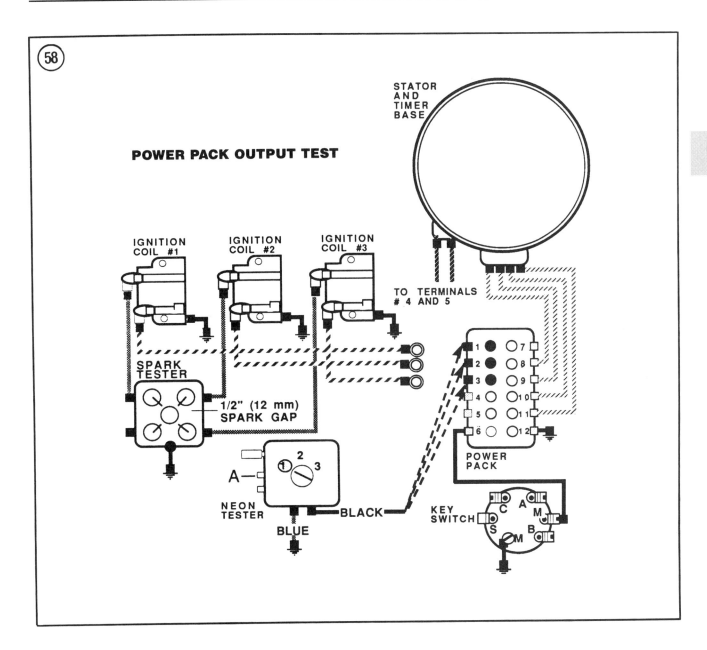

Power Pack Output Test

Refer to **Figure 58** for this procedure.

1. Mark each ignition coil primary lead with the cylinder number, then disconnect from terminals 1, 2 and 3 of the power pack.
2. Connect the black test lead to terminal 1 and the blue test lead to a good engine ground.
3. Set neon light switch on position 1 and depress load button A. Crank the engine with the ignition switch while watching the tester neon light.
4. Move the black test lead to terminal 2 and repeat Step 3.
5. Move the black test lead to terminal 3 and repeat Step 3:

 a. If the test light emits a steady glow on all outputs, replace the ignition coils.
 b. If the test light is steady, dim or intermittent on 1 or 2 outputs, replace the power pack.
 c. If there is no light on any output, check the power pack ground at terminal 12. If

the ground is good, perform the *Key Switch Test* in this chapter. If the switch is good, replace the power pack.

6. Reconnect the coil leads at the power pack.

Key Switch Test

Refer to **Figure 59** for this procedure.

1. With the spark tester installed as shown in **Figure 59**, disconnect the key switch lead at power pack terminal 6.
2. Crank the engine and watch the spark tester. If there is no spark or a spark at only one test gap, the key switch is good. If there is a spark at all coils, check the key switch lead for defects. If the lead is good, replace the key switch.

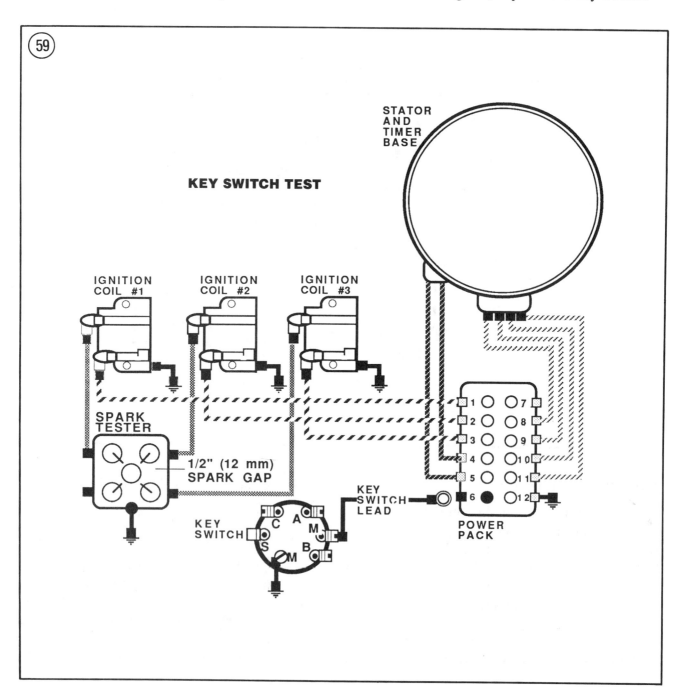

TROUBLESHOOTING

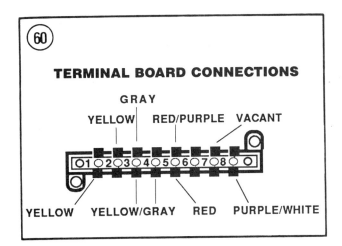

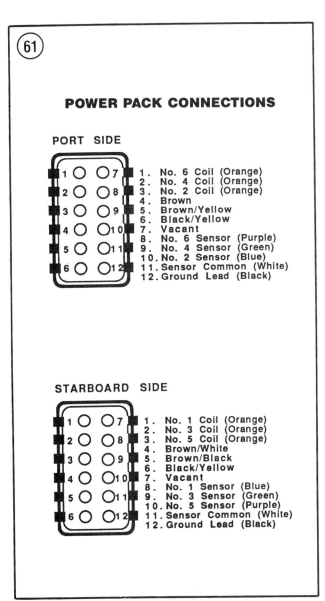

Sensor Coil Resistance Check

1. Disconnect the sensor leads from the power pack terminals 8, 9, 10 and 11.
2. Connect one ohmmeter test lead to the disconnected black/white lead. Alternately connect the other ohmmeter test lead to the white/black wires and note the reading. If it is not 7.5-9.5 ohms at room temperature (70° F), replace the sensor coil and timer base assembly.
3. Set the ohmmeter on the high scale. Connect the black test lead to a good engine ground and the red test lead alternately to the disconnected sensor leads. The ohmmeter should read infinity at each connection. If it does not, a sensor coil is grounded. Replace the sensor coil and timer base assembly.

Charge Coil Resistance Check

1. Disconnect the charge coil leads at power pack terminals 4 and 5.
2. Set the ohmmeter on the high scale and connect the test leads between the disconnected charge coil leads.
3. If the ohmmeter reading is not 870-930 ohms (1973) or 555-705 ohms (1974-1978) at room temperature (70° F), replace the charge coil and stator assembly.

1976-1978 CD 6 IGNITION TROUBLESHOOTING

The CD 6 ignition used on V6 engines consists of 2 identical ignition systems—one on the port side and one on the starboard side of the engine. The following procedures must be performed on each ignition system.

Figure 60 shows the port side terminal board connections; the starboard side is a duplicate. Although many terminal boards have the wire colors embossed either on the board itself or on the cover (if used), not all models will use all of the wires and connections shown. Double check your connections by referring to **Figure 60**.

Figure 61 shows the port and starboard power pack terminals. Refer to the diagram when reconnecting power pack leads to avoid any misconnections that can cause damage to the power pack or CD 6 ignition.

The 6 sensor coils are potted in the timer base (**Figure 62**) and are replaced as an assembly. The 2 charge coils are located in the potted stator assembly (**Figure 63**). If defective, the charge coils and stator are replaced as an assembly. Coat all electrical terminal connections (except the ground stud, nut and terminal motor cable-to-bypass cover) with OMC Black Neoprene Dip after reconnecting the leads to the terminal board or power pack.

An S-80 or M80 neon test light (available from your Johnson or Evinrude dealer) is required for this procedure.

Ignition Output Test

Refer to **Figure 64** for this procedure.
1. Connect 2 spark testers as shown in **Figure 64**. Set the tester air gaps to 7/16 in.

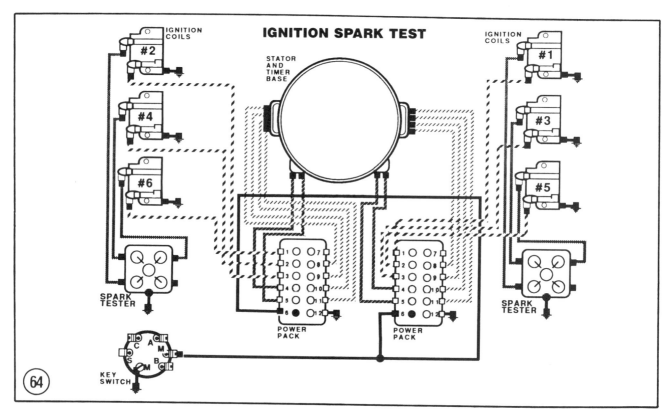

TROUBLESHOOTING

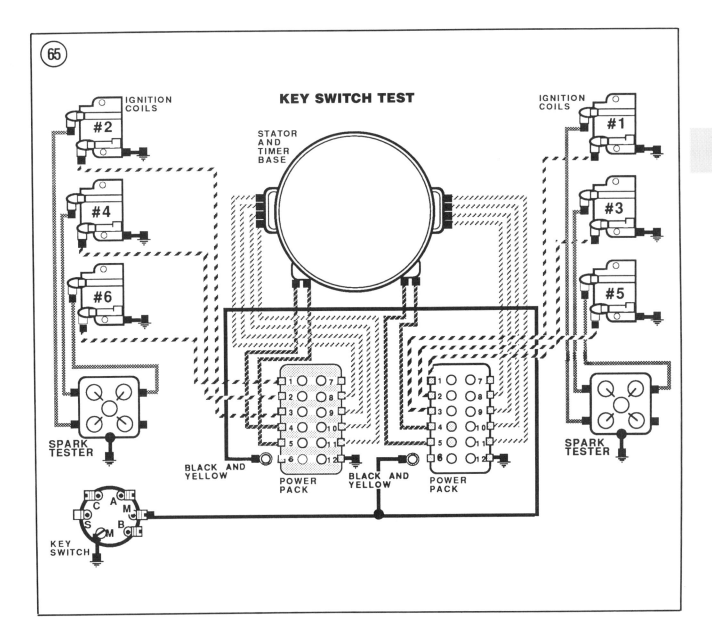

2. Crank the engine with the ignition switch while watching the spark tester. If strong, steady sparks firing one coil at a time appear at the spark tester, the ignition coil output is good.

3. If the sparks from 1 or 2 coils on a side are weak or erratic, perform the *Sensor Coil Output Test* in this chapter.

4. If there is no spark or weak/erratic sparks at all 3 coils on a side, perform the *Charge Coil Output Test* in this chapter.

5. If there is no spark from any of the 6 coils, perform the *Key Switch Test* in this chapter.

Key Switch Test

Refer to **Figure 65** for this procedure.

1. With the spark testers installed as shown in **Figure 65**, disconnect the key switch lead at power pack terminal 6.

2. Crank the engine and watch the spark tester. If there is no spark or a spark at only one test gap, the key switch is good. If there is a spark at all coils, check the key switch lead for defects. If the lead is good, replace the key switch.

CHAPTER THREE

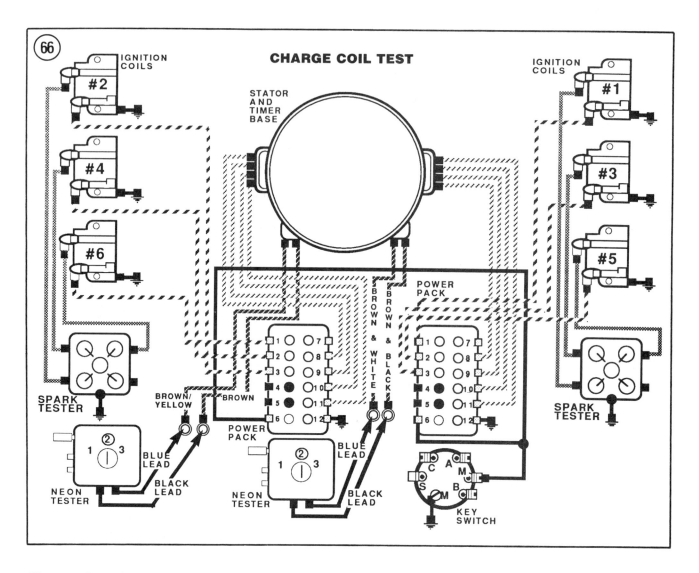

Charge Coil Output Test

Refer to **Figure 66** for this procedure.

1. Disconnect the charge coil leads at power pack terminals 4 and 5.
2. Connect the neon tester leads between the 2 disconnected charge coil leads.
3. Set neon light switch on position 2. Depress load button B and crank the engine with the ignition switch while watching the tester neon light:
   a. If the light glows steadily, perform the *Sensor Coil Output Test* in this chapter.
   b. If the light glows intermittently, check the charge coil output leads for an open or a short to ground and make sure the power pack ground lead (terminal 12) is good. If leads are good, replace the charge coil and stator assembly.
   c. If the test light does not glow, replace the charge coil and stator assembly.
4. Reconnect the charge coil leads at the power pack.

Sensor Coil Output Test

Refer to **Figure 67** for this procedure.

1. With the spark testers installed as shown in **Figure 67**, disconnect the sensor leads at terminals 8, 9, 10 and 11 of the power pack.
2. Connect the black test lead to terminal 8 and the blue test lead to terminal 11. Set neon light switch on position 3.

TROUBLESHOOTING

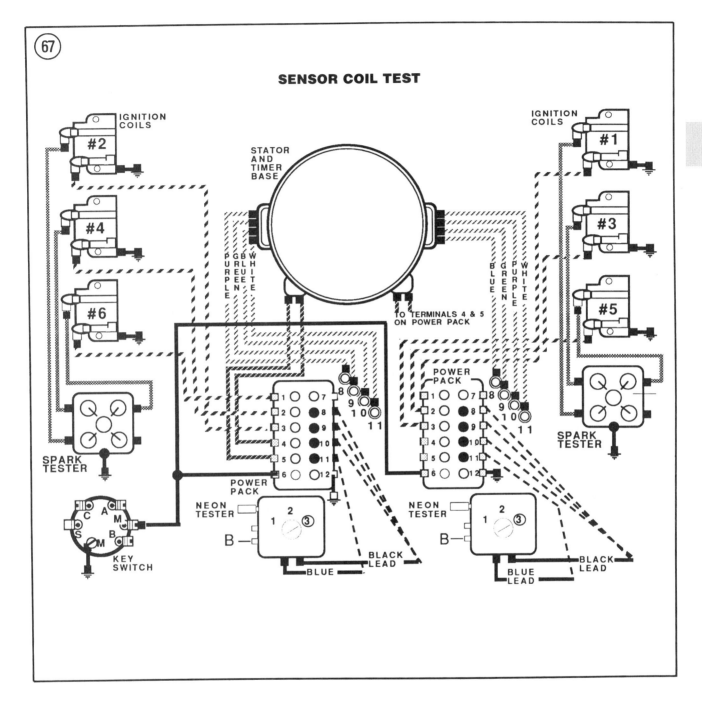

3. Crank the engine with the ignition switch while watching the spark tester and rapidly tapping tester load button B. There should be a spark at the No. 1 (starboard) and No. 6 (port) ignition coils.

4. Move the black test lead to terminal 9 and repeat Step 3. There should be a spark at the No. 3 (starboard) and No. 4 (port) ignition coils.

5. Move the black test lead to terminal 10 and repeat Step 3. There should be a spark at the No. 5 (starboard) and No. 2 (port) ignition coils.

6. If there is no spark as described in Steps 3-5, perform the *Power Pack Output Test* in this chapter. If there is a spark at any coil other than specified in Steps 3-5, replace the power pack.

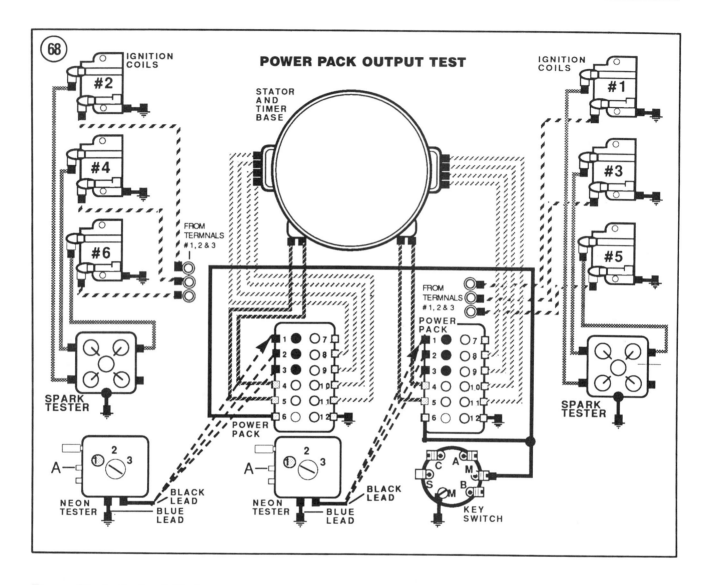

Power Pack Output Test

Refer to **Figure 68** for this procedure.

1. Mark each ignition coil primary lead with the cylinder number, then disconnect from terminals 1, 2 and 3 of the power pack.

2. Connect the black test lead to terminal 1 and the blue test lead to a good engine ground.

3. Set neon light switch on position 1 and depress load button A. Crank the engine with the ignition switch while watching the tester neon light.

4. Move the black test lead to terminal 2 and repeat Step 3.

5. Move the black test lead to terminal 3 and repeat Step 3:
   a. If the test light emits a steady glow on all outputs, replace the ignition coils.
   b. If the test light is steady, dim or intermittent on 1 or 2 outputs, replace the power pack.
   c. If there is no light from the output of both power packs, perform the *Key Switch Test* in this chapter.
   d. If there is no light on all outputs of one power pack, check the power pack ground at terminal 12. If the ground is good, replace the power pack.

6. Reconnect the coil leads at the power pack.

TROUBLESHOOTING

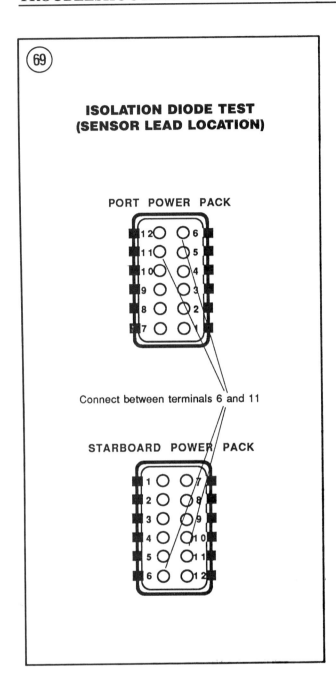

69

ISOLATION DIODE TEST
(SENSOR LEAD LOCATION)

PORT POWER PACK

Connect between terminals 6 and 11

STARBOARD POWER PACK

Power Pack Isolation Diode Test

A malfunctioning isolation diode in one or both power packs can cause a rough running engine. If the diode is not performing its functions, the power packs will interfere with each other. Refer to **Figure 69** for this procedure.

1. Disconnect the leads from power pack terminals 6 and 11.

2. With an ohmmeter on the high scale, connect the test leads between the power pack terminals and note the meter reading.

3. Reverse the test leads and note the reading. If the diode is good, one reading will be high and the other will be low. If both readings are high, the diode is open. If both readings are low, the diode is shorted. Replace the power pack if both readings are high or low.

4. Reconnect the power pack leads and repeat the procedure to check the other power pack.

Sensor Coil Resistance Test

1. Disconnect the sensor leads from the power pack terminals 8, 9, 10 and 11.

2. Connect one ohmmeter test lead to the disconnected white lead. Alternately connect the other ohmmeter test lead to each of the remaining wires and note the reading. If it is not 11-17 ohms (1976-1977) or 13-23 ohms (1978) at room temperature (70° F), replace the sensor coil and timer base assembly.

3. Set the ohmmeter on the high scale. Connect the black test lead to a good engine ground and the red test lead alternately to the disconnected sensor leads. The ohmmeter should read infinity at each connection. If it does not, one or more of the sensor coils are grounded. Replace the sensor coil and timer base assembly.

4. Reconnect the sensor leads to the power pack terminals and repeat the procedure to check the sensor leads at the other power pack.

Charge Coil Resistance Test

1. Disconnect the charge coil leads at power pack terminals 4 and 5.

2. Set the ohmmeter on the high scale and connect the test leads between the disconnected charge coil leads.

3. If the ohmmeter reading is not 220-330 ohms at room temperature (70° F), replace the charge coil and stator assembly.
4. Reconnect the charge coil leads to the power pack terminals and repeat the procedure to check the charge coil leads at the other power pack.

1979-1984 CD 3 AND 1979-1984 CD 6 IGNITION TROUBLESHOOTING

The CD 3 ignition is used on 3-cylinder engines. The CD 6 ignition is used on V6 engines and consists of 2 identical ignition systems—one on the port side and one on the starboard side of the engine. The CD 6 troubleshooting procedures must be performed on each ignition system.

One 4-wire connector joins the sensor coil leads to the power pack. A second 4-wire connector joins the power pack to the ignition coils and ignition switch. A 2-wire connector joins the charge coil to the power pack.

Since the CD 6 ignition (V6 engine) uses 2 power packs, there is one set of connectors on each side of a V6 engine. A retaining clip holds the connectors in place. The clip must be removed to unhook the connector plugs.

Jumper leads are required for troubleshooting. Fabricate 4 leads using 8 inch lengths of 16-gauge wire. Connect a pin (OMC part No. 511469) at one end and a socket (OMC part No. 581656) with one inch of tubing (OMC part No. 519628) at the other. Ohmmeter readings should be made when the engine is cold. Readings taken on a hot engine will show increased resistance caused by engine heat and may result in unnecessary parts replacement without solving the basic problem.

Output tests should be made with a Stevens or Electro-Specialties CD voltmeter.

Key Switch Elimination Test

1. Connect a spark tester to the engine. Set the tester air gap to 7/16 in. for V6 engines and 1/2 in. for 3-cylinder engines.

2. Disconnect the 4-wire connector between the power pack and ignition coil. Insert jumper wires between the connector terminals A, B and C. See **Figure 70**.

3. Crank the engine and watch the spark tester. If there is no spark or a spark at only one gap, remove the jumper wires, reconnect the connector plugs and continue with the test procedures.

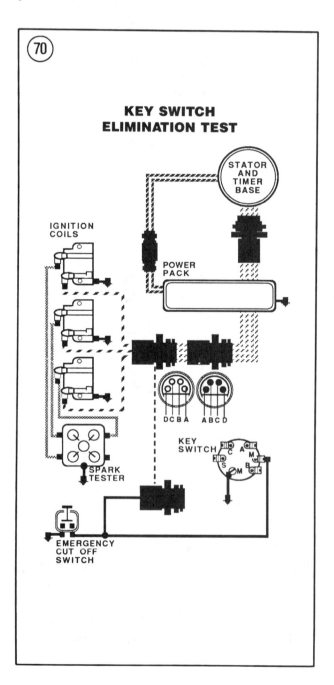

TROUBLESHOOTING

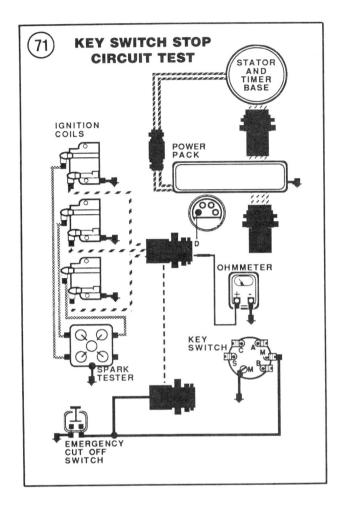

Figure 71. KEY SWITCH STOP CIRCUIT TEST

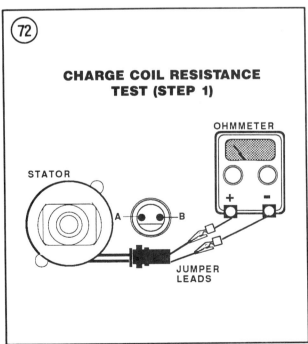

Figure 72. CHARGE COIL RESISTANCE TEST (STEP 1)

4. If a spark jumps alternately at all gaps, the problem is probably in the key switch circuit or emergency ignition cutoff switch, if so equipped.

Key Switch Stop Circuit Test

1. Disconnect the 4-wire connector between the power pack and ignition coil.
2. With the ohmmeter on the high scale, connect the red test lead to the D terminal in the ignition coil end of the connector. Connect the red test lead to a good engine ground. See **Figure 71**.
3. Turn the key switch ON. If equipped with an emergency ignition cutoff switch, put the cap end and lanyard assembly on the switch. The meter should read infinity (open circuit).
4. If the meter does not read infinity in Step 3, disconnect the black/yellow wire from the M terminal on the key switch (and the emergency ignition cutoff switch, if so equipped). If the meter reads infinity with the wire disconnected, the problem is in the key switch or emergency ignition cutoff switch, if so equipped. If the meter does not read infinity, replace the black/yellow wire.
5. Turn the key switch OFF. The meter should indicate a closed circuit. If it does not, look for an open in the key switch, key switch ground lead, black/yellow wire or the emergency ignition cutoff switch, if so equipped.
6. If the engine will not stop when the key switch is turned OFF but the stop circuit passes the test, check the power pack ground. If the ground is good, replace the power pack.

Charge Coil Resistance Test

1. Disconnect the 2-wire connector between the power pack and charge coil(s).
2. Insert jumper wires in terminals A and B of the connector stator end. See **Figure 72**. If the meter does not read 475-625 ohms, replace the charge coil(s).

3. Set the ohmmeter on the high scale. Ground the black test lead at the timer base. Connect the red test lead first to the A terminal jumper wire, then to the B terminal jumper wire and note the meter reading. See **Figure 73**. If the needle moves during this step, the charge coil(s) or leads are shorted. Repair the lead if possible; replace the charge coil(s) if the lead is good.

Sensor Coil Resistance Test

1. Disconnect the 4-wire connector between the power pack and timer base.
2. Insert jumper wires in terminals A and B of the timer base connector end. See **Figure 74**. Insert jumper wires in terminals C and D.
3. With the ohmmeter on the low scale, connect the test leads between A and D jumper wires, between B and D jumper wires and between terminals C and D, in that order. If the meter does not read between 9-21 ohms, replace the sensor coil.
4. Set the ohmmeter on the high scale. Ground the black test lead at the timer base. With the red test lead connected to a jumper wire, insert the jumper wire in terminals A, B, C and D in that order, noting the meter reading. See **Figure 75**. If the needle moves during this step, the sensor coil or leads are shorted. Repair the lead if possible; replace the sensor coil if the lead is good.
5. V6 only—Connect the black test lead to the common lead on the port side sensor connector terminal D and the red test lead to the common lead on the starboard side sensor connector terminal D. If the meter does not read infinity, replace the timer base.

Charge Coil Output Test

Refer to **Figure 76** for this procedure.
1. Disconnect the 2-wire connector between the power pack and stator.
2. Set the CD voltmeter switches to NEGATIVE and 500. Connect a jumper wire to the red test

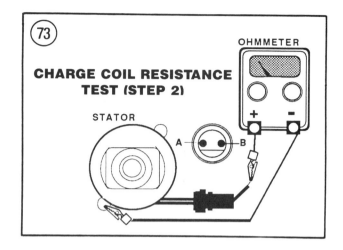

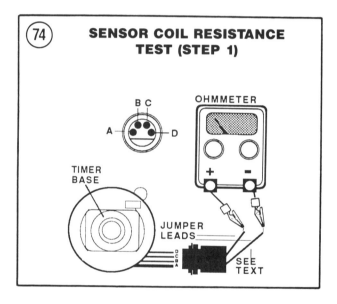

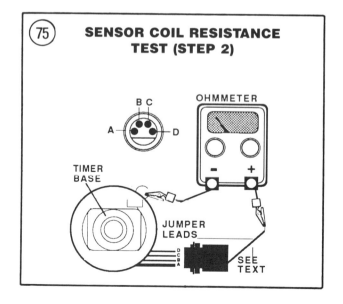

TROUBLESHOOTING

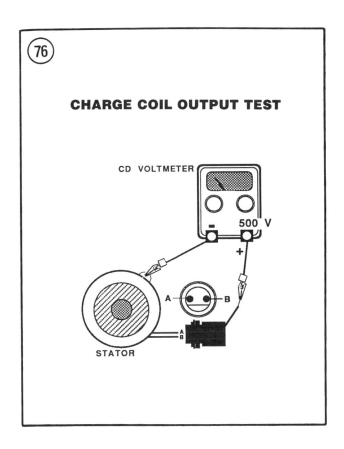

lead. Insert the red test lead jumper wire in cavity A of the stator connector. Ground the black test lead at the timer base. Crank the engine and note the meter reading.

3. Move the red test lead jumper wire to cavity B of the connector. Crank the engine again and note the meter reading. If the needle moves during this step, the charge coil(s) or leads are shorted. Repair the lead if possible; replace the charge coil(s) if the lead is good.

4. Leave the red test lead in cavity B of the connector. Remove the black test lead from the timer base. Connect it to a jumper wire and insert the wire in cavity A of the connector.

5. Crank the engine and note the meter reading. If it is less than 160 volts (V6) or 220 volts (3-cylinder), replace the charge coil.

Sensor Coil Output Test

Refer to **Figure 77** for this procedure.

1. Disconnect the 4-wire connector between the power pack and timer base.

2. Set the CD voltmeter switches to S and 5. Ground the black test lead at the timer base. Connect a jumper wire to the red test lead and probe terminals A, B, C and D in that order while cranking the engine. If the needle moves during this step, the sensor coil or leads are shorted. Repair the leads if possible; replace the sensor coil if the leads are good.

3. Remove the black test lead from the timer base. Connect it to a jumper wire and insert the wire in cavity D of the timer base 4-wire connector. Probe cavity A of the timer base 4-wire connector with the red test lead jumper wire while cranking the engine. Note the meter reading.

4. Move the red test lead jumper wire to cavity B, crank the engine and note the meter reading. Repeat this step with the red test jumper wire in cavity C.

5. If the meter reads less than 0.25 volt (V6) or 0.4 volt (3-cylinder) during Step 3 or Step 4,

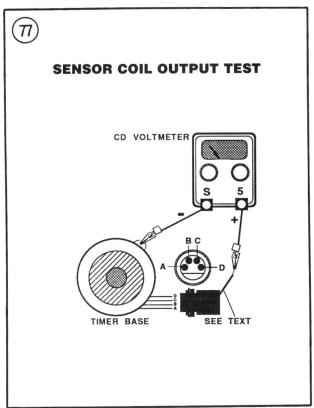

either a sensor coil is defective or engine cranking rpm is too low. The use of a fully charged battery during this procedure will eliminate the possibility of insufficient cranking rpm unless the engine is new and has not been properly broken in or there is a mechanical problem within the power head.

Power Pack Output Test

Refer to **Figure 78** for this procedure.
1. Disconnect the 4-wire connector between the power pack and the ignition coils.
2. Set the CD voltmeter switches to NEGATIVE and 500.
3. Insert jumper wires between terminals A, B and C of the connectors. Connect the black test lead to a good engine ground.
4. Connect the red test lead to the metal portion of the jumper wire at the power pack connector A cavity and crank the engine. Note the meter reading.
5. Repeat Step 4 with the red test lead connected to the B cavity jumper wire, then the C cavity jumper wire. Note each meter reading.
6. If the meter reads at least 170 volts (V6) or at least 230 volts (3-cylinder), check the ignition coil(s). If the meter does not read the specified voltage on one or more outputs, remove the jumper wires from those cavities. Probe the cavities directly with the red test lead while cranking the engine. If there is no change in the meter reading, replace the power pack.

Power Pack Isolation Diode Test (V6 Only)

A malfunctioning isolation diode in one or both power packs on V6 engines can cause a rough running engine. If the diode is not performing its function, the power packs will interfere with each other. Refer to **Figure 79** for this procedure.

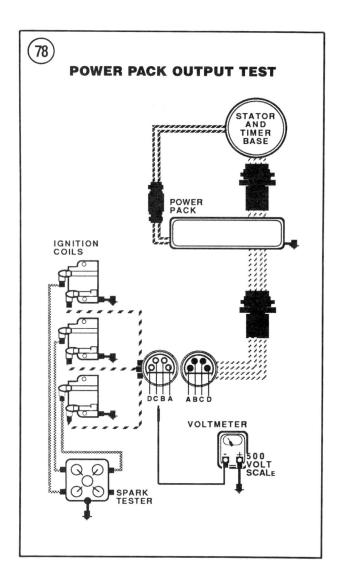

1. Disconnect the 4-wire power pack connectors to the timer base and ignition coils.
2. With an ohmmeter on the high scale, connect the test leads between the D terminals of each connector as shown in **Figure 79** and note the meter reading.
3. Reverse the test leads and note the reading. If the diode is good, one reading will be high and the other will be low. If both readings are high, the diode is open. If both readings are low, the diode is shorted. Replace the power pack if both readings are high or low.
4. Reconnect the power pack connectors and repeat the procedure to check the other power pack.

TROUBLESHOOTING

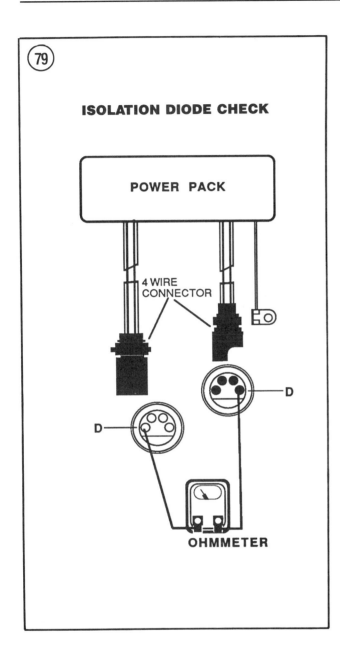

79 ISOLATION DIODE CHECK

Ignition Coil Resistance Test

1. Disconnect the high tension lead at the No. 1 ignition coil.
2. Disconnect the 4-wire connector between the power pack and ignition coils. Insert a jumper lead in terminal A of the ignition coil end of the connector.
3. Connect the ohmmeter red test lead to the A terminal jumper lead. Connect the black test lead to a good engine ground. The meter should read 0.1 ± 0.05 ohms.
4. Set the ohmmeter on the high scale. Move the black test lead to the ignition coil high tension terminal. The meter should read 225-325 ohms.
5. Repeat Steps 1-4 to test the 2 remaining coils. Use terminal B to test the No. 2 coil and terminal C to test the No. 3 coil.
6. On V6 engines, repeat the entire procedure to test the coils in the other ignition system.
7. Replace any coil that does not meet the readings specified in Step 3 or Step 4.

1985-1988 CD 3 IGNITION TROUBLESHOOTING

This CD ignition is used on 3-cylinder engines. A 3-wire connector plug connects the charge coil and key switch leads to the power pack. A 4-wire connector plug connects the timer base (sensor coil) leads to the power pack. Ignition coils connect to the power pack by individual leads. No timing adjustments are required with this ignition. Correct timing will be maintained as long as the wires are properly positioned in the connectors.

Jumper leads are required for troubleshooting. Fabricate 4 leads using 8 inch lengths of 16-gauge wire. Connect a pin (OMC part No. 511469) at one end and a socket (OMC part No. 581656) with one inch of tubing (OMC part No. 519628) at the other. Ohmmeter readings should be made when the engine is cold. Readings taken on a hot engine will show increased resistance caused by engine heat and result in unnecessary parts replacement without solving the basic problem.

Output tests should be made with a Stevens or Electro-Specialties CD voltmeter and CD adapter.

Key Switch Elimination Test

Refer to **Figure 80** for this procedure.
1. Connect a spark tester as described in this chapter. Set the tester air gap to 1/2 in.

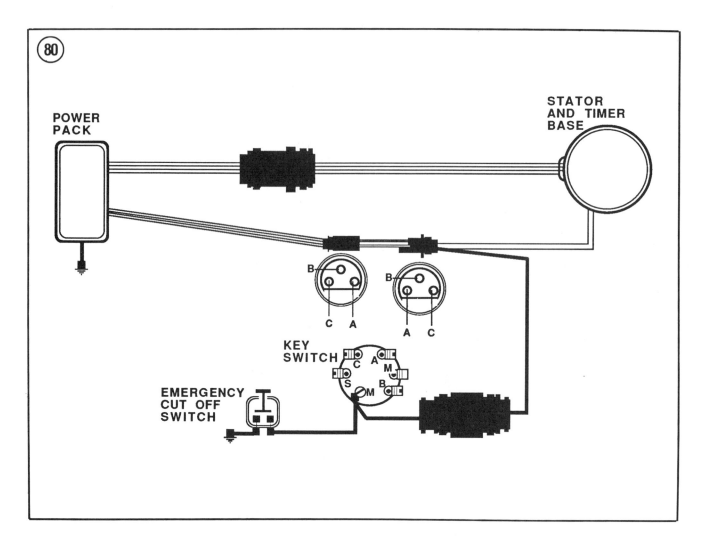

2. Separate the power pack-to-charge coil 3-wire connector. Insert jumper wires between the A and B terminals of the connector.
3. Crank the engine while watching the spark tester. If there is no spark or a spark at only one gap, remove the jumper wires and reconnect the connector plugs. Continue testing to locate the problem.
4. If a spark now jumps all gaps alternately, the problem is in the key switch circuit or the emergency ignition cutoff switch.

Key Switch/Stop Circuit Test (Except WML and TEL Models)

Refer to **Figure 81** for this procedure.

1. Separate the power pack-to-charge coil 3-wire connector.
2. Install the cap/lanyard assembly on the emergency ignition cutoff switch, if so equipped.
3. Insert a jumper wire in the C terminal of the charge coil end connector. Connect the red ohmmeter test lead to the jumper wire and the black test lead to a good ground. Turn the key switch ON. The ohmmeter should read infinity.
4. If there is meter needle movement in Step 3, disconnect the black/yellow wire at the key switch M terminal with the ohmmeter still connected.
5. If the meter reads infinity with the wire disconnected in Step 4, test the key switch. If the meter reads zero, remove the emergency ignition cutoff switch from the circuit.

TROUBLESHOOTING

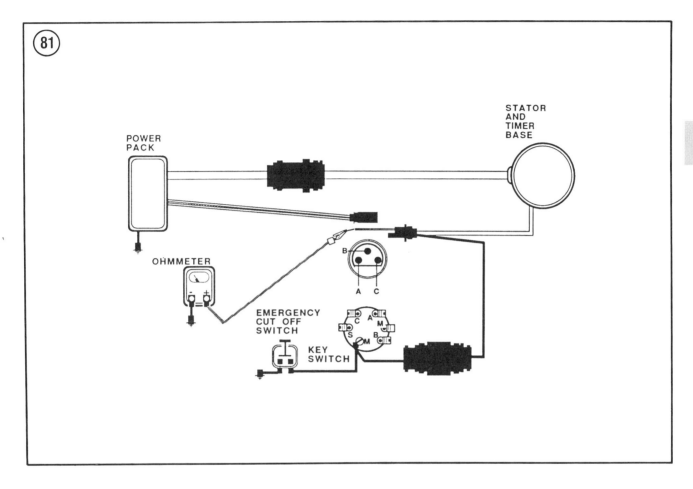

a. If the ohmmeter reads infinity, replace the cutoff switch.
b. If the ohmmeter reads zero, check for a defect in the black/yellow wire between the key switch and 3-wire connector. Repair or replace as required.

Stop Button Circuit Test (WML and TEL Models)

Refer to **Figure 82** for this procedure.
1. Separate the power pack-to-charge coil 3-wire connector.
2. Install the cap/lanyard assembly on the emergency ignition cutoff switch, if so equipped.
3. Insert a jumper wire in the C terminal of the charge coil end connector. Connect the red ohmmeter test lead to the jumper wire and the black test lead to a good ground. The ohmmeter should read infinity.

4. If the meter reading is not as specified in Step 3, remove the emergency ignition cutoff switch from the stop button circuit (if so equipped).
  a. If the ohmmeter reads infinity, replace the cutoff switch.
  b. If the ohmmeter reads zero, replace the stop button.
5. Depress the stop button. The meter should read zero. If not, replace the stop button.
6. Remove the cap/lanyard assembly from the cutoff switch, if so equipped. The meter should read zero. If not, replace the cutoff switch.

Sensor Coil Resistance Test

1. Separate the power pack-to-timer base 4-wire connector. Insert jumper wires in terminals A, B, C and D of the timer base end of the connector.

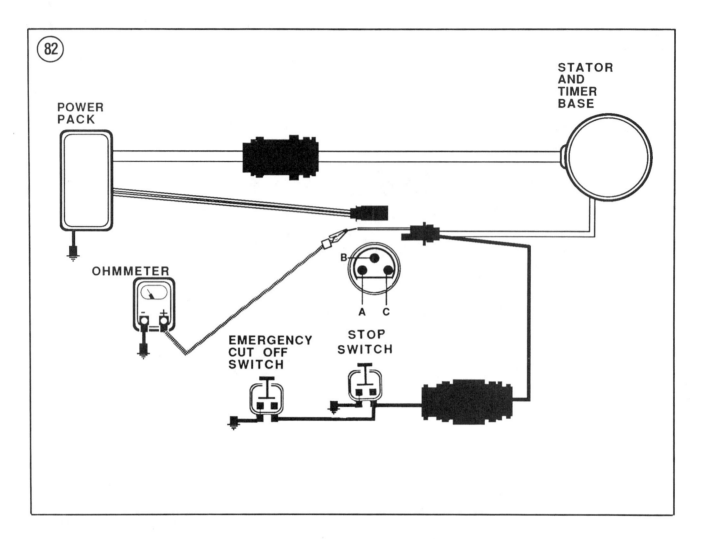

2. With an ohmmeter on the low scale, connect its test leads to the A and D terminal jumper wires, then the B and D terminal wires and finally the C and D terminal wires, noting each reading. See **Figure 83**. If the reading is not 17 ± 5 ohms at each connection, replace the timer base.

3. Set the ohmmeter on the high scale. Ground the black test lead at the armature plate and connect the red test lead alternately to the A, B, C and D terminal jumper wires (**Figure 84**). The ohmmeter needle should not move. If it does, the sensor coils are grounded. Check for a grounded sensor coil lead before replacing the timer base.

4. Remove the jumper wires and perform the *Charge Coil Resistance Test*.

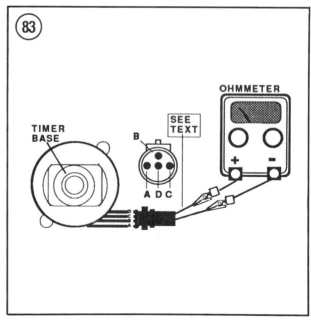

TROUBLESHOOTING

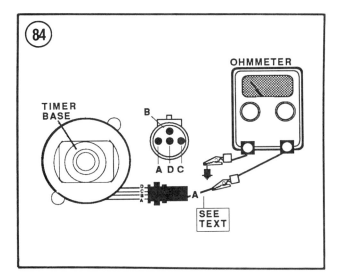

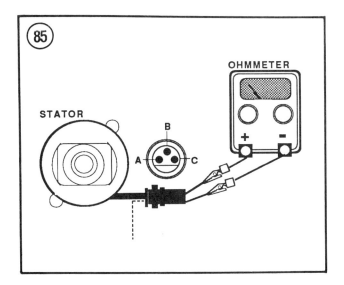

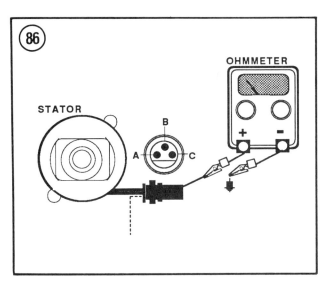

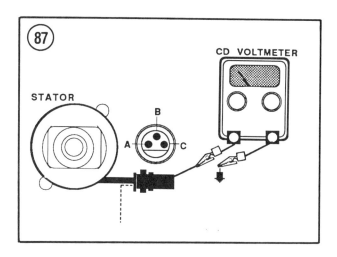

Charge Coil Resistance Test

1. With the 3-terminal connector plug disconnected, insert jumper wires in terminals A and B of the charge coil end of the connector.
2. Connect an ohmmeter between the jumper wires (**Figure 85**) and note the reading. If it is not 560 ± 25 ohms, replace the stator assembly.
3. Set the ohmmeter on the high scale. Ground the black test lead and connect the red test lead alternately to the A and B terminal jumper wires (**Figure 86**). The ohmmeter needle should not move. If it does, the charge coils are grounded. Check for a grounded charge coil lead before replacing the stator assembly.

Charge Coil Output Test

1. Disconnect the 3-wire connector. Set the CD voltmeter switches to NEGATIVE and 500. Insert the red test lead in cavity A of the charge coil end of the connector. Connect the black test lead to a good engine ground. See **Figure 87**.
2. Crank the engine and note the meter reading.
3. Move the red test lead to cavity B of the connector and crank the engine again. Note the meter reading.
4. There should be no meter reading in Step 2 or Step 3. If there is, check for a grounded charge coil lead before replacing the stator assembly.

5. Leave the red test lead in cavity B of the connector. Remove the black test lead from ground and insert it in cavity A of the connector. See **Figure 88**.

6. Crank the engine and note the meter reading. If it is less than 250 volts, check for a problem in component connectors and/or wiring. If no problem is found, replace the stator assembly.

Sensor Coil Output Test

1. Disconnect the 4-wire connector. Set the CD voltmeter switches to S and 5; if using a Merc-O-Tronic Model 781 voltmeter, set the switches to POSITIVE and 5.
2. Insert the red test lead in cavity A of the timer base end of the connector. Connect the black test lead to a good engine ground. See **Figure 89**.
3. Crank the engine and note the meter reading.
4. Move the red test lead to cavity B, C and D of the connector, cranking the engine at each connection and noting the meter reading.
5. There should be no meter reading in Step 3 or Step 4. If there is, check for a grounded sensor coil lead before replacing the timer base.
6. Return the red test lead to cavity A of the connector. Remove the black test lead from ground and insert it in cavity B of the connector. See **Figure 90**.
7. Crank the engine and note the meter reading.
8. Move the red test lead to cavity B, then cavity C of the connector, cranking the engine at each connection and noting the meter reading. See **Figure 90**.
9. If any reading in Step 7 or Step 8 is less than 0.3 volts, check for a problem in the component connectors and/or wiring. If no problem is found, replace the timer base.

Power Pack Output Test

Refer to **Figure 91** for this procedure.
1. Disconnect the primary lead from each ignition coil. Install a terminal extender on the primary terminal of each coil, then reconnect the primary lead to the extender terminal.

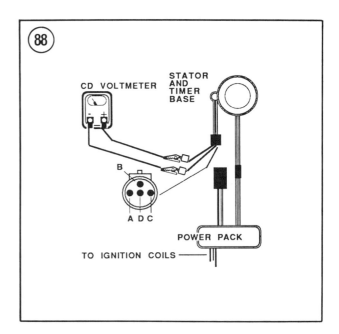

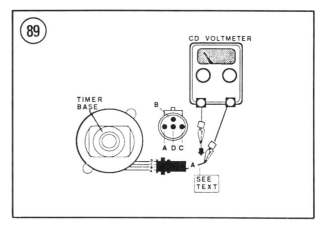

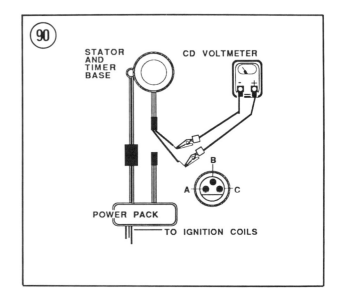

TROUBLESHOOTING

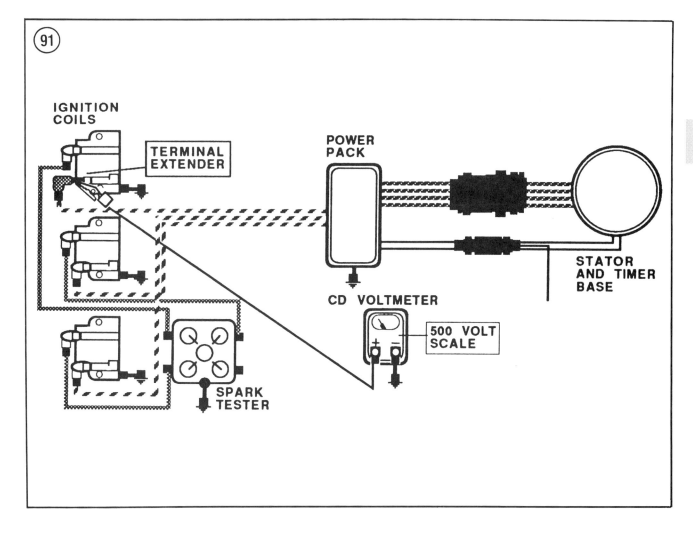

2. Set the CD voltmeter switches to NEGATIVE and 500.

3. Connect the red test lead to the No. 1 coil terminal extender (top coil) and ground the black test lead.

4. Crank the engine and note the meter. It should read 230 volts or more.

5. If the meter reading is less than 230 volts in Step 4, disconnect the primary lead from the No. 1 coil terminal extender. Connect the red test lead directly to the spring clip in the primary lead boot and repeat Step 4.
   a. If the meter now reads 230 volts or more, test the ignition coil as described in this chapter.
   b. If the meter still reads less than 230 volts, check the spring clip and primary lead wire for defects. If none are found, replace the power pack.

6. Repeat Steps 3-5 to test the No. 2 (center) and No. 3 (bottom) ignition coils.

7. Remove the terminal extenders from the coil terminals with a clockwise pulling motion. Reconnect the primary leads to the coil terminals as follows:
   a. Orange/blue lead—Top coil.
   b. Orange lead—Center coil.
   c. Orange/green lead—Bottom coil.

Ignition Coil Resistance Test

1. Disconnect the high tension and primary leads at each ignition coil.

2. Connect the ohmmeter red test lead to the primary terminal of the No. 1 coil. Connect the black test lead to a good engine ground. The meter should read 0.1 ± 0.05 ohms.

3. Set the ohmmeter on the high scale. Move the black test lead to the ignition coil high tension terminal. The meter should read 225-325 ohms.

4. If the readings are not as specified in Step 2 or Step 3, replace the No. 1 ignition coil.

5. Repeat Steps 2-4 to test the No. 2 and No. 3 ignition coils.

6. With the ohmmeter on the low scale, connect the red test lead to one end of the high tension lead and the black test lead to the other end to check wire resistance. Resistance should be near zero; if not, replace the high tension lead. Repeat this step to test the 2 other high tension leads.

1989-1990 CD 3 IGNITION TROUBLESHOOTING

This CD ignition is used on 3-cylinder engines. A 5-wire connector plug connects the charge coil and key switch leads to the power pack. A 4-wire connector plug connects the timer base (sensor coil) leads to the power pack. Ignition coils connect to the power pack by individual leads.

Jumper leads are required for troubleshooting. Fabricate 4 leads using 8 inch lengths of 16-gauge wire. Connect a pin (OMC part no. 511469) with one inch of tubing (OMC part No. 510628) at one end and a socket (OMC part No. 581656) with one inch of tubing (OMC part No. 510628) at the other. Ohmmeter readings should be made when the engine is cold. Readings taken on a hot engine will show increased resistance caused by engine heat and result in unnecessary parts replacement without solving the basic problem.

Running output tests should be made with a Stevens peak-reading voltmeter or equivalent and a Stevens adapter TS-77.

S.L.O.W. Operation And Testing

The S.L.O.W. (speed limiting overheat warning) circuit is activated when the engine temperature exceeds 203° F. When activated, the S.L.O.W. mode will limit the engine speed to approximately 2500 rpm. To be deactivated, the engine must be cooled to 162° F and the engine must be stopped.

A blocking diode located in the engine wiring harness is used to isolate the S.L.O.W. warning system from other warning horn signals.

1. Install the engine in a test tank with the proper test wheel.

2. Connect a tachometer according to manufacturer's instructions.

3. Disconnect temperature sensor tan lead at connector (**Figure 47**, typical).

4. Start engine and run at approximately 3500 rpm.

5. Connect engine wiring tan lead to a clean engine ground.

6. Note engine rpm. Engine rpm should reduce to approximately 2500 rpm.

7. Stop the engine. If engine rpm does not reduce, disconnect 5-wire connector located between flywheel stator and power pack.

8. Calibrate an ohmmeter to R×100 scale.

9. Connect black ohmmeter lead to a good engine ground. Alternately connect red

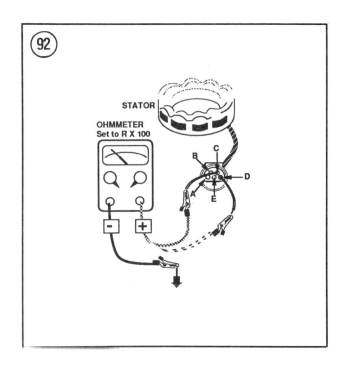

TROUBLESHOOTING

ohmmeter lead to "C" and "D" terminal ends in connector leading to flywheel stator. See **Figure 92**. No continuity or infinite resistance should be noted at both connections.

10. Connect ohmmeter leads between "C" and "D" terminal ends in connector leading to flywheel stator. See **Figure 93**. Ohmmeter should read between 455-505 ohms. If so, replace power pack.

11. Blocking Diode Test—Calibrate an ohmmeter to R×100 scale. Connect one ohmmeter lead to engine wiring tan lead for temperature sensor and other ohmmeter lead to tan wire connector in red engine harness connector. Note ohmmeter reading, then reverse leads or push "polarity" button on meter if so equipped and note reading. Continuity (near zero resistance) should be noted in one direction and no continuity or infinite resistance should be noted in other direction. If not, replace blocking diode in engine wiring harness.

Ignition System Output Test

1. Remove spark plug leads from spark plugs.
2. Mount a spark tester on the engine and connect a tester lead to each spark plug lead. Set the spark tester air gap to 1/2 in. See **Figure 48**, typical.
3. Connect ignition system emergency cutoff clip and lanyard if so equipped.
4. Crank the engine over while watching the spark tester. If a spark jumps at each air gap, test the ignition system as outlined under *Running Output Test* in this chapter.
5. If a spark jumps at *only* one or two air gaps, first make sure all connections are good, then retest. If only one or two sparks are noted, test ignition system as outlined under *Sensor Coil Test* in this chapter.
6. If *no* spark jumps at any of the spark tester's air gaps, first make sure all connections are good, then retest. If no spark is noted, test ignition system as outlined under *Key Switch Elimination Test* in this chapter.

Key Switch Elimination Test

Refer to **Figure 94** for this procedure.
1. Connect a spark tester as outlined under *Ignition System Output Test*.
2. Separate the power pack-to-flywheel stator 5-wire connector. Insert jumper wires between the A, B, C and D terminals of the connector.
3. Crank the engine while watching the spark tester. If there is no spark or a spark at only one or two air gaps, remove the jumper wires and reconnect the connector plugs. Continue testing to locate the problem.
4. If a spark jumps all air gaps in sequence, the problem is in the key switch circuit or the emergency ignition cutoff switch.

Key Switch/Stop Circuit Test

Refer to **Figure 95** for this procedure.
1. Separate the power pack-to-flywheel stator 5-wire connector.
2. Install the cap/lanyard assembly on the emergency ignition cutoff switch, if so equipped.
3. Insert a jumper wire in the E terminal of the flywheel stator end connector. Connect the red

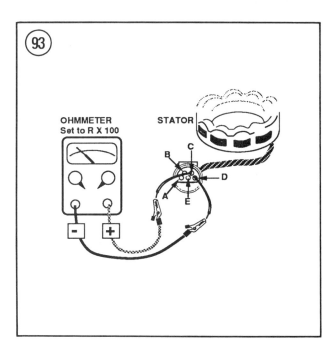

ohmmeter test lead to the jumper wire and the black test lead to a good ground. Turn the key switch ON. The ohmmeter should read infinity.

4. If there is meter needle movement in Step 3, disconnect the black/yellow wire at the key switch M terminal with the ohmmeter still connected.

5. If the meter reads infinity with the wire disconnected in Step 4, test the key switch. If the meter reads zero, remove the emergency ignition cutoff switch from the circuit.

   a. If the ohmmeter reads infinity, replace the cutoff switch.

   b. If the ohmmeter reads zero, check for a defect in the black/yellow wire between the key switch and 5-wire connector. Repair or replace as required.

Sensor Coil Output Test

1. Separate the power pack-to-timer base 4-wire connector. Insert jumper wires in all 4 terminals of the timer base end of the connector.

2. Connect black lead of a peak-reading voltmeter to terminal D of the sensor coil end of the connector.

3. Set the peak-reading voltmeter to POS and 5.

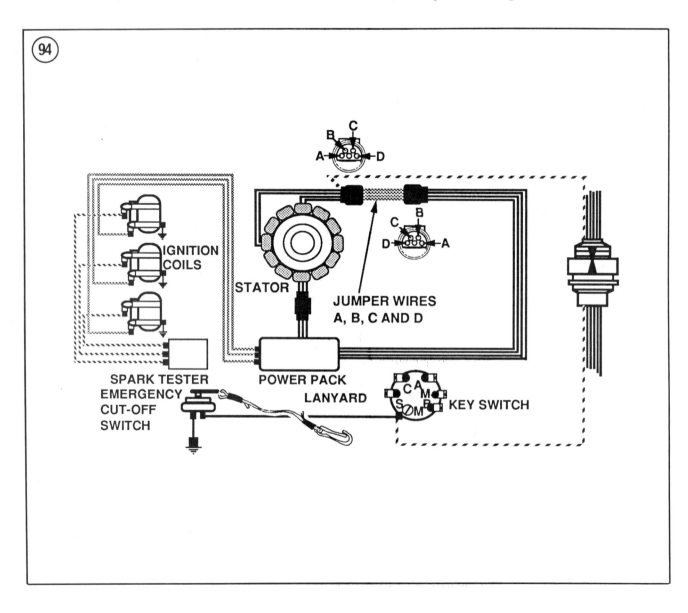

TROUBLESHOOTING

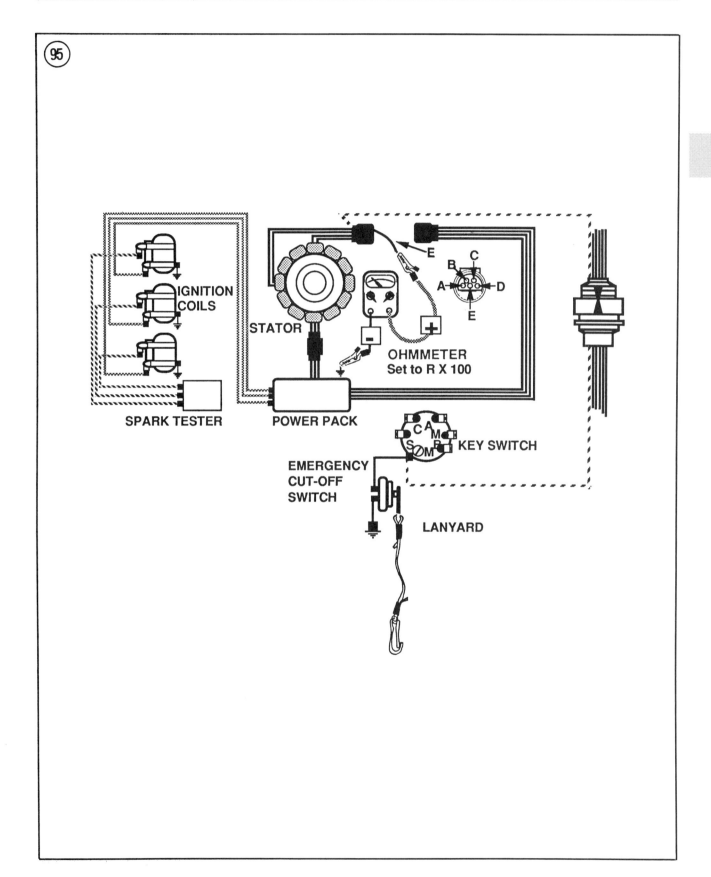

4. Connect red lead of peak-reading voltmeter to terminal A of the sensor coil end of the connector (See **Figure 96**).

5. Crank the engine and note the meter reading.

6. Move the red test lead to cavity B and C of the connector, cranking the engine at each connection and noting the meter reading.

7. If any reading in Step 5 or Step 6 is less than 0.3 volts, check for a problem in the component connectors and/or wiring. If no problem is found, perform *Sensor Coil Resistance Test* as outlined in this chapter.

Sensor Coil Resistance Test

1. Separate the power pack-to-timer base 4-wire connector. Insert jumper wires in all 4 terminals of the timer base end of the connector.

2. With an ohmmeter on the low scale, connect its test leads to the A and D terminal jumper leads, then the B and D terminal jumper leads and finally to the C and D terminal jumper leads while noting each reading. See **Figure 97**. If the reading is not 11 ±3 ohms at each connection, replace the timer base.

3. Set the ohmmeter on the high scale. Ground the black test lead and connect the red test lead alternately to the A, B, C and D terminal jumper wires (**Figure 98**). The ohmmeter needle should not move. If it does, the sensor coils or leads are grounded. Check for a grounded sensor coil lead before replacing the timer base.

Charge Coil Output Test

1. Separate the power pack-to-flywheel stator 5-wire connector. Insert jumper wires in terminals A and B of the flywheel stator end of the connector.

2. Connect black lead of a peak-reading voltmeter to terminal A of the flywheel stator end of the connector.

3. Set the peak-reading voltmeter to POS and 500.

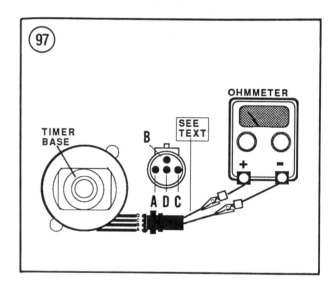

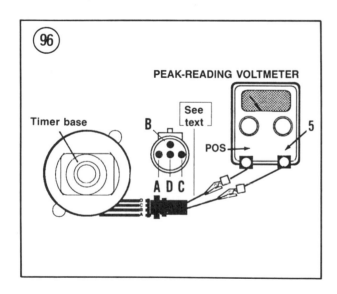

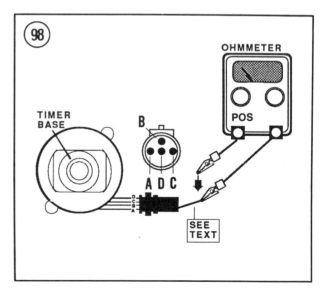

TROUBLESHOOTING

4. Connect red lead of peak-reading voltmeter to terminal B of the flywheel stator end of the connector (See **Figure 99**).
5. Crank the engine and note the meter reading.
6. If reading in Step 5 is less than 250 volts, check for a problem in the component connectors and/or wiring. If no problem is found, perform *Charge Coil Resistance Test* as outlined in this chapter.

Charge Coil Resistance Test

1. Separate the power pack-to-flywheel stator 5-wire connector. Insert jumper wires in terminals A and B of the flywheel stator end of the connector.
2. Connect an ohmmeter between the jumper wires (**Figure 100**) and note the reading. If it is not 480 ±25 ohms, replace the flywheel stator assembly.
3. Set the ohmmeter on the high scale. Ground the black test lead and connect the red test lead alternately to the A and B terminal jumper wires (**Figure 101**). The ohmmeter needle should not move. If it does, the charge coils are grounded. Check for a grounded charge coil lead before replacing the flywheel stator assembly.

Power Pack Output Test

WARNING
Disconnect spark plug leads to prevent accidental starting.

1. Remove primary leads from ignition coils.
2. Connect No. 1 cylinder ignition coil primary lead to the red lead of Stevens load adapter No. PL-88 and the black lead of load adapter to a good engine ground. See **Figure 50**, typical.
3. Connect the red lead of a peak-reading voltmeter to the connector end of the red lead

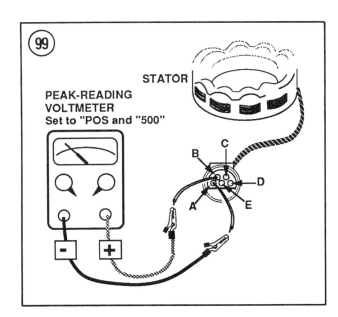

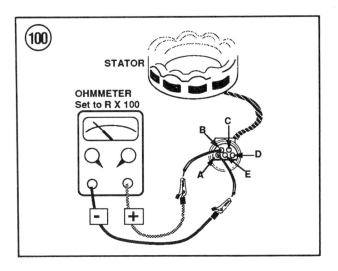

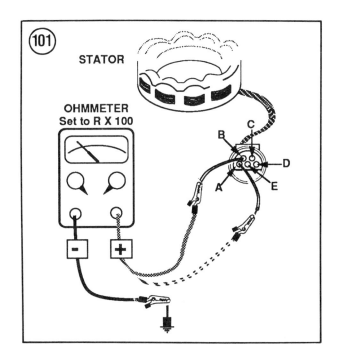

on Stevens load adapter No. PL-88. Connect the black lead of the peak-reading voltmeter to a good engine ground.
4. Set the peak-reading voltmeter on POS and 500.
5. Crank the engine over while noting the peak-reading voltmeter. The meter should show at least 230 volts.
6. Repeat Steps 2-5 for No. 2 and No. 3 cylinders ignition coil primary lead.
7. If all primary leads show at least 230 volts, then refer to *Ignition Coil Resistance Test* in this chapter.
8. If only one or two primary leads show at least 230 volts, the power pack must be replaced.

Running Output Test

1. Remove primary leads from ignition coils and connect Stevens terminal extenders No. TS-77 to ignition coils, then reconnect primary leads.
2. Connect the red lead of a peak-reading voltmeter to the No. 1 cylinder ignition coil terminal extender. Connect the black lead of the peak-reading voltmeter to a good engine ground.
3. Set the peak-reading voltmeter on POS and 500.
4. Make sure the outboard motor is mounted in a suitable test tank or testing fixture with a suitable loading device (test wheel or dynometer).
5. Start the engine and operate at the rpm ignition trouble is suspected.
6. The peak-reading voltmeter should show at least 250 volts.
7. Repeat Steps 2-6 to test No. 2 and No. 3 cylinders.
8. If one or more of the cylinders shows less than 250 volts, test charge coil as outlined in this chapter.
9. If one or more of the cylinders shows no output, test sensor coil as outlined in this chapter.
10. Remove ignition coil terminal extenders, then reconnect primary leads.

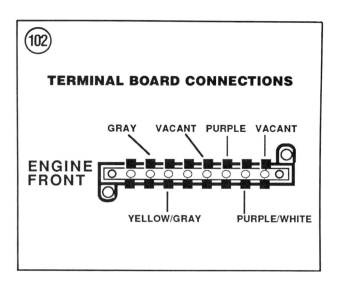

Ignition Coil Resistance Test

1. Disconnect the primary and high tension (secondary) coil leads at the ignition coil to be tested.
2. With an ohmmeter set on the low scale, connect the red test lead to the coil primary terminal. Connect the black test lead to a good engine ground (if the coil is mounted) or the coil ground tab (if the coil is unmounted). The meter should read 0.1 ± 0.05 ohms.
3. Set the ohmmeter on the high scale. Move the black test lead to the ignition coil high tension terminal. The meter should read 225-325 ohms.
4. If the readings are not as specified in Step 2 or Step 3, replace the ignition coil.
5. Repeat Steps 2-4 to test the remaining ignition coils.

1973-1977 CD 4 IGNITION TROUBLESHOOTING

The CD 4 ignition is used on V4 engines covered in this manual.

Figure 102 shows the terminal board connections. Although many terminal boards have the wire colors embossed either on the board itself or on the cover (if used), not all models will use all of the wires and connections shown. Double check your connections by referring to **Figure 102**.

TROUBLESHOOTING

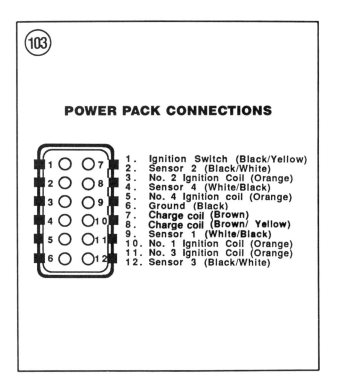

103

POWER PACK CONNECTIONS

1. Ignition Switch (Black/Yellow)
2. Sensor 2 (Black/White)
3. No. 2 Ignition Coil (Orange)
4. Sensor 4 (White/Black)
5. No. 4 Ignition coil (Orange)
6. Ground (Black)
7. Charge coil (Brown)
8. Charge coil (Brown/Yellow)
9. Sensor 1 (White/Black)
10. No. 1 Ignition Coil (Orange)
11. No. 3 Ignition Coil (Orange)
12. Sensor 3 (Black/White)

Figure 103 shows the power pack terminals. Refer to the diagram when reconnecting power pack leads to avoid any misconnections that can cause damage to the power pack or CD 4 ignition.

The sensor coils (**Figure 62**) and timer base are replaced as an assembly. The charge coil is part of the potted stator assembly (**Figure 63**). If defective, replace the charge coil and stator as an assembly.

Coat all electrical terminal connections (except the ground stud, nut and terminal motor cable-to-bypass cover) with OMC Black Neoprene Dip after reconnecting the leads to the terminal board or power pack.

An S-80 or M80 neon test light (available from your Johnson or Evinrude dealer) is required for this procedure.

Ignition Output Test

Refer to **Figure 104** for this procedure.
1. Connect a spark tester as shown in **Figure 104**. Set the tester air gap to 1/2 in. (1973) or 7/16 in. (1974-1977).
2. Crank the engine with the ignition switch while watching the spark tester. If strong, steady sparks firing one coil at a time appear at the spark tester, the ignition coil output is good.
3. If the sparks from 1 or 2 coils are weak or erratic, perform the *Sensor Coil Output Test* in this chapter.
4. If the sparks from all 4 coils are weak or erratic, perform the *Charge Coil Output Test* in this chapter.
5. If there is no spark from any of the 4 coils, perform the *Sensor Coil Output Test* in this chapter.

Sensor Coil Output Test

Refer to **Figure 105** for this procedure.
1. With the spark tester installed as shown in **Figure 105**, disconnect the sensor leads at terminals 2, 4, 9 and 12 of the power pack.
2. Disconnect the ignition coil leads at terminals 3 and 5.
3. Connect the black test lead to terminal 9 and the blue test lead to terminal 12. Set neon light switch on position 3.
4. Crank the engine with the ignition switch while watching the spark tester and rapidly tapping tester load button B. There should be a spark at the No. 1 ignition coil.
5. Reverse the test leads and repeat Step 4. There should be a spark at the No. 3 ignition coil.
6. If both coils fire at the same time in Step 4 or Step 5, replace the power pack. If neither coil fires in Step 4 or Step 5, perform the *Charge Coil Output Test* in this chapter.
7. Reconnect the ignition coil leads to terminals 3 and 5. Disconnect the ignition coil leads at terminals 10 and 11.
8. Connect the black test lead to terminal 2 and the blue test lead to terminal 4.
9. Repeat Step 4 and Step 5. There should be a spark at the No. 2 and then the No. 4 ignition coil. If both coils fire at the same time when repeating Step 4 or Step 5, replace the power

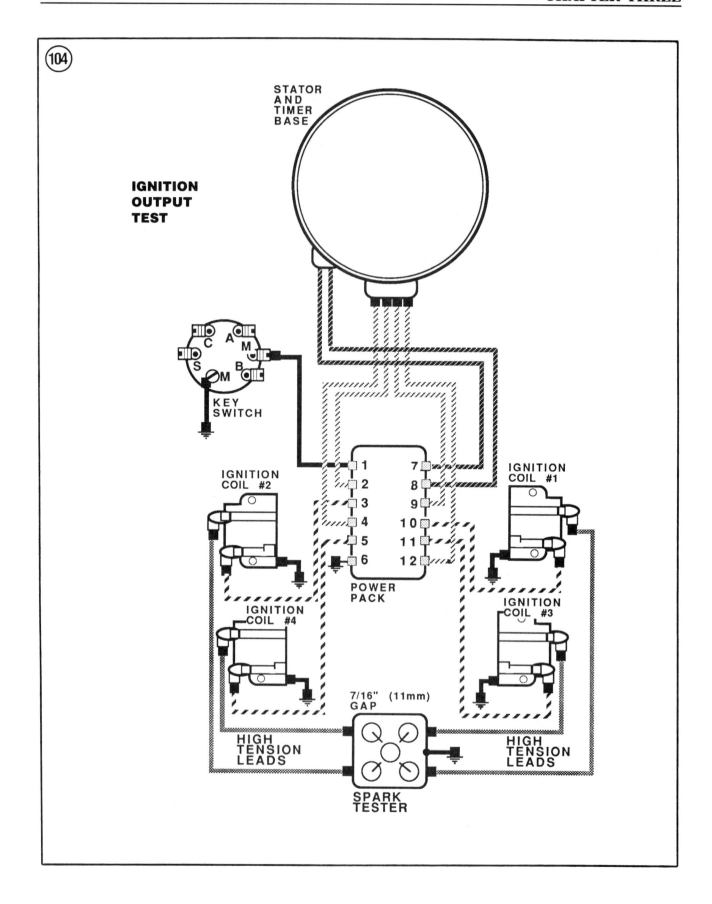

TROUBLESHOOTING

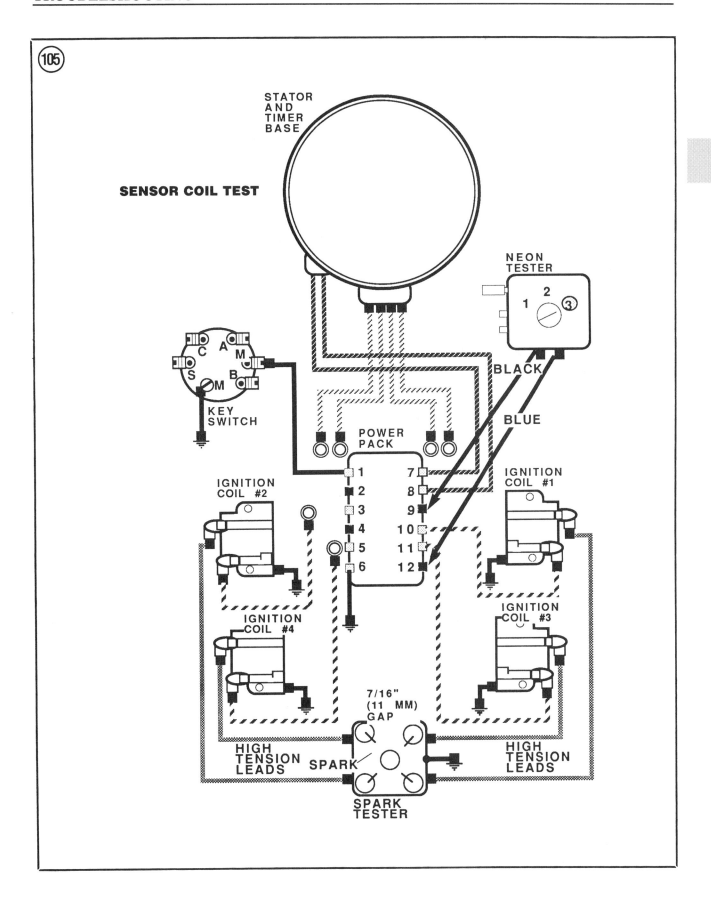

pack. If neither coil fires when repeating Step 4 or Step 5, perform the *Charge Coil Output Test* in this chapter.

Charge Coil Output Test

Refer to **Figure 106** for this procedure.
1. Disconnect the charge coil leads at power pack terminals 7 and 8.
2. Connect the black test lead to the charge coil lead disconnected from terminal 7. Connect the blue test lead to the charge coil lead disconnected from terminal 8.
3. Set neon light switch on position 2. Depress load button B and crank the engine with the ignition switch while watching the tester neon light:
   a. If the light glows steadily, perform the *Power Pack Output Test* in this chapter.
   b. If the light glows intermittently, check the charge coil output leads for an open or a short to ground and make sure that the power pack ground lead (terminal 12) is good. If leads are good, replace the charge coil and stator assembly.
   c. If the light does not glow, replace the charge coil and stator assembly.
4. Reconnect the charge coil leads at the power pack.

Power Pack Output Test

Refer to **Figure 107** for this procedure.
1. Mark each ignition coil primary lead with the cylinder number, then disconnect from terminals 3, 5, 10 and 11 of the power pack.
2. Connect the black test lead to terminal 10 and the blue test lead to a good engine ground.
3. Set neon light switch on position 1 and depress load button A. Crank the engine with the ignition switch while watching the tester neon light.
4. Move the black test lead to terminal 11 and repeat Step 3.
5. Move the black test lead to terminal 3 and repeat Step 3.
6. Move the black test lead to terminal 5 and repeat Step 3.
7. If the test light emits a steady glow on all outputs, replace the ignition coils. If the test light is steady, dim or intermittent on 1, 2 or 3 outputs, replace the power pack. If there is no light on any output, perform the *Key Switch Test* in this chapter.
8. Reconnect the coil leads at the power pack.

Key Switch Test

Refer to **Figure 108** for this procedure.
1. With the spark tester installed as shown in **Figure 108**, disconnect the key switch lead at power pack terminal 1.
2. Crank the engine and watch the spark tester. If there is no spark, the key switch is good. If there is a spark at 1, 2 or 3 coils, check the key switch lead for defects. If the lead is good, replace the key switch.

Sensor Coil Resistance Test

1. Disconnect the sensor leads from the power pack terminals 2, 4, 9 and 12.
2. Connect the ohmmeter test leads between the leads disconnected from terminals 2 and 4, then between the leads from terminals 9 and 12. Note the ohmmeter reading. If it is not 7.5-9.5 ohms (1973) or 6.5-10.5 ohms (1974-1977) at room temperature (70° F), replace the sensor coil and timer base assembly.
3. Set the ohmmeter on the high scale. Connect the black test lead to a good engine ground and the red test lead alternately to the disconnected sensor leads. The ohmmeter should read infinity at each connection. If it does not, one or more of the sensor coils are grounded. Replace the sensor coil and timer base assembly.

Charge Coil Resistance Test

1. Disconnect the charge coil leads at power pack terminals 7 and 8.

TROUBLESHOOTING

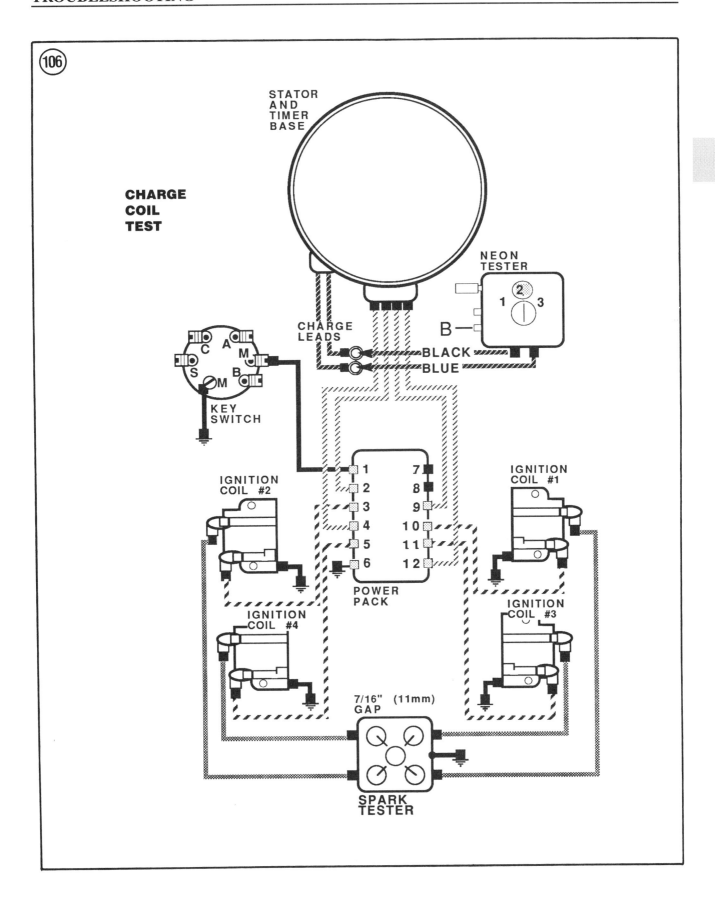

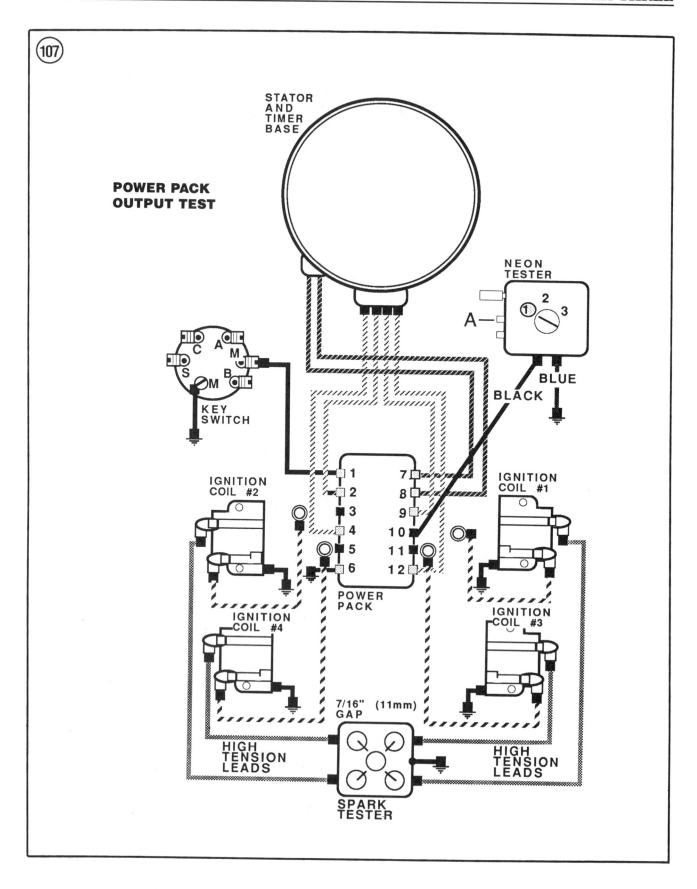

TROUBLESHOOTING

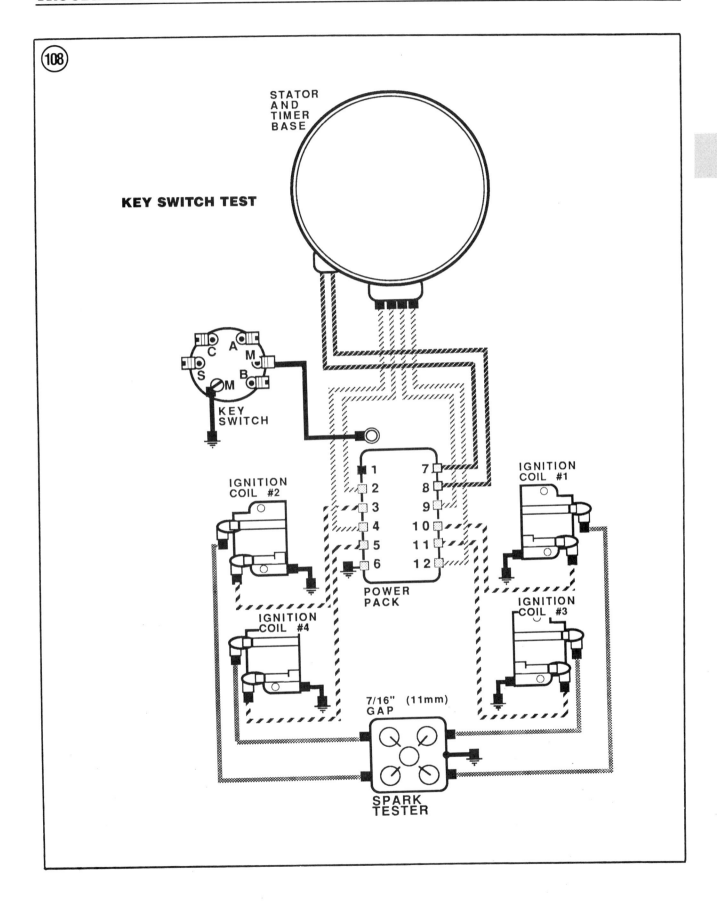

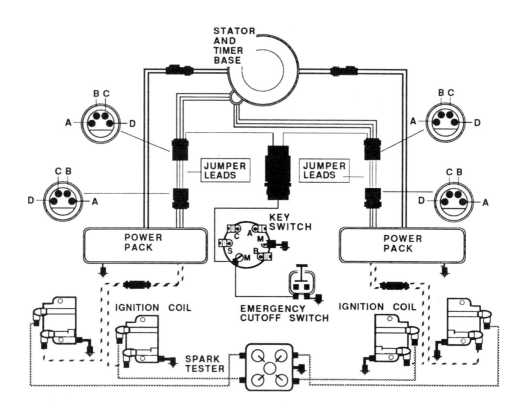

KEY SWITCH ELIMINATION TEST

TROUBLESHOOTING

2. Set the ohmmeter on the high scale and connect the test leads between the disconnected charge coil leads.

3. If the ohmmeter reading is not 835-985 ohms (1973) or 555-705 ohms (1974-1977) at room temperature (70° F), replace the charge coil and stator assembly.

1978-1984 CD 4 IGNITION TROUBLESHOOTING

The CD 4 ignition used on these V4 engines covered in this manual consists of 2 identical ignition systems—one on the port side and one on the starboard side of the engine. The troubleshooting procedures must be performed on each ignition system.

Each power pack is connected to the sensor coil and ignition switch leads by a 4-wire connector. A 2-wire connector joins each power pack to the ignition coils. A second 2-wire connector joins the charge coil to each power pack.

Since the CD 4 ignition used from 1978-on contains 2 power packs, there is one set of connectors on each side of the engine. A retaining clip holds the connectors in place. The clip must be removed to unhook the connector plugs.

Jumper leads are required for troubleshooting. Fabricate 4 leads using 8 inch lengths of 16-gauge wire. Connect a pin (OMC part No. 511469) at one end and a socket (OMC part No. 581656) with one inch of tubing (OMC part No. 519628) at the other. Ohmmeter readings should be made when the engine is cold. Readings taken on a hot engine will show increased resistance caused by engine heat and may result in unnecessary parts replacement without solving the basic problem.

Output tests should be made with a Stevens or Electro-Specialties CD voltmeter.

Key Switch Elimination Test

1. Connect a spark tester to the engine. Set the tester air gap to 1/2 in.

2. Disconnect the 4-wire connector between the power pack and timer base. Insert jumper wires between the connector terminals A, B and C. See **Figure 109**.

3. Crank the engine and watch the spark tester. If there is no spark or a spark at only one gap, remove the jumper wires, reconnect the connector plugs and continue with the test procedures.

4. If a spark jumps alternately at all gaps, the problem is probably in the key switch circuit.

Key Switch Stop Circuit Test

1. Disconnect the 4-wire connector between the power pack and timer base.

2. With the ohmmeter on the high scale, connect the red test lead to the D terminal in the ignition coil end of the connector. Connect the red test lead to a good engine ground. See **Figure 110**.

3. Turn the key switch ON. If equipped with an emergency ignition cutoff switch, put the cap end and lanyard assembly on the switch. The meter should read infinity (open circuit).

4. If the meter does not read infinity in Step 3, disconnect the black/yellow wire from the M terminal on the key switch. If the meter reads infinity with the wire disconnected, the problem is in the key switch. If the meter now indicates a closed circuit, the problem is in the black/yellow wire between the key switch and connector.

5. If the meter does not read infinity with the key switch wire disconnected, disconnect the emergency ignition cutoff switch, if so equipped. If the meter now reads infinity, the problem is in the cutoff switch. If it indicates a closed circuit, the problem is in the black/yellow wire between the key switch and connector.

Charge Coil Resistance Test

1. Disconnect the 2-wire connector between the power pack and charge coils.

2. Insert jumper wires in terminals A and B of the connector stator end. See **Figure 111**. If the meter does not read 485-635 ohms, replace the charge coil.

3. Set the ohmmeter on the high scale. Ground the black test lead at the timer base. Connect the red test lead first to the A terminal jumper wire, then to the B terminal jumper wire and note the meter reading. See **Figure 112**. If the needle moves during this step, the charge coil or leads are shorted. Repair the lead if possible; replace the charge coil if the lead is good.

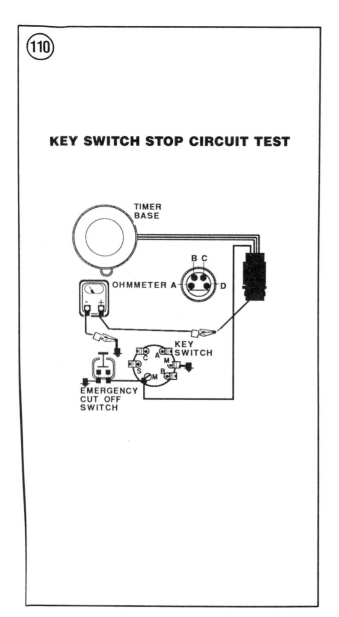

KEY SWITCH STOP CIRCUIT TEST

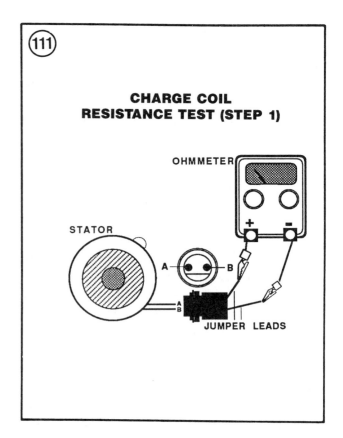

CHARGE COIL RESISTANCE TEST (STEP 1)

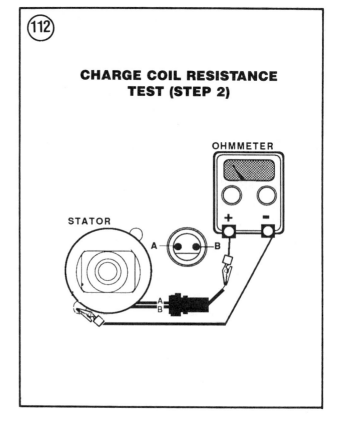

CHARGE COIL RESISTANCE TEST (STEP 2)

TROUBLESHOOTING

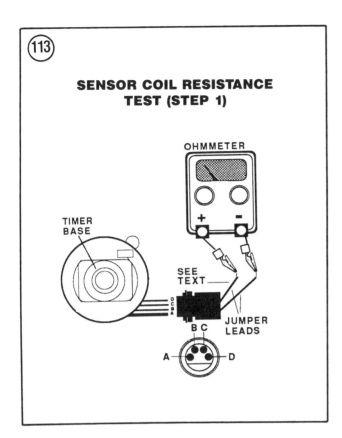

113 SENSOR COIL RESISTANCE TEST (STEP 1)

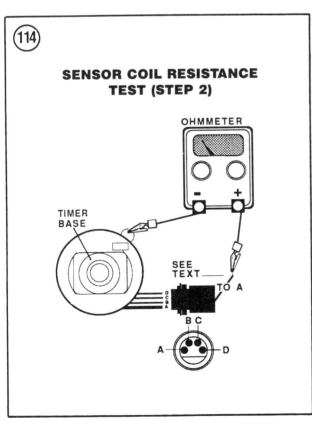

114 SENSOR COIL RESISTANCE TEST (STEP 2)

Sensor Coil Resistance Test

1. Disconnect the 4-wire connector between the power pack and timer base.
2. Insert jumper wires in terminals A, B and C of the timer base connector end. See **Figure 113**.
3. With the ohmmeter on the low scale, connect the test leads between the A and B jumper wires, then between terminals A and C. If the meter does not read between 30-50 ohms, replace the sensor coil.
4. Set the ohmmeter on the high scale. Ground the black test lead at the timer base. With the red test lead connected to a jumper wire, insert the jumper wire in terminals A, B and C in that order, noting the meter reading. See **Figure 114**. If the needle moves during this step, the sensor coil or leads are shorted. Repair the lead if possible; replace the sensor coil if the lead is good.
5. Connect the black test lead to the common lead on the port side sensor connector terminal A. Connect the red test lead to terminal A on the starboard side sensor connector. If the meter does not read infinity, replace the timer base.

Charge Coil Output Test

Refer to **Figure 115** for this procedure.
1. Disconnect the 2-wire connector between the power pack and stator.
2. Set the CD voltmeter switches to NEGATIVE and 500. Connect a jumper wire to the red test lead. Insert the red test lead jumper wire in cavity A of the stator connector. Ground the black test lead at the timer base. Crank the engine and note the meter reading.
3. Move the red test lead jumper wire to cavity B of the connector. Crank the engine again and note the meter reading. If the needle moves during this step, the charge coil or leads are shorted. Repair the lead if possible; replace the charge coil if the lead is good.
4. Leave the red test lead in cavity B of the connector. Remove the black test lead from the

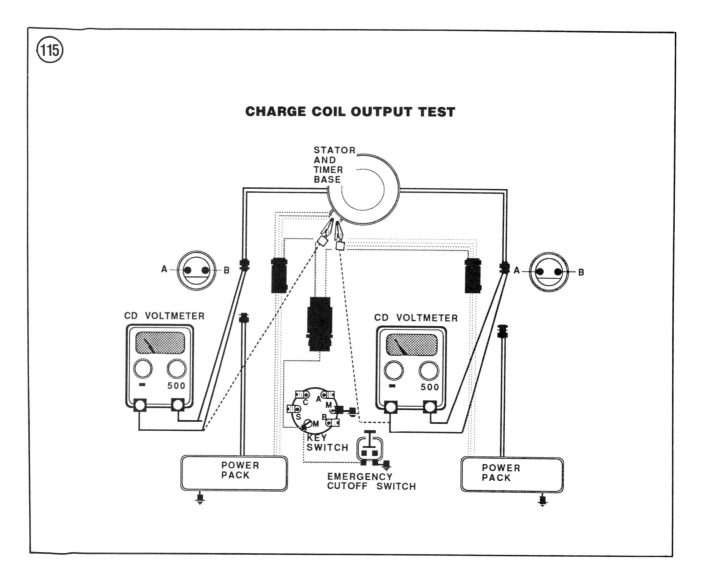

timer base. Connect it to a jumper wire and insert the wire in cavity A of the connector.
5. Crank the engine and note the meter reading. If it is less than 160 volts, replace the charge coil.
6. Repeat the procedure to check the other charge coil.

Sensor Coil Output Test

Refer to **Figure 116** for this procedure.
1. Disconnect the 4-wire connector between the power pack and timer base.
2. Set the CD voltmeter switches to S and 5. Ground the black test lead at the timer base. Connect a jumper wire to the red test lead and probe terminals A and B while cranking the engine. If the needle moves during this step, the sensor coil or leads are shorted. Repair the leads if possible; replace the sensor coil if the leads are good.
3. Remove the black test lead from the timer base. Connect it to a jumper wire and insert the wire in cavity A of the timer base 4-wire connector. Probe cavity B of the timer base 4-wire connector with the red test lead jumper wire while cranking the engine. Note the meter reading.
4. Move the red test lead jumper wire to cavity C, crank the engine and note the meter reading.

TROUBLESHOOTING

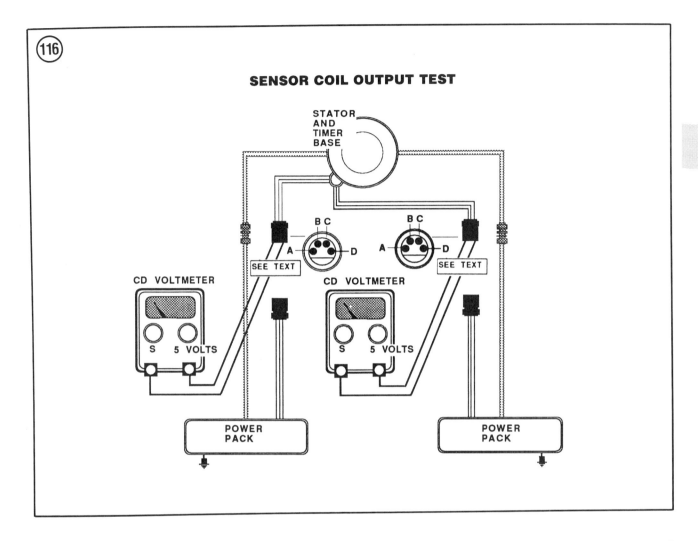

5. If the meter reads less than 0.3 volt during Step 3 or Step 4, either a sensor coil is defective or engine cranking rpm is too low. The use of a fully charged battery during this procedure will eliminate the possibility of insufficient cranking rpm unless the engine is new and has not been properly broken in or there is a mechanical problem within the power head.

Power Pack Output Test

Refer to **Figure 117** for this procedure.
1. Disconnect the 2-wire connector between the power pack and the ignition coils.
2. Set the CD voltmeter switches to NEGATIVE and 500.
3. Insert jumper wires between terminals A of the connectors. Connect the black test lead to a good engine ground.
4. Connect the red test lead to the metal portion of the jumper wire at the power pack connector B cavity and crank the engine. Note the meter reading.
5. Repeat Step 4 with the red test lead connected to the A cavity jumper wire. Note the meter reading.
6. If the meter reads at least 170 volts, check the ignition coil(s). If the meter does not read the specified voltage on any output, remove the jumper wires from those cavities. Probe the cavities directly with the red test lead while cranking the engine. If there is no change in the meter reading, replace the power pack.

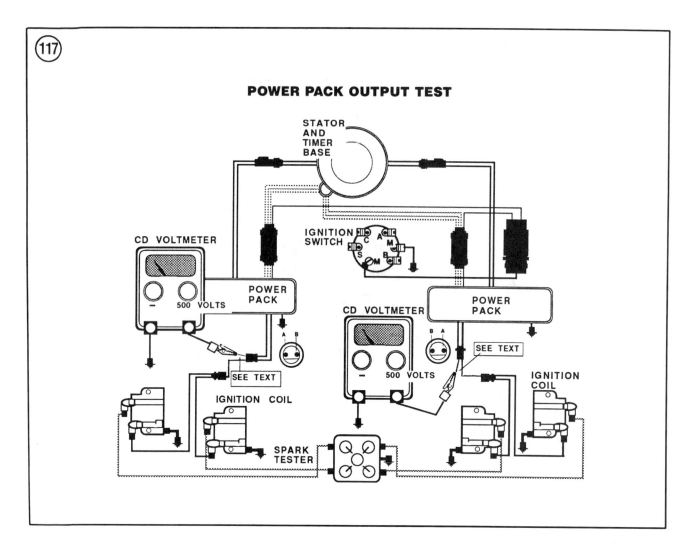

117 POWER PACK OUTPUT TEST

Power Pack Isolation Diode Test

A malfunctioning isolation diode in one or both power packs on V4 engines can cause the engine to run roughly. If the diode is not performing its function, the power packs will interfere with each other.

Refer to **Figure 118** for this procedure.

1. Disconnect the 4-wire power pack connectors to the timer base.
2. With an ohmmeter on the high scale, connect the test leads between the A and D terminals as shown in **Figure 118** and note the meter reading.
3. Reverse the test leads and note the reading. If the diode is good, one reading will be high and the other will be low. If both readings are high, the diode is open. If both readings are low, the diode is shorted. Replace the power pack if both readings are high or low.
4. Reconnect the power pack connectors and repeat the procedure to check the other power pack.

Ignition Coil Resistance Test

1. Disconnect the high tension lead at the No. 1 ignition coil.
2. Disconnect the 2-wire connector between the power pack and ignition coils. Insert a jumper lead in terminal A of the ignition coil end of the connector.
3. Connect the ohmmeter red test lead to the A terminal jumper lead. Connect the black test lead

TROUBLESHOOTING

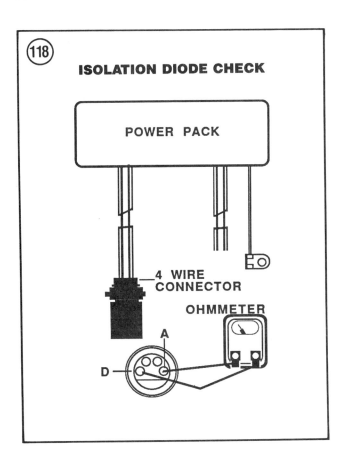

118 ISOLATION DIODE CHECK

to a good engine ground. The meter should read 0.1 ± 0.05 ohms.
4. Set the ohmmeter on the high scale. Move the black test lead to the ignition coil high tension terminal. The meter should read 250-300 ohms.
5. Repeat Steps 1-4 to test the remaining coil in the ignition system. Use terminal B to test the coil.
6. Repeat the procedure to test both coils in the other ignition.
7. Replace any coil that does not meet the readings specified in Step 3 or Step 4.

1987 88 HP, 1985-1987 90 HP, 1985 115 HP AND 1986-1987 110 HP CD 4 IGNITION TROUBLESHOOTING

The CD ignition used on these V4 engines consists of 2 identical ignition systems—one on the port side and one on the starboard side of the engine. The troubleshooting procedures must be performed on each ignition system.

Each power pack is connected to the timer base and ignition switch by a 4-wire connector. A 2-wire connector joins each power pack to the charge coil. Individual leads connect the power packs to the ignition coils.

Since this ignition contains 2 power packs, there is one set of connectors on each side of the engine. Amphenol connectors are used, with retaining clips provided to hold the connectors in place. The clips must be removed to disconnect the connector plugs.

Jumper leads are required for troubleshooting. Fabricate 4 leads using 8 inch lengths of 16-gauge wire. Connect a pin (OMC part No. 511469) at one end and a socket (OMC part No. 581656) with one inch of tubing (OMC part No. 519628) at the other. Ohmmeter readings should be made when the engine is cold. Readings taken on a hot engine will show increased resistance caused by engine heat and result in unnecessary parts replacement without solving the basic problem.

Output tests should be made with a Stevens or Electro-Specialties CD voltmeter and CD adapter.

Key Switch Elimination Test

Refer to **Figure 119** for this procedure.
1. Connect a spark tester as shown in **Figure 119**. Set the tester air gap to 1/2 in.
2. Separate the power pack-to-timer base 4-wire connector. Insert jumper wires between the A, B and C terminals of the connector.
3. Crank the engine while watching the spark tester. If there is no spark or a spark at only one gap, remove the jumper wires and reconnect the connector plugs. Continue testing to locate the problem.
4. If a spark jumps all gaps alternately, the problem is in the key switch circuit or the emergency ignition cutoff switch.

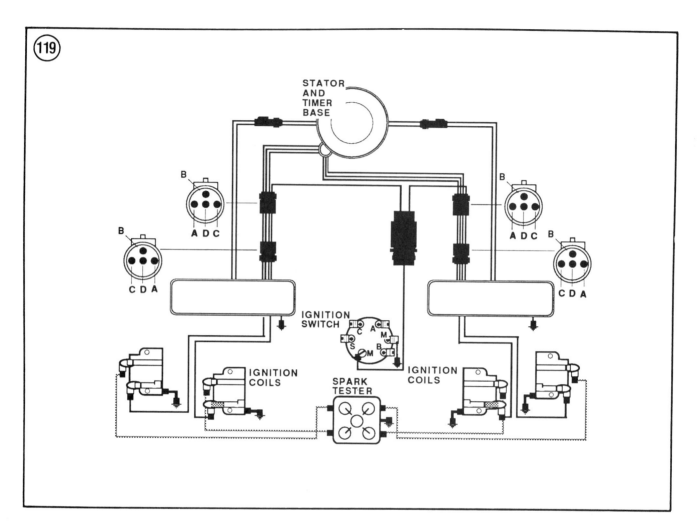

Key Switch/Stop Circuit Test

Refer to **Figure 120** for this procedure.

1. Separate the power pack-to-timer base 4-wire connector.
2. Install the cap/lanyard assembly on the emergency ignition cutoff switch, if so equipped.
3. Insert a jumper wire in the D terminal of the timer base end connector. Connect the red ohmmeter test lead to the jumper wire and the black test lead to a good ground. Turn the key switch ON. The ohmmeter should read infinity.
4. If there is meter needle movement in Step 3, disconnect the black/yellow wire at the key switch M terminal with the ohmmeter still connected.
5. If the meter reads infinity with the wire disconnected in Step 4, test the key switch. If

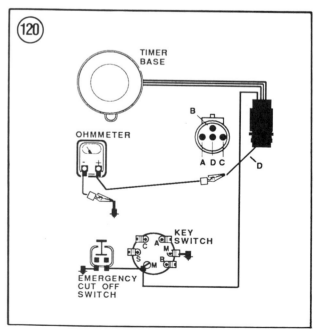

TROUBLESHOOTING

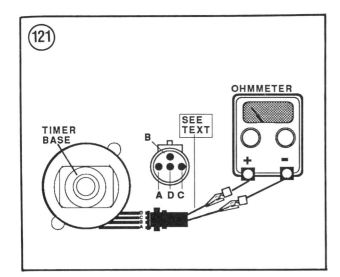

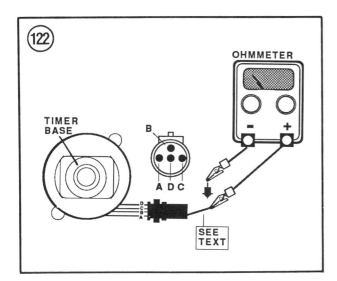

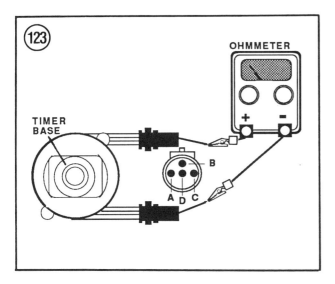

the meter reads zero, remove the emergency ignition cutoff switch from the circuit.
  a. If the ohmmeter reads infinity, replace the cutoff switch.
  b. If the ohmmeter reads zero, check for a defect in the black/yellow wire between the key switch and 4-wire connector. Repair or replace as required.

Sensor Coil Resistance Test

1. Separate the power pack-to-timer base 4-wire connector. Insert jumper wires in terminals A, B and C of the timer base end of the connector.
2. With an ohmmeter on the low scale, connect its test leads to the A and B terminal jumper wires, then the A and C terminal wires, noting each reading. See **Figure 121**. If the reading is not 40 ± 10 ohms at each connection, replace the timer base.
3. Set the ohmmeter on the high scale. Ground the black test lead and connect the red test lead alternately to the A, B and C terminal jumper wires (**Figure 122**). The ohmmeter needle should not move. If it does, the sensor coils or leads are grounded. Check for a grounded sensor coil lead before replacing the timer base.
4. With the ohmmeter on the high scale, connect the black test lead to the common lead on the sensor terminal A or jumper wire (port side). See **Figure 123**. Connect the red test lead to the same point on the starboard side. If the meter does not read infinity, replace the timer base.
5. Remove the jumper wires and perform the *Charge Coil Resistance Test*.

Charge Coil Resistance Test

1. With the 2-terminal connector plug disconnected, insert jumper wires in terminals A and B of the charge coil end of the connector.
2. Connect an ohmmeter between the jumper wires (**Figure 124**) and note the reading. If it is not 560 ± 25 ohms, replace the stator assembly.

3. Set the ohmmeter on the high scale. Ground the black test lead and connect the red test lead alternately to the A and B terminal jumper wires (**Figure 125**). The ohmmeter needle should not move. If it does, the charge coils are grounded. Check for a grounded charge coil lead before replacing the stator assembly.

Charge Coil Output Test

1. Disconnect the 2-wire connector. Set the CD voltmeter switches to NEGATIVE and 500. Insert the red test lead in cavity A of the charge coil end of the connector. Connect the black test lead to a good engine ground. See **Figure 126**.
2. Crank the engine and note the meter reading.
3. Move the red test lead to cavity B of the connector and crank the engine again. Note the meter reading.
4. There should be no meter reading in Step 2 or Step 3. If there is, check for a grounded charge coil lead before replacing the stator assembly.
5. Leave the red test lead in cavity B of the connector. Remove the black test lead from ground and insert it in cavity A of the connector. See **Figure 126**.
6. Crank the engine and note the meter reading. If it is less than 150 volts, check for a problem in component connectors and/or wiring. If no problem is found, replace the stator assembly.

Sensor Coil Output Test

1. Disconnect the 4-wire connector. Set the CD voltmeter switches to S and 5; if using a Merc-O-Tronic Model 781 voltmeter, set the switches to POSITIVE and 5.
2. Insert the red test lead in cavity A of the timer base end of the connector. Connect the black test lead to a good engine ground. See **Figure 127**.
3. Crank the engine and note the meter reading.
4. Move the red test lead to cavity B and C of the connector, cranking the engine at each connection and noting the meter reading.

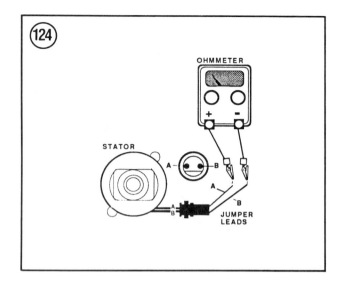

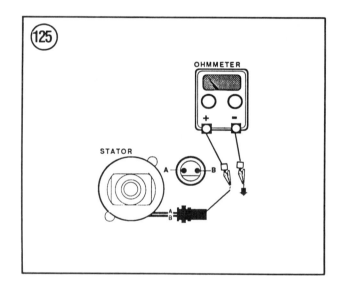

5. There should be no meter reading in Step 3 or Step 4. If there is, check for a grounded sensor coil lead before replacing the timer base.
6. Return the red test lead to cavity B of the connector. Remove the black test lead from ground and insert it in cavity A of the connector.
7. Crank the engine and note the meter reading.
8. Move the red test lead to cavity C of the connector, cranking the engine at each connection and noting the meter reading.
9. If any reading in Step 7 or Step 8 is less than 0.3 volts, check for a problem in the component connectors and/or wiring. If no problem is found, replace the timer base.

TROUBLESHOOTING

119

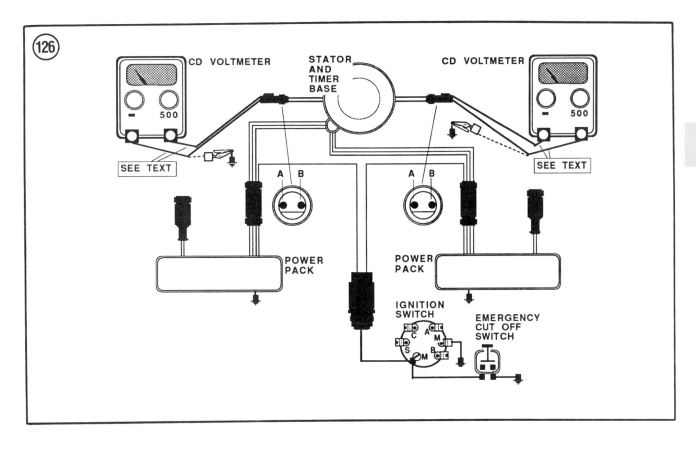

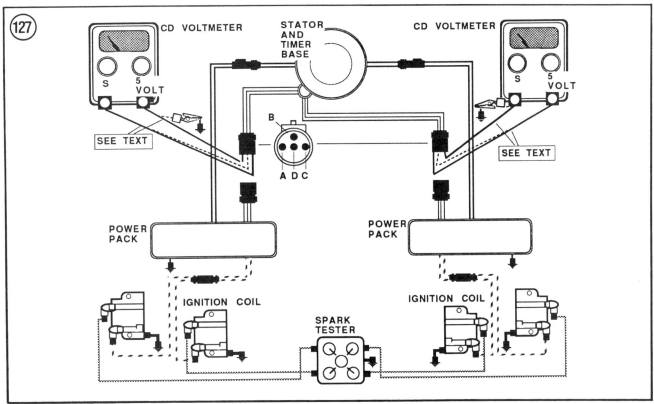

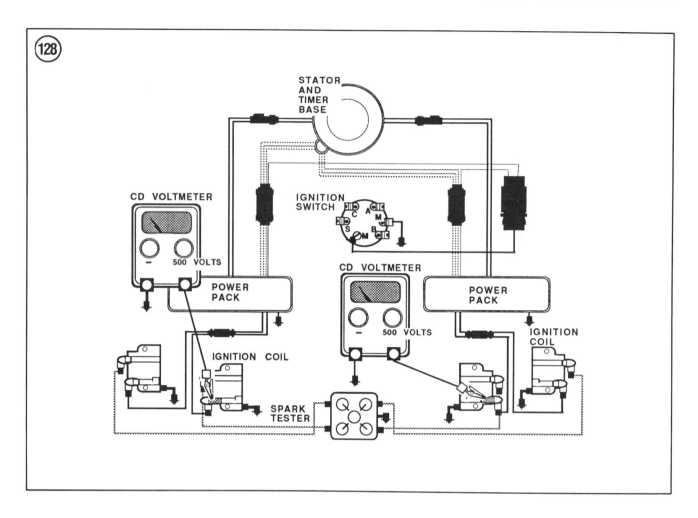

Power Pack Output Test

Refer to **Figure 128** for this procedure.

1. Disconnect the primary lead from each ignition coil. Install a terminal extender on the primary terminal of each coil, then reconnect the primary lead to the extender terminal.
2. Set the CD voltmeter switches to NEGATIVE and 500.
3. Connect the red test lead to the No. 1 coil terminal extender (top coil) and ground the black test lead.
4. Crank the engine and note the meter. It should read 150 volts or more.
5. If the meter reading is less than 150 volts in Step 4, disconnect the primary lead from the No. 1 coil terminal extender. Connect the red test lead directly to the spring clip in the primary lead boot and repeat Step 4.

   a. If the meter now reads 150 volts or more, test the ignition coil as described in this chapter.
   b. If the meter still reads less than 150 volts, check the spring clip and primary lead wire for defects. If none are found, replace the power pack.

6. Repeat Steps 3-5 to test the No. 2 (bottom) ignition coil.
7. Remove the terminal extenders from the coil terminals with a clockwise pulling motion. Reconnect the primary leads to the coil terminals. Make sure the orange/blue lead connects to the No. 1 coil.

Ignition Coil Resistance Test

See *1985-1988 CD 3 Ignition Troubleshooting* in this chapter.

TROUBLESHOOTING

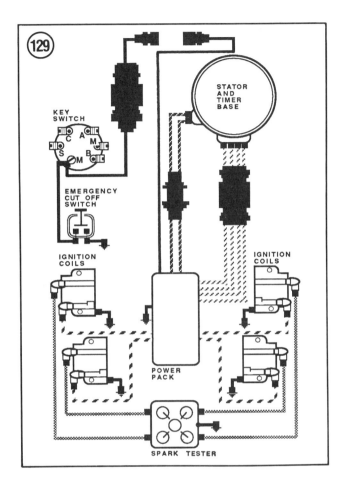

1988 88 HP, 90 HP AND 110 HP CD 4 IGNITION TROUBLESHOOTING

The CD ignition used on these V4 engines consists of a stator containing one charge coil and a timer base containing 4 sensor coils feeding voltage to a single power pack.

The power pack is connected to the sensor coils by a 5-wire connector. A 2-wire connector joins the power pack to the charge coil. A single-wire connector joins the power pack to the ignition switch. Ignition coils are connected to the power pack by individual leads. Amphenol connectors are used, with retaining clips to hold the connectors in place. The clips must be removed to disconnect the connector plugs.

Jumper leads are required for troubleshooting. Fabricate 5 leads using 8 inch lengths of 16-gauge wire. Connect a pin (OMC part No. 511469) at one end with one inch of tubing (OMC part No.

510628) and a socket (OMC part No. 581656) with one inch of tubing (OMC part No. 510628) at the other. Ohmmeter readings should be made when the engine is cold. Readings taken on a hot engine will show increased resistance caused by engine heat and result in unnecessary parts replacement without solving the basic problem.

Output tests should be made with a Stevens Model CD77, Merc-O-Tronic Model 781 or Electro-Specialties Model PRV-1 peak-reading voltmeter.

Ignition System Output Test

1. Remove spark plug leads from spark plugs.
2. Mount a spark tester on the engine and connect a tester lead to each spark plug lead. Set the spark tester air gap to 1/2 in. See **Figure 48**, typical.
3. Connect ignition system emergency cutoff clip and lanyard if so equipped.
4. Crank the engine over while watching the spark tester. If a spark jumps at each air gap, test the ignition system as outlined under *Running Output Test* in this chapter.
5. If a spark jumps at *only* one or two air gaps, first make sure all connections are good, then retest. If only one or two sparks are noted, test ignition system as outlined under *Sensor Coil Test* in this chapter.
6. If *no* spark jumps at any of the spark tester's air gaps, first make sure all connections are good, then retest. If no spark is noted, test ignition system as outlined under *Key Switch Elimination Test* in this chapter.

Key Switch Elimination Test

Refer to **Figure 129** for this procedure.
1. Connect a spark tester as shown in **Figure 129**. Set the tester air gap to ½ in.
2. Disconnect the single wire connector between the power pack and key switch.
3. Crank the engine while watching the spark tester. If there is no spark or a spark at only one

gap, remove the jumper wires and reconnect the connector plugs. Continue testing to locate the problem.

4. If a spark jumps all gaps alternately, the problem is in the key switch circuit or the emergency ignition cutoff switch.

Key Switch/Stop Circuit Test

Refer to **Figure 130** for this procedure.

1. Separate the single wire connector between the power pack and key switch (A, **Figure 130**).
2. Install the cap/lanyard assembly on the emergency ignition cutoff switch (B, **Figure 130**), if so equipped.
3. Insert a jumper wire in the key switch terminal end of the connector. Connect the red ohmmeter test lead to the jumper wire and the black test lead to a good ground. Turn the key switch ON. The ohmmeter should read infinity.
4. If there is meter needle movement in Step 3, disconnect the black/yellow wire at the key switch M terminal with the ohmmeter still connected.
5. If the meter reads infinity with the wire disconnected in Step 4, test the key switch. If the meter reads zero, remove the emergency ignition cutoff switch from the circuit.
   a. If the ohmmeter reads infinity, replace the cutoff switch.
   b. If the ohmmeter reads zero, check for a defect in the black/yellow wire between the key switch and single wire connector. Repair or replace as required.

Sensor Coil Resistance Test

1. Disconnect the 5-wire connector between the power pack and timer base. Insert jumper wires in each of the 5 terminals in the timer base connector end.
2. With an ohmmeter on the low scale, connect its test leads between the A and E terminal jumper wires, then between B and E, C and E

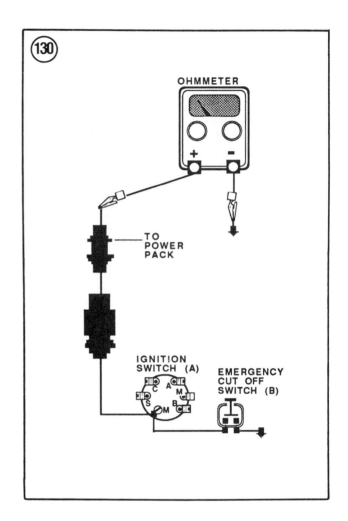

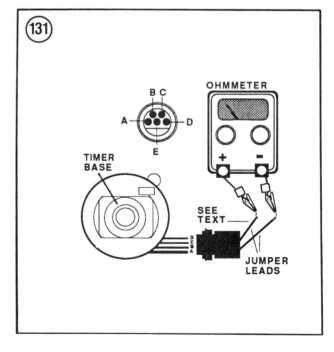

TROUBLESHOOTING

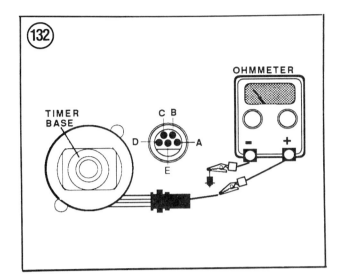

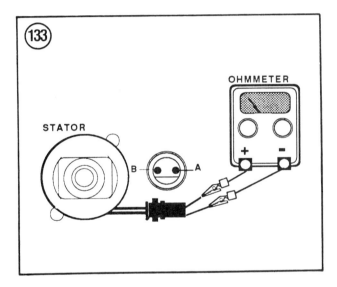

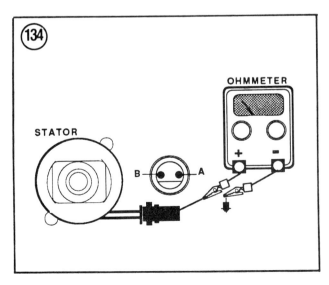

and D and E, noting each reading. See **Figure 131**. If the reading is not 40 ±10 ohms at each connection, replace the timer base.

3. Set the ohmmeter on the high scale. Ground the black test lead at the timer base and connect the red test lead alternately to each of the 5 terminal jumper wires (**Figure 132**), noting each reading. The ohmmeter needle should not move. If it does, the sensor coils or leads are grounded. Check for a grounded sensor coil lead before replacing the timer base.

4. Remove the jumper wires and perform the *Charge Coil Resistance Test*.

Charge Coil Resistance Test

1. With the 2-terminal connector plug disconnected, insert jumper wires in terminals A and B of the stator end of the connector.

2. Connect an ohmmeter between the jumper wires (**Figure 133**) and note the reading. If it is not 560 ±50 ohms on 6 amp alternator charging system or 500 ±50 ohms on 9 amp alternator charging system, replace the stator assembly.

3. Set the ohmmeter on the high scale. Ground the black test lead and connect the red test lead alternately to the A and B terminal jumper wires (**Figure 134**). The ohmmeter needle should not move. If it does, the charge coils are grounded. Check for a grounded charge coil lead before replacing the stator assembly.

Charge Coil Output Test

1. Disconnect the 2-wire connector between the power pack and stator. Set the CD voltmeter switches to NEGATIVE and 500. Insert the red test lead in cavity A of the charge coil end of the connector. Connect the black test lead to a good engine ground. See **Figure 135**.

2. Crank the engine and note the meter reading.

3. Move the red test lead to cavity B of the connector and crank the engine again. Note the meter reading.

4. There should be no meter reading in Step 2 or Step 3. If there is, check for a grounded charge coil lead before replacing the stator assembly.
5. Leave the red test lead in cavity B of the connector. Remove the black test lead from the ground and insert it in cavity A of the connector. See **Figure 135**.
6. Crank the engine and note the meter reading. If it is less than 150 volts, check for a problem in component connectors and/or wiring. If no problem is found, replace the stator assembly.

Sensor Coil Output Test

1. Disconnect the 5-wire connector between the power pack and timer base. Set the CD voltmeter switches to S and 5; if using a Merc-O-Tronic Model 781 voltmeter, set the switches to POS and 5. See **Figure 136**.
2. Connect the black test lead to a good engine ground. Connect a jumper wire to the red test lead and alternately probe all 5 connector terminals while cranking the engine.

3. There should be no meter reading in Step 2. If there is, check for a grounded sensor coil lead before replacing the timer base.
4. Remove the black test lead from ground. Connect it to a jumper wire and insert the wire in cavity E of the timer base 5-wire connector. Probe cavities A, B, C and D of the connector with the red test lead jumper wire while cranking the engine. Note the meter reading at each cavity.
5. If any reading in Step 4 is less than 0.3 volts, check for a problem in the component connectors and/or wiring. If no problem is found, replace the timer base.

Power Pack Output Test

Refer to **Figure 137** for this procedure.
1. Disconnect the primary lead from each ignition coil. Install a terminal extender on the primary terminal of each coil, then reconnect the primary lead to the extender terminal.
2. Mount a spark tester on the engine and connect a tester lead to each spark plug lead. Set the spark tester air gap to 1/2 in. See **Figure 48**, typical.

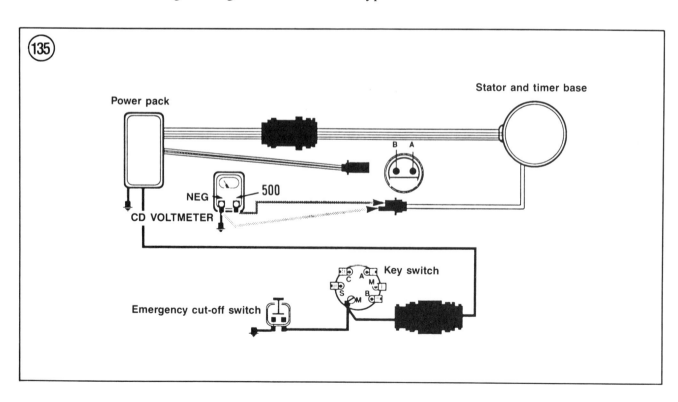

TROUBLESHOOTING

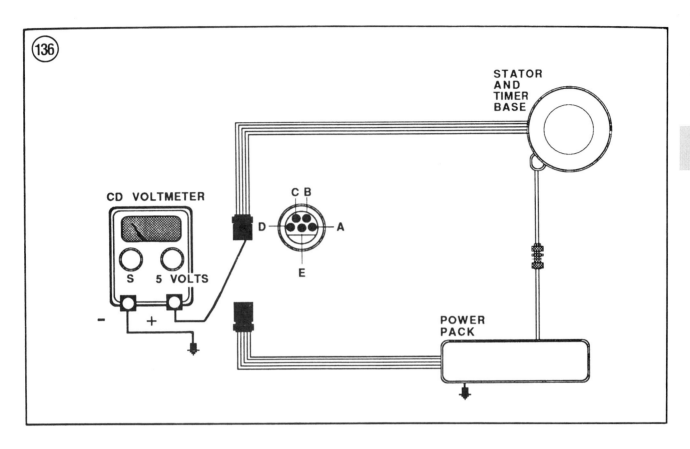

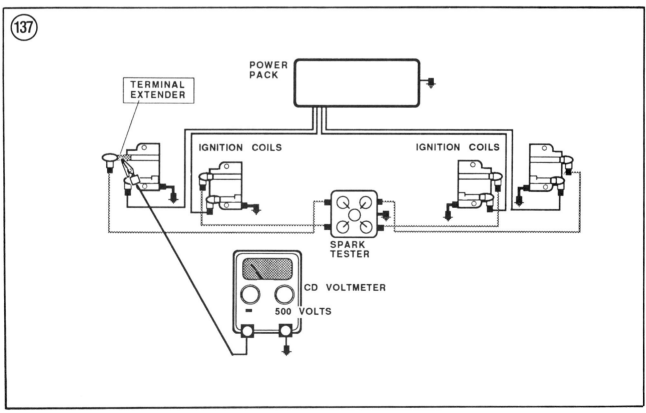

3. Set the CD voltmeter switches to NEGATIVE and 500.
4. Connect the red test lead to the No. 1 coil terminal extender (top coil) and ground the black test lead.
5. Crank the engine and note the meter. It should read 150 volts or more.
6. If the meter reading is less than 150 volts in Step 5, disconnect the primary lead from the No. 1 coil terminal extender. Connect the red test lead directly to the spring clip in the primary lead boot and repeat Step 5.
   a. If the meter now reads 150 volts or more, test the ignition coil as described in this chapter.
   b. If the meter still reads less than 150 volts, check the spring clip and primary lead wire for defects. If none are found, replace the power pack.
7. Repeat Steps 4-6 to test each remaining ignition coil.
8. Remove the terminal extenders from the coil terminals with a clockwise pulling motion. Reconnect the primary leads to the coil terminals. Make sure the leads are reconnected as follows:
   a. Orange/blue lead—Top starboard coil.
   b. Orange/green lead—Bottom starboard coil.
   c. Orange/purple lead—Top port coil.
   d. Orange/pink lead—Bottom port coil.

Ignition Coil Resistance Test

1. Disconnect the primary and high tension (secondary) coil leads at the ignition coil to be tested.
2. With an ohmmeter set on the low scale, connect the red test lead to the coil primary terminal. Connect the black test lead to a good engine ground (if the coil is mounted) or the coil ground tab (if the coil is unmounted). The meter should read 0.1 ±0.05 ohms.
3. Set the ohmmeter on the high scale. Move the black test lead to the ignition coil high tension terminal. The meter should read 225-325 ohms.
4. If the readings are not as specified in Step 2 or Step 3, replace the ignition coil.
5. Repeat Steps 2-4 to test the remaining ignition coils.

1989-1990 88 HP, 90 HP, 110 HP AND 115 HP CD 4 IGNITION TROUBLESHOOTING

The CD ignition used on these V4 engines consists of a stator containing one charge coil and a timer base containing 4 sensor coils feeding voltage to a single power pack.

The power pack is connected to the sensor coils by a 5-wire connector. A 2-wire connector joins the power pack to the charge coil. A single-wire connector joins the power pack to the ignition switch. Ignition coils are connected to the power pack by individual leads. Amphenol connectors are used, with retaining clips to hold the connectors in place. The clips must be removed to disconnect the connector plugs.

Jumper leads are required for troubleshooting. Fabricate 5 leads using 8 inch lengths of 16-gauge wire. Connect a pin (OMC part No. 511469) at one end with one inch of tubing (OMC part No. 510628) and a socket (OMC part No. 581656) with one inch of tubing (OMC part No. 510628) at the other. Ohmmeter readings should be made when the engine is cold. Readings taken on a hot engine will show increased resistance caused by engine heat and result in unnecessary parts replacement without solving the basic problem.

Output tests should be made with a Stevens Model CD77, Merc-O-Tronic Model 781 or Electro-Specialties Model PRV-1 peak-reading voltmeter.

Ignition System Output Test

1. Remove spark plug leads from spark plugs.
2. Mount a spark tester on the engine and connect a tester lead to each spark plug lead. Set the spark tester air gap to 1/2 in. See **Figure 48**, typical.
3. Connect ignition system emergency cutoff clip and lanyard if so equipped.

TROUBLESHOOTING

4. Crank the engine over while watching the spark tester. If a spark jumps at each air gap, test the ignition system as outlined under *Running Output Test* in this chapter.

5. If a spark jumps at *only* one or two air gaps, first make sure all connections are good, then retest. If only one or two sparks are noted, test ignition system as outlined under *Sensor Coil Test* in this chapter.

6. If *no* spark jumps at any of the spark tester's air gaps, first make sure all connections are good, then retest. If no spark is noted, test ignition system as outlined under *Key Switch Elimination Test* in this chapter.

Key Switch Elimination Test

Refer to **Figure 129** for this procedure.
1. Connect a spark tester as shown in **Figure 129**. Set the tester air gap to ½ in.
2. Disconnect the single wire connector between the power pack and key switch.
3. Crank the engine while watching the spark tester. If there is no spark or a spark at only one gap, remove the jumper wires and reconnect the connector plugs. Continue testing to locate the problem.
4. If a spark jumps all gaps alternately, the problem is in the key switch circuit or the emergency ignition cutoff switch.

Key Switch/Stop Circuit Test

Refer to **Figure 130** for this procedure.
1. Separate the single wire connector between the power pack and key switch (A, **Figure 130**).
2. Install the cap/lanyard assembly on the emergency ignition cutoff switch (B, **Figure 130**), if so equipped.
3. Insert a jumper wire in the key switch terminal end of the connector. Connect the red ohmmeter test lead to the jumper wire and the black test lead to a good ground. Turn the key switch ON. The ohmmeter should read infinity.

4. If there is meter needle movement in Step 3, disconnect the black/yellow wire at the key switch M terminal with the ohmmeter still connected.
5. If the meter reads infinity with the wire disconnected in Step 4, test the key switch. If the meter reads zero, remove the emergency ignition cutoff switch from the circuit.
   a. If the ohmmeter reads infinity, replace the cutoff switch.
   b. If the ohmmeter reads zero, check for a defect in the black/yellow wire between the key switch and single wire connector. Repair or replace as required.

Sensor Coil Resistance Test

1. Disconnect the 5-wire connector between the power pack and timer base. Insert jumper wires in each of the 5 terminals in the timer base connector end.
2. With an ohmmeter on the low scale, connect its test leads between the A and E terminal jumper wires, then between B and E, C and E and D and E, noting each reading. See **Figure 131**. If the reading is not 40 ±10 ohms at each connection, replace the timer base.
3. Set the ohmmeter on the high scale. Ground the black test lead at the timer base and connect the red test lead alternately to each of the 5 terminal jumper wires (**Figure 132**), noting each reading. The ohmmeter needle should not move. If it does, the sensor coils or leads are grounded. Check for a grounded sensor coil lead before replacing the timer base.
4. Remove the jumper wires and perform the *Charge Coil Resistance Test*.

Charge Coil Resistance Test

1. With the 2-terminal connector plug disconnected, insert jumper wires in terminals A and B of the stator end of the connector.
2. Connect an ohmmeter between the jumper wires (**Figure 133**) and note the reading. If it is

not 560 ±25 ohms on 6 amp alternator charging system or 480 ±25 ohms on 9 amp alternator charging system, replace the stator assembly.

3. Set the ohmmeter on the high scale. Ground the black test lead and connect the red test lead alternately to the A and B terminal jumper wires (**Figure 134**). The ohmmeter needle should not move. If it does, the charge coils are grounded. Check for a grounded charge coil lead before replacing the stator assembly.

Charge Coil Output Test

1. Disconnect the 2-wire connector between the power pack and stator. Set the peak-reading voltmeter switches to POS and 500. Insert the red test lead in cavity A of the charge coil end of the connector. Connect the black test lead to a good engine ground. See **Figure 138**.
2. Crank the engine and note the meter reading.
3. Move the red test lead to cavity B of the connector and crank the engine again. Note the meter reading.
4. There should be no meter reading in Step 2 or Step 3. If there is, check for a grounded charge coil lead before replacing the stator assembly.

5. Leave the red test lead in cavity B of the connector. Remove the black test lead from the ground and insert it in cavity A of the connector. See **Figure 138**.
6. Crank the engine and note the meter reading. If it is less than 150 volts, check for a problem in component connectors and/or wiring. If no problem is found, replace the stator assembly.

Sensor Coil Output Test

1. Disconnect the 5-wire connector between the power pack and timer base. Set the peak-reading voltmeter switches to POS and 5. See **Figure 139**.
2. Connect the black test lead to a good engine ground. Connect a jumper wire to the red test lead and alternately probe all 5 connector terminals while cranking the engine.
3. There should be no meter reading in Step 2. If there is, check for a grounded sensor coil lead before replacing the timer base.
4. Remove the black test lead from ground. Connect it to a jumper wire and insert the wire in cavity E of the timer base 5-wire connector. Probe cavities A, B, C and D of the connector

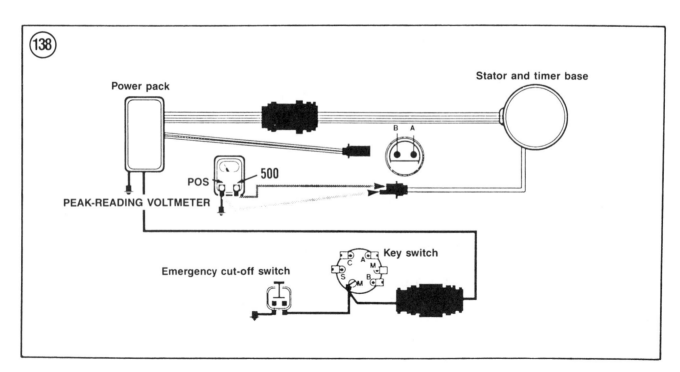

TROUBLESHOOTING

with the red test lead jumper wire while cranking the engine. Note the meter reading at each cavity.

5. If any reading in Step 4 is less than 0.3 volts, check for a problem in the component connectors and/or wiring. If no problem is found, replace the timer base.

Power Pack Output Test

WARNING
Disconnect spark plug leads to prevent accidental starting.

1. Remove primary leads from ignition coils.
2. Connect No. 1 cylinder ignition coil primary lead to the red lead of Stevens load adapter No. PL-88 and the black lead of load adapter to a good engine ground. See **Figure 50**, typical.
3. Connect the red lead of a peak-reading voltmeter to the connector end of the red lead on Stevens load adapter No. PL-88. Connect the black lead of the peak-reading voltmeter to a good engine ground.

4. Set the peak-reading voltmeter on POS and 500.
5. Crank the engine over while noting the peak-reading voltmeter. The meter should show at least 150 volts.
6. Repeat Steps 2-5 for No. 2, No. 3 and No. 4 cylinders ignition coil primary lead.
7. If all primary leads show at least 150 volts, then refer to *Ignition Coil Resistance Test* in this chapter.
8. If only one, two or three primary leads show at least 150 volts, the power pack must be replaced.

Running Output Test

1. Remove primary leads from ignition coils and connect Stevens terminal extenders No. TS-77 to ignition coils, then reconnect primary leads.
2. Connect the red lead of a peak-reading voltmeter to the No. 1 cylinder ignition coil

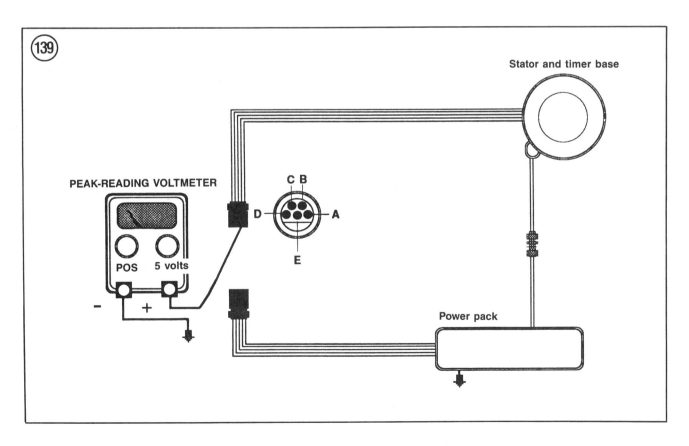

terminal extender. Connect the black lead of the peak-reading voltmeter to a good engine ground.
3. Set the peak-reading voltmeter on POS and 500.
4. Make sure the outboard motor is mounted in a suitable test tank or testing fixture with a suitable loading device (test wheel or dynometer).
5. Start the engine and operate at the rpm ignition trouble is suspected.
6. The peak-reading voltmeter should show at least 230 volts.
7. Repeat Steps 2-6 to test No. 2, No. 3 and No. 4 cylinders.
8. If one or more of the cylinders shows less than 230 volts, test charge coil as outlined in this chapter.
9. If one or more of the cylinders shows no output, test sensor coil as outlined in this chapter.
10. Remove ignition coil terminal extenders, then reconnect primary leads.

Ignition Coil Resistance Test

1. Disconnect the primary and high tension (secondary) coil leads at the ignition coil to be tested.
2. With an ohmmeter set on the low scale, connect the red test lead to the coil primary terminal. Connect the black test lead to a good engine ground (if the coil is mounted) or the coil ground tab (if the coil is unmounted). The meter should read 0.1 ±0.05 ohms.
3. Set the ohmmeter on the high scale. Move the black test lead to the ignition coil high tension terminal. The meter should read 225-325 ohms.
4. If the readings are not as specified in Step 2 or Step 3, replace the ignition coil.
5. Repeat Steps 2-4 to test the remaining ignition coils.

1985-1987 120 AND 140 HP CD 4 IGNITION TROUBLESHOOTING

The CD ignition used on these V4 engines consists of a stator containing one charge and 4 sensor coils feeding voltage to a single power pack. If the power pack is designated "CDL," it contains a built-in rpm limiter which prevents engine speed from exceeding 6,100 rpm.

The power pack is connected to the sensor coils by a 5-wire connector. A 2-wire connector joins the power pack to the charge coil. A single-wire connector joins the power pack to the ignition switch. Ignition coils are connencted to the power pack by individual leads. Amphenol connectors are used, with retaining clips to hold the connectors in place. The clips must be remoed to disconnect the connector plugs.

Jumper leads are required for troubleshooting. Fabricate 5 leads using 8 inch lengths of 16-gauge wire. Connect a pin (OMC part No. 511469) at one end and a socket (OMC part No. 581656) with one inch of tubing (OMC part No. 519628)

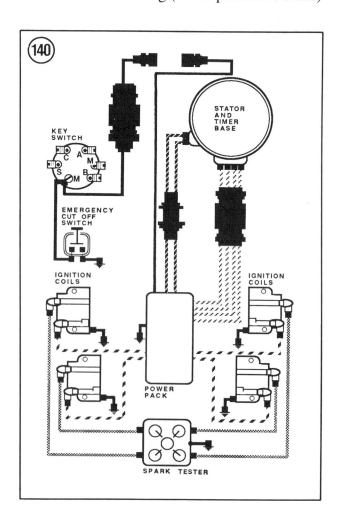

TROUBLESHOOTING

at the other. Ohmmeter readings should be made when the engine is cold. Readings taken on a hot engine will show increased resistance caused by engine heat and result in unnecessary parts replacement without solving the basic problem.

Output tests should be made with a Stevens Model CD77, Merc-O-Tronic Model 781 or Electro-Specialties Model PRV-1 CD voltmeter.

Key Switch Elimination Test

Refer to **Figure 140** for this procedure.
1. Connect a spark tester as shown in **Figure 140**. Set the tester air gap to 1/2 in.
2. Disconnect the single wire connector between the power pack and key switch.
3. Crank the engine while watching the spark tester. If there is no spark or a spark at only one gap, remove the jumper wires and reconnect the connector plugs. Continue testing to locate the problem.
4. If a spark jumps all gaps alternately, the problem is in the key switch circuit or the emergency ignition cutoff switch.

Key Switch/Stop Circuit Test

Refer to **Figure 141** for this procedure.
1. Separate the single wire connector between the power pack and key switch (A, **Figure 141**).
2. Install the cap/lanyard assembly on the emergency ignition cutoff switch (B, **Figure 141**), if so equipped.
3. Insert a jumper wire in the key switch terminal end of the connector. Connect the red ohmmeter test lead to the jumper wire and the black test lead to a good ground. Turn the key switch ON. The ohmmeter should read infinity.
4. If there is meter needle movement in Step 3, disconnect the black/yellow wire at the key switch M terminal with the ohmmeter still connected.
5. If the meter reads infinity with the wire disconnected in Step 4, test the key switch. If the meter reads zero, remove the emergency ignition cutoff switch from the circuit.
   a. If the ohmmeter reads infinity, replace the cutoff switch.
   b. If the ohmmeter reads zero, check for a defect in the black/yellow wire between the key switch and single wire connector. Repair or replace as required.

Sensor Coil Resistance Test

1. Disconnect the 5-wire connector between the power pack and timer base. Insert jumper wires in each of the 5 terminals in the timer base connector end.
2. With an ohmmeter on the low scale, connect its test leads between the A and E terminal jumper wires, then between B and E, C and E

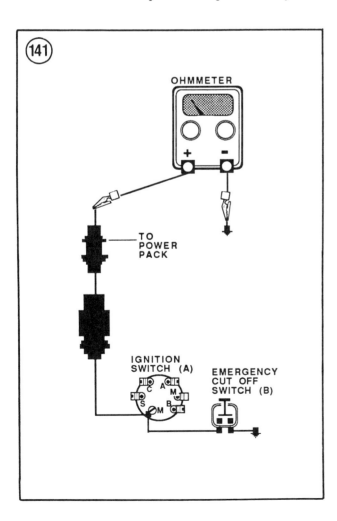

and D and E, noting each reading. See **Figure 142**. If the reading is not 40 ± 10 ohms at each connection, replace the timer base.

3. Set the ohmmeter on the high scale. Ground the black test lead at the timer base and connect the red test lead alternately to each of the 5 terminal jumper wires (**Figure 143**), noting each reading. The ohmmeter needle should not move. If it does, the sensor coils or leads are grounded. Check for a grounded sensor coil lead before replacing the timer base.

4. Remove the jumper wires and perform the *Charge Coil Resistance Test*.

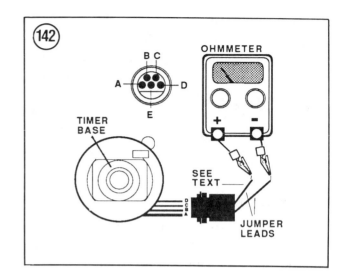

Charge Coil Resistance Test

1. With the 2-terminal connector plug disconnected, insert jumper wires in terminals A and B of the stator end of the connector.

2. Connect an ohmmeter between the jumper wires (**Figure 144**) and note the reading. If it is not 560 ± 25 ohms, replace the stator assembly.

3. Set the ohmmeter on the high scale. Ground the black test lead and connect the red test lead alternately to the A and B terminal jumper wires (**Figure 145**). The ohmmeter needle should not move. If it does, the charge coils are grounded. Check for a grounded charge coil lead before replacing the stator assembly.

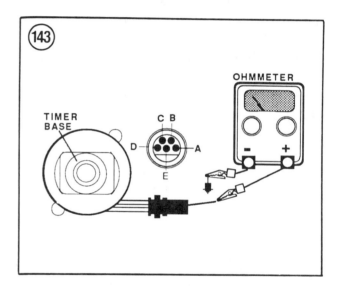

Charge Coil Output Test

1. Disconnect the 2-wire connector between the power pack and stator. Set the CD voltmeter switches to NEGATIVE and 500. Insert the red test lead in cavity A of the charge coil end of the connector. Connect the black test lead to a good engine ground. See **Figure 146**.

2. Crank the engine and note the meter reading.

3. Move the red test lead to cavity B of the connector and crank the engine again. Note the meter reading.

4. There should be no meter reading in Step 2 or Step 3. If there is, check for a grounded charge coil lead before replacing the stator assembly.

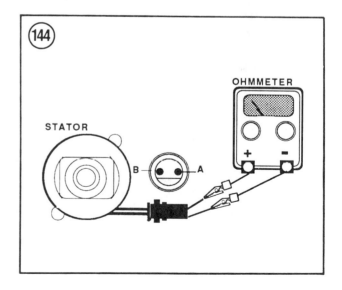

TROUBLESHOOTING

5. Leave the red test lead in cavity B of the connector. Remove the black test lead from ground and insert it in cavity A of the connector. See **Figure 146**.

6. Crank the engine and note the meter reading. If it is less than 175 volts, check for a problem in component connectors and/or wiring. If no problem is found, replace the stator assembly.

Sensor Coil Output Test

1. Disconnect the 5-wire connector between the power pack and timer base. Set the CD voltmeter switches to S and 5; if using a Merc-O-Tronic Model 781 voltmeter, set the switches to POSITIVE and 5. See **Figure 147**.

2. Connect the black test lead to a good engine ground. Connect a jumper wire to the red test lead and alternately probe all 5 connector terminals while cranking the engine.

3. There should be no meter reading in Step 2. If there is, check for a grounded sensor coil lead before replacing the timer base.

4. Remove the black test lead from ground. Connect it to a jumper wire and insert the wire in cavity E of the timer base 5-wire connector. Probe cavities A, B, C and D of the connector with the red test lead jumper wire while cranking the engine. Note the meter reading at each cavity.

5. If any reading in Step 4 is less than 0.3 volts, check for a problem in the component connectors and/or wiring. If no problem is found, replace the timer base.

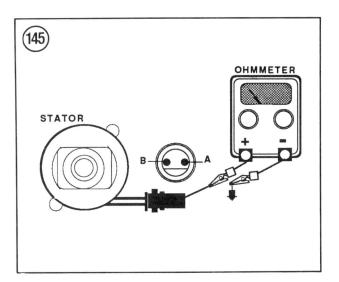

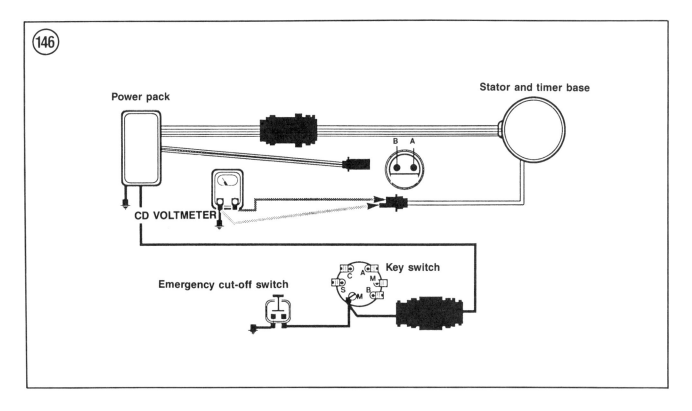

Power Pack Output Test

Refer to **Figure 148** for this procedure.

1. Disconnect the primary lead from each ignition coil. Install a terminal extender on the primary terminal of each coil, then reconnect the primary lead to the extender terminal.
2. Set the CD voltmeter switches to NEGATIVE and 500.
3. Connect the red test lead to the No. 1 coil terminal extender (top coil) and ground the black test lead.
4. Crank the engine and note the meter. It should read 175 volts or more.
5. If the meter reading is less than 175 volts in Step 4, disconnect the primary lead from the No. 1 coil terminal extender. Connect the red test lead directly to the spring clip in the primary lead boot and repeat Step 4.
   a. If the meter now reads 175 volts or more, test the ignition coil as described in this chapter.
   b. If the meter still reads less than 175 volts, check the spring clip and primary lead wire for defects. If none are found, replace the power pack.
6. Repeat Steps 3-5 to test each remaining ignition coil.
7. Remove the terminal extenders from the coil terminals with a clockwise pulling motion. Reconnect the primary leads to the coil terminals. Make sure the leads are reconnected as follows:
   a. Orange/blue lead—Top starboard coil.
   b. Orange/green lead—Bottom starboard coil.
   c. Orange/purple lead—Top port coil.
   d. Orange/pink lead—Bottom port coil.

Ignition Coil Resistance Test

1. Disconnect the primary and high tension (secondary) coil leads at the ignition coil to be tested.
2. With an ohmmeter set on the low scale, connect the red test lead to the coil primary

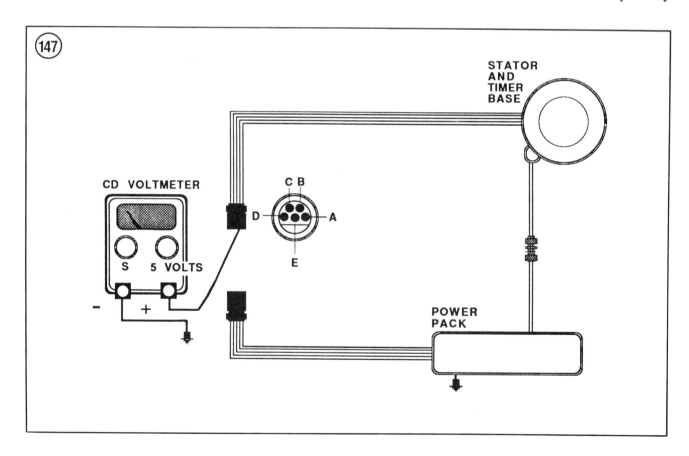

TROUBLESHOOTING

terminal. Connect the black test lead to a good engine ground (if the coil is mounted) or the coil ground tab (if the coil is unmounted). The meter should read 0.1 ±0.05 ohms.

3. Set the ohmmeter on the high scale. Move the black test lead to the ignition coil high tension terminal. The meter should read 225-325 ohms.

4. If the readings are not as specified in Step 2 or Step 3, replace the ignition coil.

5. Repeat Steps 2-4 to test the remaining ignition coils.

1988-1990 120 AND 140 HP CD 4 IGNITION TROUBLESHOOTING

The CD ignition used on these V4 engines consists of a stator containing one charge and one power coil and a timer base containing 8 sensor coils feeding voltage to a single power pack. Two sensor coils are used per cylinder. One coil for normal operation and one coil for automatic ignition timing advancement when QuikStart™ is activated.

The power pack is connected to the sensor coils by a 5-wire and a 4-wire connector. A 2-wire connector joins the power pack to the charge coil. A terminal block joins the power pack to the power coil. A single-wire connector joins the power pack to the ignition switch. Ignition coils are connected to the power pack by individual leads. Amphenol connectors are used. Special tools are required to service terminals in connector plugs.

Jumper leads are required for troubleshooting. Fabricate at least 2 and as many as 9 leads using 8 inch lengths of 16-gauge wire. Connect a pin (OMC part No. 511469) at one end with one inch of tubing (OMC part No. 510628) and a socket (OMC part No. 581656) with one inch of tubing (OMC part No. 510628) at the other. Ohmmeter readings should be made when the engine is cold. Readings taken on a hot engine will show

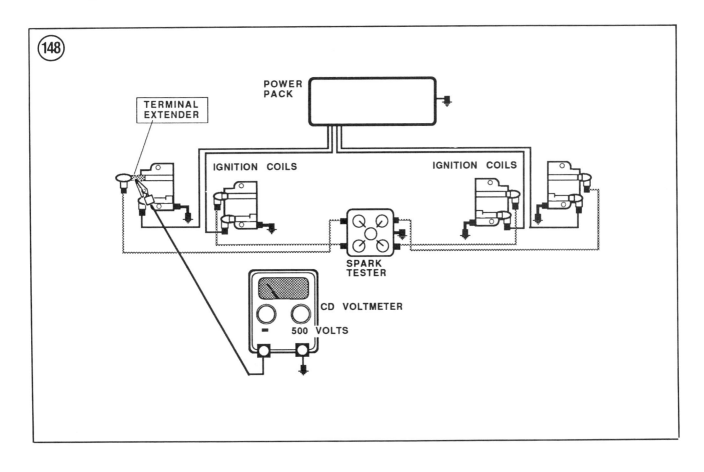

increased resistance caused by engine heat and result in unnecessary parts replacement without solving the basic problem.

Output tests should be made with a Stevens Model CD77, Merc-O-Tronic Model 781 or Electro-Specialties Model PRV-1 peak-reading voltmeter.

S.L.O.W. Operation And Testing

The S.L.O.W. (speed limiting overheat warning) circuit is activated when the engine temperature exceeds 203° F. When activated, the S.L.O.W. mode will limit the engine speed to approximately 2500 rpm. To be deactivated, the engine must be cooled to 162° F and the engine must be stopped.

A blocking diode located in the engine wiring harness is used to isolate the S.L.O.W. warning system from other warning horn signals.

1. Install the engine in a test tank with the proper test wheel.
2. Connect a tachometer according to manufacturer's instructions.
3. Disconnect temperature sensor tan lead at connector on port cylinder head (**Figure 149**).
4. Start engine and run at approximately 3500 rpm.
5. Connect engine wiring tan lead to a clean engine ground.
6. Note engine rpm. Engine rpm should reduce to approximately 2500 rpm.
7. Stop the engine. Repeat Steps 3-6 for temperature sensor located in starboard cylinder head.
8. Stop the engine. If engine rpm reduces only when one temperature sensor lead is grounded, check wiring harness and connectors. If engine rpm does not reduce when either temperature sensor lead is grounded, loosen power pack (**Figure 150**) mounting fasteners and remove terminal ends of orange and orange/black wires from terminal block which lead to power coil contained in flywheel stator.
9. Calibrate an ohmmeter to R×100 scale.
10. Connect black ohmmeter lead to a good engine ground. Alternately connect red ohmmeter lead to terminal ends of orange and orange/black wires leading to power coil. See **Figure 151**. No continuity or infinite resistance should be noted at both terminal ends.
11. Connect ohmmeter leads between orange and orange/black wire terminal ends leading to power

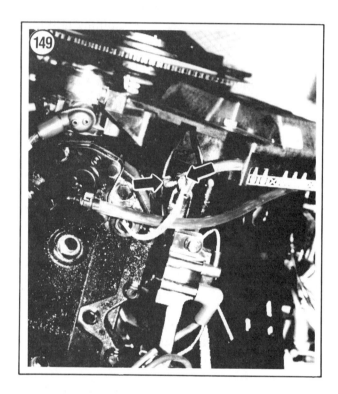

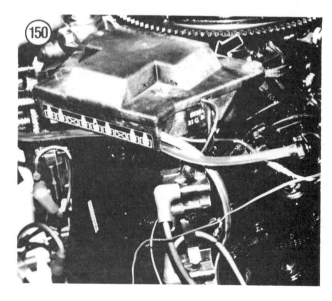

TROUBLESHOOTING

coil. See **Figure 152**. Ohmmeter should read between 86-106 ohms. If so, replace power pack.

12. Blocking Diode Test—Calibrate an ohmmeter to R×100 scale. Connect one ohmmeter lead to either port or starboard engine wiring tan lead for temperature sensor and other ohmmeter lead to tan wire terminal in red engine harness connector. Note ohmmeter reading, then reverse leads or push "polarity" button on meter if so equipped and note reading. Continuity (near zero resistance) should be noted in one direction and no continuity or infinite resistance should

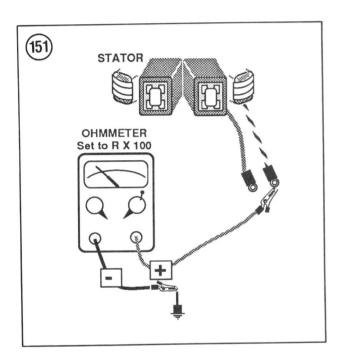

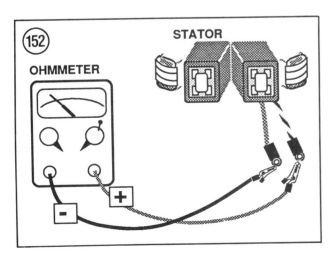

be noted in other direction. If not, replace engine wiring harness.

QuikStart™ Operation And Testing

The QuikStart™ circuit is activated each time the engine is started. QuikStart™ is an electronic feature that will automatically advance ignition timing during engine starting (all engine temperatures) and warm-up (engine temperature below 96° F) when engine speed is below approximately 1100 rpm. If the engine temperature is 96° F or more, the ignition timing will remain advanced for only approximately 5 seconds after the engine starts before returning to normal setting. When engine speed is approximately 1100 rpm or more, the QuikStart™ circuit will deactivate and return ignition timing to normal setting.

1. Install the engine in a test tank with the proper test wheel.

NOTE
Make sure engine synchronization and linkage adjustments are correctly set as outlined in Chapter Five.

2. Operate the engine until temperature is above 96° F.
3. Stop the engine and connect a tachometer according to manufacturer's instructions.
4. Disconnect temperature sensor white/black lead at connector on port cylinder head (**Figure 149**).
5. Connect a timing light to the No. 1 cylinder according to the manufacturer's instructions.
6. Start engine and shift into FORWARD gear. *Do not* operate engine above 900 rpm.
7. Observe flywheel with timing light. Pointer should align with No. 1 cylinder TDC mark.
8. Connect temperature sensor white/black lead to respective power pack lead.
9. Observe flywheel with timing light. Pointer should move to the right of the TDC mark approximately 1 in. indicating ignition timing has returned to normal setting.

10. Stop the engine and repeat Steps 4-9 for remaining cylinders.

11. If only one, two or three of the cylinders react correctly, then replace the timer base.

12. If no cylinder reacts correctly, test power coil located in flywheel stator as outlined under *S.L.O.W. Operation And Testing* in this chapter.

NOTE
If QuikStart™ remains activated when the engine is operated above 1100 rpm, replace the power pack if no electrical wiring or connections are found faulty.

Ignition System Output Test

1. Remove power pack (**Figure 150**) top cover and disconnect yellow/red wire at connector. Ground power pack with an external connection.

2. Disconnect temperature sensor white/black lead at connector on port cylinder head (**Figure 149**).

3. Connect terminal end of power pack white/black lead to a clean engine ground.

4. Remove spark plug leads from spark plugs.

5. Mount a spark tester on the engine and connect a tester lead to each spark plug lead. Set the spark tester air gap to 7/16 in. See **Figure 48**, typical.

6. Connect ignition system emergency cutoff clip and lanyard if so equipped.

7. Crank the engine over while watching the spark tester. If a spark jumps at each air gap, test the ignition system as outlined under *Running Output Test* in this chapter.

8. If a spark jumps at *only* one, two or three air gaps, first make sure all connections are good, then retest. If only one, two or three sparks are noted, test ignition system as outlined under *Sensor Coil Test* in this chapter.

9. If *no* spark jumps at any of the spark tester's air gaps, first make sure all connections are good, then retest. If no spark is noted, test ignition system as outlined under *Key Switch Elimination Test* in this chapter.

Key Switch Elimination Test

Refer to **Figure 153** for this procedure.

1. Connect a spark tester as shown in **Figure 48**, typical. Set the tester air gap to 7/16 in.

2. Disconnect the black/yellow wire at the connector between the power pack and key switch (A, **Figure 153**).

3. Crank the engine while watching the spark tester. If there is no spark or a spark at only one, two or three gaps, test charge coil and sensor coils as outlined in this chapter.

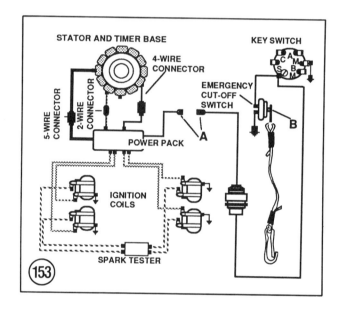

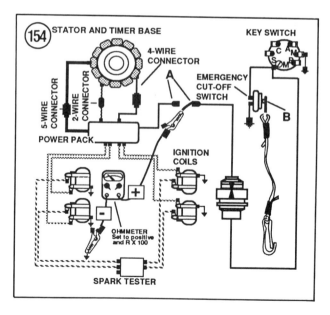

TROUBLESHOOTING

4. If a spark jumps all gaps alternately, the problem is in the key switch circuit or the emergency ignition cutoff switch.

Key Switch/Stop Circuit Test

Refer to **Figure 154** for this procedure.

1. Separate the black/yellow wire at the connector between the power pack and key switch (A, **Figure 154**).

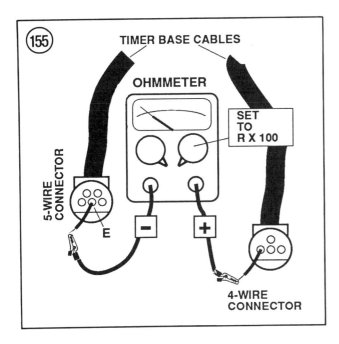

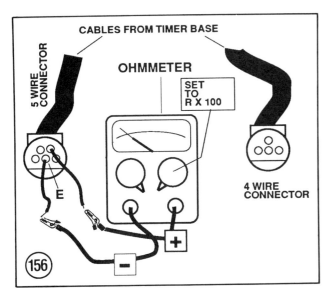

2. Install the cap/lanyard assembly on the emergency ignition cutoff switch (B, **Figure 154**), if so equipped.
3. Insert a jumper wire in the key switch terminal end of the connector. Connect the red ohmmeter test lead to the jumper wire and the black test lead to a good ground. Turn the key switch ON. The ohmmeter should read infinity.
4. If there is meter needle movement in Step 3, disconnect the black/yellow wire at the key switch M terminal with the ohmmeter still connected.
5. If the meter reads infinity with the wire disconnected in Step 4, test the key switch. If the meter reads zero, remove the emergency ignition cutoff switch from the circuit.
   a. If the ohmmeter reads infinity, replace the cutoff switch.
   b. If the ohmmeter reads zero, check for a defect in the black/yellow wire between the key switch and single wire connector. Repair or replace as required.

Sensor Coil Resistance Test

1. Disconnect the 5-wire connector and the 4-wire connector between the power pack and timer base. If 9 jumper wires were fabricated, insert a jumper wire in each of the terminals in the 5- and 4-wire timer base connector ends.
2. With an ohmmeter on the Rx100 scale, connect its black test lead to the E terminal jumper wire in the port 5-wire connector and the red test lead alternately between each of the 4 jumper wires in the starboard 4-wire connector while noting each reading. See **Figure 155**. If the reading is not 130 ±30 ohms at each connection, replace the timer base.
3. Leave the ohmmeter black test lead connected to the E terminal jumper wire in the port 5-wire connector and alternately connect the red test lead between each of the 4 remaining jumper wires in the port 5-wire connector while noting each reading. See **Figure 156**. If the reading is

not 45 ±10 ohms at each connection, replace the timer base.

4. Set the ohmmeter on the high scale. Ground the black test lead at the timer base and connect the red test lead alternately to each of the terminal jumper wires in the 5- and 4-wire timer base connector ends while noting each reading. The ohmmeter needle should not move. If it does, the sensor coils or leads are grounded. Check for a grounded sensor coil lead before replacing the timer base.

5. Remove the jumper wires and perform the *Charge Coil Resistance Test*.

Charge Coil Resistance Test

1. With the 2-terminal connector plug disconnected, insert jumper wires in terminals A and B of the stator end of the connector.
2. Connect an ohmmeter between the jumper wires (**Figure 157**) and note the reading. If it is not 480 ±25 ohms, replace the stator assembly.
3. Set the ohmmeter on the high scale. Ground the black test lead and connect the red test lead alternately to the A and B terminal jumper wires (**Figure 158**). The ohmmeter needle should not move. If it does, the charge coils are grounded.

Check for a grounded charge coil lead before replacing the stator assembly.

Charge Coil Output Test

1. Disconnect the 2-wire connector between the power pack and stator. Set the peak-reading

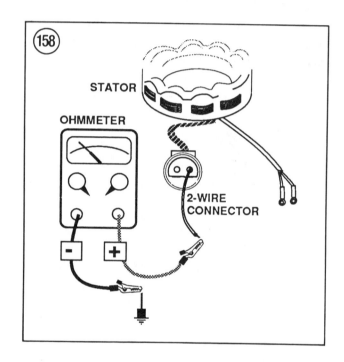

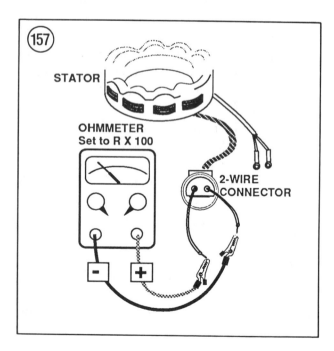

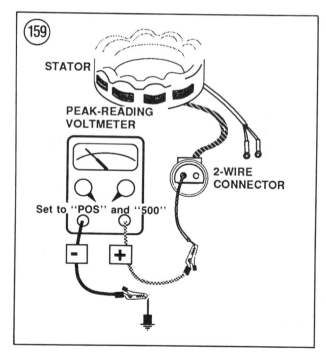

TROUBLESHOOTING

voltmeter switches to POS and 500. Insert the red test lead in cavity A of the charge coil end of the connector. Connect the black test lead to a good engine ground. See **Figure 159**.

2. Crank the engine and note the meter reading.

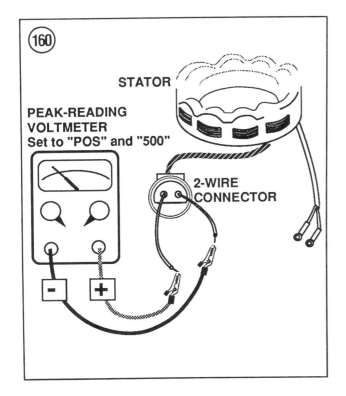

3. Move the red test lead to cavity B of the connector and crank the engine again. Note the meter reading.

4. There should be no meter reading in Step 2 or Step 3. If there is, check for a grounded charge coil lead before replacing the stator assembly.

5. Leave the red test lead in cavity B of the connector. Remove the black test lead from the ground and insert it in cavity A of the connector. See **Figure 160**.

6. Crank the engine and note the meter reading. If it is less than 175 volts, check for a problem in component connectors and/or wiring. If no problem is found, replace the stator assembly.

Sensor Coil Output Test

1. Disconnect the 5-wire connector and the 4-wire connector between the power pack and timer base. If 9 jumper wires were fabricated, insert a jumper wire in each of the terminals in the 5- and 4-wire timer base connector ends. Set the peak-reading voltmeter switches to POS and 5. See **Figure 161**.

2. Connect the black test lead to a good engine ground. Connect the red test lead alternately to each of the terminal jumper wires in the 5- and 4-wire timer base connector ends while cranking the engine and noting each reading.

3. There should be no meter reading in Step 2. If there is, check for a grounded sensor coil lead before replacing the timer base.

4. Remove the black test lead from ground. Connect the black test lead to the E terminal jumper wire in the port 5-wire connector and alternately connect the red test lead between each of the 4 remaining jumper wires in the port 5-wire connector and the 4 jumper wires in the starboard 4-wire connector while cranking the engine and noting each reading. See **Figure 162**.

5. If any reading in Step 4 is less than 0.5 volts, check for a problem in the component connectors and/or wiring. If no problem is found, replace the timer base.

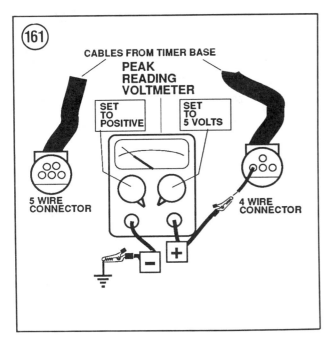

Power Pack Output Test

WARNING
Disconnect spark plug leads to prevent accidental starting.

1. Remove primary leads from ignition coils.
2. Connect No. 1 cylinder ignition coil primary lead to the red lead of Stevens load adapter No. PL-88 and the black lead of load adapter to a good engine ground. See **Figure 50**, typical.
3. Connect the red lead of a peak-reading voltmeter to the connector end of the red lead on Stevens load adapter No. PL-88. Connect the black lead of the peak-reading voltmeter to a good engine ground.
4. Set the peak-reading voltmeter on POS and 500.
5. Crank the engine over while noting the peak-reading voltmeter. The meter should show at least 150 volts.
6. Repeat Steps 2-5 for No. 2, No. 3 and No. 4 cylinders ignition coil primary lead.
7. If all primary leads show at least 150 volts, then refer to *Ignition Coil Resistance Test* in this chapter.
8. If only one, two or three primary leads show at least 150 volts, the power pack must be replaced.

Running Output Test

1. Remove primary leads from ignition coils and connect Stevens terminal extenders No. TS-77 to ignition coils, then reconnect primary leads.
2. Connect the red lead of a peak-reading voltmeter to the No. 1 cylinder ignition coil terminal extender. Connect the black lead of the peak-reading voltmeter to a good engine ground.
3. Set the peak-reading voltmeter on POS and 500.
4. Make sure the outboard motor is mounted in a suitable test tank or testing fixture with a suitable loading device (test wheel or dynamometer).
5. Start the engine and operate at the rpm ignition trouble is suspected.
6. The peak-reading voltmeter should show at least 180 volts.
7. Repeat Steps 2-6 to test No. 2, No. 3 and No. 4 cylinders.
8. If one or more of the cylinders shows less than 180 volts, test charge coil as outlined in this chapter.
9. If one or more of the cylinders shows no output, test sensor coil as outlined in this chapter.
10. Remove ignition coil terminal extenders, then reconnect primary leads.

Ignition Coil Resistance Test

1. Disconnect the primary and high tension (secondary) coil leads at the ignition coil to be tested.
2. With an ohmmeter set on the low scale, connect the red test lead to the coil primary terminal. Connect the black test lead to a good engine ground (if the coil is mounted) or the coil

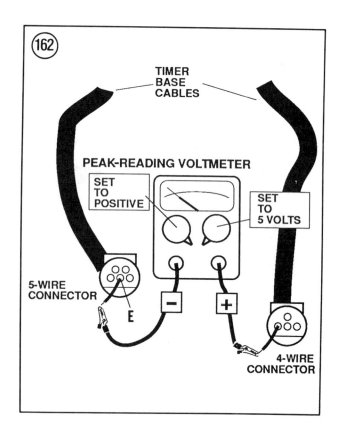

TROUBLESHOOTING

ground tab (if the coil is unmounted). The meter should read 0.1 ±0.05 ohms.

3. Set the ohmmeter on the high scale. Move the black test lead to the ignition coil high tension terminal. The meter should read 225-325 ohms.
4. If the readings are not as specified in Step 2 or Step 3, replace the ignition coil.
5. Repeat Steps 2-4 to test the remaining ignition coils.

1985 V6 AND 1986-1988 150-175 HP CD 6 IGNITION TROUBLESHOOTING

The CD ignition used on these V6 engines consists of 2 identical ignition systems—one on the port side and one on the starboard side of the engine. The troubleshooting procedures must be performed on each ignition system.

Each power pack is connected to the timer base by a 4-wire connector. A 2-wire connector joins each power pack to the charge coil. A single wire connector is used between the key switch and power pack. Individual leads connect the power packs to the ignition coils.

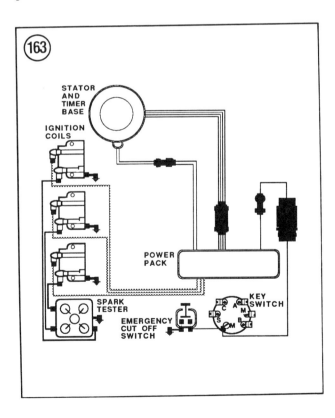

Since this ignition contains 2 power packs, there is one set of connectors on each side of the engine. Amphenol connectors are used, with retaining clips provided to hold the connectors in place. The clips must be removed to disconnect the connector plugs.

Jumper leads are required for troubleshooting. Fabricate 4 leads using 8 inch lengths of 16-gauge wire. Connect a pin (OMC part No. 511469) at one end and a socket (OMC part No. 581656) with one inch of tubing (OMC part No. 519628) at the other. Ohmmeter readings should be made when the engine is cold. Readings taken on a hot engine will show increased resistance caused by engine heat and result in unnecessary parts replacement without solving the basic problem.

Output tests should be made with a Stevens Model CD77, Merc-O-Tronic Model 781 or Electro-Specialties Model PRV-1 CD voltmeter.

Key Switch Elimination Test

Refer to **Figure 163** for this procedure.
1. Connect a spark tester as shown in **Figure 163**. Set the tester air gap to 7/16 in.
2. Disconnect the single wire connector between the power pack and key switch.
3. Crank the engine while watching the spark tester. If there is no spark or a spark at only one gap, remove the jumper wires and reconnect the connector plugs. Continue testing to locate the problem.
4. If a spark jumps all gaps alternately, the problem is in the key switch circuit or the emergency ignition cutoff switch.

Key Switch/Stop Circuit Test

Refer to **Figure 164** for this procedure.
1. Separate the single wire connector between the power pack and key switch.
2. Install the cap/lanyard assembly on the emergency ignition cutoff switch, if so equipped.
3. Insert a jumper wire in the key switch terminal end of the connector. Connect the red ohmmeter

test lead to the jumper wire and the black test lead to a good ground. Turn the key switch ON. The ohmmeter should read infinity.

4. If there is meter needle movement in Step 3, disconnect the black/yellow wire at the key switch M terminal with the ohmmeter still connected.

5. If the meter reads infinity with the wire disconnected in Step 4, test the key switch. If the meter reads zero, remove the emergency ignition cutoff switch from the circuit.

   a. If the ohmmeter reads infinity, replace the cutoff switch.

   b. If the ohmmeter reads zero, check for a defect in the black/yellow wire between the key switch and single wire connector. Repair or replace as required.

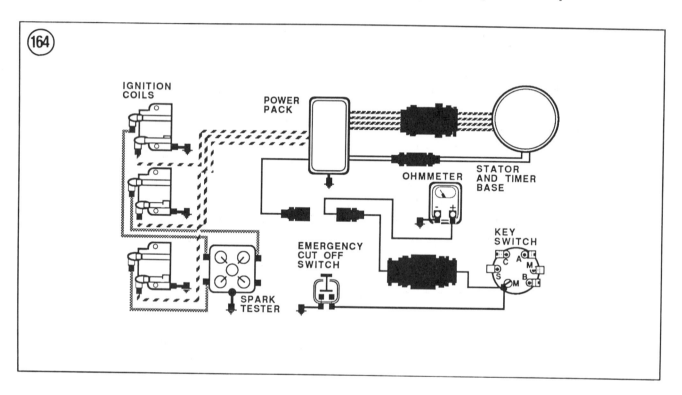

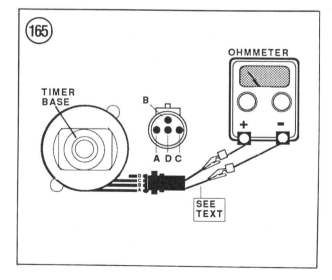

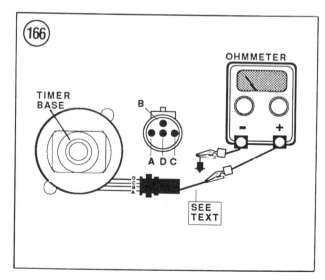

TROUBLESHOOTING

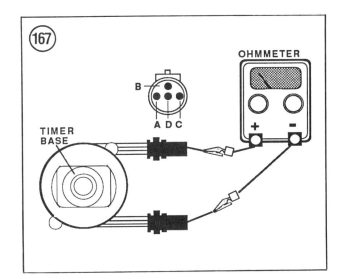

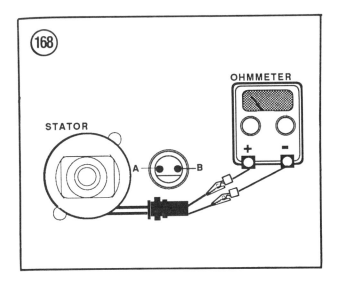

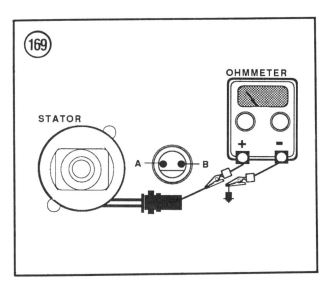

Sensor Coil Resistance Test

1. Disconnect the 5-wire connector between the power pack and timer base. Insert jumper wires in each of the 4 terminals in the timer base connector end.

2. With an ohmmeter on the low scale, connect its test leads between the A and D terminal jumper wires, then between B and D and C and D, noting each reading. See **Figure 165**. If the reading is not 17 ± 5 ohms at each connection, replace the timer base.

3. Set the ohmmeter on the high scale. Ground the black test lead at the timer base and connect the red test lead alternately to each of the 4 terminal jumper wires (**Figure 166**), noting each reading. The ohmmeter needle should not move. If it does, the sensor coils or leads are grounded. Check for a grounded sensor coil lead before replacing the timer base.

4. With the ohmmeter on the high scale, connect the black test lead to the common lead on the sensor terminal D or jumper wire (port side). See **Figure 167**. Connect the red test lead to the same point on the starboard side. If the meter does not read infinity, replace the timer base.

5. Remove the jumper wires and perform the *Charge Coil Resistance Test*.

Charge Coil Resistance Test

1. With the 2-terminal connector plug disconnected, insert jumper wires in terminals A and B of the stator end of the connector.

2. Connect an ohmmeter between the jumper wires (**Figure 168**) and note the reading. If it is as follows, replace the stator assembly:
   a. 6, 9 (prior to 1988) or 10 amp system—560 ± 25 ohms.
   b. 1988 9 amp system—500 ± 50 ohms.
   c. 1985 35 amp system—600 ± 25 ohms.
   d. 1986-1988 35 amp system—970 ± 15 ohms.

3. Set the ohmmeter on the high scale. Ground the black test lead and connect the red test lead alternately to the A and B terminal jumper wires (**Figure 169**). The ohmmeter needle should not

move. If it does, the charge coils are grounded. Check for a grounded charge coil lead before replacing the stator assembly.

Charge Coil Output Test

1. Disconnect the 2-wire connector between the power pack and stator. Set the CD voltmeter switches to NEGATIVE and 500. Insert the red test lead in cavity A of the charge coil end of the connector. Connect the black test lead to a good engine ground. See **Figure 170**.
2. Crank the engine and note the meter reading.
3. Move the red test lead to cavity B of the connector and crank the engine again. Note the meter reading.
4. There should be no meter reading in Step 2 or Step 3. If there is, check for a grounded charge coil lead before replacing the stator assembly.
5. Leave the red test lead in cavity B of the connector. Remove the black test lead from ground and insert it in cavity A of the connector. See **Figure 171**.
6. Crank the engine and note the meter reading. If it is less than 200 volts, check for a problem in component connectors and/or wiring. If no problem is found, replace the stator assembly.

Sensor Coil Output Test

1. Disconnect the 4-wire connector between the power pack and timer base. Set the CD voltmeter

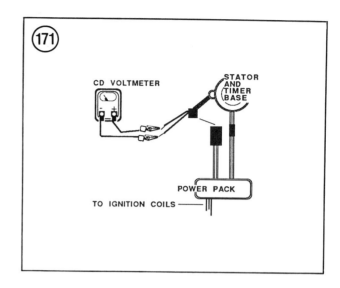

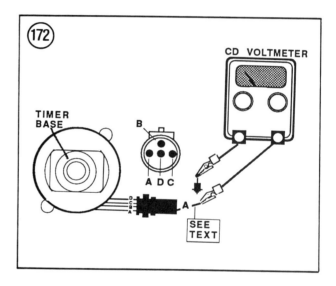

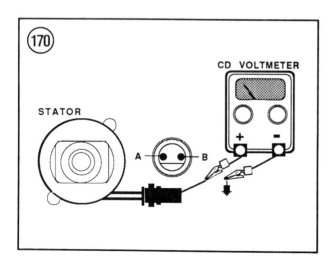

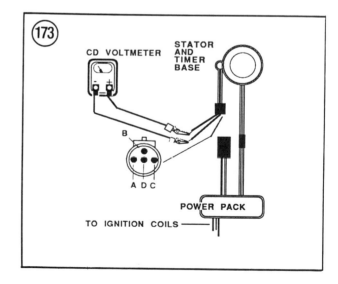

TROUBLESHOOTING

switches to S and 5; if using a Merc-O-Tronic Model 781 voltmeter, set the switches to POSITIVE and 5.

2. Connect the black test lead to a good engine ground. Connect a jumper wire to the red test lead and alternately probe all 4 connector terminals while cranking the engine. See **Figure 172**.

3. There should be no meter reading in Step 2. If there is, check for a grounded sensor coil lead before replacing the timer base.

4. Remove the black test lead from ground. Connect it to a jumper wire and insert the wire in cavity A of the timer base 4-wire connector. Probe cavities B and C of the connector with the red test lead jumper wire while cranking the engine. Note the meter reading at each cavity. See **Figure 173**.

5. If any reading in Step 4 is less than 0.2 volts, check for a problem in the component connectors and/or wiring. If no problem is found, replace the timer base.

Power Pack Output Test

Refer to **Figure 174** for this procedure.

1. Disconnect the primary lead from each ignition coil. Install a terminal extender on the primary terminal of each coil, then reconnect the primary lead to the extender terminal.

2. Set the CD voltmeter switches to NEGATIVE and 500.

3. Connect the red test lead to the No. 1 coil terminal extender (top coil) and ground the black test lead.

4. Crank the engine and note the meter. It should read 175 volts or more.

5. If the meter reading is less than 175 volts in Step 4, disconnect the primary lead from the No. 1 coil terminal extender. Connect the red test lead directly to the spring clip in the primary lead boot and repeat Step 4.

   a. If the meter now reads 175 volts or more, test the ignition coil as described in this chapter.
   b. If the meter still reads less than 175 volts, check the spring clip and primary lead wire for defects. If none are found, replace the power pack.

6. Repeat Steps 3-5 to test each remaining ignition coil.

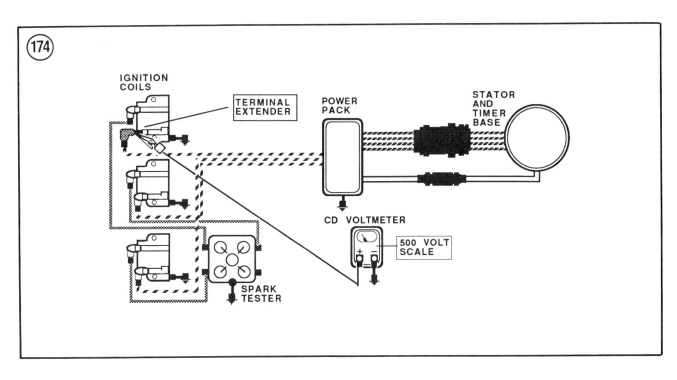

7. Remove the terminal extenders from the coil terminals with a clockwise pulling motion. Reconnect the primary leads to the coil terminals. Make sure the leads are reconnected as follows:
   a. Orange/blue lead—Top coil.
   b. Orange lead—Center coil.
   c. Orange/green lead—Bottom coil.

Ignition Coil Resistance Test

See *1985-1988 CD 3 Ignition Troubleshooting* in this chapter.

1989-1990 150 AND 175 HP CD 6 IGNITION TROUBLESHOOTING

The CD ignition used on these V6 engines consists of a stator containing two charge coils and one power coil (35 amp models) and a timer base containing 6 single-wound sensor coils on 9 amp models and 6 double-wound sensor coils on 35 amp models feeding voltage to a single power pack. On 35 amp models, one double-wound sensor coil is used per cylinder. One winding of the coil is for normal operation and the other winding is for automatic ignition timing advancement when QuikStart™ is activated.

The power pack is connected to the sensor coils by a 6-wire connector on 9 amp models and by two 4-wire connectors on 35 amp models. A 4-wire connector joins the power pack to the charge coils on 9 amp models and a 6-wire connector joins the power pack to the power coil and charge coils on 35 amp models. A single-wire connector joins the power pack to the ignition switch on 9 amp models and a 5-wire connector joins the power pack to the ignition switch on 35 amp models. Ignition coils are connected to the power pack by individual leads. Amphenol connectors are used. Special tools are required to service terminals in connector plugs.

Jumper leads are required for troubleshooting. Fabricate at least 2 and as many as 8 leads using 8 inch lengths of 16-gauge wire. Connect a pin (OMC part No. 511469) at one end with one inch of tubing (OMC part No. 510628) and a socket (OMC part No. 581656) with one inch of tubing (OMC part No. 510628) at the other. Ohmmeter readings should be made when the engine is cold. Readings taken on a hot engine will show increased resistance caused by engine heat and result in unnecessary parts replacement without solving the basic problem.

Output tests should be made with a Stevens Model CD77, Merc-O-Tronic Model 781 or Electro-Specialties Model PRV-1 peak-reading voltmeter.

S.L.O.W. Operation And Testing
35 Amp Models

The S.L.O.W. (speed limiting overheat warning) circuit is activated when the engine temperature exceeds 203° F. When activated, the S.L.O.W. mode will limit the engine speed to approximately 2500 rpm. To be deactivated, the engine must be cooled to 162° F and the engine must be stopped.

A blocking diode located in the engine wiring harness is used to isolate the S.L.O.W. warning system from other warning horn signals.

1. Install the engine in a test tank with the proper test wheel.
2. Connect a tachometer according to manufacturer's instructions.
3. Disconnect temperature sensor tan lead at connector on port cylinder head (**Figure 149**, typical).

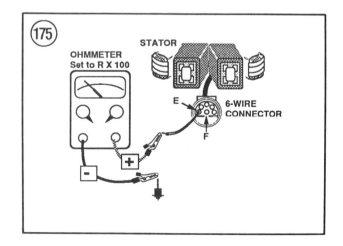

TROUBLESHOOTING

4. Start engine and run at approximately 3500 rpm.
5. Connect engine wiring tan lead to a clean engine ground.
6. Note engine rpm. Engine rpm should reduce to approximately 2500 rpm.
7. Stop the engine. Repeat Steps 3-6 for temperature sensor located in starboard cylinder head.
8. Stop the engine. If engine rpm reduces only when one temperature sensor lead is grounded, check wiring harness and connectors. If engine rpm does not reduce when either temperature sensor lead is grounded, separate 6-wire connector between flywheel stator and power pack to test power coil in stator.
9. Calibrate an ohmmeter to R×100 scale.
10. Connect black ohmmeter lead to a good engine ground. Insert a jumper wire in terminals E and F of 6-wire connector leading to flywheel stator. Alternately connect red ohmmeter lead to jumper wires in terminals E and F and note meter readings. See **Figure 175**. No continuity or infinite resistance should be noted at both terminal ends.
11. Connect ohmmeter leads to jumper wires in stator connector terminals E and F and note meter readings. See **Figure 176**. Ohmmeter should read between 86-106 ohms. If so, replace power pack.

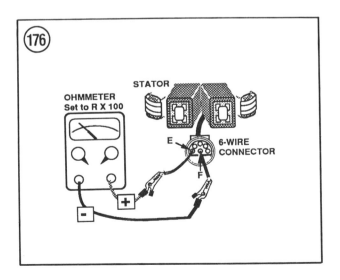

12. Blocking Diode Test—Calibrate an ohmmeter to R×100 scale. Connect one ohmmeter lead to either port or starboard engine wiring tan lead for temperature sensor and other ohmmeter lead to tan wire terminal in red engine harness connector. Note ohmmeter reading, then reverse leads or push "polarity" button on meter if so equipped and note reading. Continuity (near zero resistance) should be noted in one direction and no continuity or infinite resistance should be noted in other direction. If not, replace engine wiring harness blocking diode.

QuikStart™ Operation And Testing
35 Amp Models

The QuikStart™ circuit is activated each time the engine is started. QuikStart™ is an electronic feature that will automatically advance ignition timing during engine starting (all engine temperatures) and warm-up (engine temperature below 96° F) when engine speed is below approximately 1100 rpm. If the engine temperature is 96° F or more, the ignition timing will remain advanced for only approximately 5 seconds after the engine starts before returning to normal setting. When engine speed is approximately 1100 rpm or more, the QuikStart™ circuit will deactivate and return ignition timing to normal setting.

1. Install the engine in a test tank with the proper test wheel.

NOTE
Make sure engine synchronization and linkage adjustments are correctly set as outlined in Chapter Five.

2. Operate the engine until temperature is above 96° F.
3. Stop the engine and connect a tachometer according to manufacturer's instructions.
4. Disconnect temperature sensor white/black lead at connector on port cylinder head (**Figure 149**, typical).

5. Connect a timing light to the No. 1 cylinder according to the manufacturer's instructions.
6. Start engine and shift into FORWARD gear. *Do not* operate engine above 900 rpm.
7. Observe flywheel with timing light. Pointer should align with No. 1 cylinder TDC mark.
8. Connect temperature sensor white/black lead to respective power pack lead.
9. Observe flywheel with timing light. Pointer should move to the right of the TDC mark approximately 1 in. indicating ignition timing has returned to normal setting.
10. Stop the engine and repeat Steps 4-9 for remaining cylinders.
11. If only five or less of the cylinders react correctly, then replace the timer base.
12. If no cylinder reacts correctly, test power coil located in flywheel stator as outlined under *S.L.O.W. Operation And Testing* in this chapter.

NOTE
If QuikStart™ remains activated when the engine is operated above 1100 rpm, replace the power pack if no electrical wiring or connections are found faulty.

Ignition System Output Test

1. Remove spark plug leads from spark plugs.
2. Mount a spark tester on the engine and connect a tester lead to each spark plug lead. Set the spark tester air gap to 7/16 in. See **Figure 48**, typical.
3. Connect ignition system emergency cutoff clip and lanyard if so equipped.
4. 35 Amp Models—Separate 5-wire connector between power pack and engine harness. Install jumper wires between terminals A and D in both connectors. Install a jumper wire in terminal E of power pack connector and attach opposite end of jumper wire to a good engine ground. See **Figure 177**.
5. Crank the engine over while watching the spark tester. If a spark jumps at each air gap, test the ignition system as outlined under *Running Output Test* in this chapter.

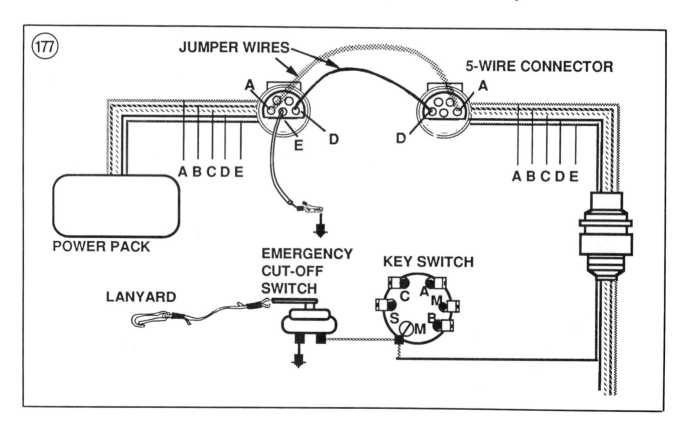

TROUBLESHOOTING

6. If a spark jumps at *only* five or less air gaps, first make sure all connections are good, then retest. If only five or less sparks are noted, test ignition system as outlined under *Charge Coil Test* in this chapter.

7. If *no* spark jumps at any of the spark tester's air gaps, first make sure all connections are good, then retest. If no spark is noted, test ignition system as outlined under *Key Switch Elimination Test* in this chapter.

Key Switch Elimination Test

1. Connect a spark tester as shown in **Figure 48**, typical. Set the tester air gap to 7/16 in.

2A. 9 Amp Models—Disconnect the black/yellow wire at the connector between the power pack and key switch (A, **Figure 178**).

2B. 35 Amp Models—Separate 5-wire connector between power pack and engine harness. Install a jumper wire in terminal E of power pack connector and attach opposite end of jumper wire to a good engine ground. See **Figure 179**.

3. Crank the engine while watching the spark tester. If there is no spark or a spark at five or less gaps, test charge coil and sensor coils as outlined in this chapter.

4. If a spark jumps all gaps alternately, the problem is in the key switch circuit or the emergency ignition cutoff switch.

Key Switch/Stop Circuit Test
9 Amp Models

1. Separate the black/yellow wire at the connector between the power pack and key switch (A, **Figure 180**).

2. Install the cap/lanyard assembly on the emergency ignition cutoff switch (B, **Figure 180**), if so equipped.

3. Insert a jumper wire in the key switch terminal end of the connector. Connect the red ohmmeter test lead to the jumper wire and the black test lead to a good ground. Turn the key switch ON. The ohmmeter should read infinity.

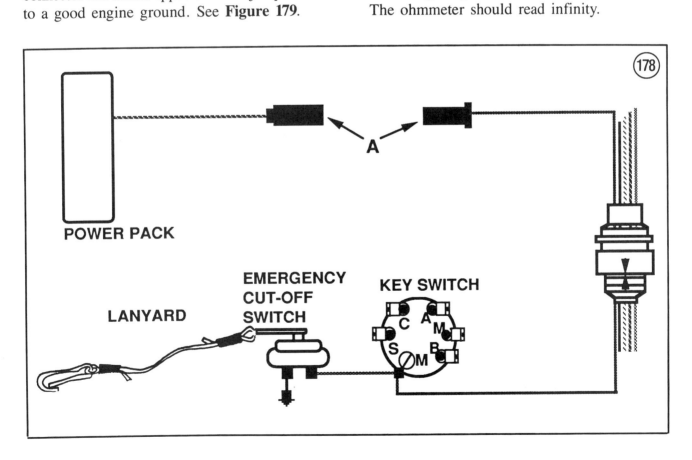

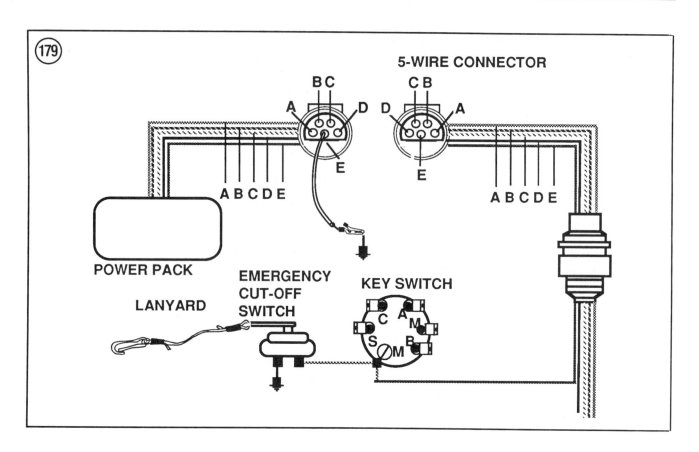

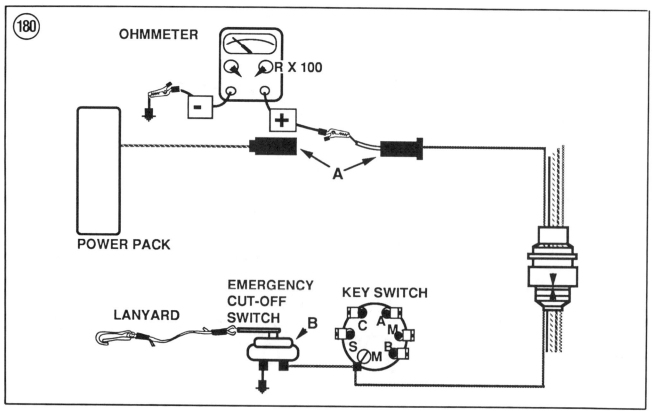

TROUBLESHOOTING

4. If there is meter needle movement in Step 3, disconnect the black/yellow wire at the key switch M terminal with the ohmmeter still connected.

5. If the meter reads infinity with the wire disconnected in Step 4, test the key switch. If the meter reads zero, remove the emergency ignition cutoff switch from the circuit.
   a. If the ohmmeter reads infinity, replace the cutoff switch.
   b. If the ohmmeter reads zero, check for a defect in the black/yellow wire between the key switch and single wire connector. Repair or replace as required.

35 Amp Models

1. Separate 5-wire connector between power pack and engine harness.

2. Install the cap/lanyard assembly on the emergency ignition cutoff switch, if so equipped.

3. Install a jumper wire in terminal A of engine harness connector. See **Figure 181**.

4. Connect the red ohmmeter test lead to the jumper wire and the black test lead to a good ground. Turn the key switch ON. The ohmmeter should read infinity.

5. If there is meter needle movement in Step 4, disconnect the black/yellow wire at the key switch M terminal with the ohmmeter still connected.

6. If the meter reads infinity with the wire disconnected in Step 5, test the key switch. If the meter reads zero, remove the emergency ignition cutoff switch from the circuit.
   a. If the ohmmeter reads infinity, replace the cutoff switch.
   b. If the ohmmeter reads zero, check for a defect in the black/yellow wire between the key switch and single wire connector. Repair or replace as required.

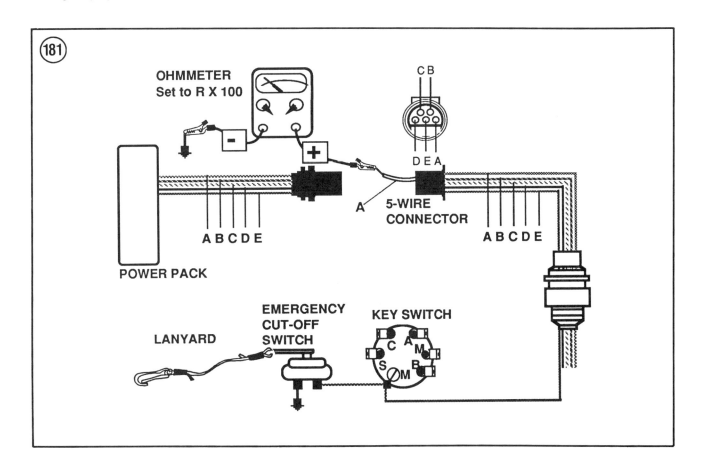

Sensor Coil Resistance Test

9 Amp Models

1. Disconnect the 6-wire connector between the power pack and timer base. If 6 jumper wires were fabricated, insert a jumper wire in each of the terminals in the 6-wire timer base connector end.
2. Disconnect timer base ground wire.
3. With an ohmmeter on the R×100 scale, connect its black test lead to the terminal end of the timer base ground wire and the red test lead alternately between each of the 6 jumper wires in the timer base connector end while noting each reading. See **Figure 182**. If the reading is not 40 ±10 ohms at each connection, replace the timer base.
4. Disconnect ohmmeter black test lead from timer base ground wire. Isolate timer base ground wire and connect ohmmeter black test lead to timer base or a good engine ground.
5. Set the ohmmeter on the high scale. Alternately connect the red test lead between each of the 6 jumper wires in the timer base connector end while noting each reading. See **Figure 183**. The ohmmeter needle should not move. If it does, the sensor coils or leads are grounded. Check for a grounded sensor coil lead before replacing the timer base.
5. Remove the jumper wires and perform the *Charge Coil Resistance Test*.

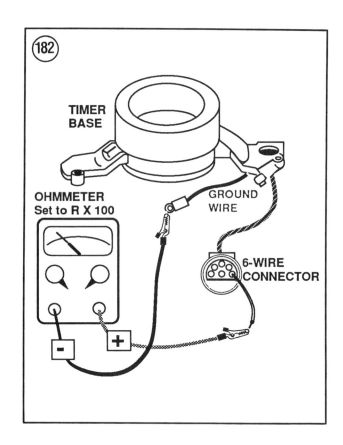

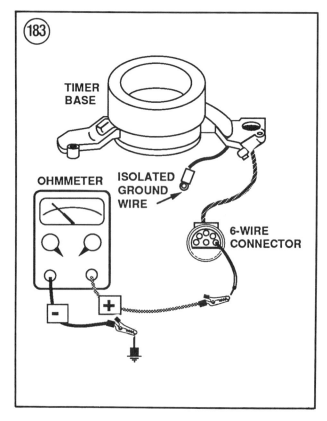

35 Amp Models

1. Disconnect the two 4-wire connectors between the power pack and timer base. If 8 jumper wires were fabricated, insert a jumper wire in each of the terminals in the two 4-wire timer base connector ends.

NOTE
Verify ohmmeter polarity. For the following test, the red test lead represents the positive lead.

TROUBLESHOOTING

2. With an ohmmeter on the R×100 scale, connect its red test lead to the D terminal jumper wire in the port 4-wire connector and the black test lead alternately between each of the 3 remaining jumper wires in the port 4-wire connector and between terminals A, B and C jumper wires in the starboard 4-wire connector while noting each reading. See **Figure 184**.

NOTE
Do to internal timer base components, readings obtained in Step 2 may be slightly higher or lower than those specified as individual meter impedance may be affected. If so noted, the readings should be consistent.

3. Recommended readings in Step 2 are:
   a. 360 ±30 ohms when using a Stevens AT-101.
   b. 970 ±100 ohms when using a Mercotronics M-700.
4. Leave the red test lead connected to the D terminal jumper wire in the port 4-wire connector and attach the black test lead to the D terminal jumper wire in the starboard 4-wire connector. See **Figure 185**. The meter should read 230 ±30 ohms.
5. Set the ohmmeter on the high scale. Ground the black test lead at the timer base or a good engine ground and connect the red test lead alternately to each of the terminal jumper wires in the two 4-wire timer base connector ends while noting each reading. The ohmmeter needle should not move. If it does, the sensor coils or leads are grounded. Check for a grounded sensor coil lead before replacing the timer base.
6. If the timer base does not pass all of the above test, replace the timer base.
7. Remove the jumper wires and perform the *Charge Coil Resistance Test*.

Charge Coil Resistance Test
9 Amp Models

1. With the 4-terminal connector plug disconnected, insert jumper wires in terminals A, B, C and D of the stator end of the connector.
2. Connect an ohmmeter between the jumper wires in terminals A and B and note the reading. Connect an ohmmeter between the jumper wires in terminals C and D and note the reading. See **Figure 186**. If a reading of 480 ±25 ohms is not obtained in both test, replace the stator assembly.
3. Set the ohmmeter on the high scale. Ground the black test lead and connect the red test lead alternately to the A, B, C and D terminal jumper

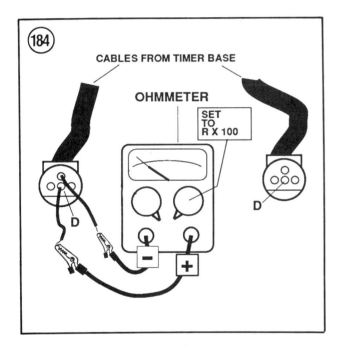

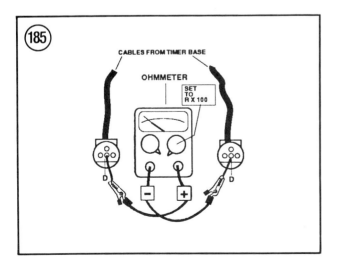

wires (**Figure 187**). The ohmmeter needle should not move. If it does, the charge coils are grounded. Check for a grounded charge coil lead before replacing the stator assembly.

35 Amp Models

1. With the 6-terminal connector plug disconnected, insert jumper wires in terminals A, B, C and D of the stator end of the connector. See **Figure 188**.
2. Connect an ohmmeter between the jumper wires in terminals A and B and note the reading (**Figure 188**). Connect an ohmmeter between the jumper wires in terminals C and D and note the reading. If a reading of 985 ±25 ohms is not obtained in both test, replace the stator assembly.
3. Set the ohmmeter on the high scale. Ground the black test lead and connect the red test lead alternately to the A, B, C and D terminal jumper wires (**Figure 189**). The ohmmeter needle should

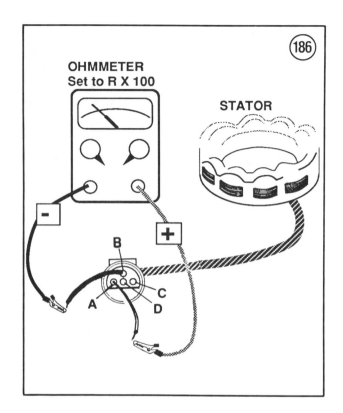

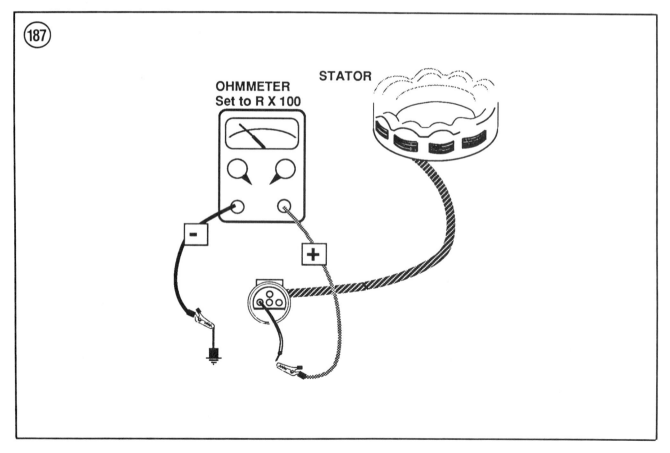

TROUBLESHOOTING

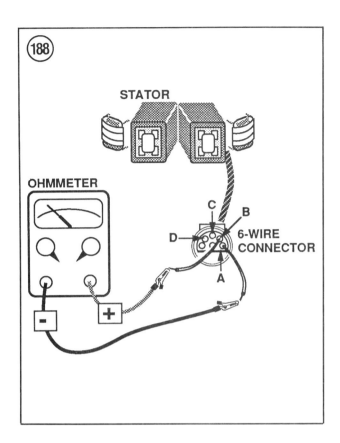

not move. If it does, the charge coils are grounded. Check for a grounded charge coil lead before replacing the stator assembly.

Charge Coil Output Test
9 Amp Models

1. With the 4-terminal connector plug disconnected, insert jumper wires in terminals A, B, C and D of the stator end of the connector. Set the peak-reading voltmeter switches to POS and 500.
2. Connect the red test lead to the jumper wire in terminal A and the black test lead to a good engine ground. See **Figure 190**.
3. Crank the engine and note the meter reading.
4. Move the red test lead to jumper wires in terminals B, C and D and crank the engine while noting the meter reading for each terminal connection.
5. There should be no meter reading in Steps 2-4. If there is, check for a grounded charge coil lead before replacing the stator assembly.

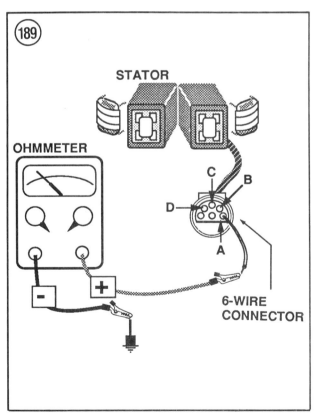

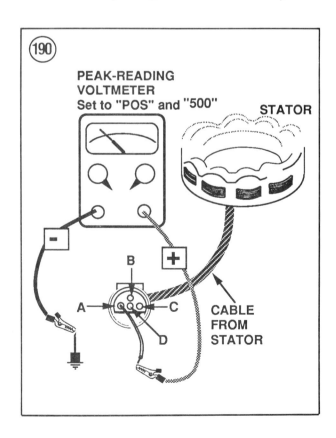

6. Connect the black test lead to the jumper wire in terminal A and the red test lead to the jumper wire in terminal B. See **Figure 191**.

7. Crank the engine and note the meter reading.

8. Move the black test lead to terminal C and the red test lead to terminal D. Crank the engine and note the meter reading. If a reading of 200 volts or higher is not obtained in both test, first check for a problem in component connectors and/or wiring. If no problem is found, replace the stator assembly.

35 Amp Models

1. With the 6-terminal connector plug disconnected, insert jumper wires in terminals A, B, C and D of the stator end of the connector. See **Figure 192**. Set the peak-reading voltmeter switches to POS and 500.

2. Connect the red test lead to the jumper wire in terminal A and the black test lead to a good engine ground. See **Figure 192**.

3. Crank the engine and note the meter reading.

4. Move the red test lead to jumper wires in terminals B, C and D and crank the engine while noting the meter reading for each terminal connection.

5. There should be no meter reading in Steps 2-4. If there is, check for a grounded charge coil lead before replacing the stator assembly.

6. Connect the black test lead to the jumper wire in terminal A and the red test lead to the jumper wire in terminal B. See **Figure 193**.

7. Crank the engine and note the meter reading.

8. Move the black test lead to terminal C and the red test lead to terminal D. Crank the engine and note the meter reading. If a reading of 200 volts or higher is not obtained in both test, first check for a problem in component connectors and/or wiring. If no problem is found, replace the stator assembly.

Sensor Coil Output Test
9 Amp Models

1. Disconnect the 6-wire connector between the power pack and timer base. If 6 jumper wires

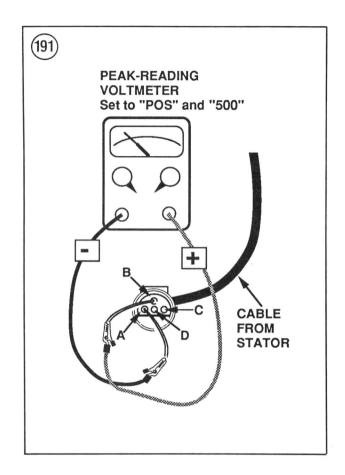

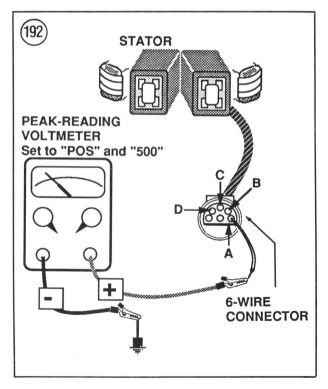

TROUBLESHOOTING

were fabricated, insert a jumper wire in each of the terminals in the 6-wire timer base connector end.

2. Disconnect timer base ground wire and isolate.

3. Set the peak-reading voltmeter switches to POS and 5.

4. Connect the red test lead to the jumper wire in terminal A and the black test lead to a good engine ground. See **Figure 194**.

5. Crank the engine and note the meter reading.

6. Move the red test lead to jumper wires in 5 remaining terminals and crank the engine while noting the meter reading for each terminal connection.

7. There should be no meter reading in Steps 4-6. If there is, check for a grounded charge coil lead before replacing the stator assembly.

6. Connect the black test lead to the timer base ground wire and the red test lead to the jumper wire in terminal A. See **Figure 195**.

7. Crank the engine and note the meter reading.

8. Move the red test lead to jumper wires in 5 remaining terminals and crank the engine while noting the meter reading for each terminal connection. If a reading of 0.2 volts or higher is not obtained in all 6 test, first check for a problem in component connectors and/or wiring. If no problem is found, replace the stator assembly.

35 Amp Models

1. Disconnect the two 4-wire connectors between the power pack and timer base. If 8

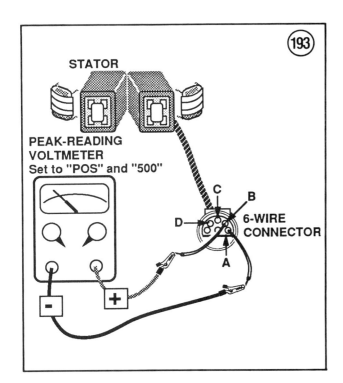

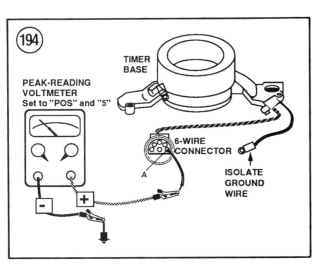

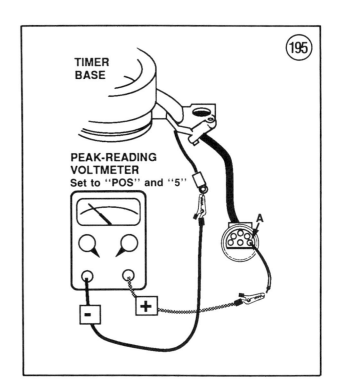

jumper wires were fabricated, insert a jumper wire in each of the terminals in the two 4-wire timer base connector ends.

2. Set the peak-reading voltmeter switches to POS and 5.

3. Ground the black test lead at the timer base or a good engine ground and connect the red test lead alternately to each of the terminal jumper wires in the two 4-wire timer base connector ends. See **Figure 196**. Crank the engine while noting each reading. There should be no meter reading. If there is, the sensor coils or leads are grounded. Check for a grounded sensor coil lead.

4. Connect the black test lead to the D terminal jumper wire in the port 4-wire connector and the red test lead alternately between each of the 3 remaining jumper wires in the port 4-wire connector and between terminals A, B and C jumper wires in the starboard 4-wire connector. See **Figure 197**. Crank the engine while noting the meter reading for each terminal connection. If a reading of 0.2 volts or higher is not obtained in all 6 test, first check for a problem in component connectors and/or wiring.

5. Connect a jumper wire between the D terminal of the port timer base and engine connectors and the D terminal of the starboard timer base and engine connectors. Ground the black test lead at the timer base or a good engine ground and connect the red test lead alternately to each of the 3 remaining jumper wires in the port and starboard 4-wire timer base connectors. See **Figure 198**. Crank the engine while noting the meter reading for each terminal connection. If a reading of 1.2 volts or higher is not obtained in all 6 tests, first check for a problem in component connectors and/or wiring.

6. If the timer base does not pass all of the above test, replace the timer base.

Power Pack Output Test

WARNING
Disconnect spark plug leads to prevent accidental starting.

1. Remove primary leads from ignition coils.

2. Connect No. 1 cylinder ignition coil primary lead to the red lead of Stevens load adapter No. PL-88 and the black lead of load adapter to a good engine ground. See **Figure 50**, typical.

3. Connect the red lead of a peak-reading voltmeter to the connector end of the red lead on Stevens load adapter No. PL-88. Connect the black lead of the peak-reading voltmeter to a good engine ground.

4. Set the peak-reading voltmeter on POS and 500.

5. Crank the engine over while noting the peak-reading voltmeter. The meter should show at least 175 volts.

6. Repeat Steps 2-5 for remaining 5 cylinders ignition coil primary lead.

7. If all primary leads show at least 175 volts, then refer to *Ignition Coil Resistance Test* in this chapter.

8. If not all of the primary leads show at least 175 volts, the power pack must be replaced.

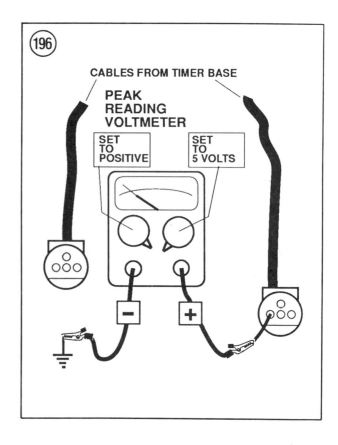

TROUBLESHOOTING

Running Output Test

1. Remove primary leads from ignition coils and connect Stevens terminal extenders No. TS-77 to ignition coils, then reconnect primary leads.
2. Connect the red lead of a peak-reading voltmeter to the No. 1 cylinder ignition coil terminal extender. Connect the black lead of the peak-reading voltmeter to a good engine ground.
3. Set the peak-reading voltmeter on POS and 500.

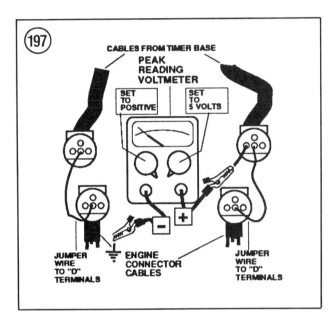

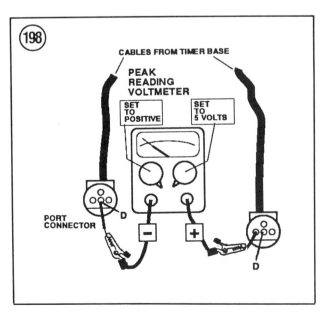

4. Make sure the outboard motor is mounted in a suitable test tank or testing fixture with a suitable loading device (test wheel or dynamometer).
5. Start the engine and operate at the rpm ignition trouble is suspected.
6. The peak-reading voltmeter should show at least 250 volts.
7. Repeat Steps 2-6 to test 5 remaining cylinders.
8. If one or more of the cylinders shows less than 250 volts, test charge coil as outlined in this chapter.
9. If one or more of the cylinders shows no output, test sensor coil as outlined in this chapter.
10. Remove ignition coil terminal extenders, then reconnect primary leads.

Ignition Coil Resistance Test

1. Disconnect the primary and high tension (secondary) coil leads at the ignition coil to be tested.
2. With an ohmmeter set on the low scale, connect the red test lead to the coil primary terminal. Connect the black test lead to a good engine ground (if the coil is mounted) or the coil ground tab (if the coil is unmounted). The meter should read 0.1 ± 0.05 ohms.
3. Set the ohmmeter on the high scale. Move the black test lead to the ignition coil high tension terminal. The meter should read 225-325 ohms.
4. If the readings are not as specified in Step 2 or Step 3, replace the ignition coil.
5. Repeat Steps 2-4 to test the remaining ignition coils.

1986-1987 200-225 HP CD IGNITION TROUBLESHOOTING

The CD ignition used on these V6 engines consists of 2 identical ignition systems—one on the port side and one on the starboard side of the engine. The troubleshooting procedures must be performed on each ignition system.

Each power pack is connected to the timer base by a 4-wire connector. A 2-wire connector joins each power pack to the charge coil. A single wire connector is used between the key switch and each power pack. A single wire connector and isolation diode connect the starboard power pack (cylinders 1-3-5) to the shift interrupter switch. Individual leads connect the power packs to the ignition coils.

Since this ignition contains 2 power packs, there is one set of connectors on each side of the engine. Amphenol connectors are used, with retaining clips provided to hold the connectors in place. The clips must be removed to disconnect the connector plugs.

Jumper leads are required for troubleshooting. Fabricate 4 leads using 8 inch lengths of 16-gauge wire. Connect a pin (OMC part No. 511469) at one end and a socket (OMC part No. 581656) with one inch of tubing (OMC part No. 519628) at the other. Ohmmeter readings should be made when the engine is cold. Readings taken on a hot engine will show increased resistance caused by engine heat and result in unnecessary parts replacement without solving the basic problem.

Output tests should be made with a Stevens Model CD77, Merc-O-Tronic Model 781 or Electro-Specialties Model PRV-1 CD voltmeter.

Key Switch Elimination Test

Refer to **Figure 199** for this procedure.
1. Connect a spark tester as shown in **Figure 199**. Set the tester air gap to 7/16 in.
2. Disconnect the single wire connector between each power pack and key switch (arrows, **Figure 199**).

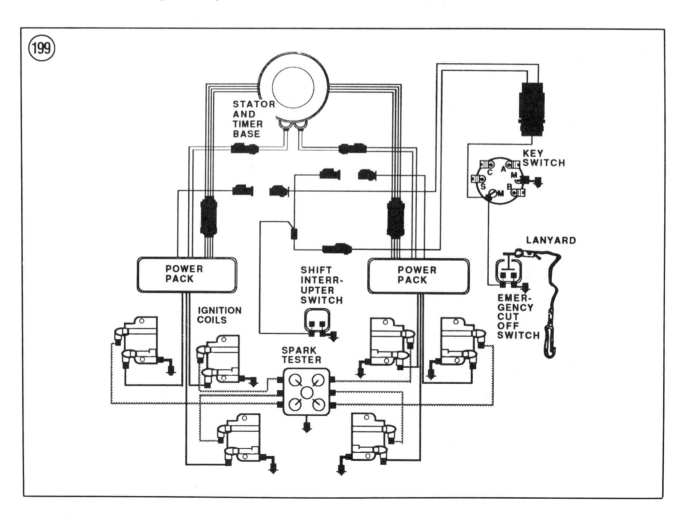

TROUBLESHOOTING

3. Crank the engine while watching the spark tester. If there is no spark or a spark at only one gap, remove the jumper wires and reconnect the connector plugs. Continue testing to locate the problem.

4. If a spark jumps all gaps alternately, the problem is in the key switch circuit, wiring harness or the emergency ignition cutoff switch.

Key Switch/Stop Circuit Test

Refer to **Figure 200** for this procedure.

1. Separate the single wire connector between each power pack and key switch (**Figure 199**).
2. Install the cap/lanyard assembly on the emergency ignition cutoff switch.
3. Connect the red ohmmeter test lead to the terminal in the key switch end of the connector (arrow, **Figure 200**). Connect the black test lead to a good ground. Turn the key switch ON. The ohmmeter should read infinity.
4. If there is meter needle movement in Step 3, disconnect the black/yellow wire at the key switch M terminal with the ohmmeter still connected.
5. If the meter reads infinity with the wire disconnected in Step 4, test the key switch. If the meter reads zero, remove the emergency ignition cutoff switch from the circuit.

   a. If the ohmmeter reads infinity, replace the cutoff switch.

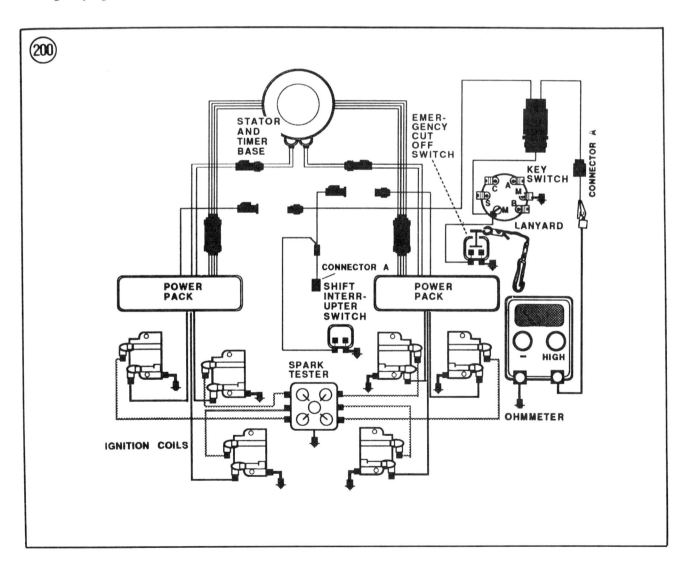

b. If the ohmmeter reads zero, check for a defect in the black/yellow wire between the key switch and single wire connector. Repair or replace as required.

Shift Switch Elimination Test

Refer to **Figure 201** for this procedure.
1. Connect a spark tester as shown in **Figure 201**. Set the tester air gap to 7/16 in.
2. Disconnect the shift interrupter switch from the starboard power pack and wiring harness at the points shown in **Figure 201**.
3. Crank the engine while watching the spark tester. If there is no spark at gaps 1, 3 and 5, continue testing to locate the problem.
4. If a spark jumps alternately at gaps 1, 3 and 5, the problem is in the shift switch. Test the switch resistance as described in this chapter.

Shift Switch Resistance Test

1. Set the ohmmeter on its high scale and connect it between the power pack end of the switch harness and a good engine ground. If the meter does not read infinity, replace the switch and harness.
2. If the meter reads infinity in Step 1, depress the switch while watching the meter scale. The meter should read zero. If it does not, replace the switch and harness.

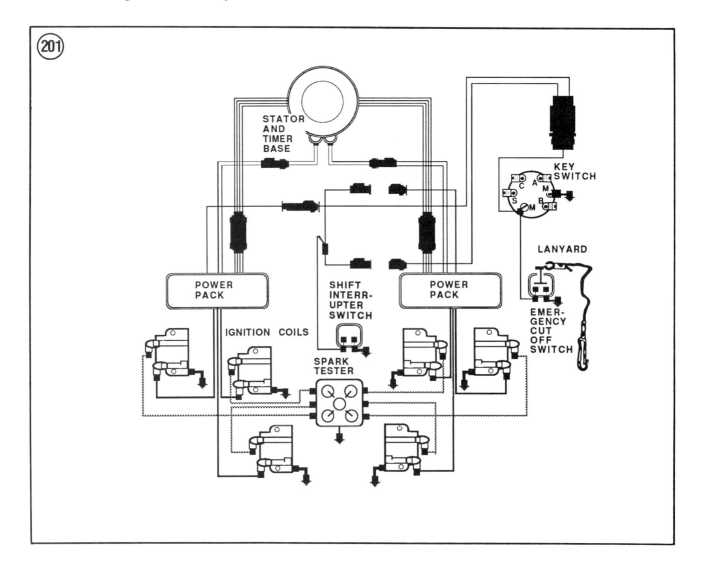

TROUBLESHOOTING

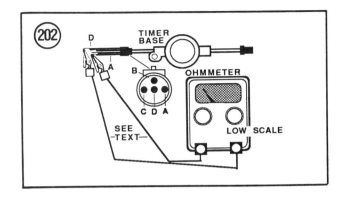

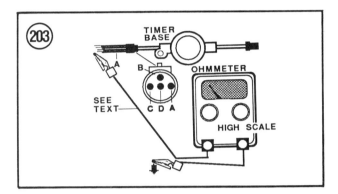

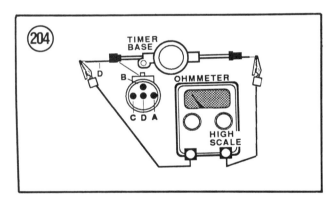

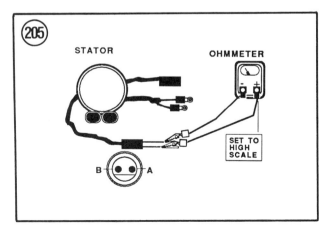

3. Connect the ohmmeter between both switch harness single pin connectors. Note the reading, then reverse the ohmmeter leads and note the second reading. One reading should be high and the other reading should be low. If 2 high or 2 low readings are obtained, replace the switch and harness.

Sensor Coil Resistance Test

1. Disconnect the 4-wire connector between the power pack and timer base. Insert jumper wires in each of the 4 terminals in the timer base connector end.

2. With an ohmmeter on the low scale, connect its test leads between the A and D terminal jumper wires, then between B and D and C and D, noting each reading. See **Figure 202**. If the reading is not 40 ± 5 ohms at each connection, replace the timer base.

3. Set the ohmmeter on the high scale. Ground the black test lead at the timer base and connect the red test lead alternately to each of the 4 terminal jumper wires (**Figure 203**), noting each reading. The ohmmeter needle should not move. If it does, the sensor coils or leads are grounded. Check for a grounded sensor coil lead before replacing the timer base.

4. With the ohmmeter on the high scale, connect the black test lead to the common lead on the sensor terminal D or jumper wire (port side). See **Figure 204**. Connect the red test lead to the same point on the starboard side. If the meter does not read infinity, replace the timer base.

5. Remove the jumper wires and perform the *Charge Coil Resistance Test*.

Charge Coil Resistance Test

1. With the 2-terminal connector plug disconnected, insert jumper wires in terminals A and B of the stator end of the connector.

2. Connect an ohmmeter between the jumper wires (**Figure 205**) and note the reading. If it is not 970 ± 15 ohms, replace the stator assembly.

3. Set the ohmmeter on the high scale. Ground the black test lead and connect the red test lead alternately to the A and B terminal jumper wires (**Figure 206**). The ohmmeter needle should not move. If it does, the charge coils are grounded. Check for a grounded charge coil lead before replacing the stator assembly.

Charge Coil Output Test

1. Disconnect the 2-wire connector between the power pack and stator. Set the CD voltmeter switches to NEGATIVE and 500. Insert the red test lead in cavity A of the charge coil end of the connector. Connect the black test lead to a good engine ground. See **Figure 207**.
2. Crank the engine and note the meter reading.
3. Move the red test lead to cavity B of the connector and crank the engine again. Note the meter reading.
4. There should be no meter reading in Step 2 or Step 3. If there is, check for a grounded charge coil lead before replacing the stator assembly.
5. Leave the red test lead in cavity B of the connector. Remove the black test lead from ground and insert it in cavity A of the connector. See **Figure 208**.
6. Crank the engine and note the meter reading. If it is less than 130 volts, check for a problem in component connectors and/or wiring. If no problem is found, replace the stator assembly.

Sensor Coil Output Test

1. Disconnect the 4-wire connector between the power pack and timer base. Set the CD voltmeter switches to S and 5; if using a Merc-O-Tronic Model 781 voltmeter, set the switches to POSITIVE and 5. See **Figure 209**.
2. Connect the black test lead to a good engine ground. Connect a jumper wire to the red test lead and alternately probe all 4 connector terminals while cranking the engine. See **Figure 209**.

3. There should be no meter reading in Step 2. If there is, check for a grounded sensor coil lead before replacing the timer base.
4. Remove the black test lead from ground. Connect it to a jumper wire and insert the wire

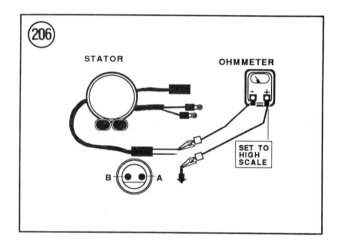

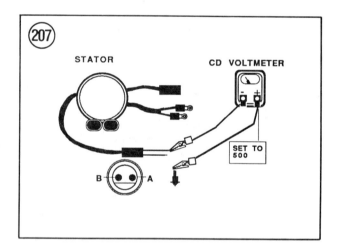

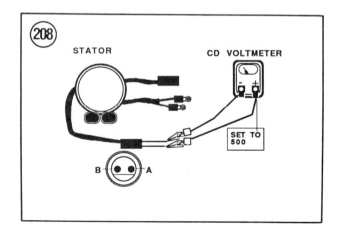

TROUBLESHOOTING

in cavity A of the timer base 4-wire connector. Probe cavities B and C of the connector with the red test lead jumper wire while cranking the engine. Note the meter reading at each cavity. See **Figure 210**.

5. If any reading in Step 4 is less than 0.3 volts, check for a problem in the component connectors and/or wiring. If no problem is found, replace the timer base.

Power Pack Output Test

1. Disconnect the primary lead from each ignition coil. Install a terminal extender on the primary terminal of each coil, then reconnect the primary lead to the extender terminal.
2. Set the CD voltmeter switches to NEGATIVE and 500.

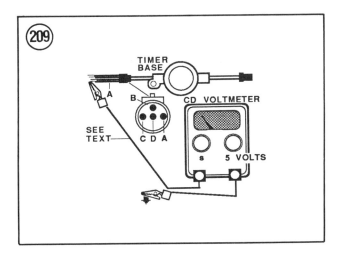

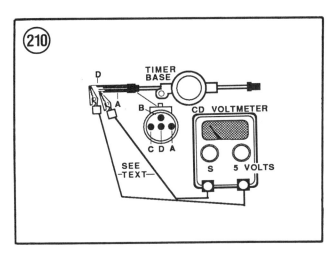

3. Connect the red test lead to the No. 1 coil terminal extender (top coil) and ground the black test lead.
4. Crank the engine and note the meter. It should read 100 volts or more.
5. If the meter reading is less than 100 volts in Step 4, disconnect the primary lead from the No. 1 coil terminal extender. Connect the red test lead directly to the spring clip in the primary lead boot and repeat Step 4.
   a. If the meter now reads 100 volts or more, test the ignition coil as described in this chapter.
   b. If the meter still reads less than 100 volts, check the spring clip and primary lead wire for defects. If none are found, replace the power pack.
6. Repeat Steps 3-5 to test each remaining ignition coil.
7. Remove the terminal extenders from the coil terminals with a clockwise pulling motion. Reconnect the primary leads to the coil terminals. Make sure the leads are reconnected as follows:
   a. Orange/blue lead—Top coil.
   b. Orange lead—Center coil.
   c. Orange/green lead—Bottom coil.

Ignition Coil Resistance Test

See *1985-1988 CD 3 Ignition Troubleshooting* in this chapter.

1988-1990 200-225 HP CD IGNITION TROUBLESHOOTING

The CD ignition used on these V6 engines consists of a stator containing two charge coils and one power coil and a timer base containing 6 double-wound sensor coils feeding voltage to a single power pack. One double-wound sensor coil is used per cylinder. One winding of the coil is for normal operation and the other winding is for automatic ignition timing advancement when QuikStart™ is activated.

The power pack is connected to the sensor coils by two 4-wire connectors. Two 2-wire connectors join the power pack to the charge coils and a terminal block located adjacent to the power pack connects the power pack to the power coil. Two single-wire connectors join the power pack to the ignition switch. Two single-wire connectors and an isolation diode connect the power pack (starboard side cylinders, 1-3-5) to the shift interrupter switch. Ignition coils are connected to the power pack by individual leads. Amphenol connectors are used. Special tools are required to service terminals in connector plugs.

Jumper leads are required for troubleshooting. Fabricate at least 2 and as many as 8 leads using 8 inch lengths of 16-gauge wire. Connect a pin (OMC part No. 511469) at one end with one inch of tubing (OMC part No. 510628) and a socket (OMC part No. 581656) with one inch of tubing (OMC part No. 510628) at the other. Ohmmeter readings should be made when the engine is cold. Readings taken on a hot engine will show increased resistance caused by engine heat and result in unnecessary parts replacement without solving the basic problem.

Output tests should be made with a Stevens Model CD77, Merc-O-Tronic Model 781 or Electro-Specialties Model PRV-1 peak-reading voltmeter.

S.L.O.W. Operation And Testing

The S.L.O.W. (speed limiting overheat warning) circuit is activated when the engine temperature exceeds 203° F. When activated, the S.L.O.W. mode will limit the engine speed to approximately 2500 rpm. To be deactivated, the engine must be cooled to 162° F and the engine must be stopped.

A blocking diode located in the engine wiring harness is used to isolate the S.L.O.W. warning system from other warning horn signals.

1. Install the engine in a test tank with the proper test wheel.
2. Connect a tachometer according to manufacturer's instructions.
3. Disconnect temperature sensor tan lead at connector on port cylinder head (**Figure 149**, typical).
4. Start engine and run at approximately 3500 rpm.
5. Connect engine wiring tan lead to a clean engine ground.
6. Note engine rpm. Engine rpm should reduce to approximately 2500 rpm.
7. Stop the engine. Repeat Steps 3-6 for temperature sensor located in starboard cylinder head.
8. Stop the engine. If engine rpm reduces only when one temperature sensor lead is grounded, check wiring harness and connectors. If engine rpm does not reduce when either temperature sensor lead is grounded, loosen power pack (**Figure 150**) mounting fasteners and remove terminal ends of orange and orange/black wires from terminal block which lead to power coil contained in flywheel stator.
9. Calibrate an ohmmeter to R×100 scale.
10. Connect black ohmmeter lead to a good engine ground. Alternately connect red ohmmeter lead to terminal ends of orange and

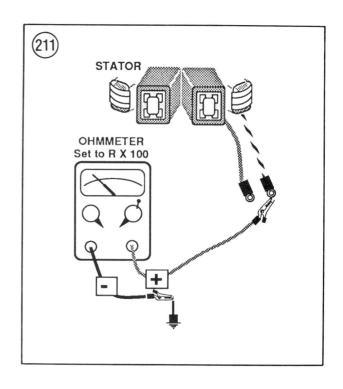

TROUBLESHOOTING

orange/black wires leading to power coil. See **Figure 211**. No continuity or infinite resistance should be noted at both terminal ends.

11. Connect ohmmeter leads between orange and orange/black wire terminal ends leading to power coil. See **Figure 212**. Ohmmeter should read between 86-106 ohms. If so, replace power pack.

12. Blocking Diode Test—Calibrate an ohmmeter to R×100 scale. Connect one ohmmeter lead to either port or starboard engine wiring tan lead for temperature sensor and other ohmmeter lead to tan wire terminal in red engine harness connector. Note ohmmeter reading, then reverse leads or push "polarity" button on meter if so equipped and note reading. Continuity (near zero resistance) should be noted in one direction and no continuity or infinite resistance should be noted in other direction. If not, replace engine wiring harness.

QuikStart™ Operation And Testing

The QuikStart™ circuit is activated each time the engine is started. QuikStart™ is an electronic feature that will automatically advance ignition timing during engine starting (all engine temperatures) and warm-up (engine temperature below 96° F) when engine speed is below approximately 1100 rpm. If the engine temperature is 96° F or more, the ignition timing will remain advanced for only approximately 5 seconds after the engine starts before returning to normal setting. When engine speed is approximately 1100 rpm or more, the QuikStart™ circuit will deactivate and return ignition timing to normal setting.

1. Install the engine in a test tank with the proper test wheel.

NOTE
Make sure engine synchronization and linkage adjustments are correctly set as outlined in Chapter Five.

2. Operate the engine until temperature is above 96° F.
3. Stop the engine and connect a tachometer according to manufacturer's instructions.
4. Disconnect temperature sensor white/black lead at connector on port cylinder head (**Figure 149**, typical).
5. Connect a timing light to the No. 1 cylinder according to the manufacturer's instructions.
6. Start engine and shift into FORWARD gear. *Do not* operate engine above 900 rpm.
7. Observe flywheel with timing light. Pointer should align with No. 1 cylinder TDC mark.
8. Connect temperature sensor white/black lead to respective power pack lead.
9. Observe flywheel with timing light. Pointer should move to the right of the TDC mark approximately 1 in. indicating ignition timing has returned to normal setting.
10. Stop the engine and repeat Steps 4-9 for remaining cylinders.
11. If only five or less of the cylinders react correctly, then replace the timer base.
12. If no cylinder reacts correctly, test power coil located in flywheel stator as outlined under *S.L.O.W. Operation And Testing* in this chapter.

NOTE
If QuikStart™ remains activated when the engine is operated above 1100 rpm, replace the power pack if no electrical wiring or connections are found faulty.

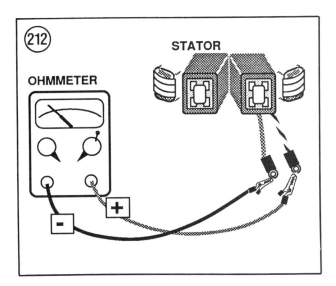

Ignition System Output Test

1. Remove power pack (**Figure 150**) top cover and disconnect yellow/red wire at connector. Ground power pack with an external connection.
2. Disconnect temperature sensor white/black lead at connector on port cylinder head (**Figure 149**).
3. Connect terminal end of power pack white/black lead to a clean engine ground.
4. Remove spark plug leads from spark plugs.
5. Mount a spark tester on the engine and connect a tester lead to each spark plug lead. Set the spark tester air gap to 7/16 in. See **Figure 48**, typical.
6. Connect ignition system emergency cutoff clip and lanyard if so equipped.
7. Crank the engine over while watching the spark tester. If a spark jumps at each air gap, test the ignition system as outlined under *Running Output Test* in this chapter.
8. If a spark jumps at *only* five or less air gaps, first make sure all connections are good, then retest. If only five or less sparks are noted, test ignition system as outlined under *Charge Coil Test* in this chapter.
9. If a spark jumps at *only* air gaps for cylinders 2, 4 and 6, first make sure all connections are good, then retest. If only sparks are noted for cylinders 2, 4 and 6, test ignition system as outlined under *Shift Switch Elimination Test* in this chapter.
10. If *no* spark jumps at any of the spark tester's air gaps, first make sure all connections are good, then retest. If no spark is noted, test ignition system as outlined under *Key Switch Elimination Test* in this chapter.

Key Switch Elimination Test

Refer to **Figure 213** for this procedure.
1. Connect a spark tester as shown in **Figure 48**, typical. Set the tester air gap to 7/16 in.
2. Disconnect the black/yellow wire at the two connectors between the power pack and key switch (A, **Figure 213**).

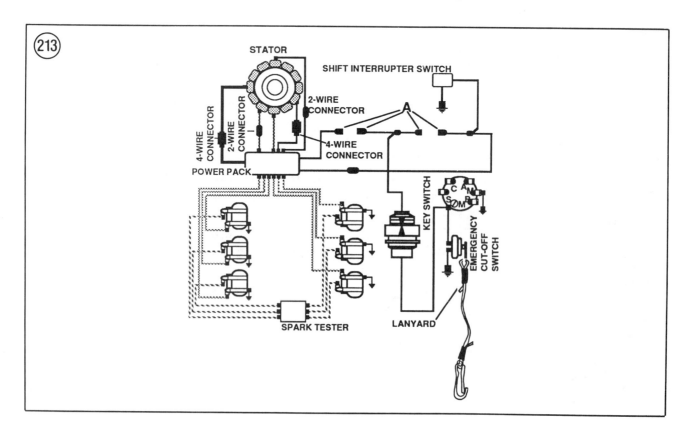

TROUBLESHOOTING

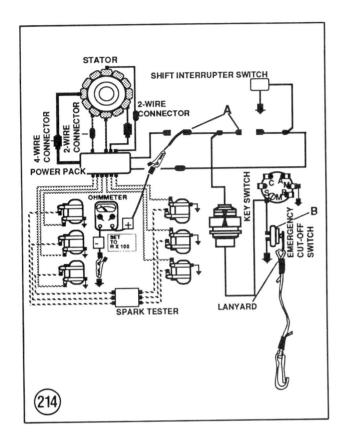

214

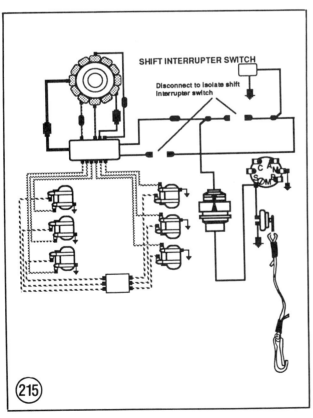

215

3. Crank the engine while watching the spark tester. If there is no spark or a spark at only five or less air gaps, test charge coil and sensor coils as outlined in this chapter.

4. If a spark jumps all gaps alternately, the problem is in the key switch circuit or the emergency ignition cutoff switch.

Key Switch/Stop Circuit Test

Refer to **Figure 214** for this procedure.

1. Separate the black/yellow wire at the two connectors between the power pack and key switch (A, **Figure 214**).
2. Install the cap/lanyard assembly on the emergency ignition cutoff switch (B, **Figure 214**), if so equipped.
3. Insert a jumper wire in the key switch terminal end of either connector. Connect the red ohmmeter test lead to the jumper wire and the black test lead to a good ground. Turn the key switch ON. The ohmmeter should read infinity.
4. If there is meter needle movement in Step 3, disconnect the black/yellow wire at the key switch M terminal with the ohmmeter still connected.
5. If the meter reads infinity with the wire disconnected in Step 4, test the key switch. If the meter reads zero, remove the emergency ignition cutoff switch from the circuit.
   a. If the ohmmeter reads infinity, replace the cutoff switch.
   b. If the ohmmeter reads zero, check for a defect in the black/yellow wire between the key switch and the two single-wire connectors. Repair or replace as required.

Shift Switch Elimination Test

Refer to **Figure 215** for this procedure.

1. Connect a spark tester as shown in **Figure 48**, typical. Set the tester air gap to 7/16 in.
2. Isolate the shift interrupter switch by disconnecting the black/yellow wire at the two

connectors between the power pack and shift interrupter switch. See **Figure 215**.

3. Crank the engine while watching the spark tester. A spark should be noted at air gaps for cylinders 1, 3 and 5. If there is no spark at air gaps, test shift interrupter switch as outlined under *Shift Switch Resistance Test* in this chapter.

Shift Switch Resistance Test

1. Isolate the shift interrupter switch as shown in **Figure 215**.
2. Set the ohmmeter on its high scale and connect it between the power pack end of the switch harness and a good engine ground. See **Figure 216**. If the ohmmeter does not read infinity, replace the switch and harness.
3. If the ohmmeter reads infinity in Step 2, depress the switch while watching the meter scale. The ohmmeter should read zero. If it does not, replace the switch and harness.
4. Connect the ohmmeter between both switch harness single pin connectors. Note the reading, then reverse the ohmmeter leads and note the second reading. One reading should be infinite resistance and the other reading should be near zero. If two infinite or two zero readings are obtained, then replace the switch and harness.

Sensor Coil Resistance Test

1. Disconnect the two 4-wire connectors between the power pack and timer base. If 8 jumper wires were fabricated, insert a jumper wire in each of the terminals in the two 4-wire timer base connector ends.

NOTE
Verify ohmmeter polarity. For the following test, the red test lead represents the positive lead.

2. With an ohmmeter on the R×100 scale, connect its red test lead to the D terminal jumper wire in the port 4-wire connector and the black test lead alternately between each of the 3 remaining jumper wires in the port 4-wire connector and between terminals A, B and C jumper wires in the starboard 4-wire connector while noting each reading. See **Figure 217**.

NOTE
*Do to internal timer base components, readings obtained in **Step 2** may be slightly higher or lower than those specified as individual meter impedance may be affected. If so noted, the readings should be consistent.*

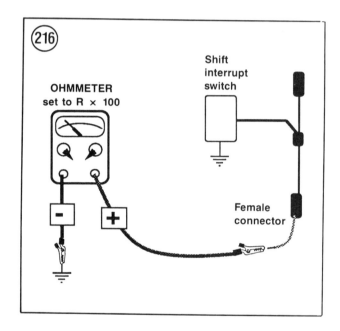

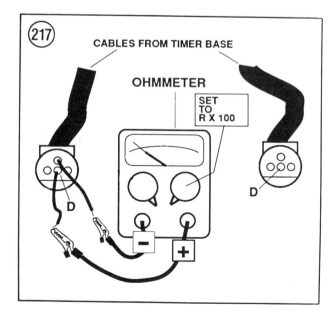

TROUBLESHOOTING

3. Recommended readings in Step 2 are:
   a. 360 ±30 ohms when using a Stevens AT-101.
   b. 970 ±100 ohms when using a Mercotronics M-700.
4. Leave the red test lead connected to the D terminal jumper wire in the port 4-wire connector and attach the black test lead to the D terminal jumper wire in the starboard 4-wire connector. See **Figure 218**. The meter should read 230 ±30 ohms.

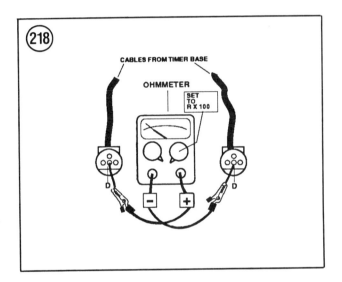

5. Set the ohmmeter on the high scale. Ground the black test lead at the timer base or a good engine ground and connect the red test lead alternately to each of the terminal jumper wires in the two 4-wire timer base connector ends while noting each reading. The ohmmeter needle should not move. If it does, the sensor coils or leads are grounded. Check for a grounded sensor coil lead before replacing the timer base.
6. If the timer base does not pass all of the above test, replace the timer base.
7. Remove the jumper wires and perform the *Charge Coil Resistance Test*.

Charge Coil Resistance Test

1. Disconnect the two 2-terminal connector plugs leading to the stator assembly. Insert jumper wires in terminals of the two connectors leading to the stator.
2. Connect an ohmmeter between the jumper wires in terminals A and B of one connector and note the reading (**Figure 219**). Connect an ohmmeter between the jumper wires in terminals A and B of the remaining connector and note the reading. If a reading of 985 ±25 ohms is not obtained in both test, replace the stator assembly.

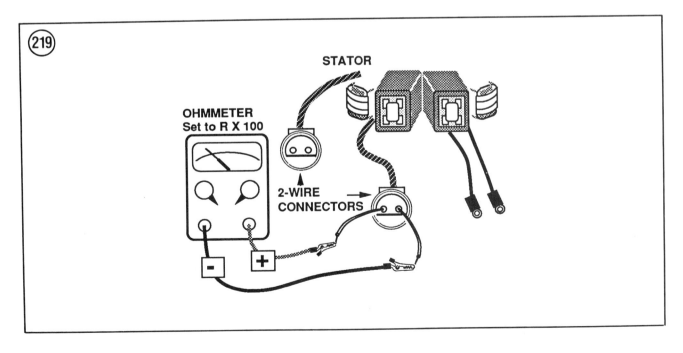

3. Set the ohmmeter on the high scale. Ground the black test lead and connect the red test lead alternately to each of the jumper wires in the two connectors (**Figure 220**). The ohmmeter needle should not move. If it does, the charge coils are grounded. Check for a grounded charge coil lead before replacing the stator assembly.

Charge Coil Output Test

1. With the two 2-terminal connector plugs leading to stator assembly disconnected, insert jumper wires in terminals of the two connectors leading to stator. Set the peak-reading voltmeter switches to POS and 500.
2. Alternately connect the red test lead to the jumper wires in terminals A and B of one stator connector and the black test lead to a good engine ground. See **Figure 221**.
3. Crank the engine for each connection and note the meter reading.
4. Move the red test lead to jumper wires in terminals A and B of the remaining stator connector and crank the engine while noting the meter reading for each terminal connection.
5. There should be no meter reading in Steps 2-4. If there is, check for a grounded charge coil lead before replacing the stator assembly.
6. Connect the black test lead to the jumper wire in terminal A and the red test lead to the jumper wire in terminal B of one stator connector. See **Figure 222**.
7. Crank the engine and note the meter reading.
8. Move the black test lead to terminal A and the red test lead to terminal B of the remaining stator connector. Crank the engine and note the meter reading. If a reading of 130 volts or higher is not obtained in both test, first check for a problem in component connectors and/or wiring. If no problem is found, replace the stator assembly.

Sensor Coil Output Test

1. Disconnect the two 4-wire connectors between the power pack and timer base. If 8 jumper wires were fabricated, insert a jumper wire in each of the terminals in the two 4-wire timer base connector ends.

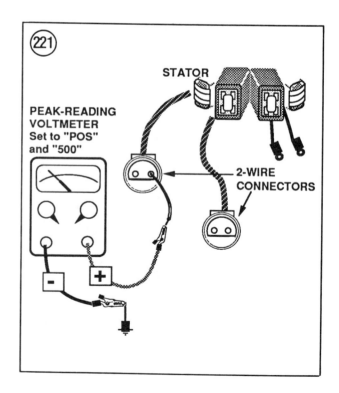

TROUBLESHOOTING

2. Set the peak-reading voltmeter switches to POS and 5.

3. Ground the black test lead at the timer base or a good engine ground and connect the red test lead alternately to each of the terminal jumper wires in the two 4-wire timer base connector ends. See **Figure 223**. Crank the engine while noting each reading. There should be no meter reading. If there is, the sensor coils or leads are grounded. Check for a grounded sensor coil lead.

4. Connect the black test lead to the D terminal jumper wire in the port 4-wire connector and the red test lead alternately between each of the 3 remaining jumper wires in the port 4-wire connector and between terminals A, B and C jumper wires in the starboard 4-wire connector. See **Figure 224**. Crank the engine while noting the meter reading for each terminal connection. If a reading of 0.2 volts or higher is not obtained in all 6 test, first check for a problem in component connectors and/or wiring.

5. Connect a jumper wire between the D terminal of the port timer base and engine connectors and the D terminal of the starboard timer base and engine connectors. Ground the black test lead at the timer base or a good engine ground and connect the red test lead alternately to each of the 3 remaining jumper wires in the port and starboard 4-wire timer base connectors. See **Figure 225**. Crank the engine while noting

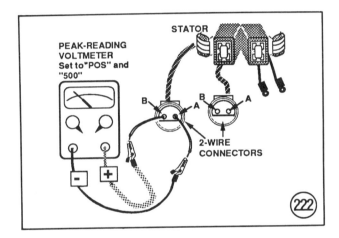

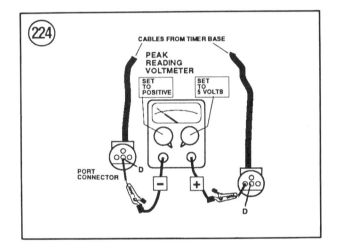

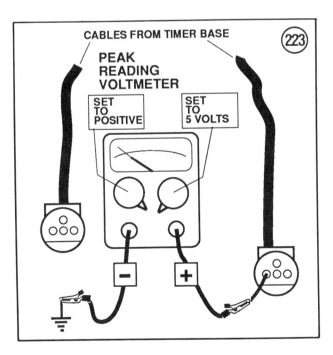

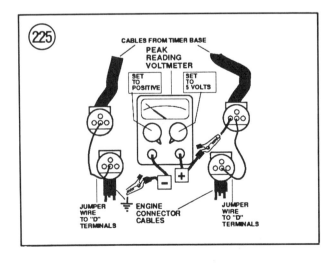

the meter reading for each terminal connection. If a reading of 1.2 volts or higher is not obtained in all 6 tests, first check for a problem in component connectors and/or wiring.
6. If the timer base does not pass all of the above test, replace the timer base.

Power Pack Output Test

WARNING
Disconnect spark plug leads to prevent accidental starting.

1. Remove primary leads from ignition coils.
2. Connect No. 1 cylinder ignition coil primary lead to the red lead of Stevens load adapter No. PL-88 and the black lead of load adapter to a good engine ground. See **Figure 50**, typical.
3. Connect the red lead of a peak-reading voltmeter to the connector end of the red lead on Stevens load adapter No. PL-88. Connect the black lead of the peak-reading voltmeter to a good engine ground.
4. Set the peak-reading voltmeter on POS and 500.
5. Crank the engine over while noting the peak-reading voltmeter. The meter should show at least 100 volts.
6. Repeat Steps 2-5 for remaining 5 cylinders ignition coil primary lead.
7. If all primary leads show at least 100 volts, then refer to *Ignition Coil Resistance Test* in this chapter.
8. If not all of the primary leads show at least 100 volts, the power pack must be replaced.

Running Output Test

1. Remove primary leads from ignition coils and connect Stevens terminal extenders No. TS-77 to ignition coils, then reconnect primary leads.
2. Connect the red lead of a peak-reading voltmeter to the No. 1 cylinder ignition coil terminal extender. Connect the black lead of the peak-reading voltmeter to a good engine ground.
3. Set the peak-reading voltmeter on POS and 500.
4. Make sure the outboard motor is mounted in a suitable test tank or testing fixture with a suitable loading device (test wheel or dynomometer).
5. Start the engine and operate at the rpm ignition trouble is suspected.
6. The peak-reading voltmeter should show at least 130 volts.
7. Repeat Steps 2-6 to test 5 remaining cylinders.
8. If one or more of the cylinders shows less than 130 volts, test charge coil as outlined in this chapter.
9. If one or more of the cylinders shows no output, test sensor coil as outlined in this chapter.
10. Remove ignition coil terminal extenders, then reconnect primary leads.

Ignition Coil Resistance Test

1. Disconnect the primary and high tension (secondary) coil leads at the ignition coil to be tested.
2. With an ohmmeter set on the low scale, connect the red test lead to the coil primary terminal. Connect the black test lead to a good engine ground (if the coil is mounted) or the coil ground tab (if the coil is unmounted). The meter should read 0.1 ±0.05 ohms.
3. Set the ohmmeter on the high scale. Move the black test lead to the ignition coil high tension terminal. The meter should read 225-325 ohms.
4. If the readings are not as specified in Step 2 or Step 3, replace the ignition coil.
5. Repeat Steps 2-4 to test the remaining ignition coils.

IGNITION AND NEUTRAL START SWITCH

The ignition and neutral start switches can be tested with a self-powered test lamp or ohmmeter. If defective, replace the ignition switch with a marine switch. Do not use an automotive ignition switch.

TROUBLESHOOTING

Ignition Switch Test

Refer to **Figure 226** for this procedure.

1. Disconnect the negative battery lead. Disconnect the positive battery lead. Disconnect the ignition switch leads.
2. Connect a test lamp or ohmmeter leads between the switch terminals marked BATT and A. With the switch in the OFF position, there should be no continuity.
3. Turn the switch to the ON position. The test lamp should light or the meter show continuity.
4. Turn the switch to the START position. The test lamp should light or the meter show continuity.
5. Hold the switch key in the START position and move the test lead from terminal A to terminal S. The test lamp should light or the meter show continuity.
6. Turn the switch off. Move the test leads to the 2 terminals marked M. The test lamp should light or the meter show continuity.
7. Turn the switch first to the START, then to the ON position. There should be no continuity in either position.
8. Turn the switch off. Move the test leads to terminal B and terminal C. Turn the switch ON. There should be no continuity. If equipped with a choke primer system, push inward on the key and the test lamp should light or the meter show continuity.
9. Repeat Step 8 with the switch in the START position. The results should be the same.

NOTE
It is possible that the switch may pass this test but still have an internal short. If the switch passes but does not function properly, have it leak-tested by a dealer.

10. Replace the switch if it fails any of the steps in this procedure.

NEUTRAL START SWITCH

The throttle cam or remote control box neutral start switch is not adjustable. If it does not prevent the motor from starting when the throttle is advanced beyond the START position, replace it.

To check the neutral start switch, disconnect the negative battery cable and connect an ohmmeter between the switch leads. There should be continuity only when the engine control is in NEUTRAL. If continuity is shown with the engine control in FORWARD or REVERSE or if the engine cranks in either gear, replace the switch.

FUEL SYSTEM

Many outboard owners automatically assume that the carburetor is at fault when the engine does not run properly. While fuel system problems are not uncommon, carburetor adjustment is seldom the answer. In many cases, adjusting the carburetor only compounds the problem by making the engine run worse.

Fuel system troubleshooting should start at the gas tank and work through the system, reserving the carburetors as the final point. The majority of fuel system problems result from an empty fuel tank, sour fuel, a plugged fuel filter or a

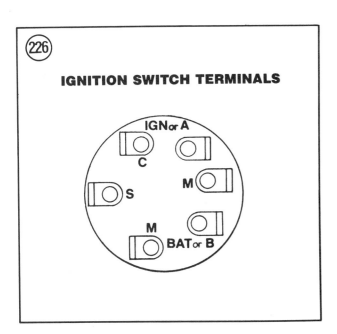

malfunctioning fuel pump. **Table 3** provides a series of symptoms and causes that can be useful in localizing fuel system problems.

Troubleshooting

As a first step, check the fuel flow. Remove the fuel tank cap and look into the tank. If there is fuel present, disconnect and ground the spark plug leads as a safety precaution. Disconnect the fuel line at the carburetor (**Figure 227**, typical) and place it in a suitable container to catch any discharged fuel. See if gas flows freely from the line when the primer bulb is squeezed.

If there is no fuel flow from the line, the fuel petcock may be shut off or blocked by rust or foreign matter, the fuel line may be stopped up or kinked or a primer bulb check valve may be defective. If a good fuel flow is present, crank the engine 10-12 times to check fuel pump operation. A pump that is operating satisfactorily will deliver a good, constant flow of fuel from the line. If the amount of flow varies from pulse to pulse, the fuel pump is probably failing.

Carburetor chokes can also present problems. A choke that sticks open will show up as a hard starting problem; one that sticks closed will result in a flooding condition.

During a hot engine shut-down, the fuel bowl temperature can rise above 200° F, causing the fuel inside to boil. While marine carburetors are vented to atmosphere to prevent this problem, there is a possibility some fuel will percolate over the high-speed nozzle.

A leaking inlet needle and seat or a defective float will allow an excessive amount of fuel into the intake manifold. Pressure in the fuel line after the engine is shut down forces fuel past the leaking needle and seat. This raises the fuel bowl level, allowing fuel to overflow into the manifold.

Excessive fuel consumption may not necessarily mean an engine or fuel system problem. Marine growth on the boat's hull, a bent or otherwise damaged propeller or a fuel line leak can cause an increase in fuel consumption. These areas should all be checked *before* blaming the carburetor.

Hesitation on Acceleration (1979-1983 70 and 75 hp)

Some 1983 75 hp engines were fitted with carburetors (part No. 393196 or part No. 393197) which are calibrated too lean in the intermediate range (1,000-2,000 rpm). This results in a hesitation on acceleration. If the ignition timing is correctly set, remove the air silencer cover and locate the part number on the carburetor. Carburetors with part No. 393196S or 393197S have already been modified. If there is no "S" in the part number, remove the intermediate orifice plug screw. Remove and discard the intermediate orifice, then reinstall the plug screw.

Another hesitation problem plaguing 1982-1983 70 and 75 hp engines is caused by a carburetor with an idle pickup tube in which the pickup hole was incorrectly positioned. The problem can also affect 1979 and 1980 models if a replacement carburetor has been installed.

To determine if the idle pickup tube is at fault, remove the carburetor (Chapter Six). Hold the throttle valve wide open and inspect the idle pickup tube from the engine side of the

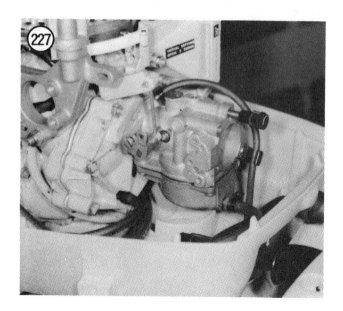

TROUBLESHOOTING

carburetor. The tiny hole in the pickup tube should be positioned within 30° of the centerline on either side. If it is not, see your Johnson or Evinrude dealer, as a warranty claim can be filed for carburetor replacement.

Lean or Poor Idle Conditions (1980 55-60 hp; 1980-1981 V4 and V6)

A 0.045 in. idle orifice was originally installed on 1980 55-60 hp carburetors. These engines had a tendency toward a lean idle which was most often diagnosed as an intermittent and light backfiring while idling.

During the 1980 model run, a production change incorporated a 0.046 in. idle orifice to richen the idle mixture. Carburetors with the larger idle orifice have an "E" stamped after the carburetor number. These engines are very sensitive and will not tolerate imprecise synchronization and linkage adjustments. Check any 1980 55-60 hp engine which exhibits this condition to make sure that:
  a. Synchronization and linkage adjustments are correct. If not, reset according to procedures in Chapter Five.
  b. Water flows from the rear exhaust relief holes during idle. If not, the inner-to-outer exhaust housing seal is leaking and should be replaced. See Chapter Eight.
  c. The carburetors have the larger idle orifice. Remove the carburetor and check the idle orifice size. If fitted with the 0.045 in. orifice, replace with a 0.046 in. orifice (part No. 327342).

The air silencer drain system metering orifice was originally drilled in the intake manifold on 1980-1981 V4 and V6 models. During the production run, the metering orifice location was changed to the drain nipple. Some models built during this change-over period have no metering orifice in either the manifold or nipple.

To determine if this is the cause of a poor idle condition with a 1980-1981 V4 or V6 engine, pinch off the silencer drain hose with a pair of pliers while the engine is idling. If the idle improves, there is no metering device installed.

To correct the condition, disconnect the drain hose from the silencer or manifold. Lubricate the inner diameter of the hose with rubbing alcohol and install a restrictor plug (part No. 311126) in the end of the hose. When the hose is reinstalled to the silencer or manifold, the restrictor plug will be forced into the hose.

Drop in High Speed Performance (1986 150 HP XP [TM] and GT [TM] Models)

A drop in high speed performance may be a result of an over-rich carburetor main circuit. To correct, replace the original No. 61C main circuit jets with No. 59C jets, part No. 317330 and increase maximum spark advance from 28° to 30° BTDC. Make sure the correct propeller is installed on the outboard motor to allow the engine to reach the top half of the recommended wide open throttle operating range.

Electric Primer System

A primer solenoid is used on some 1980-1982 and all 1983-on electric start models instead of a choke solenoid. See **Figure 228**. When the key is inserted in the ignition switch and depressed, the solenoid opens electrically and allows fuel to pass from the fuel pump directly to the intake manifold in sufficient quantity to start the engine.

The primer solenoid operation can be checked by running the engine at approximately 2,000 rpm and depressing the ignition key. If the solenoid is functioning properly, the engine will run rich and drop about 1,000 rpm until the key is released. If the solenoid is suspected of not operating properly, shut the engine off and disconnect the purple/white wire at the terminal board. Connect an ohmmeter between the purple/white wire and the black primer solenoid ground lead. The ohmmeter should read 4-6 ohms.

If the solenoid does not perform as described, remove and repair or replace it. See Chapter Six.

ENGINE TEMPERATURE AND OVERHEATING

Proper engine temperature is critical to good engine operation. An engine that runs too hot will be damaged internally. One that operates too cool will not run smoothly or efficiently.

A variety of problems can cause engine overheating. Some of the most commonly encountered are a defective thermostat, low output or defective water pump, damaged or mispositioned water passage restrictors or even engine flashing in the cylinder head casting water discharge passage that was not removed during manufacture.

NOTE
1980 Johnson and Evinrude 150, 175 and 200 hp V6 engines with overheating problems may have been built with a faulty core. Such engines have mislocated water restrictor locating ribs which must be modified. To determine if this is the problem, check the bottom of the starboard water jacket on the exhaust port side between cylinders No. 3 and No. 5 for the letter designation. If it is an "E," take the motor to your dealer for recommended OMC modification.

Troubleshooting

Engine temperature can be checked with the use of Markal Thermomelt Stiks available at your Johnson or Evinrude dealer. This heat-sensitive stick looks like a large crayon (**Figure 229**) and will melt on contact with a metal surface at a specific temperature.

Two thermomelt sticks are required to properly check a Johnson or Evinrude outboard: a 125° F (52° C) stick and a 163° F (73° C) stick. The stick should not be applied to the center of the cylinder head, as this area is normally hotter than 163° F.

The test is most efficient when carried out on a motor operating on a boat in the water. If necessary to perform the test using a test tank, run the engine at 3,000 rpm for a minimum of 5 minutes to bring it to operating temperature. Make sure inlet water temperature is below 80° F (26° C) and perform the test as follows.

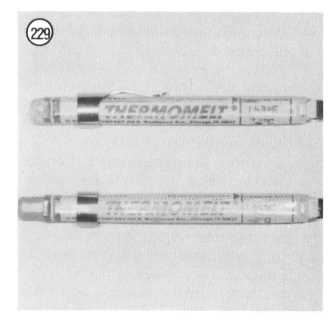

TROUBLESHOOTING

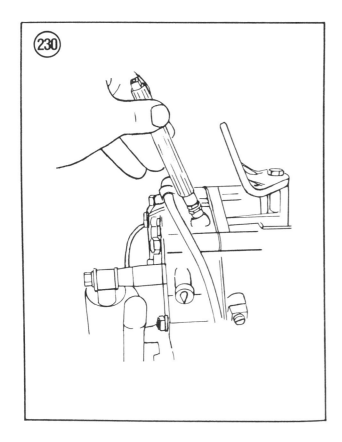

1. Mark the cylinder water jacket with each stick (**Figure 230**). The mark will appear similar to a chalk mark. Make sure sufficient material is applied to the metal surface.
2. With the engine at operating temperature and running at idle in FORWARD gear, 125° F stick mark should melt. If it does not melt on thermostat-equipped models, the thermostat is stuck open and the engine is running cold.
3. With the engine at operating temperature and running at full throttle in FORWARD gear, the 163° F stick mark should not melt. If it does, the power head is overheating. Look for a defective water pump or clogged or leaking cooling system. On thermostat-equipped models, the thermostat may be stuck closed.

Temperature Switches (Remote Control Models)

Engines equipped with remote controls have a temperature switch installed in each cylinder head. The normally open switch is connected in series with a warning horn in the remote control unit. If the engine overheats, the switch(es) close the warning circuit and the horn sounds.

WARNING
Never heat oil over an open flame since it is volatile when warmed.

Switch operation is checked by removing the cylinder head cover with the switch, then removing the switch from the cover. Connect a test lamp to the switch terminals and submerge the switch in cooking oil. See **Figure 231**. While heating the oil observe the following switch closing and opening temperatures.

1. Models prior to 1988:
   a. Tan wire—Switch contacts should close (test lamp ON) between 205-217° F (96-102° C) and contacts should open (test lamp OFF) when oil temperature cools to 168-182° F (75-83° C).

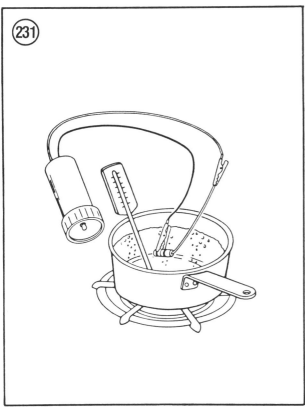

2. Models after 1987:
   a. Tan wire—Switch contacts should close (test lamp ON) between 197-209° F (92-98° C) and contacts should open (test lamp OFF) when oil temperature cools to 147-177° F (65-79° C).
   b. White/black wire—Switch contacts should close (test lamp ON) between 93-99° F (34-38° C) and contacts should open (test lamp OFF) when oil temperature cools to 86-92° F (30-34° C).

If the switch fails to pass tests outlined above, replace it with a new one.

ENGINE

Engine problems are generally symptoms of something wrong in another system, such as ignition, fuel or starting. If properly maintained and serviced, the engine should experience no problems other than those caused by age and wear.

Overheating and Lack of Lubrication

Overheating and lack of lubrication cause the majority of engine mechanical problems. Outboard motors create a great deal of heat and are not designed to operate at a standstill for any length of time. Using a spark plug of the wrong heat range can burn a piston. Incorrect ignition timing, a defective water pump or thermostat, a propeller that is too large (over-propping) or an excessively lean fuel mixture can also cause the engine to overheat.

Preignition

Preignition is the premature burning of fuel and is caused by hot spots in the combustion chamber (**Figure 232**). The fuel actually ignites before it is supposed to. Glowing deposits in the combustion chamber, inadequate cooling or overheated spark plugs can all cause preignition. This is first noticed in the form of a power loss but will eventually result in extensive damage to the internal parts of the engine because of higher combustion chamber temperatures.

Detonation

Commonly called "spark knock" or "fuel knock," detonation is the violent explosion of fuel in the combustion chamber prior to the proper time of combustion (**Figure 233**). Severe damage can result. Use of low octane gasoline is a common cause of detonation.

Even when high octane gasoline is used, detonation can still occur if the engine is improperly timed. Other causes are over-advanced ignition timing, lean fuel mixture at or near full throttle, inadequate engine cooling, cross-firing of spark plugs, excessive accumulation of deposits on piston and combustion chamber or the use of a prop that is too large (over-propping).

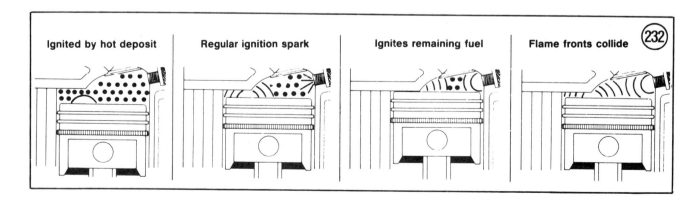

TROUBLESHOOTING

Since outboard motors are noisy, engine knock or detonation is likely to go unnoticed by owners, especially at high engine rpm when wind noise is also present. Such inaudible detonation, as it is called, is usually the cause when engine damage occurs for no apparent reason.

Poor Idling

A poor idle can be caused by improper carburetor adjustment, incorrect timing or ignition system malfunctions. Check the gas cap vent for an obstruction.

On 1982 V4 and V6 engines, plugged crankcase recirculating valves will also cause a poor or rough idle, especially when the engine is tilted down after running it in a tilted position. To check this condition, tilt the motor out to a full trim position. Remove one recirculating hose at a time from the bypass cover nipples. If fuel does not pulse from the hose, shut the engine off and check the hose for clogging. If the hose is not clogged, remove and clean the check valve. Reinstall the valve and tighten to 75-85 in.-lb. (brass) or 20-25 in.-lb. (plastic).

Misfiring

Misfiring can result from a weak spark or a dirty spark plug. Check for fuel contamination. If misfiring occurs only under heavy load, as when accelerating, it is usually caused by a defective spark plug. Run the motor at night to check for spark leaks along the plug wire and under spark plug cap or use a spark leak tester.

WARNING
Do not run engine in a dark garage to check for spark leak. There is considerable danger of carbon monoxide poisoning.

Water Leakage in Cylinder

The fastest and easiest way to check for water leakage in a cylinder is to check the spark plugs. Water will clean a spark plug. If one of the plugs on a multi-cylinder engine is clean and the others is dirty, there is most likely a water leak in the cylinder with the clean plug.

To remove all doubt, install a dirty plug in all cylinders. Run the engine in a test tank or on the boat in water for 5-10 minutes. Shut the engine off and remove the plugs. If one plug is clean and the other(s) dirty (or if all plugs are clean), a water leak in the cylinder(s) is the problem.

Flat Spots

If the engine seems to die momentarily when the throttle is opened and then recovers, check for a dirty main jet in the carburetor, water in the fuel or an excessively lean mixture.

Power Loss

Several factors can cause a lack of power and speed. Look for air leaks in the fuel line or fuel

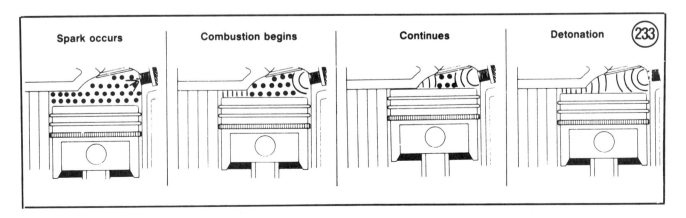

pump, a clogged fuel filter or a choke/throttle valve that does not operate properly. Check ignition timing.

A piston or cylinder that is galling, incorrect piston clearance or a worn/sticky piston ring may be responsible. Look for loose bolts, defective gaskets or leaking machined mating surfaces on the cylinder head, cylinder or crankcase. Also check the crankcase oil seal; if worn, it can allow gas to leak between cylinders.

Piston Seizure

This is caused by one or more pistons with incorrect bore clearances, piston rings with an improper end gap, the use of an oil-fuel mixture containing less than 1 part oil to 50 parts of gasoline or an oil of poor quality, a spark plug of the wrong heat range or incorrect ignition timing. Overheating from any cause may result in piston seizure.

Excessive Vibration

Excessive vibration may be caused by loose motor mounts, worn bearings or a generally poor running motor.

Engine Noises

Experience is needed to diagnose accurately in this area. Noises are difficult to differentiate and even harder to describe. Deep knocking noises usually mean main bearing failure. A slapping noise generally comes from a loose piston. A light knocking noise during acceleration may be a bad connecting rod bearing. Pinging should be corrected immediately or damage to the piston will result. A compression leak at the head-to-cylinder joint will sound like a rapid on-off squeal.

Table 1 STARTER TROUBLESHOOTING

Trouble	Cause	Remedy
Pinion does not move when starter is turned on	Blown fuse	Replace fuse.
	Pinion rusted to armature shaft	Remove, clean or replace as required.
	Series coil or shunt broken or shorted	Replace coil or shunt.
	Loose switch connections	Tighten connections.
	Rusted or dirty plunger	Clean plunger.
Pinion meshes with ring gear but starter does not run	Worn brushes or brush springs touching armature	Replace brushes or brush springs.
	Dirty or burned commutator	Clean or replace as required.
	Defective armature field coil	Replace armature.
	Worn or rusted armature shaft bearing	Replace bearing.
Starter motor runs at full speed before pinion meshes with ring gear	Worn pinion sleeve	Replace sleeve.
	Pinion does not stop in correct position	Replace pinion.

(continued)

TROUBLESHOOTING

Table 1 STARTER TROUBLESHOOTING (continued)

Trouble	Cause	Remedy
Pinion meshes with gear and motor starts but engine does not crank	Defective overrunning clutch	Replace overrunning clutch.
Starter motors does not stop when turned off after engine has started	Rusted or dirty plunger	Clean or replace plunger.
Low starter motor speed with high current draw	Armature may be dragging on pole shoes from bent shaft, worn bearings or loose pole shoes	Replace shaft or bearings and/or tighten pole shoes.
	Tight or dirty bearings	Loosen or clean bearings.
High current draw with no armature rotation	A direct ground switch, @ terminal or @ brushes or field connections	Replace defective parts.
	Frozen shaft bearings which prevent armature from rotating	Loosen, clean or replace bearings.
Starter motor has grounded armature or field winding	Field and/or armature is burned or lead is thrown out of commutator due to excess leakage	Raise grounded brushes from commutator and insulate them with cardboard. Use an ignition analyzer and test points to check between insulated terminal or starter motor and starter motor frame (remove ground connection of shunt coils on motors with this feature). If analyzer shows resistance (meter needle moves to right), there is a ground. Raise other brushes from armature and check armature and fields separately to locate ground.
Starter motor has grounded armature or field winding	Current passes through armature first, then to ground field windings	Disconnect grounded leads, then locate any abnormal grounds in starter motor.

(continued)

Table 1 STARTER TROUBLESHOOTING (continued)

Trouble	Cause	Remedy
Starter motor has grounded armature or field winding (continued)	Wiring or key switch corroded	Coat with sealer to protect against further corrosion.
	Starter solenoid	Check for resistance between: (a) positive (+) terminal of battery and large input terminal of starter solenoid, (b) large wire @ top of starter motor and negative (−) terminal of battery, and (c) small terminal of starter solenoid and positive battery terminal. Key switch must be in START position. Repair all defective parts.
	Starter motor	With a fully charged battery, connect a negative (−) jumper wire to upper terminal on side of starter motor and a positive jumper to large lower terminal of starter motor. If motor still does not operate, remove for overhaul or replacement.
Starter turns over too slowly	Low battery or poor contact at battery terminal	See "Starter does not operate."
	Poor contact at starter solenoid or starter motor	Check all terminals for looseness and tighten all nuts securely.
Starter motor fails to operate and draws no current and/or high resistance	Open circuit in fields or armature, @ connections or brushes or between brushes and commutator	Repair or adjust broken or weak brush springs, worn brushes, high insulation between commutator bars or a dirty, gummy or oily commutator.
High resistance in starter motor	Low no-load speed and a low current draw and low developed torque	Closed "open" field winding on unit which has 2 or 3 circuits in starter motor (unit in which current divides as it enters, taking 2 or 3 parallel paths).
High free speed and high current draw	Shorted fields in starter motor	Install new fields and check for improved performance. Fields normally have very low resistance, thus it is difficult to detect shorted fields, since difference in current draw between normal starter motor field windings would not be very great.
Excessive voltage drop	Cables too small	Install larger cables to accomodate high current draw.
High circuit resistance	Dirty connections	Clean connections.
Starter does not operate	Run-down battery	Check battery with hydrometer. If reading is below 1.230, recharge or replace battery.
	Poor contact at terminals	Remove terminal clamps. Scrape terminals and clamps clean and tighten bolts securely.

(continued)

TROUBLESHOOTING

Table 1 STARTER TROUBLESHOOTING (continued)

Trouble	Cause	Remedy
Starter does not operate (continued)	Starter mechanism	Disconnect positive (+) battery terminal. Rotate pinion gear in disengaged position. Pinion gear and motor should run freely by hand. If motor does not turn over easily, clean starter and replace all defective parts.
Starter spins freely but does not engage engine	Low battery or poor contact at battery terminal	See "Starter does not operate".
	Poor contact at starter solenoid or starter motor	See "Starter does not operate".
	Dirty or corroded pinion drive	Clean thoroughly and lubricate the spline underneath the pinion with Lubriplate 777.
Starter does not engage freely	Pinion or flywheel gear	Inspect mating gears for excessive wear. Replace all defective parts.
	Small anti-drift spring	If drive pinion interferes with flywheel gear after engine has started, inspect anti-drift spring located under pinion gear. Replace all defective parts. NOTE: if drive pinion tends to stay engaged in flywheel gear when starter motor is in idle position, start motor @ 1/4 throttle to allow starter pinion gear to release flywheel ring gear instantly.
Starter keeps on spinning after key is turned ON	Key not fully returned	Check that key has returned to normal ON position from START position. Replace switch if key constantly stays in START position.
	Starter solenoid	Inspect starter solenoid to see if contacts have become stuck in closed position. If starter does not stop running with small yellow lead disconnect from starter solenoid, replace starter solenoid.
	Wiring or key switch	Inspect all wires for defects. Open remote control box and inspect wiring @ switches. Repair or replace all defective parts.
Wires overheat	Battery terminals improperly connected	Check that negative marking on harness matches that of battery. If battery is connected improperly, red wire to rectifier will overheat.
	Short circuit in system	Inspect all wiring connections and wires for looseness or defects. Open remote control box and inspect wiring @ switches.

(continued)

Table 1 STARTER TROUBLESHOOTING (continued)

Trouble	Cause	Remedy
Wires overheat (continued)	Short circuit in choke solenoid	Repair or replace all defective parts. Check for high resistance. If blue choke wire heats rapidly when choke is used, choke solenoid may have internal short. Replace if defective.
	Short circuit in starter solenoid	If yellow starter solenoid lead overheats, there may be internal short (resistance) in starter solenoid. Replace if defecrive.
	Low battery voltage	Battery voltage is checked with an ampere-volt tester when battery is under a starting load. Battery must be recharged if it resisters under 9.5 volts. If battery is below specified hydrometer reading of 1.230, it will not turn engine fast enough to start it.

Table 2 IGNITION TROUBLESHOOTING

Symptom	Probable cause
Engine won't start, but fuel and spark are good	Defective or dirty spark plugs. Spark plug gap set too wide. Improper spark timing. Shorted stop button. Air leaks into fuel pump. Broken piston ring(s). Cylinder head, crankcase or cylinder sealing faulty. Worn crankcase oil seal.
Engine misfires @ idle	Incorrect spark plug gap. Defective, dirty or loose spark plugs. Spark plugs of incorrect heat range. Leaking or broken high tension wires. Weak armature magnets. Defective coil or condenser. Defective ignition switch. Spark timing out of adjustment.

(continued)

TROUBLESHOOTING

Table 2 IGNITION TROUBLESHOOTING (continued)

Symptom	Probable cause
Engine misfires @ high speed	See "Engine misfires @ idle." Coil breaks down. Coil shorts through insulation. Spark plug gap too wide. Wrong type spark plugs. Too much spark advance.
Engine backfires	Cracked spark plug insulator. Improper timing. Crossed spark plug wires. Improper ignition timing.
Engine preignition	Spark advanced too far. Incorrect type spark plug. Burned spark plug electrodes.
Engine noises (knocking at power head)	Spark advanced too far.
Ignition coil fails	Extremely high voltage. Moisture formation. Excessive heat from engine.
Spark plugs burn and foul	Incorrect type plug. Fuel mixture too rich. Inferior grade of gasoline. Overheated engine. Excessive carbon in combustion chambers.
Ignition causing high fuel consumption	Incorrect spark timing. Leaking high tension wires. Incorrect spark plug gap. Fouled spark plugs. Incorrect spark advance. Weak ignition coil. Preignition.

Table 3 FUEL SYSTEM TROUBLESHOOTING

Symptom	Probable cause
No fuel @ carburetor	No gas in tank. Air vent in gas cap not open. Air vent in gas cap clogged. Fuel tank sitting on fuel line. Fuel line fittings not properly connected to engine or fuel tank. Air leak @ fuel connection. Fuel pickup clogged. Defective fuel pump.
Flooding @ carburetor	Choke out of adjustment. High float level. Float stuck. Excessive fuel pump pressure. Float saturated beyond buoyancy.
Rough operation	Dirt or water in fuel. Reed valve open or broken. Incorrect fuel level in carburetor bowl. Carburetor loose @ mounting flange. Throttle shutter not closing completely. Throttle shutter valve installed incorrectly.
Carburetor spit-back at idle	Chipped or broken reed valve(s).
Engine misfires at high speed	Dirty carburetor. Lean carburetor adjustment. Restriction in fuel system. Low fuel pump pressure.
Engine backfires	Poor quality fuel. Air-fuel mixture too rich or too lean. Improperly adjusted carburetor.
Engine preignition	Excessive oil in fuel. Inferior grade of gasoline. Lean carburetor mixture.
Spark plugs burn and foul	Fuel mixture too rich or inferior grade of gasoline.
High gas consumption: Flooding or leaking	Cracked carburetor casting. Leaks @ line connections. Defective carburetor bowl gasket. High float level. Plugged vent hole in cover. Loose needle and seat. Defective needle valve seat gasket. Worn needle valve and seat. Float binding in bowl. High fuel pump pressure.
Overrich mixture	Choke lever stuck. High float level. High fuel pump pressure.
Abnormal speeds	Carburetor out of adjustment. Too much oil in fuel.

TROUBLESHOOTING

Table 4 STATOR RESISTANCE VALUES

Engine	Ohms
1973-1975	
50 hp	1.0-2.0
65-135 hp	
6 amp system	0.8-1.2
12 amp system	0.3-0.7
1976	
55 hp	1.0-1.6
70-75 hp	0.6-1.2
85-115 hp	
6 amp system	1.0-1.4
12 amp system	0.3-0.7
135 hp	0.15-0.35
1977	
55 hp	1.0-1.6
70-75 hp	0.6-1.2
85EL-115EL	
6 amp system	1.0-1.4
12 amp system	0.3-0.7
85TXL-115TXL	0.15-0.35
140-200 hp	0.15-0.35
1978	
55 hp	
Yellow/gray	0.25-0.45
Yellow/blue	0.1-0.3
70-75 hp	0.6-1.2
V4	
6 amp system	0.8-1.2
10 amp system	0.3-0.7
V6	0.15-0.35
1979-1984	
50, 55 and 60 hp	
Yellow/gray	0.25-0.45
Yellow/blue	0.1-0.3
70-75 hp	0.7-1.3
V4	
6 amp system	0.8-1.6
10 amp system	0.3-0.7
V6	0.3-0.7

(continued)

Table 4 STATOR RESISTANCE VALUES (continued)

Engine	Ohms
1985	
50-60 hp	0.22-0.32
70-150 hp	1.2-1.4
90-140 hp	0.60-0.64
185-235 hp	0.12-0.22
1986	
50 hp	0.22
60-150 hp	
6 amp system	1.2-1.4
9 amp system	0.60-0.64
150-200 hp	0.12-0.22
1987	
48-50 hp	0.22-0.32
60-75 hp	1.2-1.4
88-150 hp	
6 amp system	1.2-1.4
9 amp system	0.60-0.64
35 amp system	0.12-0.22
All others	
9 amp system	0.60-0.64
35 amp system	0.12-0.22
1988-1990	
48-50 hp	0.22-0.32
60-70 hp	1.2-1.4
88-115 hp	
6 amp system	1.2-1.4
9 amp system	0.65-0.75
All others	
9 amp system	0.65-0.75
35 amp system	0.12-0.22

Chapter Four

Lubrication, Maintenance and Tune-up

The modern outboard motor delivers more power and performance than ever before, with higher compression ratios, new and improved electrical systems and other design advances. Proper lubrication, maintenance and tune-ups have thus become increasingly important as ways in which you can maintain a high level of performance, extend engine life and extract the maximum economy of operation.

You can do your own lubrication, maintenance and tune-ups if you follow the correct procedures and use common sense. The following information is based on recommendations from Johnson and Evinrude that will help you keep your outboard motor operating at its peak performance level.

Tables 1-4 are at the end of the chapter.

LUBRICATION

Proper Fuel Selection

Two-stroke engines are lubricated by mixing oil with the fuel. The various components of the engine are thus lubricated as the fuel-oil mixture passes through the crankcase and cylinders. Since outboard fuel serves the dual function of producing ignition and distributing the lubrication, the use of low octane marine white gasolines should be avoided. Such gasolines also have a tendency to cause ring sticking and port plugging.

In the past, Johnson and Evinrude have recommended the use of regular leaded gasoline with a minimum posted pump octane rating of 87 (50-75 hp and V4) or 89 (V6). However, obtaining regular leaded gasoline is becoming extremely difficult and what can be found contains a very low amount of lead.

For this reason, fuel recommendations have been changed on the earlier models which may result in some engine modifications to accommodate the new recommendation. Refer to the following.

NOTE
*Gasoline containing methanol is **not** recommended for use on 1973-1986 models.*

1. 1973-1985 48-115 HP:
    a. 1973 65 HP, 1975-1978 75 HP and 1985 115 HP—Lead-free regular, lead-free premium, leaded regular, leaded premium or gasohol with a pump posted octane rating of 91 or higher. If a pump posted octane rating of 88 is used, the maximum spark advance must be reduced 4°. If a pump posted octane rating of 86 is used, consult your local dealer as engine modifications must be performed in addition to reducing maximum spark advance.
    b. All other models—Lead-free regular, lead-free premium, leaded regular, leaded premium or gasohol with a pump posted octane rating of 86 or higher.
2. 1973-1985 135-235 HP—Lead-free regular, lead-free premium, leaded regular, leaded premium or gasohol with a pump posted octane rating of 91 or higher. If a pump posted octane rating of 88 is used, the maximum spark advance must be reduced 4° except on 1983-1985 150 XP™ and GT™ models. If a pump posted octane rating of 86 is used, consult your local dealer as engine modifications must be performed in addition to reducing maximum spark advance.
3. All 1986 models—Lead-free regular, lead-free premium, leaded regular, leaded premium or gasohol with a pump posted octane rating of 86 or higher.
4. All 1987-1990 models—Lead-free regular, lead-free premium, leaded regular, leaded premium or gasohol with a pump posted octane rating of 87 or higher. A gasoline/methanol mixture may be used if the mixture contains *no more than* 5 percent methanol plus 5 percent cosolvent alcohols and meets the pump posted octane rating of 87.

Sour Fuel

Fuel should not be stored for more than 60 days (under ideal conditions). Gasoline forms gum and varnish deposits as it ages. Such fuel will cause starting problems. A fuel additive such as OMC 2+4 Fuel Conditioner should be used to prevent gum and varnish formation during storage or prolonged periods of non-use but it is always better to drain the tank in such cases. Always use fresh gasoline when mixing fuel for your outboard.

Gasohol

Some gasolines sold for marine use now contain alcohol, although this fact may not be advertised. A mixture of 10 percent ethyl alcohol or ethanol and 90 percent unleaded gasoline is called gasohol. This is considered suitable for use in Johnson and Evinrude outboards. Some gasolines, however, contain methyl alcohol or methanol. This is *not* recommended for 1973-1986 outboards and only under the provisions outlined under *Proper Fuel Selection* in this chapter for 1987-1990 models.

Fuels with an alcohol content tend to slowly absorb moisture from the air. When the moisture content of the fuel reaches approximately one percent, it combines with the alcohol and separates from the fuel. This separation does not normally occur when gasohol is used in an automobile, as the tank is generally emptied within a few days after filling it.

The problem does occur in marine use, however, because boats often remain idle between start-ups for days or even weeks. This length of time permits separation to take place. The alcohol-water mixture settles at the bottom of the fuel tank. Since outboard motors will not run on this mixture, it is necessary to drain the fuel tank, flush out the fuel system with clean gasoline and then remove, clean and reinstall the spark plugs before the engine can be started.

Continued use of fuels containing methanol can cause deterioration of fuel system components. The major danger of using gasohol in an outboard motor is that a shot of the water-alcohol mix may be picked up and sent to one

LUBRICATION, MAINTENANCE AND TUNE-UP

of the carburetors of a multicylinder engine. Since this mixture contains no oil, it will wash oil off the bore of any cylinder it enters. The other carburetor receiving good fuel-oil mixture will keep the engine running while the cylinder receiving the water-alcohol mixture can suffer internal damage.

The problem of unlabeled gasohol has become so prevalent around the United States that Miller Tools (32615 Park Lane, Garden City, MI 48135) now offers an Alcohol Detection Kit (part No. C-4846) so that owners and mechanics can determine the quality of fuel being used.

The kit cannot differentiate between types of alcohol (ethanol, methanol, etc.) nor is it considered to be absolutely accurate from a scientific standpoint, but it is accurate enough to determine whether or not there is sufficient alcohol in the fuel to cause the user to take precautions.

If gasohol or a gasoline/methanol mixture (1987-1990 models) is used with any regularity, Johnson and Evinrude recommend that the carburetor jets be changed to provide a richer mixture to maintain good performance and prevent possible engine damage. See your dealer for recommended rejetting according to your needs.

Recommended Fuel Mixture

NOTE
If engine is equipped with Variable Ratio Oiling (VRO) or OMC Economixer, read this chapter and then refer to Chapter Eleven.

Use the specified gasoline for your Johnson or Evinrude outboard and mix with Johnson or Evinrude Outboard Lubricant in the following ratios:

CAUTION
Do not, under any circumstances, use multigrade or other high detergent automotive oils or oils containing metallic additives. Such oils are harmful to 2-stroke engines. Since they do not mix properly with gasoline, do not burn as 2-cycle oils do and leave an ash residue, their use may result in piston scoring, bearing failure or other engine damage.

a. Thoroughly mix one pint of Johnson or Evinrude Outboard Lubricant with each 6 gallons of gasoline in your remote fuel tank. This provides a 50:1 mixture.
b. Operation in Canada requires mixing one U.S. pint of Johnson or Evinrude Outboard Lubricant to each 5 Imperial gallons of gasoline in the remote fuel tank.

If Johnson or Evinrude Outboard Lubricant is not available, any high-quality 2-stroke oil intended for outboard use may be substituted provided the oil meets NMMA (BIA) rating TC-W II or TC-W and specifies so on the container. Follow the manufacturer's mixing instructions on the container but do not exceed a 50:1 ratio.

CAUTION
Only 2-stroke oil meeting NMMA TC-WII certification is recommended for usage in 1990 models.

Correct Fuel Mixing

WARNING
Gasoline is an extreme fire hazard. Never use gasoline near heat, sparks or flame. Do not smoke while mixing fuel.

Mix the fuel and oil outdoors or in a well-ventilated indoor location. Using less than the specified amount of oil can result in insufficient lubrication and serious engine damage. Using more oil than specified causes spark plug fouling, erratic carburetion, excessive smoking and rapid carbon accumulation.

Cleanliness is of prime importance. Even a very small particle of dirt can cause carburetion problems. Always use fresh gasoline. Gum and varnish deposits tend to form in gasoline stored in a tank for any length of time. Use of sour fuel can result in carburetor problems and spark plug fouling.

Above 32° F (0° C)

Measure the required amounts of gasoline and oil accurately. Pour the Outboard Lubricant into the portable tank and add the fuel. Install the tank filler cap and mix the fuel by tipping the tank on its side and back to an upright position several times. See **Figure 1**.

If a built-in tank is used, insert a large metal filter funnel in the tank filler neck. Slowly pour the Outboard Lubricant into the funnel at the same time the tank is being filled with gasoline. See **Figure 2**.

Below 32° F (0° C)

Measure the required amounts of gasoline and oil accurately. Pour about one gallon of gasoline in the tank and add the required amount of Outboard Lubricant. Install the tank filler cap and shake the tank to thoroughly mix the fuel and oil. Remove the cap and add the balance of the gasoline.

If a built-in tank is used, insert a large metal filter funnel in the tank filler neck. Mix the required amount of Outboard Lubricant with one gallon of gasoline in a separate container. Slowly pour the mixture into the funnel at the same time the tank is being filled with gasoline.

Consistent Fuel Mixtures

The carburetor idle adjustment is sensitive to fuel mixture variations which result from the use of different oils and gasolines or from inaccurate measuring and mixing. This may require readjustment of the idle needle. To prevent the necessity for constant readjustment of the carburetor from one batch of fuel to the next, always be consistent. Prepare each batch of fuel exactly the same as previous ones.

Pre-mixed fuels sold at some marinas are not recommended for use in Johnson or Evinrude outboards, since the quality and consistency of pre-mixed fuels can vary greatly. The possibility of engine damage resulting from use of an incorrect fuel mixture outweighs the convenience offered by pre-mixed fuel. This is especially true if the marina includes alcohol in its pre-mixed fuel. The alcohol tends to settle at the bottom of the storage tank. See *Gasohol* in this chapter.

Lower Drive Unit Lubrication

Replace the lower drive unit lubricant after the first 20 hours of operation. Check every 50 hours of operation and top up if necessary. Drain and refill every 100 hours of operation or at least once a season. Use OMC HI-VIS Gearcase Lubricant. If not available, OMC Premium Blend Gearcase

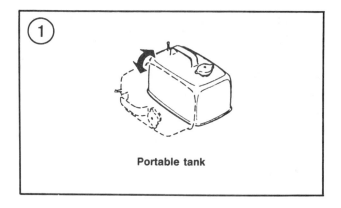

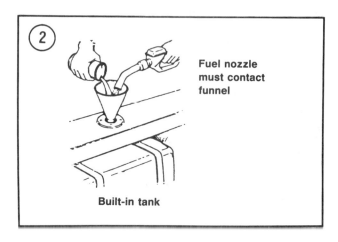

LUBRICATION, MAINTENANCE AND TUNE-UP

Lube can be used as a substitute in all except V6 gearcases.

CAUTION
Do not use regular automotive grease in the lower drive unit. Its expansion and foam characteristics are not suitable for marine use.

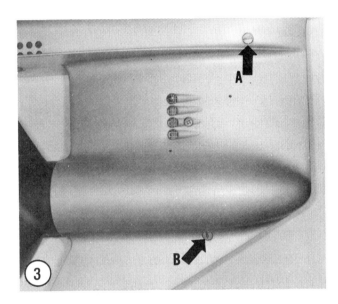

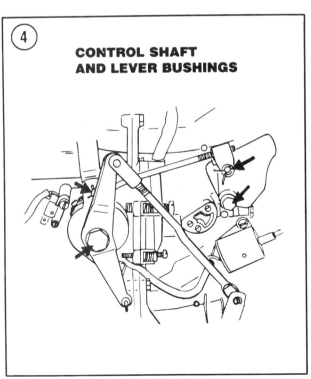

CONTROL SHAFT AND LEVER BUSHINGS

1. Disconnect the negative battery cable or the armature plate-to-power pack wiring to eliminate any accidental starting of the motor.
2. Place a suitable container under the gearcase.

CAUTION
Never lubricate the gearcase without first removing the oil level screw, as the injected lubricant displaces air which must be allowed to escape. The gearcase cannot be completely filled otherwise.

3. Locate and remove the oil level plug and washer (A, **Figure 3**, typical).
4. Locate and remove the drain/fill plug and washer (B, **Figure 3**, typical).
5. Allow the lubricant to completely drain.
6. Inject OMC HI-VIS Gearcase Lubricant into the drain/fill plug hole until excess fluid flows out the oil level plug hole.
7. Drain about one ounce of fluid to allow for lubricant expansion.
8. Install the oil level plug. Remove the lubricant tube or nozzle from drain/fill hole and install the drain/fill plug. Be sure the washers are in place under the head of each, so that water will not leak past the threads into the housing. Tighten both plugs to 60-80 in.-lb. (7-9 N•m).

Other Lubrication Points

Refer to **Figures 4-15** (typical) and **Table 1** for other lubricant points, frequency of lubrication and lubricant to be used.

CAUTION
When lubricating the steering cable on models so equipped, make sure its core is fully retracted into the cable housing. Lubricating the cable while extended can cause a hydraulic lock to occur.

Salt Water Corrosion of Gear Housing Bearing Carrier/Nut

Salt water corrosion that is allowed to build up unchecked can eventually split the gear housing and destroy the lower unit. If the motor

CHAPTER FOUR

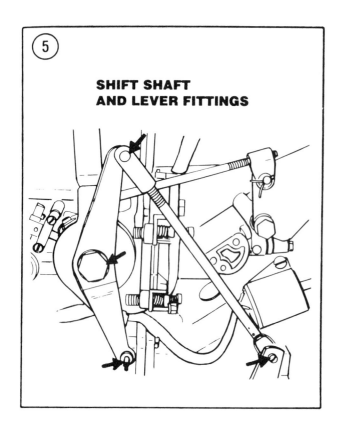

⑤ SHIFT SHAFT AND LEVER FITTINGS

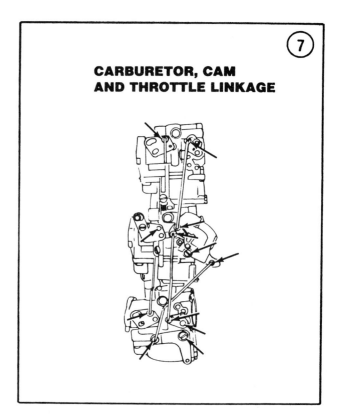

⑦ CARBURETOR, CAM AND THROTTLE LINKAGE

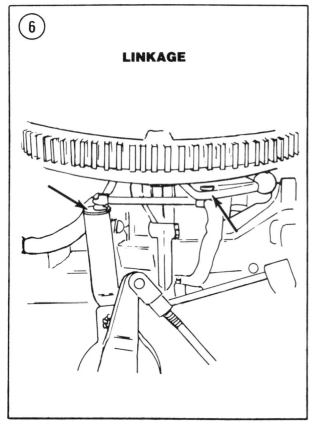

⑥ LINKAGE

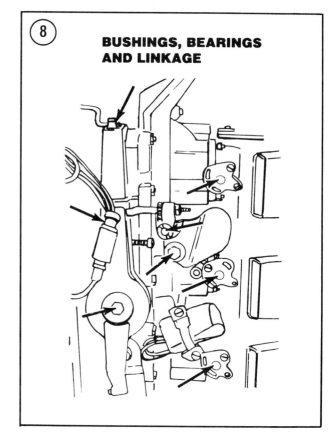

⑧ BUSHINGS, BEARINGS AND LINKAGE

LUBRICATION, MAINTENANCE AND TUNE-UP

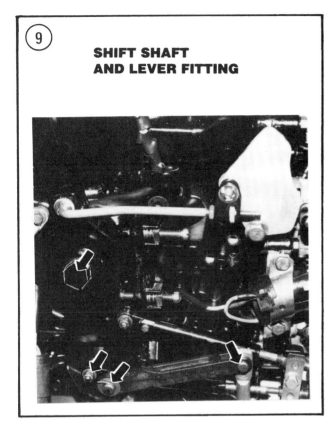

⑨ SHIFT SHAFT AND LEVER FITTING

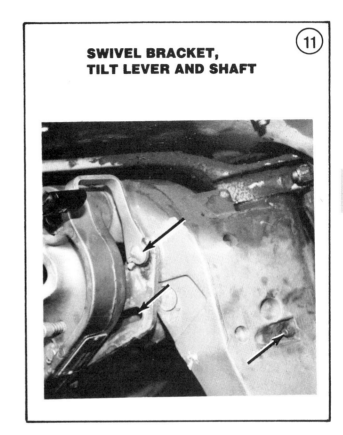

⑪ SWIVEL BRACKET, TILT LEVER AND SHAFT

⑩ TILT TUBE SHAFT

⑫ MOTOR COVER LEVER SHAFTS

is used in salt water, remove the propeller assembly and bearing housing at least once a year after the initial 20-hour inspection. Clean all corrosive deposits and dried-up lubricant from each end of the housing (**Figure 16**). Lubricate the bearing housing, O-ring and screw threads with OMC Gasket Sealing Compound. Install bearing housing and tighten screws to specifications (Chapter Nine).

STORAGE

The major consideration in preparing an outboard motor for storage is to protect it from rust, corrosion and dirt. Johnson and Evinrude recommend the following procedure.

1. If boat is equipped with a built-in fuel tank, add one ounce of OMC 2+4 fuel conditioner to fuel tank for each gallon of fuel tank capacity. Top off fuel tank with recommended fuel shown in this chapter.
2. Operate the motor in a test tank with the proper test wheel or on the boat in the water. Start the engine and allow it to warm up while allowing the stabilized fuel to circulate.
3. Stop the engine after approximately five minutes.
4. In a portable six gallon fuel tank, prepare the following storage mixture:
   a. Add five gallons of recommended fuel.
   b. Add two quarts of OMC Storage Fogging Oil.
   c. Add one pint of Evinrude or Johnson Outboard Lubricant.
   d. Add one pint of OMC 2+4 fuel conditioner.
5. Thoroughly blend mixture in fuel tank.
6. Connect storage mixture to engine.
7. Operate the motor in a test tank with the proper test wheel or on the boat in the water. Start the engine and operate at approximately 1500 rpm for five minutes.

NOTE
*On variable ratio oiling (VRO) equipped models, **do not** disconnect fuel hose with engine running in an attempt to run the carburetors dry. When a low amount of gasoline is supplied to the VRO pump, a fuel mixture with a high oil to gasoline ratio is created. The resulting mixture could cause excessive oil consumption and difficulty in restarting.*

8. Stop the engine and disconnect the storage mixture.

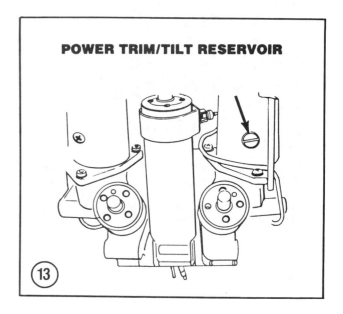

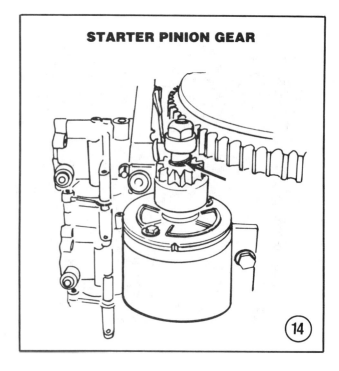

LUBRICATION, MAINTENANCE AND TUNE-UP

9. Remove the engine cover.
10. Remove the spark plugs as described in this chapter.
11. Spray a liberal amount of OMC Storage Fogging Oil through spark plug holes into each cylinder.
12. Rotate flywheel several rotations clockwise to distribute the OMC Storage Fogging Oil throughout the cylinder.

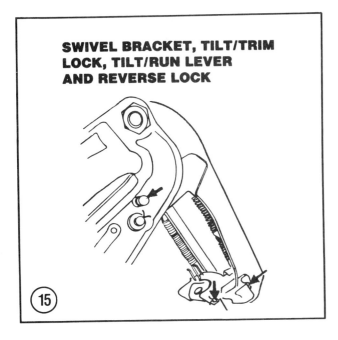

13. Remove the outboard motor from the test tank or water and rotate the flywheel several rotations clockwise to drain any water from the water pump.
14. Clean and regap or replace plugs. Leave spark plug leads disconnected.
15. If storage fuel mixture is not to be used on any other outboard motors, *safely* drain and clean fuel tank. For regular use portable tank, stabilize fuel as recommended with OMC 2+4 fuel conditioner or drain and clean fuel tank. Store tank(s) in a well-ventilated area away from heat or open flame.
16. Drain and refill gearcase as described in this chapter. Check condition of level and drain/fill plug gaskets. Replace as required.
17. Refer to **Figures 4-15** and **Table 1** as appropriate and lubricate motor at all specified points.
18. Remove and check propeller condition. Remove any burring from drive pin hole and replace drive pin if worn or bent. Look for propeller shaft seal damage from fishing line. Clean and lubricate propeller shaft with OMC Triple-Guard grease. Reinstall propeller with a new cotter pin or tab lock washer.
19. Clean all external parts of the motor with OMC All-Purpose Marine Cleaner and apply a good quality marine polish.
20. Store the outboard motor in an upright position in a dry and well-ventilated area.
21. Service the battery as follows:

   a. Disconnect the negative battery cable, then the positive battery cable.

   b. Remove all grease, corrosion and dirt from the battery surface.

   c. Check the electrolyte level in each battery cell and top up with distilled water, if necessary. Fluid level in each cell should not be higher than 3/16 in. above the perforated baffles.

   d. Lubricate the terminal bolts with grease or petroleum jelly.

> *CAUTION*
> *A discharged battery can be damaged by freezing.*

e. With the battery in a fully charged condition (specific gravity 1.260-1.275), store in a dry place where the temperature will not drop below freezing. Do not store on a concrete surface.

f. Recharge the battery every 45 days or whenever the specific gravity drops below 1.230. Before charging, cover the plates with distilled water, but not more than 3/16 in. above the perforated baffles. The charge rate should not exceed 6 amps. Discontinue charging when the specific gravity reaches 1.260 at 80° F (27° C).

g. Before placing the battery back into service after winter storage, remove the excess grease from the terminals, leaving a small amount on. Install battery in a fully charged state.

COMPLETE SUBMERSION

An outboard motor which has been lost overboard should be recovered as quickly as possible. If lost in salt water or fresh water containing sand or silt, disassemble and clean it immediately—any delay will result in rust and corrosion of internal components once it has been removed from the water.

If the motor was running when it was lost, do not attempt to start it until it has been disassembled and checked. Internal components may be out of alignment and running the motor may cause permanent damage.

The following emergency steps should be accomplished immediately if the motor was lost in fresh water.

> *CAUTION*
> *If it is not possible to disassemble and clean the motor immediately, resubmerge it in fresh water to prevent rust and corrosion formation until such time as it can be properly serviced.*

1. Remove the engine cover.
2. Remove the spark plugs as described in this chapter.
3. Disconnect connector(s) between stator charge coil(s) and power pack(s).
4. Remove the carburetor float bowl drain screws if so equipped. See Chapter Six.
5. Wash the outside of the motor with clean water to remove weeds, mud and other debris.

> *CAUTION*
> *If there is a possibility that sand or silt may have entered the power head or gearcase, do not try to start the motor or severe internal damage may occur.*

6. Drain as much water as possible from the power head by placing the motor in a horizontal position. Use the starter rope to rotate the flywheel with the spark plug holes facing downward.

> *CAUTION*
> *Do not force the motor if it does not turn over freely. This may be an indication of internal damage such as a bent connecting rod or broken piston.*

7. Pour either Evinrude or Johnson Outboard Lubricant into cylinders through spark plug holes.
8. Remove carburetors and disassemble. See Chapter Six.
9. Disassemble electric starter motor, if so equipped, and disconnect all electrical connections. Wash with clean fresh water. Spray all electrical components and connections with a water displacing electrical spray and allow to dry. Reassemble electric starter motor, if so equipped, and reconnect all electrical connections.
10. Reinstall spark plugs, carburetors and electric starter motor if so equipped.
11. Blend a fresh fuel mixture following OMC's recommended engine break-in procedure. Try starting the motor. If the motor will start, let it run at least 30 minutes following OMC's recommended engine break-in procedure.

LUBRICATION, MAINTENANCE AND TUNE-UP

12. If motor will not start in Step 11, try to diagnose the cause as fuel, electrical or mechanical, then correct. If the engine cannot be started within three hours, disassemble, clean and oil all parts thoroughly as soon as possible.

ANTI-CORROSION MAINTENANCE

1. Flush the cooling system with fresh water as described in this chapter after each use in salt water. Wash exterior with fresh water.
2. Dry exterior of motor and apply primer over any paint nicks and scratches. Use only tin anti-fouling paint; do not use paints containing mercury or copper. Do not paint sacrificial zinc anodes or trim tab.
3. Apply OMC Black Neoprene Dip to all exposed electrical connections except the positive terminal on the starter solenoid.
4. Check sacrificial anodes and replace any that are less than two-thirds their original size.

5. Lubricate more frequently than specified in **Table 1**. If used consistently in salt water, reduce lubrication intervals by one-half.

ENGINE FLUSHING

Periodic engine flushing will prevent salt or silt deposits from accumulating in the water passageways. This procedure should also be performed whenever an outboard motor is operated in salt water or polluted water.

Keep the motor in an upright position during and after flushing. This prevents water from passing into the power head through the drive shaft housing and exhaust ports during the flushing procedure. It also eliminates the possibility of residual water being trapped in the drive shaft housing or other passageways.

Johnson and Evinrude recommend that the outboard be run with a test wheel instead of the propeller when operated in a test tank or with a flush-test device. See **Figure 17** (typical). Test wheel recommendations are given in **Table 2**.

1. Remove the propeller and install the correct test wheel.
2. Attach the flushing device according to manufacturer's instructions. See **Figure 18** (typical).
3. Connect a garden hose between a water tap and the flushing device.
4. Open the water tap partially—do not use full pressure.
5. Shift into NEUTRAL, then start motor. Keep engine speed at approximately 1,500 rpm.
6. Adjust water flow so that there is a slight loss of water around the rubber cups of the flushing device.
7. Check the motor to make sure that water is being discharged from the "tell-tale" nozzle. If it is not, stop the motor immediately and determine the cause of the problem.

CAUTION
Flush the motor for at least 5 minutes if used in salt water.

8. Flush motor until discharged water is clear. Stop motor.
9. Close water tap and remove flushing device from lower unit.
10. Remove test wheel and reinstall propeller.

TUNE-UP

A tune-up consists of a series of inspections, adjustments and parts replacements to compensate for normal wear and deterioration of outboard motor components. Regular tune-ups are important for power, performance and economy. Johnson and Evinrude recommend that their outboards be serviced every 6 months or 50 hours of operation. If subjected to limited use, the engine should be tuned at least once a year.

Since proper outboard motor operation depends upon a number of interrelated system functions, a tune-up consisting of only one or two corrections will seldom give lasting results. For best results, a thorough and systematic procedure of analysis and correction is necessary.

Prior to performing a tune-up, it is a good idea to flush the motor as described in this chapter and check for satisfactory water pump operation.

The tune-up sequence recommended by Johnson and Evinrude includes the following:
   a. Compression check.
   b. Spark plug service.
   c. Lower unit and water pump check.
   d. Fuel system service.
   e. Ignition system service.
   f. Battery, starter motor and solenoid check (if so equipped).
   g. Internal wiring harness.
   h. Engine synchronization and adjustment.
   i. Performance test (on boat).

Any time the fuel or ignition systems are adjusted or defective parts replaced, the engine timing, synchronization and linkage adjustment *must* be checked. These procedures are described in Chapter Five. Perform the synchronization and linkage adjustment procedure for your engine *before* running the performance test.

Compression Check

An accurate cylinder compression check gives a good idea of the condition of the basic working parts of the engine. It is also an important first step in any tune-up, as a motor with low or unequal compression between cylinders *cannot* be satisfactorily tuned. Any compression problem discovered during this check must be corrected before continuing with the tune-up procedure.

1. With the engine warm, disconnect the spark plug wires and remove the plugs as described in this chapter.
2. Disconnect the power pack-to-armature plate connector to disable the ignition system.
3. Connect the compression tester to the top spark plug hole according to manufacturer's instructions (**Figure 19**).
4. Make sure the throttle is held wide open and crank the engine through at least 4 compression strokes. Record the gauge reading.
5. Repeat Step 3 and Step 4 on each remaining cylinder.

While minimum cylinder compression should not be less than 100 psi, the actual readings are not as important as the differences in readings when interpreting the results. A variation of more than 15 psi between 2 cylinders indicates a problem with the lower reading cylinder, such

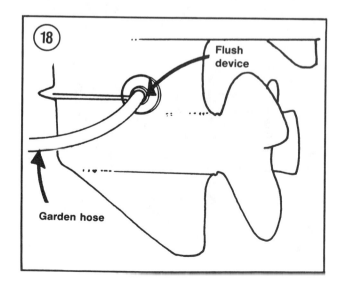

LUBRICATION, MAINTENANCE AND TUNE-UP

as worn or sticking piston rings and/or scored pistons or cylinders. In such cases, pour a tablespoon of engine oil into the suspect cylinder and repeat Step 3 and Step 4. If the compression is raised significantly (by 10 psi in an old engine), the rings are worn and should be replaced.

If the power head shows signs of overheating (discolored or scorched paint) but the compression test turns up nothing abnormal, check the cylinder(s) visually through the transfer ports for possible scoring. A cylinder can be slightly scored and still deliver a relatively good compression reading. In such a case, it is also a good idea to double-check the water pump operation as a possible cause for overheating.

Spark Plug Selection

Johnson and Evinrude outboards are equipped with Champion or AC spark plugs selected for average use conditions. Under adverse use conditions, the recommended spark plug may foul or overheat. In such cases, check the ignition and carburetion systems to make sure they are operating correctly. If no defect is found, replace the spark plug with one of a hotter or colder heat range as required. **Table 3** gives the recommended spark plugs for all models covered in this book. **Table 4** contains a cross-reference for Champion, NGK, AC, Motorcraft and Autolite spark plugs.

Spark Plug Removal

CAUTION
Whenever the spark plugs are removed, dirt around them can fall into the plug holes. This can cause engine damage that is expensive to repair.

1. Blow out any foreign matter from around the spark plugs with compressed air. Use a compressor if you have one. If you do not, use a can of compressed inert gas, available from photo stores.
2. Disconnect the spark plug wires (**Figure 20**, typical) by twisting the wire boot back and forth on the plug insulator while pulling outward. Pulling on the wire instead of the boot may cause internal damage to the wire.
3. Remove the plugs with an appropriate size spark plug socket. Keep the plugs in order so you know which cylinder they came from.
4. Examine each spark plug. See **Figure 21** for conventional gap plugs and **Figure 22** for surface gap plug. Compare plug condition with **Figure 23** (conventional gap) or **Figure 24** (surface gap). Spark plug condition indicates engine condition and can warn of developing trouble.
5. Check each plug for make and heat range. All should be of the same make and number or heat range.
6. Discard the plugs. Although they could be cleaned and reused if in good condition, they seldom last very long. New plugs are inexpensive and far more reliable.

Spark Plug Gapping (Conventional Gap Only)

New plugs should be carefully gapped to ensure a reliable, consistent spark. Use a special

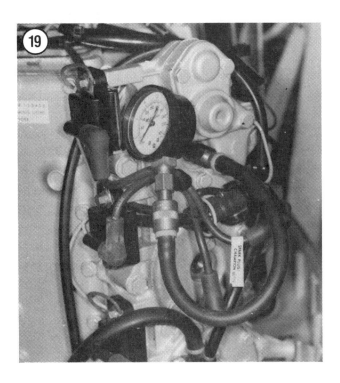

spark plug tool with a round gauge. See **Figure 25** for one common type.

1. Remove the plugs and gaskest from the boxes. Install the gaskets.

NOTE
Some plug brands may have small end pieces that must be screwed on Figure 26 before the plugs can be used.

2. Insert an appropriate round feeler gauge (see **Table 3**) between the electrodes. If the gap is correct, there will be a slight drag as the wire is pulled through. If there is no drag or if the wire will not pull through, bend the side electrode with the gapping tool (**Figure 27**) to change the gap. Remeasure with the wire gauge.

CAUTION
Never try to close the electrode gap by tapping the spark plug on a solid surface. This can damage the plug internally. Always use the gapping and adjusting tool to open or close the gap.

Spark Plug Installation

Improper installation of spark plugs is one of the most common causes of poor spark plug performance in outboard motors. The gasket on the plug must be fully compressed against a clean plug seat in order for heat transfer to take place effectively. This requires close attention to proper tightening during installation.

1. Inspect the spark plug hole threads and clean them with a thread chaser (**Figure 28**). Wipe the cylinder head seats clean before installing the new plugs.
2. Screw each plug in by hand until it seats. Very little effort is required. If force is necessary, the plug is cross-threaded. Unscrew it and try again.
3. Tighten the spark plugs. If you have a torque wrench, tighten to 17-20 ft.-lb. (24-27 N•m). If not, seat the plug finger-tight on the gasket, then tighten an additional ¼ turn with a wrench.

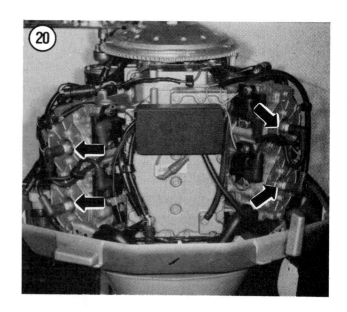

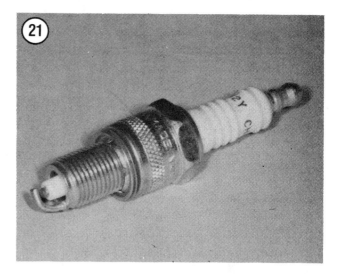

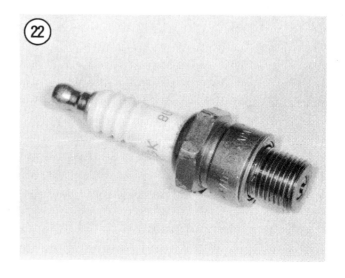

LUBRICATION, MAINTENANCE AND TUNE-UP

SPARK PLUG ANALYSIS (CONVENTIONAL GAP SPARK PLUGS)

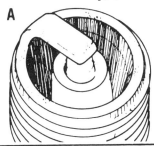
A

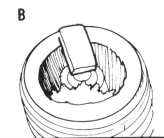
B

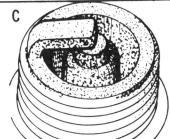
C

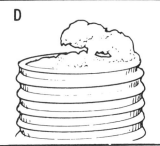
D

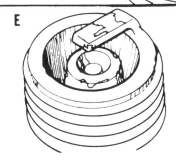
E

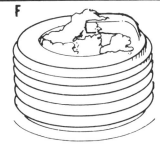
F

A. Normal—Light tan to gray color of insulator indicates correct heat range. Few deposits are present and the electrodes are not burned.

B. Core bridging—These defects are caused by excessive combustion chamber deposits striking and adhering to the firing end of the plug. In this case, they wedge or fuse between the electrode and core nose. They originate from the piston and cylinder head surfaces. Deposits are formed by one or more of the following:
   a. Excessive carbon in cylinder.
   b. Use of non-recommended oils.
   c. Immediate high-speed operation after prolonged trolling.
   d. Improper fuel-oil ratio.

C. Wet fouling—Damp or wet, black carbon coating over entire firing end of plug. Forms sludge in some engines. Caused by one or more of the following:
   a. Spark plug heat range too cold.
   b. Prolonged trolling.
   c. Low-speed carburetor adjustment too rich.
   d. Improper fuel-oil ratio.
   e. Induction manifold bleed-off passage obstructed.
   f. Worn or defective breaker points.

D. Gap bridging—Similar to core bridging, except the combustion particles are wedged or fused between the electrodes. Causes are the same.

E. Overheating—Badly worn electrodes and premature gap wear are indicative of this problem, along with a gray or white "blistered" appearance on the insulator. Caused by one or more of the following:
   a. Spark plug heat range too hot.
   b. Incorrect propeller usage, causing engine to lug.
   c. Worn or defective water pump.
   d. Restricted water intake or restriction somewhere in the cooling system.

F. Ash deposits or lead fouling—Ash deposits are light brown to white in color and result from use of fuel or oil additives. Lead fouling produces a yellowish brown discoloration and can be avoided by using unleaded fuels.

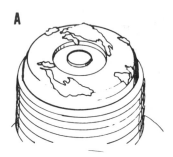

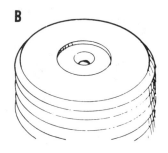

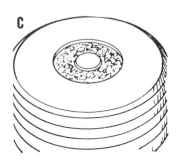

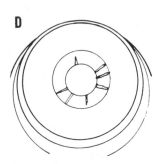

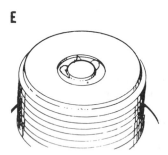

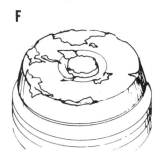

**SURFACE GAP
SPARK PLUG ANALYSIS**

A. Normal—Light tan or gray colored deposits indicate that the engine/ignition system condition is good. Electrode wear indicates normal spark rotation.

B. Worn out—Excessive electrode wear can cause hard starting or a misfire during acceleration.

C. Cold fouled—Wet oil-fuel deposits are caused by "drowning" the plug with raw fuel mix during cranking, overrich carburetion or an improper fuel-oil ratio. Weak ignition will also contribute to this condition.

D. Carbon tracking—Electrically conductive deposits on the firing end provide a low-resistance path for the voltage. Carbon tracks form and can cause misfires.

E. Concentrated arc—Multi-colored appearance is normal. It is caused by electricity consistently following the same firing path. Arc path changes with deposit conductivity and gap erosion.

F. Aluminum throw-off—Caused by preignition. This is not a plug problem but the result of engine damage. Check engine to determine cause and extent of damage.

LUBRICATION, MAINTENANCE AND TUNE-UP

4. Inspect each spark plug wire before reconnecting it to its cylinder. If insulation is damaged or deteriorated, install a new plug wire. Push wire boot onto plug terminal and make sure it seats fully.

Lower Unit and Water Pump Check

A faulty water pump or one that performs below specifications can result in extensive engine damage. Thus, it is a good idea to replace the water pump impeller, seals and gaskets once a year or whenever the lower unit is removed for service. See Chapter Nine.

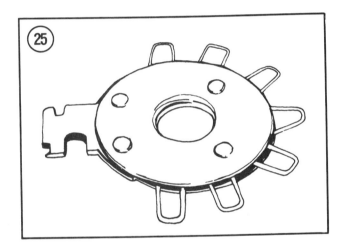

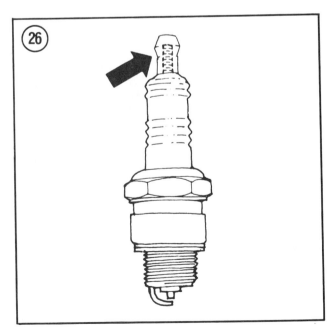

Fuel System Service

The clearance between the carburetor and choke shutter should not be greater than 0.015 in. when the choke is closed or a hard starting condition will result. When changing from one brand of gasoline to another, it may be necessary to readjust the carburetor idle mixture needle slightly (1/4 turn) to accommodate the variations in volatility.

Fuel Lines

1. Visually check all fuel lines for kinks, leaks, deterioration or other damage.
2. Disconnect fuel lines and blow out with compressed air to dislodge any contamination or foreign material.
3. Coat fuel line fittings sparingly with OMC Gasket Sealing Compound and reconnect the lines.

Engine Fuel Filter

Three types of engine fuel filters are used: an inline disposable filter installed prior to the fuel

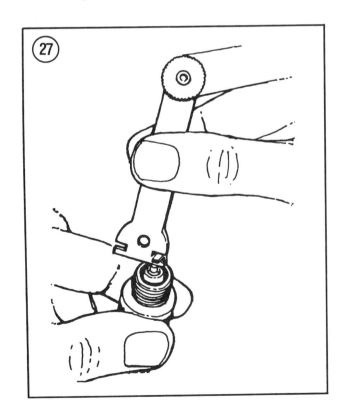

pump, an inline filter canister installed prior to the fuel pump and a filter screen located within the fuel pump.

Inline Disposable Filter

1. Remove the engine cover.
2. Remove the filter hose clamps and replace the filter.
3. Inspect fuel hose and replace if damage is noted.
4. Install hose clamps.

Inline Filter Canister

Refer to **Figure 29** for this procedure.
1. Remove the engine cover.
2. Unscrew the inline filter cover and remove the filter element.
3. Clean the filter element in solvent and blow dry with compressed air.
4. Blow any particles from the filter canister with compressed air.
5. Reinstall the filter element.
6. Install the filter cover.
7. Check filter canister for leakage by priming the fuel system with the fuel line primer bulb.

Fuel Pump Filter

1. Remove the bolt holding the filter cover to the fuel pump (**Figure 30**).
2. Remove the filter screen from the filter cover or pump housing (A, **Figure 31**).
3. Remove and discard the filter cover gasket (B, **Figure 31**).
4. Clean the screen in OMC Engine Cleaner. If screen is excessively dirty or plugged, discard it and install a new one.
5. Install the filter screen in the filter cover (A, **Figure 31**).
6. Reinstall the filter cover to the fuel pump with a new gasket and tighten the screw securely (**Figure 30**).
7. Check filter assembly for leakage by priming fuel system with fuel line primer bulb.

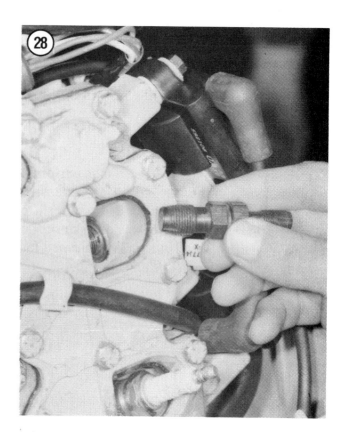

FUEL FILTER ASSEMBLY

1. Filter cover

LUBRICATION, MAINTENANCE AND TUNE-UP

Fuel Pump

The fuel pump does not generally require service during a tune-up. However, if the engine has 50 hours on it since the fuel pump was last serviced, it is a good idea to remove and disassemble the pump, inspect each part carefully for wear or damage and reassemble it with a new diaphragm and gaskets. See Chapter Six.

Fuel pump diaphragms are fragile and a defective one often produces symptoms that appear to be an ignition system problem. A common malfunction results from a tiny pinhole or crack in the diaphragm caused by an engine backfire. This defect allows gasoline to enter the crankcase and wet-foul the spark plug at idle speed, causing hard starting and engine stall at low rpm. The problem disappears at higher speeds, as fuel quantity is limited. Since the plug is not fouled by excess fuel at higher speeds, it fires normally.

Fuel Pump Pressure Test (50-75 hp)

Perform this test with the engine running in a test tank or on the boat in the water.

1. Momentarily loosen the fuel tank cap to release any pressure. Make sure fuel tank is not more than 24 in. below the fuel pump.
2. Tee a pressure gauge between the carburetor and fuel pump (**Figure 32**).
3. Connect a tachometer according to manufacturer's specifications.
4. Start the engine and note the pressure gauge. The pump pressure must be at least 1 psi at 600 rpm, 1.5 psi at 2,500-3,000 rpm and 2.5 psi at 4,500 rpm. If not, rebuild the fuel pump with a new diaphragm and gaskets. See Chapter Six.

Fuel Pump Pressure Test (V4 and V6)

Perform this test with the engine running in a test tank or on the boat in the water. Refer to **Figure 33** (V4) or **Figure 34** (V6) for this procedure.

1. Momentarily loosen the fuel tank cap to release any pressure. Make sure fuel tank is not more than 24 in. below the fuel pump.
2A. V4—Tee a pressure gauge between the carburetor and fuel pump. See **Figure 33**.

2B. V6—Tee a pressure gauge between the carburetor and lower fuel pump. See **Figure 34**.
3. Connect a tachometer according to manufacturer's specifications.
4. Remove the fuel pump filter cover(s) and make sure the filter(s) are clean.
5. Start the engine and run at full throttle for one minute to stabilize pump pressure. The pressure should be between 2½-6 psi.
6A. If the pressure is less than 2½ psi, stop the engine and pressurize the fuel system with the primer bulb. Hold the pressure and inspect the pump(s), lines and fittings for leaks. If there are no leaks, replace the fuel pump. On V6 engines, replace one pump and repeat the test. If the pressure is still too low, the other pump is the defective one.
6B. If the pressure is 2½-3½ psi with a portable fuel tank and the fuel pump filter is new or clean, the pump should be replaced, as it will cause problems as the filter screen gradually gets dirty.
6C. If the pressure is less than 2½ psi and the boat has a built-in fuel tank, connect the engine to a portable fuel tank and repeat the test. If the pressure increases when the engine is connected to the portable tank, there is a restriction somewhere in the built-in tank system.

Engine Synchronization and Linkage Adjustment

See Chapter Five.

Performance Test (On Boat)

Before performance testing the engine, make sure that the boat bottom is cleaned of all marine growth and that there is no evidence of a "hook" or "rocker" in the bottom. Any of these conditions will reduce performance considerably.

The boat should be performance tested with an average load and with the motor tilted at an angle that will allow the boat to ride on an even keel. If equipped with an adjustable trim tab, it should be properly adjusted to allow the boat to steer in either direction with equal ease.

Check engine rpm at full throttle. If not within the maximum rpm range for the motor as specified in Chapter Five, check the propeller pitch. A high pitch propeller will reduce rpm while a lower pitch prop will increase it.

Readjust the idle mixture and speed under actual operating conditions as required to obtain the best low-speed engine performance.

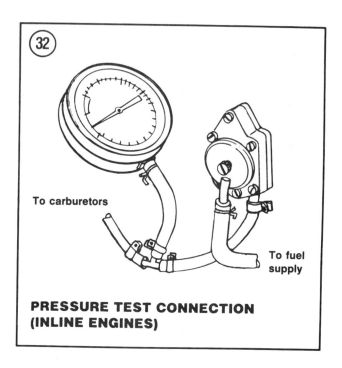

PRESSURE TEST CONNECTION (INLINE ENGINES)

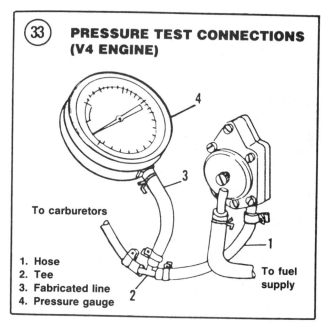

PRESSURE TEST CONNECTIONS (V4 ENGINE)

1. Hose
2. Tee
3. Fabricated line
4. Pressure gauge

LUBRICATION, MAINTENANCE AND TUNE-UP

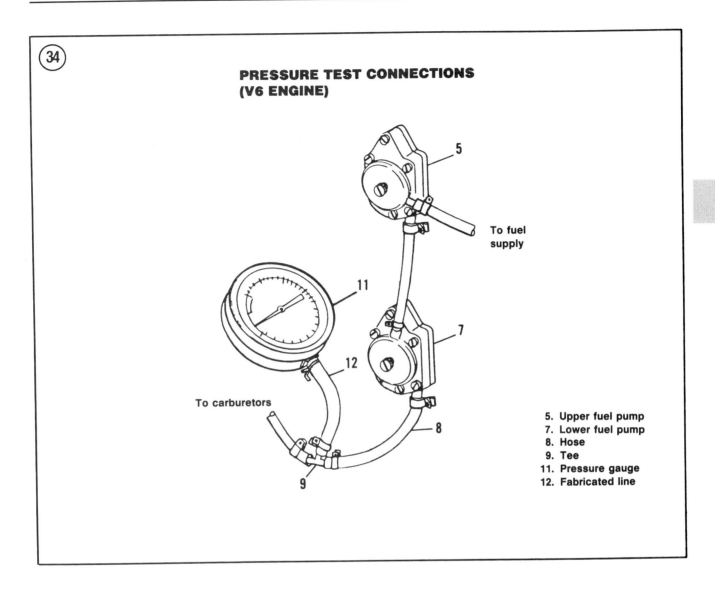

34 PRESSURE TEST CONNECTIONS (V6 ENGINE)

5. Upper fuel pump
7. Lower fuel pump
8. Hose
9. Tee
11. Pressure gauge
12. Fabricated line

Table 1 LUBRICATION AND MAINTENANCE[1]

Lubrication points	Figure no.
50 AND 60 HP	
Control shaft and control lever bushings	4
Shift shaft fittings and lever	8
Swivel bracket, tilt lock lever, tilt lever link and shaft	11
Tilt tube shaft	10
Starter pinion gear shaft	14[2]
Trailering lock	—
Carburetor and choke linkage	—
Engine cover latches	—
(continued)	

Table 1 LUBRICATION AND MAINTENANCE[1] (continued)

Lubrication points	Figure no.
70 AND 75 HP	
Shift, throttle, carburetor, choke linkages and springs; throttle cam, roller shaft, shift lever shaft and cover latch	6
Tilt tube shaft	10
Swivel bracket, tilt lock lever and reverse lock	15
Starter pinion gear shaft	14
Power trim/tilt reservoir	13[3]
Engine cover latches	—
V4 AND V6	
Control shaft bushings, bearings and shift linkage	8
Carburetor, cam and throttle linkage	7
Tilt tube shaft	10
Swivel bracket fitting, tilt/trim lock, tilt/run lever and reverse lock (manual models)	15
Swivel bracket fitting and trail lock (power trim/tilt models)	—
Starter pinion gear shaft	14[2]
Power trim/tilt reservoir	13

1. Lubricate with OMC Triple-Guard Grease every 60 days (fresh water) or 30 days (salt water) as required. Figures are representative of the models. Grease fitting locations may differ according to year and manual/power trim models.
2. Use Lubriplate 777.
3. Use OMC Power Trim and Tilt Fluid.

Table 2 TEST WHEEL RECOMMENDATIONS

Engine	Year	Test wheel	Engine rpm
48 hp	1987-1988	387635	4,900
48 hp	1989-1990	432968	5,200
50 hp	1973-1975	386950	5,300
50 hp	1977-1978	382861	4,900
50 hp	1980-1988 electric	387635	4,900
50 hp	1980-1988 manual	382861	5,200
50 hp	1989-1990	432968	5,200
55 hp	1976-1979	387635	5,300
55 hp	1980-1983	382861	5,200
60 hp	1980-1985	387635	5,600
60 hp	1986-1988	386665	5,000
60 hp	1989	386665	5,700
60 hp	1990	386665	5,000

(continued)

LUBRICATION, MAINTENANCE AND TUNE-UP

Table 2 TEST WHEEL RECOMMENDATIONS (continued)

Engine	Year	Test wheel	Engine rpm
70 hp	1974-1985	386665	5,000
70 hp	1986-on	386665	5,700
75 hp	1975-on short shaft	386950	5,200
75 hp	1975-on long shaft	386665	5,200
75 hp	1987 standard shaft	386950	5,200
85 hp	1973-1980	382861	4,800
88 hp	1987-1988	386246	4,800
88 hp	1989-1990	382861	4,800
90 hp	1981-1985	382861	4,800
90 hp	1986-1988	386246	4,800
90 hp	1989-1990	382861	4,800
100 hp	1979-1980	382861	4,800
110 hp	1986-1988	386246	4,800
110 hp	1989-1990	382861	4,800
115 hp	1973-1983	384933	4,800
115 hp	1984	386246	4,800
115 hp	1990	382861	4,800
120 hp	1985-on	386246	5,300
135 hp	1973-1976	386246	4,800
140 hp	1977-1985	386246	4,900
140 hp	1986-on	386246	5,500
150 hp	1979-1981	389931	4,800
150 hp	1982-1983	324890	4,800
150 hp	1984-1985	387388	4,800
150 hp	1986-on	387388	4,500
175 hp	1977-on	387388	4,800
200 hp	1976-1985	387388	4,800
200 hp	1986-1987	387388	5,300
200 hp	1988-1989	387388	5,500
225 hp	1986	387388	5,500
225 hp	1987-on	387388	5,700
235 hp	1978	387388	5,000
235 hp	1979-1985	387388	5,200

Table 3 RECOMMENDED SPARK PLUGS

Engine	Champion plug type	Gap (in.)
48 hp		
1987-1988	See note 1	See note 1
1989-1990	See note 2	See note 2
50 hp		
1973-1974	UL77V	—
1975-1984	L77J4	0.040
1985-1988	See note 1	See note 1
1989-1990	See note 2	See note 2
55 hp	L77J4	0.040
60 hp		
1980-1984	L77J4	0.040
1985-on	See note 1	See note 1
65 hp	UL74V	—

(continued)

Table 3 RECOMMENDED SPARK PLUGS (continued)

Engine	Champion plug type	Gap (in.)
70 & 75 hp		
1974	UL77V	—
1975-1984	L77J4	0.040
1985-on	See note 1	See note 1
85 hp		
1973-1974	UL77V	—
1975-1980	L77J4	0.040
88 hp	See note 3	See note 3
90 hp		
1981-1984	L77J4	0.040
1985-1986	See note 1	See note 1
1987-1990	See note 3	See note 3
100 hp		
1979-1980	L77J4	0.040
110 hp		
1986	See note 1	See note 1
1987-1990	See note 3	See note 3
115 hp		
1973-1974	UL77V	—
1975-1984	L77J4	0.040
1990	See note 3	See note 3
120 hp	See note 1	See note 1
135 hp	UL77V	—
140 hp		
1977-1984	UL77V	—
1985-on	See note 1	See note 1
150 hp		
1977-1986	QUL77V or UL77V	—
1987-1990	See note 3	See note 3
175 hp		
1977-1986	QUL77V or UL77V	—
1987-1990	See note 3	See note 3
185 hp	QUL77V or UL77V	—
200 hp		
1975-1986	QUL77V or UL77V	—
1987-1990	See note 3	See note 3
225 hp	See note 3	See note 3
235 hp	QUL77V or UL77V	—

1. OMC recommends the use of Champion QL77JC4 or L77JC4 plugs gapped to 0.040 in. for sustained low-speed operation. For sustained high-speed use, OMC recommends Champion QL78V or L78V (non-adjustable gap).
2. OMC recommends the use of Champion QL78C plugs gapped to 0.030 in. for sustained low-speed operation. For sustained high-speed use, OMC recommends Champion QL16V (non-adjustable gap).
3. OMC recommends the use of Champion QL77JC4 or L77JC4 plugs gapped to 0.040 in. on models prior to 1989 and 0.030 in. on 1989-1990 models for sustained low-speed operation. For sustained high-speed use, OMC recommends Champion QUL77V or UL77V (non-adjustable gap).

Table 4 SPARK PLUG CROSS-REFERENCE CHART

NGK	Champion	AC	Motorcraft	Autolite
BUHXW-1	UL77V	VB40FFM	AVZKO	2892
B9HS10	L77J4	M40FFX	AV901X-4	2634

Chapter Five

Engine Synchronization and Linkage Adjustments

If an engine is to deliver its maximum efficiency and peak performance, the ignition must be timed and the carburetor operation synchronized with the ignition. This procedure is the final step of a tune-up. It must also be performed whenever the fuel or ignition systems are serviced or adjusted.

Procedures for engine synchronization and linkage adjustment on Johnson and Evinrude outboards differ according to model and ignition system. This chapter is divided into self-contained sections dealing with particular models/ignition systems for fast and easy reference.

Each section specifies the appropriate procedure and sequence to be followed and provides the necessary tune-up data. Read the general information at the beginning of the chapter and then select the section pertaining to your outboard. **Table 1** is at the end of the chapter.

ENGINE TIMING

As engine rpm increases, the ignition system must fire the spark plugs more rapidly. Proper ignition timing synchronizes the spark plug firing with engine speed.

Ignition timing should not change during normal engine operation once it is correctly set. A timing light is required to check the timing. See **Figure 1**. The engine must be run at full throttle in forward gear. This requires the use of a test tank and test wheel, as checking engine timing while speeding across open water is neither easy nor safe.

SYNCHRONIZING

As engine speed increases, the carburetor must provide an increased amount of fuel for combustion. Synchronizing is the process of timing the carburetor operation to the ignition (and thereby the engine speed).

REQUIRED EQUIPMENT

Dynamic engine timing uses a stroboscopic timing light connected to the No. 1 spark plug wire. See **Figure 1**. As the engine is cranked or operated, the light flashes each time the spark plug fires. When the light is pointed at the

CHAPTER FIVE

moving flywheel, the mark on the flywheel appears to stand still. The flywheel mark should align with the stationary timing pointer on the engine.

A simple tool called a throttle shaft amplifier can be made with an alligator clip and a length of stiff wire (a paper clip will do). This tool will exaggerate the movement of the carburetor throttle shaft and tell you that it's moving. **Figure 2** shows the tool installed. The tool is especially useful on engines where the throttle cam and cam follower are partially hidden by the flywheel. To make the tool, enlarge the alligator clip's gripping surface by grinding out the front teeth on one side and secure the wire to the end of the clip. See **Figure 3**.

On late models where the procedure calls for aligning the throttle shaft(s) roll pin(s) in a vertical position, a toothpick inserted in the roll pin will serve the same purpose as the tool in **Figure 3**.

A tachometer connected to the engine is used to determine engine speed during idle and high-speed adjustments.

CAUTION
Never operate the engine without water circulating through the gearcase to the engine. This will damage the water pump and the gearcase and can cause engine damage.

Some form of water supply is required whenever the engine is operated during the procedure. Using a test tank is the most convenient method, although the procedures may be carried out with the boat in the water.

CAUTION
Do not use a flushing device to provide water during synchronization and linkage adjustment. Without the exhaust backpressure of a submerged gearcase, the engine will run lean. The proper test wheel must be used to put a load on the propeller shaft or engine damage can result from excessive rpm.

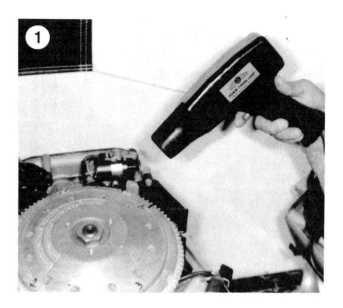

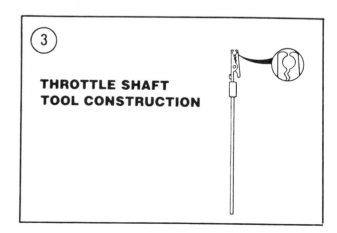

THROTTLE SHAFT TOOL CONSTRUCTION

ENGINE SYNCHRONIZATION AND LINKAGE ADJUSTMENTS

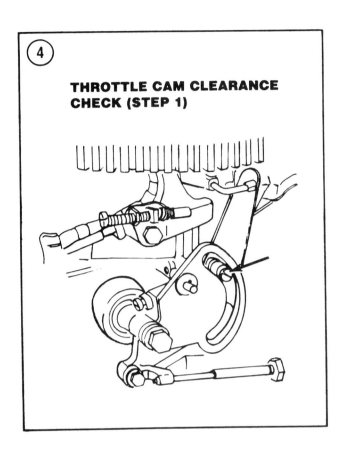

THROTTLE CAM CLEARANCE CHECK (STEP 1)

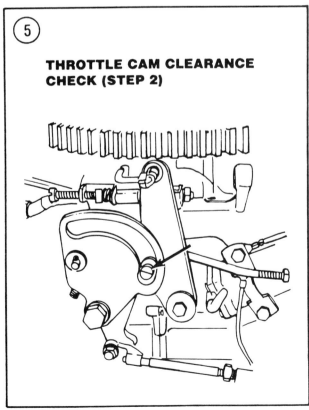

THROTTLE CAM CLEARANCE CHECK (STEP 2)

1973-1988 50 HP (TILLER MODELS)

Throttle Cable/Initial Throttle Cam Adjustment

1. Rotate the tiller handle idle speed adjustment knob counterclockwise to the minimum speed position.
2. Rotate twist grip to its full open position and check the throttle roller position in the throttle cam slot (**Figure 4**). It should be approximately 1/4 in. from the end of the slot.
3. Rotate the twist grip to its fully closed position and check the throttle roller position in the throttle cam slot (**Figure 5**). It should be approximately 1/4 in. from the end of the slot.
4. If the roller is not properly positioned in the cam slot in Step 2 or Step 3, loosen the throttle cable connector retaining screw or throttle cable jam nut (as equipped). Rotate the cable connector until the roller comes to rest about 1/4 in. from each end of the cam slot as the twist grip is opened and closed. Tighten the retaining screw or jam nut.

Throttle Valve Synchronization

1. Remove the engine cover.
2. Remove the air silencer cover.
3. Retard the throttle lever to a point where the throttle cam roller does not touch the cam.
4. Loosen the upper carburetor lever adjustment screw (1, **Figure 6**).
5. Rotate the throttle shaft partially open, then let it snap back to the closed position. Depress the adjusting link tab slightly to remove any backlash and tighten the adjustment screw.
6. Move the cam follower or lower throttle shaft while watching the throttle valves. If the throttle valves do not start to move at the same time, repeat Steps 3-5.

Throttle Cam Follower Pickup Point and Timing

1. Connect a throttle shaft amplifier tool (**Figure 2**) to the top carburetor throttle shaft.

2. Watching the amplifier tool, slowly rotate the throttle cam. As the end of the tool starts to move, check the cam and cam follower alignment. The lower embossed mark on the cam (2, **Figure 6**) should align with the center of the cam follower (3, **Figure 6**).

3. If the cam follower and cam mark do not align in Step 2, loosen the cam follower screw (**Figure 7**) and let the throttle spring close the throttle valves. Align the cam mark and follower and press on the cam follower lever (4, **Figure 6**) to maintain the alignment while tightening the screw.

4. Repeat Step 2 to check the adjustment. If incorrect, repeat Step 3, then repeat Step 2 as required.

5. When adjustment is correct, connect a timing light to the No. 1 cylinder according to manufacturer's instructions.

6. Slowly move the throttle lever until the tip of the amplifier tool starts to move. Remove the tool without disturbing the throttle lever position.

7. Start the engine and check the spark advance with the timing light, advancing the idle speed adjustment knob as required to keep the engine running. Refer to **Table 1** for proper pickup timing.

8. If the pickup timing is incorrect in Step 7, remove the throttle cam shoulder screw (1, **Figure 7**) or loosen the cam rod jam nut (1, **Figure 8**).

   a. Shoulder screw—Rotate the cam on the throttle lever link (2, **Figure 7**) clockwise (to advance) OR counterclockwise (to retard) as required. One full turn of the cam will change timing approximately 2 degrees. Install the shoulder screw.

   b. Jam nut—Turn the top of the thumbwheel (2, **Figure 8**) toward the engine (advance) or away from the engine (retard) as required to bring pickup timing within specifications. Tighten the jam nut.

9. Repeat Step 6 and Step 7 to check the adjustment. If incorrect, repeat Step 8, then repeat Step 6 and Step 7 as required.

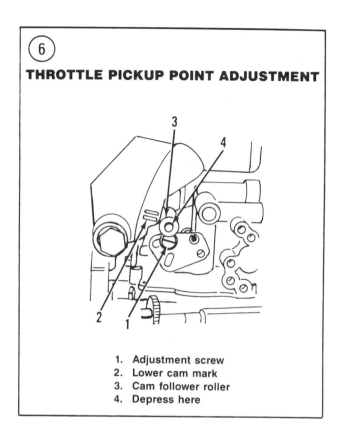

THROTTLE PICKUP POINT ADJUSTMENT

1. Adjustment screw
2. Lower cam mark
3. Cam follower roller
4. Depress here

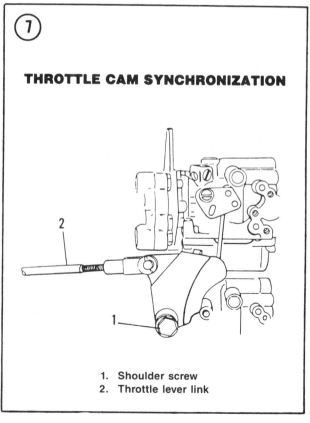

THROTTLE CAM SYNCHRONIZATION

1. Shoulder screw
2. Throttle lever link

ENGINE SYNCHRONIZATION AND LINKAGE ADJUSTMENTS

Maximum Spark Advance Adjustment

CAUTION
This procedure should be performed in a test tank with the proper test wheel installed on the engine. The use of the propeller and/or a flushing device can result in an incorrect setting and possible engine damage.

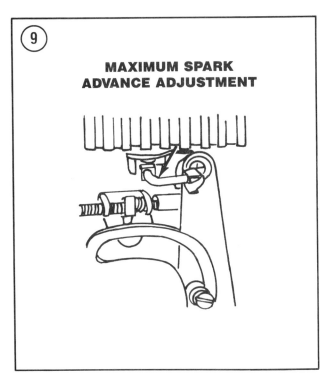

MAXIMUM SPARK ADVANCE ADJUSTMENT

1. Connect a timing light to the No. 1 cylinder according to manufacturer's instructions.
2. Connect a tachometer according to manufacturer's instructions.
3. Start the engine and run in forward gear at a minimum of 3,500 rpm (timer base fully advanced).
4. Check the timing mark position with the timing light. The specified mark (**Table 1**) on the flywheel grid should align with the engine mark.

NOTE
*Compare the specification in **Table 1** with the "Maximum Advance" specification on the engine decal. Use the decal specification if it differs.*

5. If the timing marks do not align as specified, shut the engine off. Remove the spark advance rod (**Figure 9**). Shorten the rod by bending its ends together to advance timing; lengthen it by expanding the ends to retard timing.
6. Reinstall the spark advance rod if adjustment was necessary and repeat Step 3 and Step 4. If timing mark alignment is still incorrect, repeat Step 5, then Step 3 and Step 4 as required.

Wide-open Throttle Stop Adjustment

1. Move the throttle control to the full throttle position.
2A. If the throttle shafts use roll pins:
   a. Use a toothpick as described in this chapter to check that each roll pin is exactly vertical.
   b. If adjustment is necessary, loosen the wide-open throttle stop screw locknut. Turn the stop screw until the roll pins are vertical, then tighten the locknut.
2B. If the throttle shafts do not use roll pins:
   a. Lightly depress the throttle cam roller (**Figure 10**) and check the clearance between the cam roller and cam. It should be 0.02-0.06 in.

b. If clearance is incorrect, loosen the throttle cable connector retaining screw. Rotate the cable connector until the correct clearance is obtained, then tighten the retaining screw.

Shift Lever Detent Adjustment

1. Place the shift lever in its NEUTRAL position.
2. If the lower detent spring is not completely engaged in the shift lever detent notch (**Figure 11**), loosen the detent spring screw. Move the spring until it fully engages the notch, then tighten the screw snugly.

Idle Speed Adjustment
1986

This procedure should be performed with the boat floating unrestrained in the water and the correct propeller installed.
1. Remove the engine cover.
2. Connect a tachometer according to manufacturer's instructions.
3. Start the engine and warm to normal operating temperature.
4. Shift the engine into FORWARD gear and note the idle speed on the tachometer. It should be 700-750 rpm.
5. If idle speed requires adjustment, shut the engine off as a safety precaution and rotate the idle speed screw (**Figure 12**) clockwise (to increase) or counterclockwise (to decrease) as required to being idle speed within specifications.
6. Start the engine and recheck idle speed. If not within specifications, repeat Step 5 as required.
7. When idle speed is correct, shut the engine off. Remove the tachometer and install the engine cover.

All others

This procedure should be performed with the boat floating in the water and tied securely at the slip or dock to prevent any fore or aft motion.
1. Remove the engine cover.
2. Connect a tachometer according to manufacturer's instructions.
3. Start the engine and warm to normal operating temperature.
4. Shift the engine into FORWARD gear and note the idle speed on the tachometer. It should be 750 rpm.
5. If idle speed requires adjustment, face the steering handle and rotate the idle speed

ENGINE SYNCHRONIZATION AND LINKAGE ADJUSTMENTS

adjustment knob clockwise (to increase) or counterclockwise (to decrease) as required to bring idle speed within specifications.

6. When idle speed adjustment is completed, shut the engine off. Remove the tachometer and install the engine cover.

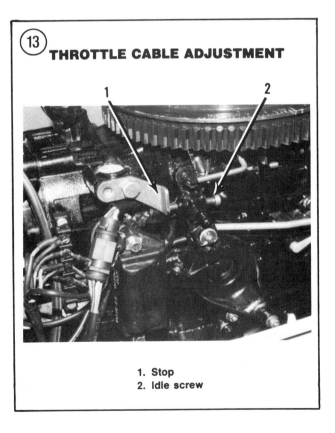

1973-1988 48, 50, 55 AND 60 HP (2-CYLINDER REMOTE CONTROL MODELS)

Throttle Cable Adjustment

The throttle cable must be adjusted to allow the throttle lever to return to the idle stop. A cable that is too loose will cause a high and unstable idle, resulting in shifting difficulties. If too tight, shifting will feel stiff and the warm-up lever will move up during a shift into NEUTRAL.

1. Slowly move the control lever back until the idle stop screw contacts its stop (**Figure 13**).
2. If the idle stop screw does not contact the stop, adjust the trunnion nut (**Figure 14**) as required.

Throttle Valve Synchronization

1. Remove the engine cover.
2. Remove the air silencer cover.
3. Retard the throttle lever to a point where the throttle cam roller does not touch the cam.

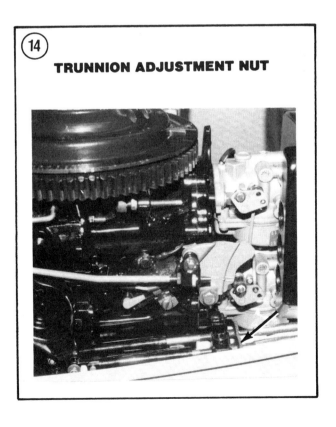

4. Loosen the upper carburetor lever adjustment screw (**Figure 15**).

5. Rotate the throttle shaft partially open, then let it snap back to the closed position. Apply finger pressure to the adjusting lever (**Figure 15**) and tighten the adjustment screw.

6. Move the cam follower while watching the throttle valves. If the throttle valves do not start to move at the same time, repeat Steps 3-5.

Throttle Cam Follower Pickup Point and Timing

1. Connect a throttle shaft amplifier tool (**Figure 2**) to the top carburetor throttle shaft.

2. Watching the amplifier tool, slowly rotate the throttle cam. As the end of the tool starts to move, check the cam and cam follower alignment. The lower embossed mark on the cam (2, **Figure 16**) should align with the center of the cam follower (3, **Figure 16**).

3. If the cam follower and cam mark do not align in Step 2, loosen the throttle arm screw (1, **Figure 16**) and let the throttle spring close the throttle valves. Align the cam mark and follower and press on the cam follower lever (4, **Figure 16**) to maintain the alignment while tightening the screw.

4. Repeat Step 2 to check the adjustment. If incorrect, repeat Step 3, then repeat Step 2 as required.

5. When adjustment is correct, connect a timing light to the No. 1 cylinder according to manufacturer's instructions.

6. Slowly move the throttle lever until the tip of the amplifier tool starts to move. Remove the tool without disturbing the throttle lever position.

7. Start the engine and check the spark advance with the timing light, advancing the idle speed adjustment knob or throttle control lever (as equipped) as required to keep the engine running. Refer to **Table 1** for proper pickup timing.

8. If the pickup timing is incorrect in Step 7, loosen the throttle arm screw (1, **Figure 16**) or

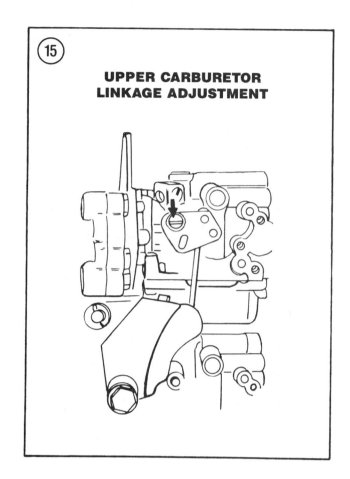

UPPER CARBURETOR LINKAGE ADJUSTMENT

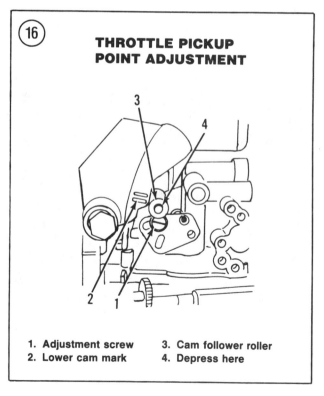

THROTTLE PICKUP POINT ADJUSTMENT

1. Adjustment screw
2. Lower cam mark
3. Cam follower roller
4. Depress here

ENGINE SYNCHRONIZATION AND LINKAGE ADJUSTMENTS

SOLENOID ADJUSTMENT

1. Solenoid body 2. Clamp screws

⑰

⑱ **FULL THROTTLE STOP ADJUSTMENT SCREW**

the control lever jam nut (1, **Figure 8**), as equipped.
   a. Shoulder screw—Align the throttle cam mark with center of the cam follower. Move the carburetor throttle arm until the throttle valves are closed. Depress the cam follower lever (4, **Figure 16**) to maintain the alignment and tighten the throttle arm screw.
   b. Jam nut—Turn the top of the thumbwheel (2, **Figure 8**) toward the engine (to advance) or away from the engine (to retard) as required to bring pickup timing within specifications.
9. Repeat Step 6 and Step 7 to check the adjustment. If incorrect, repeat Step 8, then repeat Step 6 and Step 7 as required.

Choke Valve Synchronization (Models with Choke Solenoid)

1. Loosen the choke link fastener screw.
2. Close both chokes with finger pressure on the upper and lower choke shaft levers and tighten the choke link screw.
3. Make sure the manual choke lever is in the OFF position.
4. Loosen the choke solenoid bracket screws (**Figure 17**). Move the solenoid upward until the terminal block rests against the solenoid clamp. Tighten the bracket screws.

Wide-open Throttle Stop Adjustment

1. Move the throttle control to the full throttle position and hold the throttle linkage in that position.
2A. If the throttle shafts use roll pins:
   a. Use a toothpick as described in this chapter to check that each roll pin is exactly vertical.
   b. If adjustment is necessary, adjust the wide-open throttle stop screw (**Figure 18**) until the roll pins are vertical.

2B. If the throttle shafts do not use roll pins:
  a. Measure the clearance between the cam roller and cam. It should be 0.000-0.020 in.
  b. If clearance exceeds 0.020 in., adjust the full throttle stop screw (**Figure 18**) as required to bring the clearance within specifications.

Maximum Spark Advance Adjustment

CAUTION
This procedure should be performed in a test tank with the proper test wheel installed on the engine. The use of the propeller and/or a flushing device can result in an incorrect setting and possible engine damage.

1. Connect a timing light to the No. 1 cylinder according to manufacturer's instructions.
2. Connect a tachometer according to manufacturer's instructions.
3. Start the engine and run in forward gear at a minimum of 3,500 rpm (timer base fully advanced).
4. Check the timing mark position with the timing light. The specified mark (**Table 1**) on the flywheel grid should align with the engine mark.

NOTE
Compare the specification in Table 1 with the "Maximum Advance" specification on the engine decal. Use the decal specification if it differs.

WARNING
Do not attempt to adjust the spark advance with the engine running in Step 5. The adjustment screw is located close to the moving flywheel and vibrates slightly when the engine is running. If the screwdriver slips out of the screw, serious personal injury can result.

5. If the timing marks do not align as specified, shut the engine off. Loosen the adjusting screw locknut (1, **Figure 19**). Adjust the screw clockwise (to retard) or counterclockwise (to

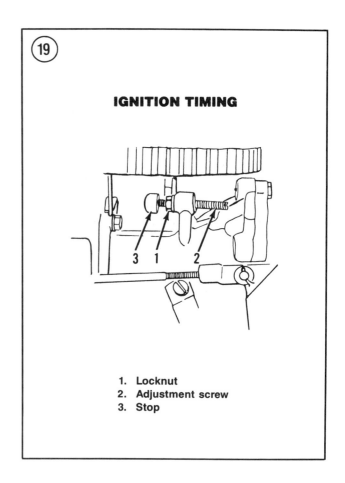

IGNITION TIMING

1. Locknut
2. Adjustment screw
3. Stop

ENGINE SYNCHRONIZATION AND LINKAGE ADJUSTMENTS

advance) as required, then tighten the locknut. One full turn in either direction changes timing approximately one degree.

6. Restart the engine and repeat Step 4. If timing mark alignment is still incorrect, repeat Step 5, then Step 3 and Step 4 as required.

7. When the timing marks align as specified, shut the engine off.

Idle Speed Adjustment
1986-on

This procedure should be performed with the boat floating unrestrained in the water and the correct propeller installed.

1. Remove the engine cover.
2. Connect a tachometer according to manufacturer's instructions.
3. Start the engine and warm to normal operating temperature.
4. Shift the engine into FORWARD gear and note the idle speed on the tachometer. It should be 700-750 rpm.
5. If idle speed requires adjustment, shut the engine off as a safety precaution and rotate the idle speed screw (**Figure 20**) clockwise (to increase) or counterclockwise (to decrease) as required to being idle speed within specifications.
6. Start the engine and recheck idle speed. If not within specifications, repeat Step 5 as required.
7. When idle speed is correct, shut the engine off. Remove the tachometer and install the engine cover.

All others

This procedure should be performed with the boat floating in the water and tied securely at the slip or dock to prevent any fore or aft motion.

1. Remove the engine cover.
2. Connect a tachometer according to manufacturer's instructions.
3. Start the engine and warm to normal operating temperature.
4. Shift the engine into REVERSE gear and note the idle speed on the tachometer. It should be 700 rpm.
5. If idle speed requires adjustment, adjust the idle speed screw (**Figure 21**) to set the idle speed to specifications.
6. When idle speed adjustment is completed, shut the engine off. Remove the tachometer and install the engine cover.

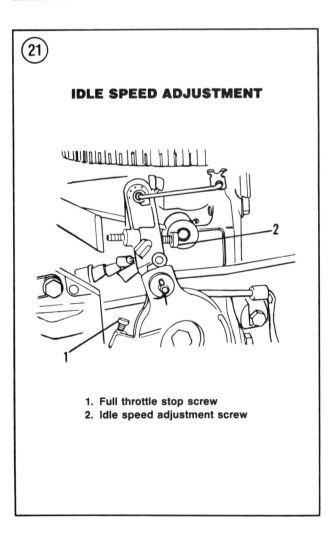

IDLE SPEED ADJUSTMENT

1. Full throttle stop screw
2. Idle speed adjustment screw

1989-1990 48 AND 50 HP MODELS

Throttle Valve Synchronization

1. Remove the engine cover.
2. Retard the throttle lever to a point where the throttle cam roller does not touch the cam.

3. Loosen the upper carburetor lever adjustment screw (1, **Figure 22**).
4. Rotate the throttle shaft partially open, then let it snap back to the closed position. Apply finger pressure to the adjusting lever (1, **Figure 22**) and tighten the adjustment screw.
5. Move the cam follower while watching the throttle valves. If the throttle valves do not start to move at the same time, repeat Steps 3-5.

Throttle Cam Follower Pickup Point

1. Loosen cam follower roller adjustment screw (2, **Figure 22**).
2. Slowly rotate throttle cam (10, **Figure 22**) toward cam follower roller (7, **Figure 22**).
3. When pickup mark (3, **Figure 22**) on throttle cam intersects center of cam follower roller, the throttle cam and cam follower roller should just touch. Correctly adjust components, then tighten cam follower roller adjustment screw (2, **Figure 22**) to retain adjustment.
4. Move idle speed screw (4, **Figure 22**) against stop mounted on power head.
5. Spark lever cam follower (5, **Figure 22**) should be positioned between marks (6, **Figure 22**). If not, loosen locknut and adjust idle speed screw.
6. With linkage positioned as stated in Step 5, a 0.010 in. gap (8, **Figure 22**) between throttle cam and cam follower roller should be noted.
7. If gap is incorrect, pry control rod connector (9, **Figure 22**) from throttle cam ball socket and rotate connector as required to obtain 0.010 in. gap.

Wide-open Throttle Stop Adjustment

1. Move the throttle control to the full throttle position and hold the throttle linkage in that position.
2. Note the position of the pin installed in each carburetor throttle shaft. The pins must be vertical.

3. If the pins are not correctly located in Step 2, loosen locknut (1, **Figure 23**) and rotate wide-open throttle stop screw (2, **Figure 23**) until the roll pins are vertical.

Maximum Spark Advance Adjustment

CAUTION
This procedure should be performed in a test tank with the proper test wheel installed on the engine. The use of the propeller and/or a flushing device can result in an incorrect setting and possible engine damage.

1. Connect a timing light to the No. 1 cylinder according to manufacturer's instructions.

1. Adjustment screw
2. Adjustment screw
3. Pickup mark
4. Idle speed screw
5. Spark lever cam follower
6. Alignment marks
7. Cam follower roller
8. Gap (0.010 in.)
9. Control rod connector
10. Throttle cam

ENGINE SYNCHRONIZATION AND LINKAGE ADJUSTMENTS

2. Connect a tachometer according to manufacturer's instructions.
3. Start the engine and run in forward gear at a minimum of 5,000 rpm.
4. Check the timing mark position with the timing light. The timing mark must align with the 19 degree ±1 degree mark on the flywheel grid.

WARNING
Do not attempt to adjust the spark advance with the engine running in Step 5. Attempting to do so could result in serious personal injury.

5. If the timing marks do not align as specified, shut the engine off. Pry connector end of maximum spark advance control rod (3, **Figure 23**) from spark lever ball socket, then rotate connector end clockwise (to advance) or counterclockwise (to retard) as required. Reattach connector end to ball socket. Two full turns in either direction changes timing approximately one degree.
6. Restart the engine and repeat Step 4. If timing mark alignment is still incorrect, repeat Step 5, then Step 3 and Step 4 as required.
7. When the timing marks align as specified, shut the engine off.

Idle Speed Adjustment

This procedure should be performed with the boat floating unrestrained in the water and the correct propeller installed.
1. Remove the engine cover.
2. Connect a tachometer according to manufacturer's instructions.
3. Start the engine and warm to normal operating temperature.
4. Shift the engine into FORWARD gear and note the idle speed on the tachometer. It should be 725-775 rpm.
5. If idle speed requires adjustment, shut the engine off as a safety precaution. Loosen locknut and rotate the idle speed screw (4, **Figure 22**) clockwise (to increase) or counterclockwise (to decrease) as required to bring idle speed within specifications.
6. Start the engine and recheck idle speed. If not within specifications, repeat Step 5 as required.
7. When idle speed is correct, shut the engine off. Remove the tachometer.

NOTE
If idle speed is adjusted, cam follower pickup point must be rechecked and adjusted, if needed, as outlined in this chapter. With idle speed and cam follower pickup point correctly adjusted, idle timing should be 3 degrees ±2 degrees ATDC with timing light properly connected to No. 1 cylinder.

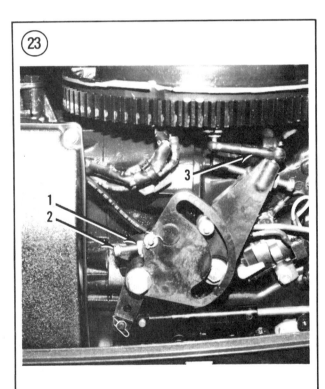

1. Locknut
2. Wide-open throttle stop screw
3. Maximum spark advance control rod

60, 65, 70 AND 75 HP (3-CYLINDER MODELS)

Timing Pointer Alignment

1. Disconnect the spark plug leads. Remove the spark plugs.
2. Install piston stop tool No. 384887 in the No. 1 spark plug hole (**Figure 24**).
3. Rotate the flywheel clockwise until the TDC mark is approximately 1½ in. beyond the pointer. Depress the piston stop plunger until it touches the piston and then tighten the locknut.
4. Mark the flywheel rim directly under the point. This is point "A," **Figure 25**.
5. Continue rotating the flywheel clockwise until the piston contacts the piston stop tool again. Mark the flywheel again directly under the pointer. This is point "B," **Figure 25**.
6. Remove the piston stop tool and measure the distance between the marks made in Step 4 and Step 5. The mid-point between the 2 marks (point "C," **Figure 25**) should fall directly on the TDC mark cast into the flywheel if timing pointer alignment is correct.
7. If the mid-point does not fall on the flywheel TDC mark, the pointer is out of alignment. Rotate the flywheel to align the mid-point with the timing pointer. Hold the flywheel in this position and loosen the pointer screws. Move the pointer to align with the TDC mark cast into the flywheel and tighten the pointer screws.

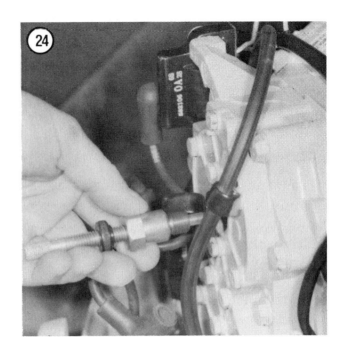

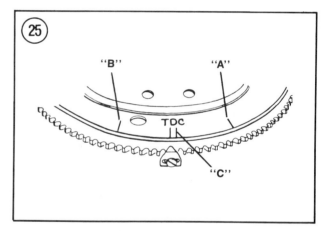

Throttle Cable Adjustment

The throttle cable must be adjusted to allow the throttle lever to return to the idle stop. A cable that is too loose will cause a high and unstable idle, resulting in shifting difficulties. If too tight, shifting will feel stiff and the warm-up lever will move up during a shift into NEUTRAL.

1. Slowly move the control lever back until the throttle lever contacts the idle stop screw (A, **Figure 26**).

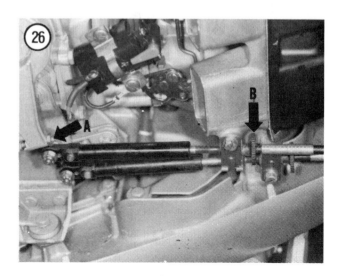

ENGINE SYNCHRONIZATION AND LINKAGE ADJUSTMENTS

2. If the throttle lever does not contact the stop screw, adjust the trunnion nut (B, **Figure 26**) as required.

Throttle Valve Synchronization

1. Remove the engine cover.
2. Remove the air silencer cover.
3. Retard the throttle lever to a point where the throttle cam roller does not touch the cam.
4. Loosen the upper and lower carburetor lever adjustment screws (A and B, **Figure 27**). The throttle return spring will close the throttle valves.
5. Rotate the throttle shaft partially open, then let it snap back to the closed position. Apply upward finger pressure on the upper adjusting link tab to remove any backlash and tighten the adjustment screw, then repeat to adjust the lower adjusting link.
6. Move the cam follower while watching the throttle valves. If the throttle valves do not start to move at the same time, repeat Steps 3-5.

Throttle Cam Follower Pickup Point and Timing

1. Connect a throttle shaft amplifier tool (**Figure 2**) to the top carburetor throttle shaft.
2. Watching the amplifier tool, slowly rotate the throttle cam. As the end of the tool starts to move, check the cam and cam follower alignment. The embossed mark (65 hp) or lower embossed mark (all others) on the cam should align with the center of the cam follower. See **Figure 28** (typical).

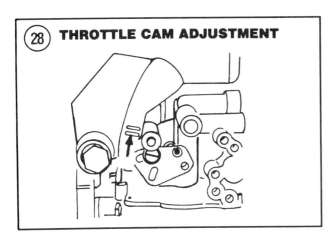

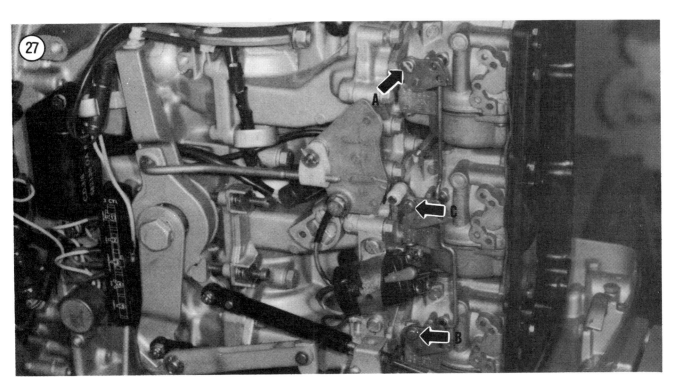

3. If the cam follower and cam mark do not align in Step 2, manually align the throttle cam mark with the center of the cam follower. Loosen the center carburetor throttle arm screw (C, **Figure 27**) and let the throttle shaft return spring close the throttle valves. Tighten the throttle arm screw.

4. Repeat Step 2 to check the adjustment. If incorrect, repeat Step 3, then repeat Step 2 as required.

5. When adjustment is correct, connect a timing light to the No. 1 cylinder according to manufacturer's instructions.

6. Slowly move the throttle lever until the tip of the amplifier tool starts to move. Remove the tool without disturbing the throttle lever position.

7. Start the engine and check the spark advance with the timing light, advancing the idle speed screw as required to keep the engine running. Refer to **Table 1** for proper pickup timing.

8. If the pickup timing is incorrect in Step 7, disconnect the throttle cam yoke rod (C, **Figure 29**) from the throttle control lever or loosen the control lever jam nut (1, **Figure 8**), as equipped.
   a. Throttle cam yoke rod—Align the throttle cam mark with center of the cam follower. Move the carburetor throttle arm until the throttle valves are closed. Rotate the cam yoke rod as required to maintain the alignment and reattach the rod to the throttle control lever.
   b. Jam nut—Turn the top of the thumbwheel (2, **Figure 8**) toward the engine (to advance) or away from the engine (to retrd) as required to bring pickup timing within specifications.

9. Repeat Step 6 and Step 7 to check the adjustment. If incorrect, repeat Step 8, then repeat Step 6 and Step 7 as required.

FULL SPARK ADJUSTMENT

1. Adjustment screw
2. Locknut

Choke Valve Synchronization (Models with Choke Solenoid)

1. Make sure the manual choke lever is in the OFF position.

ENGINE SYNCHRONIZATION AND LINKAGE ADJUSTMENTS

2. Loosen the choke solenoid bracket screws (A, **Figure 29**). Move the solenoid back until it touches the engine boss, then tighten the bracket screws.

3. Loosen the upper and lower choke lever retaining screws. See B, **Figure 29**.

4. Close both chokes with finger pressure on the upper and lower choke shaft levers and tighten the retaining screws.

Maximum Spark Advance Adjustment

CAUTION
This procedure should be performed in a test tank with the proper test wheel installed on the engine. The use of the propeller and/or a flushing device can result in an incorrect setting and possible engine damage.

1. Connect a timing light to the No. 1 cylinder according to manufacturer's instructions.

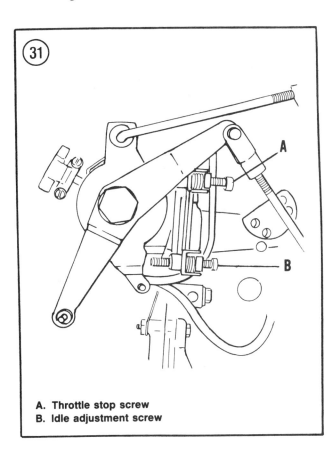

A. Throttle stop screw
B. Idle adjustment screw

2. Connect a tachometer according to manufacturer's instructions.

3. Start the engine and run in forward gear as follows:
   a. 60 hp—5,000-5,400 rpm.
   b. 65 hp—4,300-4,600 rpm.
   c. 1974-1983 70 hp—4,600-5,000 rpm.
   d. 1975-1983 75 hp—4,800-5,200 rpm.
   e. 1984-on 70 and 75 hp—5,000-5,400 rpm.

4. Check the timing mark position with the timing light. The specified mark (**Table 1**) on the flywheel grid should align with the engine mark.

NOTE
*Compare the specification in **Table 1** with the "Maximum Advance" specification on the engine decal. Use the decal specification if it differs.*

WARNING
Do not attempt to adjust the spark advance with the engine running in Step 5. The adjustment screw on many engines is located close to the moving flywheel and vibrates slightly when the engine is running. If the screwdriver slips out of the screw, serious personal injury can result.

5. If the timing marks do not align as specified, shut the engine off. Loosen the adjusting screw locknut. **Figure 30** shows the adjusting screw and locknut location with the flywheel removed for clarity. Adjust the screw clockwise (to retard) or counterclockwise (to advance) as required, then tighten the locknut. One full turn in either direction changes timing approximately one degree.

6. Restart the engine and repeat Step 4. If timing mark alignment is still incorrect, repeat Step 5, then Step 3 and Step 4 as required.

7. When the timing marks align as specified, open the throttle to the full throttle position and hold it there.

8. Loosen the throttle stop screw locknut and adjust the stop screw (A, **Figure 31**) to allow the

throttle valves to open fully without loading the throttle shaft. Tighten the locknut.

Idle Speed Adjustment

This procedure should be performed with the boat floating unrestrained in the water and the correct propeller installed.

1. Remove the engine cover.
2. Connect a tachometer according to manufacturer's instructions.
3. Start the engine and warm to normal operating temperature.
4A. 65 hp—With the engine running, shift into REVERSE and adjust the idle screw (B, **Figure 31**) to set the idle speed at 650 rpm.
4B. All others—With the engine running, shift into FORWARD and adjust the idle screw (B, **Figure 31**) to set the idle speed at 700-750 rpm.
5. Shut the engine off. Remove the timing light and tachometer. Install the engine cover.

65 HP (3-CYLINDER TILLER MODELS)

Timing Pointer Alignment

See *60, 65, 70 and 75 hp Timing Pointer Alignment* in this chapter.

Throttle Cable/Initial Throttle Cam Adjustment

1. Rotate the tiller handle idle speed adjustment knob counterclockwise to the minimum speed position.
2. Rotate twist grip to its full open position and check the throttle roller position in the throttle cam slot (**Figure 32**). It should be approximately 1/4 in. from the end of the slot.
3. Rotate the twist grip to its fully closed position and check the throttle roller position in the throttle cam slot (**Figure 33**). It should be approximately 1/4 in. from the end of the slot.
4. If the roller is not properly positioned in the cam slot in Step 2 or Step 3, loosen the throttle cable connector retaining screw or throttle cable

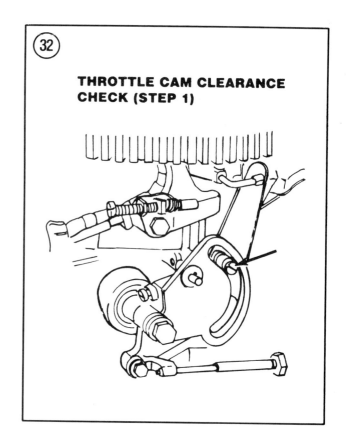

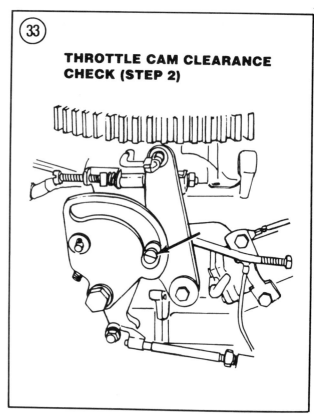

ENGINE SYNCHRONIZATION AND LINKAGE ADJUSTMENTS

jam nut (as equipped). Rotate the cable connector until the roller comes to rest about ¼ in. from each end of the cam slot as the twist grip is opened and closed. Tighten the retaining screw or jam nut.

Throttle Valve Synchronization

1. Remove the engine cover.
2. Remove the air silencer cover.
3. Retard the throttle lever to a point where the throttle cam roller does not touch the cam.
4. Loosen the upper and lower carburetor lever adjustment screws (**Figure 34**).
5. Rotate the throttle shaft partially open, then let it snap back to the closed position. Apply upward finger pressure on the upper adjusting link tab to remove any backlash and tighten the adjustment screw, then repeat to adjust the lower adjusting link.
6. Move the cam follower while watching the throttle valves. If the throttle valves do not start to move at the same time, repeat Steps 3-5.

Throttle Cam Follower Pickup Point and Timing

1. Connect a throttle shaft amplifier tool (**Figure 2**) to the top carburetor throttle shaft. See A, **Figure 35**.
2. Watching the amplifier tool, slowly rotate the throttle cam. As the end of the tool starts to move, check the cam and cam follower alignment. The lower embossed mark on the cam (B, **Figure 35**) should align with the center of the cam follower.
3. If the cam follower and cam mark do not align in Step 2, loosen the cam follower screw (below the roller) and let the throttle spring close the throttle valves. Align the cam mark and follower and press on the cam follower lever to maintain the alignment while tightening the screw.
4. Repeat Step 2 to check the adjustment. If incorrect, repeat Step 3, then repeat Step 2 as required.

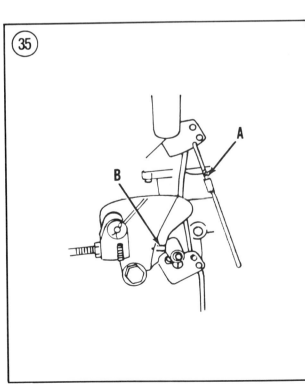

5. When adjustment is correct, connect a timing light to the No. 1 cylinder according to manufacturer's instructions.

6. Slowly move the throttle lever until the tip of the amplifier tool starts to move. Remove the tool without disturbing the throttle lever position.

7. Start the engine and check the spark advance with the timing light, advancing the idle speed adjustment knob as required to keep the engine running. Refer to **Table 1** for proper pickup timing.

8. If the pickup timing is incorrect in Step 7, loosen the cam rod jam nut and turn the thumbwheel (**Figure 36**) toward the engine (advance) or away from the engine (retard) as required to bring pickup timing within specifications.

9. Repeat Step 6 and Step 7 to check the adjustment. If incorrect, repeat Step 8, then repeat Step 6 and Step 7 as required. Tighten the cam rod jam nut.

Maximum Spark Advance Adjustment

CAUTION
This procedure should be performed in a test tank with the proper test wheel installed on the engine. The use of the propeller and/or a flushing device can result in an incorrect setting and possible engine damage.

1. Connect a timing light to the No. 1 cylinder according to manufacturer's instructions.

2. Connect a tachometer according to manufacturer's instructions.

3. Start the engine and run in forward gear at a minimum of 5,000 rpm (timer base fully advanced).

4. Check the timing mark position with the timing light. The specified mark (**Table 1**) on the flywheel grid should align with the engine mark.

NOTE
*Compare the specification in **Table 1** with the "Maximum Advance" specification on the engine decal. Use the decal specification if it differs.*

5. If the timing marks do not align as specified, shut the engine off. Remove the spark advance rod (**Figure 37**). Shorten the rod by bending its

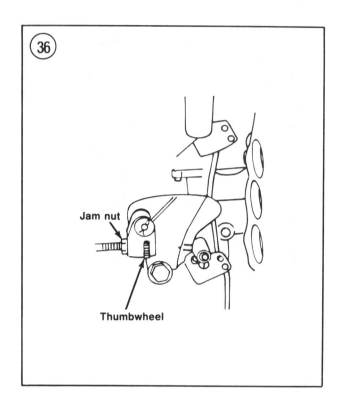

ENGINE SYNCHRONIZATION AND LINKAGE ADJUSTMENTS

ends together to advance timing; lengthen it by expanding the ends to retard timing.

6. Reinstall the spark advance rod if adjustment was necessary and repeat Step 3 and Step 4. If timing mark alignment is still incorrect, repeat Step 5, then Step 3 and Step 4 as required.

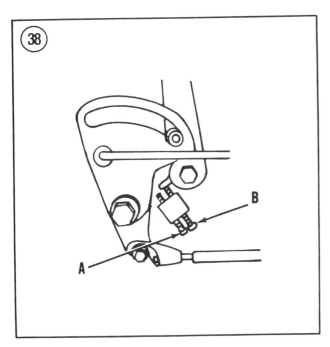

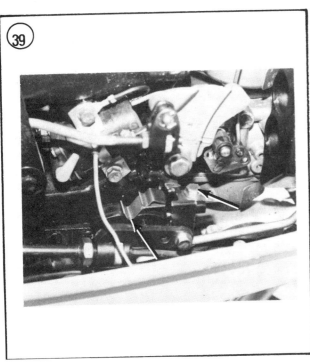

Wide-open Throttle Stop Adjustment

1. Move the throttle control to the full throttle position.
2. Use a toothpick as described in this chapter to check that each roll pin is exactly vertical.
3. If adjustment is necessary, turn the wide-open throttle stop screw (A, **Figure 38**) until the roll pins are vertical.

Shift Lever Detent Adjustment

1. Place the shift lever in its NEUTRAL position.
2. If the lower detent spring is not completely engaged in the shift lever detent notch (**Figure 39**), loosen the detent spring screw. Move the spring until it fully engages the notch, then tighten the screw snugly.

Idle Speed Adjustment

This procedure should be performed with the boat floating unrestrained in the water and the correct propeller installed.

1. Remove the engine cover.
2. Connect a tachometer according to manufacturer's instructions.
3. Start the engine and warm to normal operating temperature.
4. Shift the engine into FORWARD gear and note the idle speed on the tachometer. It should be 700-750 rpm.
5. If idle speed requires adjustment, shut the engine off as a safety precaution and rotate the idle speed screw (B, **Figure 38**) clockwise (to increase) or counterclockwise (to decrease) as required to being idle speed within specifications.
6. Start the engine and recheck idle speed. If not within specifications, repeat Step 5 as required.
7. When idle speed is correct, shut the engine off. Remove the tachometer and install the engine cover.

V4 ENGINES
(EXCEPT 1985-ON 120-140 HP AND 1986 ON 88-115 HP)

Timing Pointer Alignment

See *60, 65, 70 and 75 hp Timing Pointer Alignment* in this chapter.

Throttle Valve Synchronization

1. Remove the engine cover.
2. Remove the air silencer cover.
3. Remove the starboard air silencer extension.
4. Retard the throttle lever to a point where the throttle cam roller does not touch the cam.
5. Loosen the lower carburetor lever adjustment screw (**Figure 40**). The throttle shaft return spring should close the throttle valve.

THROTTLE AND CHOKE LINKAGE ADJUSTMENT

1. Upper carburetor throttle arm
2. Roller must not touch
3. Throttle link
4. Lower carburetor throttle arm
5. Choke links and connector

CHOKE VALVE SYNCHRONIZATION

1. Choke lever
2. Depress here
3. Choke link screw
4. Choke link

CHOKE SOLENOID INSTALLATION (1973-1978 AIR INTAKE MOUNTING)

1. Spring
3. Plunger
4. Yoke
5. Ground wire

ENGINE SYNCHRONIZATION AND LINKAGE ADJUSTMENTS

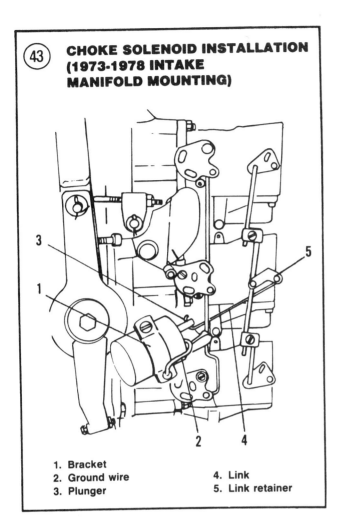

CHOKE SOLENOID INSTALLATION (1973-1978 INTAKE MANIFOLD MOUNTING)

1. Bracket
2. Ground wire
3. Plunger
4. Link
5. Link retainer

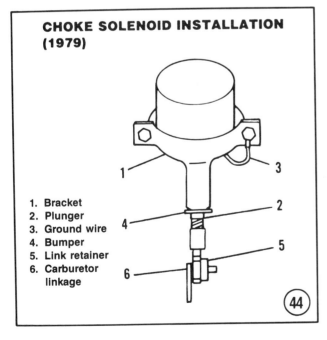

CHOKE SOLENOID INSTALLATION (1979)

1. Bracket
2. Plunger
3. Ground wire
4. Bumper
5. Link retainer
6. Carburetor linkage

6. Rotate the throttle shafts partially open, then let them snap back to the closed position. Apply a slight upward pressure on the adjusting link tab to remove any backlash and tighten the adjustment screw.

7. Move the cam follower while watching the throttle valves. If the throttle valves do not start to move at the same time, repeat Steps 4-6.

Choke Valve Synchronization (Models with Choke Solenoid)

1. Loosen the choke rod fastener lockscrew (1, **Figure 41**).
2A. 1973-1978—Make sure the manual choke lever is in the ON position.
2B. 1979—Make sure the manual choke lever is in the OFF position.
3A. 1973-1977—Close both chokes with finger pressure on the upper and lower choke shaft levers and tighten the lockscrew.
3B. 1978-1979—Depress lower choke link at point 2, **Figure 41** and tighten the lockscrew.
4. Loosen the choke solenoid bracket screws enough to move the solenoid:
   a. 1973-1978—Push solenoid plunger into solenoid to completely close the choke valves and hold the plunger in that position. Move solenoid until plunger bottoms, then move solenoid back 1/32 in. on plunger and tighten bracket screws. Make sure solenoid key does not bind in plunger slot. See **Figure 42** and **Figure 43**.
   b. 1979—Press upward on the plastic choke clevis until the choke valves are completely closed. The plunger should bottom in the solenoid. Move the solenoid up 1/16-1/8 in. and tighten the bracket screws. See **Figure 44**.

Throttle Pickup Point Adjustment

1. Connect a throttle shaft amplifier tool (**Figure 2**) to the top carburetor throttle shaft.

2. Watching the amplifier tool, slowly rotate the throttle cam. As the end of the tool starts to move, check the cam and cam follower alignment. The short or top embossed mark on the cam (1, **Figure 45**) should align with the center of the cam follower (3, **Figure 45**).

3. If the cam follower and cam mark do not align in Step 2, loosen the idle speed screw (4, **Figure 45**) and let the throttle return spring close the throttle valves.

4. Align the specified throttle cam mark with the center of the cam follower and tighten the throttle arm screw.

5. Repeat Step 2 to check the adjustment. If incorrect, repeat Step 3 and Step 4, then repeat Step 2 as required.

Throttle Cable Adjustment

The throttle cable must be adjusted to allow the idle adjustment screw to contact the idle stop. A cable that is too loose will cause a high and unstable idle, resulting in shifting difficulties. If too tight, shifting will feel stiff and the warm-up lever will move up during a shift into NEUTRAL.

1. Slowly move the control lever back until the idle stop screw contacts its stop (**Figure 46**).

2. If the idle stop screw does not contact the stop, adjust the trunnion nut (**Figure 46**) as required.

Throttle and Timer Linkage Synchronization

1. Connect a timing light to the No. 1 cylinder according to manufacturer's instructions.

2. Start the engine and point the timing light at the timing pointer and flywheel.

3. Slowly open the throttle with the control box warm-up lever until the specified throttle pickup point (**Table 1**) aligns with the timing pointer.

4. Leaving the throttle in this position, shut the engine off.

5. Adjust the throttle cam yoke (5, **Figure 45**) as required to align the short or top embossed

THROTTLE CAM ADJUSTMENT

1. Throttle cam top mark
2. Cam follower screw
3. Cam follower
4. Idle speed adjustment screw
5. Throttle cam yoke
6. Advance stop adjustment screw

ENGINE SYNCHRONIZATION AND LINKAGE ADJUSTMENTS

mark on the throttle cam with the center of the cam roller.

6. Start the engine and repeat Step 3. If adjustment is still incorrect, repeat Step 4 and Step 5, then Step 3.

Maximum Spark Advance Adjustment

CAUTION
This procedure should be performed in a test tank with the proper test wheel installed on the engine. The use of the propeller and/or a flushing device can result in an incorrect setting and possible engine damage.

1. Connect a timing light to the No. 1 cylinder according to manufacturer's instructions.

THROTTLE CABLE ADJUSTMENT
1. Idle stop screw
2. Throttle lever
3. Throttle cable casing guide
4. Throttle trunnion adjustment nut

(46)

2. Connect a tachometer according to manufacturer's instructions.
3. Start the engine and open the throttle as follows:
   a. 1973-1981—Until the timer base rests firmly against its stop.
   b. 1982-1983—To a minimum engine speed of 4,000 rpm.
   c. All others—To a minimum engine speed of 5,000 rpm.

NOTE
Compare the specification in Step 4 with the Maximum Advance specification on the engine decal. Use the decal specification if it differs.

4. Check the timing mark position with the timing light. The specified mark (**Table 1**) on the flywheel grid should align with the timing pointer.

WARNING
Do not attempt to adjust the spark advance with the engine running in Step 5. The adjustment screw is located close to or under the moving flywheel and vibrates slightly when the engine is running. If the screwdriver slips out of the screw, serious personal injury can result.

5. If the timing marks do not align as specified, shut the engine off and loosen the adjusting screw locknut (**Figure 45**). Adjust the screw clockwise (to retard) or counterclockwise (to advance) as required, then tighten the locknut. One full turn in either direction changes timing approximately one degree.
6. Restart the engine and repeat Step 4. If timing mark alignment is still incorrect, repeat Step 5, then Step 3 and Step 4 as required.

Wide-open Throttle Stop Adjustment

1. Open the throttle to the full throttle position. The upper carburetor throttle shaft roll pin should contact its stop. Hold the throttle lever in that position.
2. Loosen the throttle stop screw locknut and adjust the stop screw (**Figure 47**) to allow the throttle valves to open fully without loading the throttle shaft. Tighten the throttle stop locknut.
3. Insert a strip of tracing paper between the roll pin and throttle stop. There should be a slight drag (0.003 in. clearance) on the paper when it is removed if the adjustment is correct.

Idle Speed Adjustment

This procedure should be performed with the boat floating in the water (propeller installed) and tied securely at the slip or dock to prevent any fore or aft motion.

1. Remove the engine cover.
2. Connect a tachometer according to manufacturer's instructions.
3. Start the engine and warm to normal operating temperature.
4. Adjust the idle adjustment screw (**Figure 48**) as required to set the engine speed to 650 rpm in FORWARD gear.

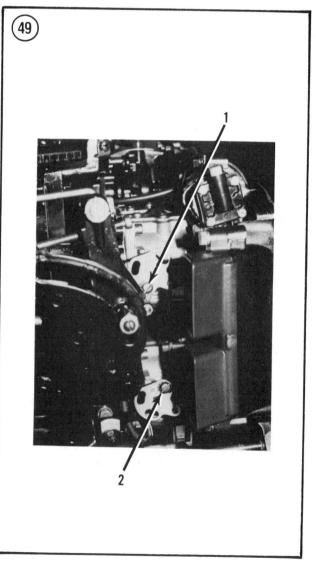

ENGINE SYNCHRONIZATION AND LINKAGE ADJUSTMENTS

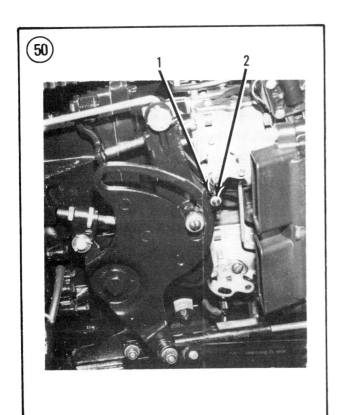

V4 ENGINES
(1986-ON 88-115 HP)

Timing Pointer Alignment

See *60, 65, 70 and 75 hp Timing Pointer Alignment* in this chapter.

Throttle Valve Synchronization

1. Remove the engine cover.
2. Remove the air silencer cover.
3. Remove the starboard air silencer extension, if so equipped.
4. Loosen the cam follower screw (1, **Figure 49**) Move the cam follower away from the throttle cam.
5. Loosen the lower carburetor adjusting lever screw (2, **Figure 49**).
6. Rotate the throttle shafts partially open, then let them snap back to the closed position. Apply a slight downward pressure on the adjusting link tab to remove any backlash and tighten the adjustment screw.
7. Move the cam follower while watching the throttle valves. If the throttle valves do not start to move at the same time, repeat Steps 4-6.

Cam Follower Pickup Adjustment

1. Connect a throttle shaft amplifier tool (**Figure 2**) to the top carburetor throttle shaft.
2. Watching the amplifier tool, slowly rotate the throttle cam. As the end of the tool starts to move, check the cam and cam follower alignment. The embossed mark on the cam (1, **Figure 50**) should align with the center of the cam follower (2, **Figure 50**).
3. If the cam follower and cam mark do not align in Step 2, loosen the cam follower screw (**Figure 51**) and let the throttle return spring close the throttle valves.
4. Align the throttle cam mark with the center of the cam follower and tighten the cam follower screw.

5. Repeat Step 2 to check the adjustment. If incorrect, repeat Step 3 and Step 4, then repeat Step 2 as required.

Maximum Spark Advance Adjustment

CAUTION
This procedure should be performed in a test tank with the proper test wheel installed on the engine. The use of the propeller and/or a flushing device can result in an incorrect setting and possible engine damage.

1. Connect a timing light to the No. 1 cylinder according to manufacturer's instructions.
2. Connect a tachometer according to manufacturer's instructions.

NOTE
If the engine cannot reach a minimum speed of 5,000 rpm in Step 3, rotate the throttle cam link thumbwheel until the required speed is obtained.

3. Start the engine and open the throttle to a minimum 5,000 rpm with the timer base fully advanced.

NOTE
Compare the specification in Step 4 with the Maximum Advance specification on the engine decal. Use the decal specification if it differs.

4. Check the timing mark position with the timing light. The specified mark (**Table 1**) on the flywheel grid should align with the timing pointer.

WARNING
Do not attempt to adjust the spark advance with the engine running in Step 5. The adjustment screw is located close to the moving flywheel and vibrates slightly when the engine is running. If the screwdriver slips out of the screw, serious personal injury can result.

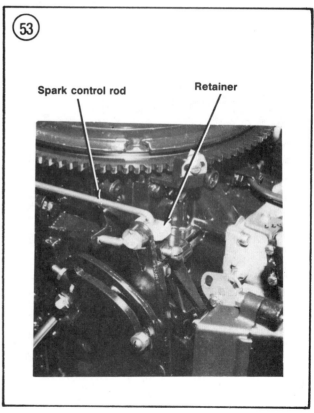

ENGINE SYNCHRONIZATION AND LINKAGE ADJUSTMENTS

5A. 1986-1988 Models—If the timing marks do not align as specified, shut the engine off.

a. If timing is within ±2° of the specified setting, loosen the eccentric screw as shown in **Figure 52**. Rotate the eccentric, tighten the screw and recheck timing.

b. If timing varies more than ±2° from the specified setting, reposition the spark control rod in another hole in its retainer. See **Figure 53**. Move the rod forward (to advance) or backward (to retard) as required. Movement between 2 adjacent holes will change timing by approximately 4°. Recheck timing. If necessary, fine-tune by rotating the eccentric as described above.

5B. 1989-1990 Models—If the timing marks do not align as specified, shut the engine off. Loosen jam nut at retainer on spark control rod (**Figure 53**, typical). Rotate thumbwheel in center of retainer toward the engine (to advance) or away from the engine (to retard) as required to adjust maximum spark advance. Tighten jam nut snugly to secure adjustment.

6. Restart the engine and repeat Step 3 and Step 4. If timing mark alignment is still incorrect, repeat Step 5A or Step 5B, then Step 3 and Step 4 as required.

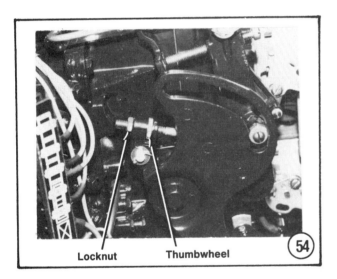

Cam Follower Pickup Timing

1. Connect a timing light to the No. 1 cylinder according to manufacturer's instructions.

2. Start the engine and check the timing mark position with the timing light. Move the spark advance lever until the specified mark (**Table 1**) on the flywheel grid aligns with the engine mark. At this point, the embossed mark on the throttle cam should align with the center of the cam follower roller.

3. If the throttle cam mark and cam follower roller do not align as specified in Step 2, loosen the throttle cam link locknut and rotate thumbwheel (**Figure 54**) away from the engine (to advance) or toward the engine (to retard) as required to bring pickup timing within specifications (**Table 1**). Tighten the locknut snugly.

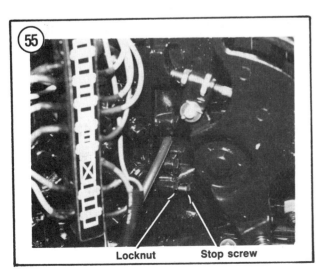

Wide-open Throttle Stop Adjustment

1. Open the throttle to the full throttle position. The carburetor throttle shafts roll pins should be exactly vertical.

2. If the roll pins are not vertical, loosen the wide-open stop screw locknut and adjust the stop screw (**Figure 55**) to allow the throttle valves to

open fully without loading the throttle shaft. Tighten the locknut.

Idle Speed Adjustment

This procedure should be performed with the boat floating in the water (propeller installed) and unrestrained, not tied securely at the slip or dock.
1. Remove the engine cover.
2. Connect a tachometer according to manufacturer's instructions.
3. Start the engine and warm to normal operating temperature.
4. Adjust the idle speed screw (**Figure 56**) as required to set the engine speed to 600-700 rpm in FORWARD gear.

V4 ENGINES
(1985-ON 120-140 HP)

Timing Pointer Alignment

See *60, 65, 70 and 75 hp Timing Pointer Alignment* in this chapter.

Throttle Valve Synchronization
1985-1988 Models

1. Remove the engine cover.
2. Loosen the cam follower screw (C, **Figure 57**). Move the cam follower away from the throttle cam.
3. Loosen the throttle shaft synchronizing screw (D, **Figure 57**) two full turns. The throttle plates should close.
4. Rotate the throttle shaft to partially open the throttle plates, then let them snap back to the closed position. Make sure they are closed and retighten the synchronizing screw. Leave the cam follower screw loose.

1989-1990 Models

1. Remove the engine cover.
2. Loosen the cam follower screw (C, **Figure 58**). Move the cam follower away from the throttle cam.
3. Loosen the 2 carburetor link adjustment screws (S, **Figure 58**) *no more than* one-half turn.
4. Allow all throttle plates to seat within bores.
5. Tighten the 2 carburetor link adjustment screws (S, **Figure 58**) while supporting rear of link.

Throttle Cam Pickup Point Adjustment

Refer to **Figure 58** and **Figure 59** for this procedure.
1. 1988-1989 Models—Loosen lock ring and rotate adjustment knob counterclockwise until no internal spring pressure is noted (**Figure 58**).
2. With the cam follower screw loose, hold the follower against the throttle cam and adjust the throttle arm stop screw (D, **Figure 59**) until the cam and follower contact the center of the cam follower roller intersecting the index line (I, **Figure 58**) on the cam.
3. When the cam index line intersects the center of the cam follower roller, tighten the cam follower screw.

ENGINE SYNCHRONIZATION AND LINKAGE ADJUSTMENTS

THROTTLE VALVE SYNCHRONIZATION

A. Throttle cam
B. Cam follower roller
C. Cam follower screw
D. Throttle shaft synchronizing screw

Wide-open Throttle Stop Adjustment

Refer to **Figure 59** for this procedure.

1. Move the throttle arm and linkage toward the wide-open throttle position and loosen the locknut.
2. Adjust the wide-open throttle stop screw until the carburetor throttle plates open fully without loading the throttle shaft.
3. Continue adjusting the stop screw until the cam follower WOT index line faces directly forward and is perpendicular with the air silencer base. Hold the screw from moving and tighten the locknut.

Throttle Cable Adjustment

The throttle cable must be adjusted to allow the idle adjustment screw to contact the idle stop. A cable that is too loose will cause a high and unstable idle, resulting in shifting difficulties. If too tight, shifting will feel stiff and the warm-up lever will move up during a shift into NEUTRAL.

Refer to **Figure 59** for this procedure.

1. Remove the throttle cable trunnion cover.
2. With the throttle cable in the air silencer trunnion pocket, adjust the cable nut until the throttle arm stop screw touches the crankcase firmly.
3. Install the trunnion cover and tighten the cover nut securely.

Maximum Spark Advance Adjustment

CAUTION
This procedure should be performed in a test tank with the proper test wheel installed on the engine. The use of the propeller and/or a flushing device can result in an incorrect setting and possible engine damage.

Refer to **Figure 60** for this procedure.

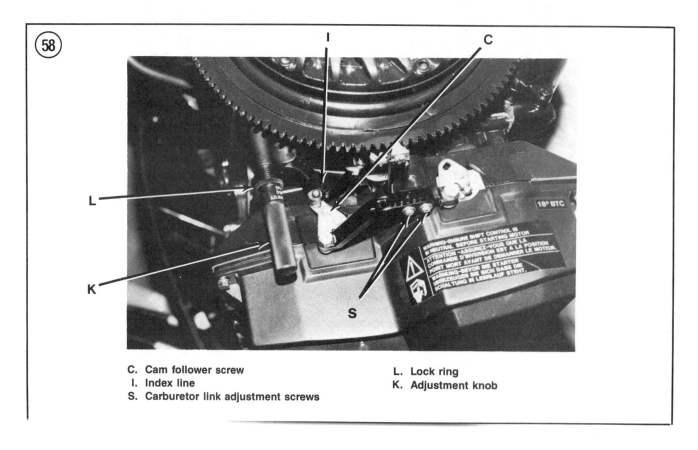

C. Cam follower screw
I. Index line
S. Carburetor link adjustment screws
L. Lock ring
K. Adjustment knob

ENGINE SYNCHRONIZATION AND LINKAGE ADJUSTMENTS

A. Cam follower screw
B. Cam follower roller
C. Throttle cam
D. Throttle arm stop screw
E. WOT stop screw
F. WOT index mark
G. Throttle cable nut
H. Trunnion cover screw
I. Locknut

MAXIMUM SPARK ADVANCE ADJUSTMENT

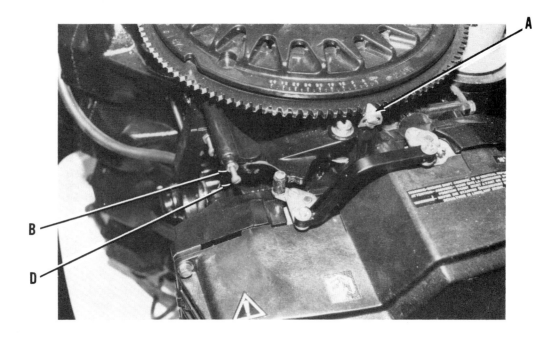

A. Timing pointer
B. Locknut
C. Full advance timing screw
D. Idle speed screw

ENGINE SYNCHRONIZATION AND LINKAGE ADJUSTMENTS

1. Connect a timing light to the No. 1 cylinder according to manufacturer's instructions.
2. Connect a tachometer according to manufacturer's instructions.
3. Start the engine and warm to normal operating temperature.
4. Open the throttle to obtain an engine speed of 4,500-5,000 rpm.

NOTE
Compare the specification in Step 5 with the Maximum Advance specification on the engine decal. Use the decal specification if it differs.

5. Check the timing mark position with the timing light. The specified mark (**Table 1**) on the flywheel grid should align with the timing pointer.

WARNING
Do not attempt to adjust the spark advance with the engine running in Step 5. The adjustment screw is located close to or under the moving flywheel and vibrates slightly when the engine is running. If the screwdriver slips out of the screw, serious personal injury can result.

6. If the timing marks do not align as specified, shut the engine off and loosen the full advance timing screw locknut. Adjust the screw clockwise (to retard) or counterclockwise (to advance) as required, then tighten the locknut. One full turn in either direction changes timing approximately one degree.
7. Restart the engine and repeat Step 5. If timing mark alignment is still incorrect, repeat Step 6, then Step 4 and Step 5 as required.

Idle Speed Adjustment
1985-1987 Models

This procedure should be performed with the boat floating in the water (propeller installed) and unrestrained, not tied securely at the slip or dock. Refer to **Figure 57**.
1. Remove the engine cover.
2. Connect a tachometer according to manufacturer's instructions.
3. Start the engine and warm to normal operating temperature.
4. Loosen the idle speed screw locknut and adjust the screw as required to set the engine speed to 650 rpm in FORWARD gear.
5. When idle speed is correct, shut the engine off. Hold the screw from moving and tighten the locknut.
6. Remove the tachometer and install the engine cover.

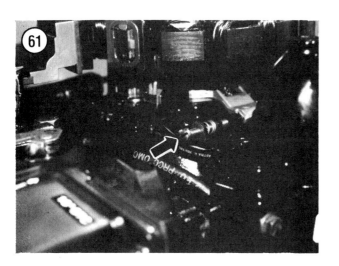

Idle Timing and Idle Speed Adjustment
1988-1989 Models

CAUTION
This procedure should be performed in a test tank with the proper test wheel installed or with the boat floating in the water (propeller installed) and unrestrained, not tied securely at the slip or dock. The use of the propeller in a test tank and/or a flushing device can result in an incorrect setting and possible engine damage.

Refer to **Figure 58**, **Figure 59** and **Figure 61** for this procedure.

1. Loosen lock ring and rotate lock ring and adjustment knob clockwise until completely seated (**Figure 58**).
2. Start the engine and warm to normal operating temperature.
3. Shift the engine into FORWARD gear and note the idle timing while making sure the throttle arm stop screw is firmly seated against crankcase (**Figure 59**).
4. The idle timing should be 8° ATDC.
5. If idle timing requires adjustment, shut the engine off as a safety precaution. Rotate idle timing screw clockwise (to advance) or counterclockwise (to retard) as required (**Figure 61**).
6. Repeat Steps 2-4. Repeat Step 5 if needed.
7. Start the engine and shift into FORWARD gear. Adjust throttle arm stop screw until engine idles at 950 rpm (**Figure 59**).
8. Stop the engine. With the throttle arm stop screw (D, **Figure 59**) firmly seated against the crankcase, turn adjustment knob (K, **Figure 58**) counterclockwise until the timer base just starts to move away from the idle timing screw (**Figure 61**).
9. Rotate lock ring against adjustment knob to secure setting (**Figure 58**).
10. With the throttle arm stop screw (D, **Figure 59**) lightly seated against the crankcase, rotate the stop screw counterclockwise until the cam index line (I, **Figure 58**) intersects the center of the cam follower roller (B, **Figure 59**). Continue to rotate stop screw one complete turn counterclockwise from this point.

NOTE
*After performing the previous adjustments, the engine should idle between 575-700 rpm in gear and the cam follower roller should not touch the throttle cam when the engine is in NEUTRAL. Refer to **Throttle Cam Pickup Point Adjustment** in this chapter if incorrect adjustment is noted.*

V6 ENGINES
(1975-1985)

Timing Pointer Alignment

See *60, 65, 70 and 75 hp Timing Pointer Alignment* in this chapter.

Throttle Cable Adjustment

The throttle cable must be adjusted to allow the idle adjustment screw to contact the idle stop. A cable that is too loose will cause a high and unstable idle, resulting in shifting difficulties. If too tight, shifting will feel stiff and the warm-up lever will move up during a shift into NEUTRAL.

1. Slowly move the remote control lever back until the idle stop screw contacts its stop.
2. If the idle stop screw does not contact the stop, adjust the trunnion nut (**Figure 62**) as required.

Throttle Valve Synchronization

1. Remove the engine cover.
2. Remove the air silencer cover.
3. Retard the throttle lever to a point where the throttle cam roller does not touch the cam.
4. Loosen the upper and lower carburetor lever adjustment screws (**Figure 63**). The throttle shaft return springs should close the throttle valves.

ENGINE SYNCHRONIZATION AND LINKAGE ADJUSTMENTS

5. Rotate the throttle shafts partially open, then let them snap back to the closed position. Apply a slight downward pressure on the adjusting link tab to remove any backlash and tighten the adjusting screws.

6. Move the cam follower while watching the throttle valves. If the throttle valves do not start to move at the same time, repeat Steps 3-5.

Choke Valve Synchronization
(Models with Choke Solenoid)

1. Loosen the choke rod fastener lockscrews (**Figure 63**).
2. 1976-1978—Make sure the manual choke lever is in the ON position.
3. Hold the center choke closed, push the upper choke rod up to close the choke and tighten the lockscrew.
4. Hold the center choke closed, push the lower choke rod down to close the choke and tighten the lockscrew.
5. Loosen the choke solenoid bracket screws enough to move the solenoid:
    a. 1976-1977—Close the choke valves with the manual choke lever. Push the solenoid plunger into the solenoid until the plunger bottoms. Holding the plunger in that position, move the solenoid forward 0.016 in. and tighten the bracket screws. See **Figure 63**.
    b. 1978-1980—Push the solenoid plunger into the solenoid until the plunger bottoms. Holding the plunger in that position, slide the solenoid and bracket forward until the choke valves are cloed and there is no slack in the plunger-to-choke lever linkage. Mark the solenoid at the forward edge of

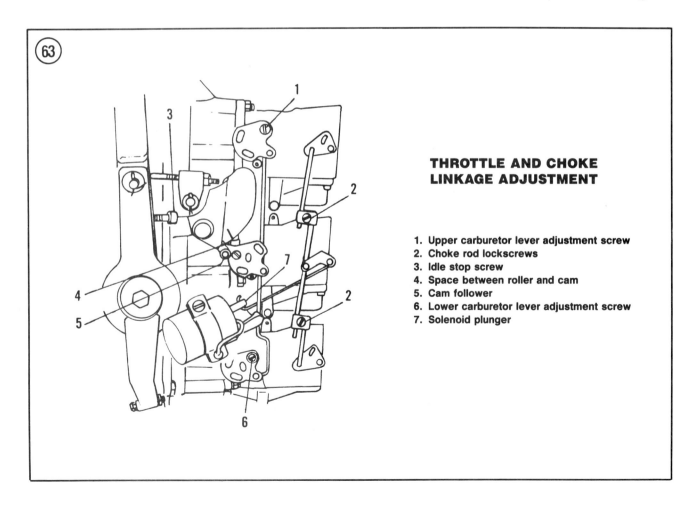

THROTTLE AND CHOKE LINKAGE ADJUSTMENT

1. Upper carburetor lever adjustment screw
2. Choke rod lockscrews
3. Idle stop screw
4. Space between roller and cam
5. Cam follower
6. Lower carburetor lever adjustment screw
7. Solenoid plunger

the bracket, hold the bracket in place and move the solenoid forward 0.016 in. Tighten the solenoid bracket fasteners.

Throttle Pickup Point Adjustment

1. Connect a throttle shaft amplifier tool (**Figure 2**) to the top carburetor throttle shaft.
2. Watching the amplifier tool, slowly rotate the throttle cam. As the end of the tool starts to move, check the cam and cam follower alignment. The bottom embossed mark on the cam (1, **Figure 64**) should align with the center of the cam follower (2, **Figure 64**).
3. If the cam follower and cam mark do not align in Step 2, loosen the cam follower screw (3, **Figure 64**) and rotate the throttle shafts to close the throttle valves.
4. Align the bottom throttle cam mark with the center of the cam follower and tighten the cam follower screw.
5. Repeat Step 2 to check the adjustment. If incorrect, repeat Step 3 and Step 4, then repeat Step 2 as required.
6. On 1976-1977 engines, check alignment of the throttle cam "START" mark and remote control warm-up lever. If the "START" mark on the cam does not align with the center of the cam roller when the control box warm-up lever is moved to the START position, adjust the control box stop screw as required. See **Figure 65**.

THROTTLE PICKUP POINT ADJUSTMENT

1. Lower mark
2. Cam follower
3. Cam follower screw
4. Full throttle adjustment screw
5. Throttle cam yoke

ENGINE SYNCHRONIZATION AND LINKAGE ADJUSTMENTS

65

WARM-UP LEVER ADJUSTMENT

1. Adjustment stop screw
2. Increase friction
3. Decrease friction

66

Throttle and Timer Linkage Synchronization

1. Connect a timing light to the No. 1 cylinder according to manufacturer's instructions.
2. Start the engine and point the timing light at the timing pointer and flywheel.
3. Slowly open the throttle with the control box warm-up lever until the specified throttle pickup points (**Table 1**) align with the timing pointer.
4. Leaving the throttle in this position, shut the engine off.
5. Adjust the throttle cam yoke (5, **Figure 64**) as required to align the bottom embossed mark on the throttle cam with the center of the cam roller.
6. Start the engine and repeat Step 3. The throttle cam mark should align with the center of the cam roller if yoke adjustment is correct. If mark and roller do not align in this step, repeat Step 4 and Step 5, then Step 3.

Maximum Spark Advance Adjustment

CAUTION
This procedure should be performed in a test tank with the proper test wheel installed on the engine. The use of the propeller and/or a flushing device can result in an incorrect setting and possible engine damage.

1. Connect a timing light to the No. 1 cylinder according to manufacturer's instructions.
2. Connect a tachometer according to manufacturer's instructions.
3. Loosen the maximum advance screw locknut and adjust the screw to extend 3/8-7/16 in. beyond the timing pointer bracket. See A, **Figure 66**.
4. Start the engine and run at 4,300-4,600 rpm (1976-1983) or a minimum of 5,000 rpm (1984-1985) in FORWARD gear.

NOTE
Compare the specification in Step 5 with the Maximum Advance specification on

the engine decal. Use the decal specification if it differs.

5. Check the timing mark position with the timing light. Note the amount of advance shown by the flywheel grid marked "No. 1" and timing pointer alignment.

6. Stop the engine and connect the timing light to the No. 2 cylinder.

7. Repeat Step 4 and Step 5, noting the amount of advance shown by the flywheel grid marked "No. 2" and the timing pointer alignment.

NOTE
*Some early production 1981 V6 engines were equipped with an incorrect spark advance decal on the spark advance lever. Refer to **Table 1** for the correct 1981 model specifications in Step 8. If the spepcification differs from that on the decal, use **Table 1**.*

8. Refer to specifications (**Table 1**) or the engine decal for desired maximum advance.

WARNING
Do not attempt to adjust the spark advance with the engine running in Step 9. The adjustment screw is located close to or under the moving flywheel and vibrates slightly when the engine is running. If the screwdriver slips out of the screw, serious personal injury can result.

9. Shut the engine off and loosen the adjusting screw locknut (A, **Figure 66**). Adjust the screw clockwise to retard or counterclockwise to advance the timing of the cylinder which showed the greatest amount of advance in Step 5 and Step 7 to specifications (**Table 1**), then tighten the locknut. One full turn in either direction changes timing approximately one degree.

10. Restart the engine and repeat Steps 4-7 as required. If timing mark alignment is still incorrect, repeat Step 9, then Steps 4-7 as required.

Wide-open Throttle Stop Adjustment

1. Open the throttle to the full throttle position. The upper carburetor throttle shaft roll pin should contact its stop. Hold the throttle lever in that position.
2. Loosen the throttle stop screw locknut and adjust the stop screw (**Figure 67**) to allow the throttle valves to open fully without loading the throttle shaft. Tighten the throttle stop locknut.
3. Insert a strip of tracing paper between the roll pin and throttle stop. There should be a slight drag (0.003 in. clearance) on the paper when it is removed if the adjustment is correct.

Idle Speed Adjustment

This procedure should be performed with the boat floating in the water (propeller installed) and tied securely at the slip or dock to prevent any fore or aft motion.

1. Remove the engine cover.
2. Connect a tachometer according to manufacturer's instructions.
3. Start the engine and warm to normal operating temperature.

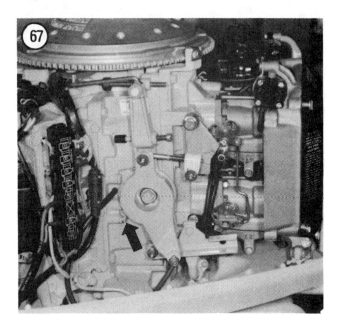

ENGINE SYNCHRONIZATION AND LINKAGE ADJUSTMENTS

4. Adjust the idle adjustment screw (B, **Figure 66**) as required to set the engine speed to 650 rpm in FORWARD gear.

V6 ENGINES
(1986-ON 150-175 HP)

Timing Pointer Alignment

See *60, 65, 70 and 75 hp Timing Pointer Alignment* in this chapter.

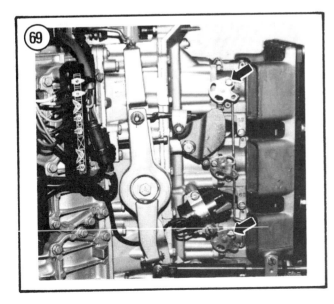

Throttle Valve Synchronization

1. Remove the engine cover.
2. Remove the air silencer cover.
3. Loosen the cam follower screw (**Figure 68**) Move the cam follower away from the throttle cam.
4. Loosen the upper and lower carburetor lever adjustment screws (**Figure 69**). The throttle shaft return springs should close the throttle valves.
5. Rotate the throttle shafts partially open, then let them snap back to the closed position. Apply a slight downward pressure on the lower adjusting link tab to remove any backlash and tighten the adjusting screw, then repeat to tighten the upper link screw.
6. Move the cam follower while watching the throttle valves. If the throttle valves do not start to move at the same time, repeat Steps 3-5.

Cam Follower Pickup Adjustment

1. Connect a throttle shaft amplifier tool (**Figure 2**) to the top carburetor throttle shaft.
2. Watching the amplifier tool, slowly rotate the throttle cam. As the end of the tool starts to move, check the cam and cam follower alignment. The short embossed mark on the cam should align with the center of the cam follower.
3. If the cam follower and cam mark do not align in Step 2, loosen the cam follower screw (**Figure 68**) and let the throttle return spring close the throttle valves.
4. Align the throttle cam mark with the center of the cam follower and tighten the cam follower screw.
5. Repeat Step 2 to check the adjustment. If incorrect, repeat Step 3 and Step 4, then repeat Step 2 as required.

Cam Follower Pickup Timing

1. Connect a timing light to the No. 1 cylinder according to manufacturer's instructions.
2. Start the engine and check the timing mark position with the timing light. The specified mark

(**Table 1**) on the flywheel grid should align with the flywheel pointer.

3. If the timing marks do not align as specified in Step 2, loosen the throttle cam link locknut and rotate thumbwheel away from the engine (to advance) or toward the engine (to retard) as required to bring pickup timing within specifications (**Table 1**). Tighten the locknut snugly.

Wide-open Throttle Stop Adjustment

1. Open the throttle to the full throttle position. The carburetor throttle shafts roll pins should be exactly vertical.

2. If the roll pins are not vertical, loosen the wide-open stop screw locknut and adjust the stop screw (**Figure 70**) to allow the throttle valves to open fully without loading the throttle shaft. Tighten the locknut.

Maximum Spark Advance Adjustment

CAUTION
This procedure should be performed in a test tank with the proper test wheel installed on the engine. The use of the propeller and/or a flushing device can result in an incorrect setting and possible engine damage.

1. Connect a timing light to the No. 1 cylinder according to manufacturer's instructions.
2. Connect a tachometer according to manufacturer's instructions.
3. Loosen the maximum advance screw locknut and adjust the screw to extend 3/8-7/16 in. beyond the timing pointer bracket. See A, **Figure 66**.
4. Start the engine and run at a minimum of 5,000 rpm in FORWARD gear.

NOTE
Compare the specification in Step 5 with the Maximum Advance specification on the engine decal. Use the decal specification if it differs.

5. Check the timing mark position with the timing light. Note the amount of advance shown by the flywheel grid marked "No. 1" and timing pointer alignment.
6. Stop the engine and connect the timing light to the No. 2 cylinder.
7. Repeat Step 4 and Step 5, noting the amount of advance shown by the flywheel grid marked "No. 2" and the timing pointer alignment.
8. Refer to specifications (**Table 1**) or the engine decal for desired maximum advance.

WARNING
Do not attempt to adjust the spark advance with the engine running in Step 9. The adjustment screw is located close to or under the moving flywheel and vibrates slightly when the engine is running. If the screwdriver slips out of the screw, serious personal injury can result.

9. Shut the engine off and loosen the adjusting screw locknut (A, **Figure 66**). Adjust the screw clockwise to retard or counterclockwise to advance the timing of the cylinder which showed the greatest amount of advance in Step 5 and Step

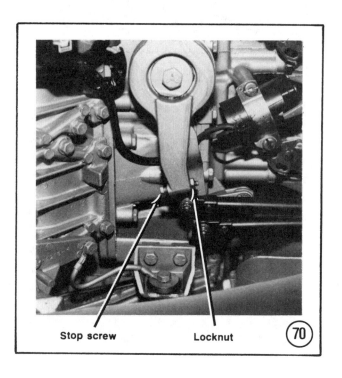

ENGINE SYNCHRONIZATION AND LINKAGE ADJUSTMENTS

7 to obtain the specified advance (**Table 1**), then tighten the locknut. One full turn in either direction changes timing approximately one degree.

10. Restart the engine and repeat Steps 4-7 as required. If timing mark alignment is still incorrect, repeat Step 9, then Steps 4-7 as required.

Idle Speed Adjustment

This procedure should be performed with the boat floating in the water (propeller installed) and unrestrained, not tied securely at the slip or dock.

1. Remove the engine cover.

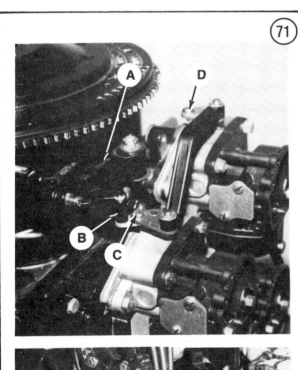

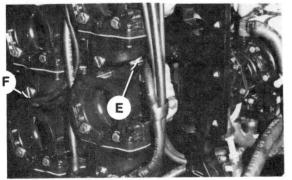

2. Connect a tachometer according to manufacturer's instructions.
3. Start the engine and warm to normal operating temperature.
4. Adjust the idle speed screw (B, **Figure 66**) as required to set the engine speed to 600-700 rpm in FORWARD gear.

V6 ENGINES (1986-ON 200-225 HP)

Timing Pointer Alignment

See *60, 65, 70 and 75 hp Timing Pointer Alignment* in this chapter.

Throttle Valve Synchronization
1986-1988 Models

1. Remove the engine cover.
2. Loosen the cam follower screw (C, **Figure 71**). Move the cam follower away from the throttle cam.
3. Loosen the throttle shaft synchronizing screw (D, **Figure 71**) two full turns. The throttle plates should close.
4. Remove the air silencer cover and baffle.
5. Loosen (do not remove) the top screw on the port throttle shaft connector (E, **Figure 71**) and the bottom screw on the starboard throttle shaft connector (F, **Figure 71**).
6. Rotate the throttle shaft to partially open the throttle plates, then let them snap back to the closed position. Make sure they are closed and retighten all screws (except the cam follower screw) in the reverse order in which they were loosened. Leave the cam follower screw loose.

1989-1990 Models

1. Remove the engine cover.
2. Loosen the cam follower screw (C, **Figure 71**). Move the cam follower away from the throttle cam.
3. Loosen the 2 carburetor link adjustment screws (S, **Figure 58**) *no more than* one-half turn.

4. Allow all throttle plates to seat within bores.
5. Remove the air silencer cover and baffle.
6. Loosen (do not remove) the bottom screw on the port throttle shaft connector (E, **Figure 71**) and the top screw on the starboard throttle shaft connector (F, **Figure 71**).
7. Rotate the throttle shaft to partially open the throttle plates, then let them snap back to the closed position. Make sure they are closed and retighten the loosened screws in the throttle shaft connectors.
8. Tighten the 2 carburetor link adjustment screws (S, **Figure 58**) while supporting rear of link.
9. Leave the cam follower screw loose.

Throttle Cam Pickup Point Adjustment

Refer to **Figure 58** and **Figure 72** for this procedure.
1. 1988-1989 Models—Loosen lock ring and rotate adjustment knob counterclockwise until no internal spring pressure is noted (**Figure 58**).
2. With the cam follower screw loose, hold the follower against the throttle cam and adjust the throttle arm stop screw (**Figure 72**) until the cam and follower contact the center of the cam follower roller intersecting the index line (I, **Figure 58**) on the cam.
3. When the cam index line intersects the center of the cam follower roller, tighten the cam follower screw.

Wide-open Throttle Stop Adjustment

Refer to **Figure 73** for this procedure.
1. Move the throttle arm and linkage toward the wide-open throttle position and loosen the locknut.
2. Adjust the wide-open throttle stop screw until the carburetor throttle plates open fully without loading the throttle shaft.
3. Continue adjusting the stop screw until the cam follower WOT index line faces directly forward and is perpendicular with the air silencer base. Hold the screw from moving and tighten the locknut.

Throttle Cable Adjustment

The throttle cable must be adjusted to allow the idle adjustment screw to contact the idle stop. A cable that is too loose will cause a high and unstable idle, resulting in shifting difficulties. If too tight, shifting will feel stiff and the warm-up lever will move up during a shift into NEUTRAL.
Refer to **Figure 73** for this procedure.
1. Remove the throttle cable trunnion cover.
2. With the throttle cable in the air silencer trunnion pocket, adjust the cable nut until the throttle arm stop screw touches the crankcase firmly.
3. Install the trunnion cover and tighten the cover nut securely.

Maximum Spark Advance Adjustment

CAUTION
This procedure should be performed in a test tank with the proper test wheel installed on the engine. The use of the

ENGINE SYNCHRONIZATION AND LINKAGE ADJUSTMENTS

propeller and/or a flushing device can result in an incorrect setting and possible engine damage.

Refer to **Figure 60** for this procedure.

1. Connect a timing light to the No. 1 cylinder according to manufacturer's instructions.
2. Connect a tachometer according to manufacturer's instructions.
3. Start the engine and warm to normal operating temperature.
4. Open the throttle to obtain an engine speed of 4,500-5,000 rpm.

NOTE
Compare the specification in Step 5 with the Maximum Advance specification on the engine decal. Use the decal specification if it differs.

1. Wide-open throttle stop screw
2. Throttle cable nut

5. Check the timing mark position with the timing light. The specified mark (**Table 1**) on the flywheel grid should align with the timing pointer.

WARNING
Do not attempt to adjust the spark advance with the engine running in Step 5. The adjustment screw is located close to or under the moving flywheel and vibrates slightly when the engine is running. If the screwdriver slips out of the screw, serious personal injury can result.

6. If the timing marks do not align as specified, shut the engine off and loosen the full advance timing screw locknut. Adjust the screw clockwise (to retard) or counterclockwise (to advance) as required, then tighten the locknut. One full turn in either direction changes timing approximately one degree.
7. Restart the engine and repeat Step 5. If timing mark alignment is still incorrect, repeat Step 6, then Step 4 and Step 5 as required.

Idle Speed Adjustment
1986-1987 Models

This procedure should be performed with the boat floating in the water (propeller installed) and unrestrained, not tied securely at the slip or dock. Refer to **Figure 60**.

1. Remove the engine cover.
2. Connect a tachometer according to manufacturer's instructions.
3. Start the engine and warm to normal operating temperature.
4. Loosen the idle speed screw locknut and adjust the screw as required to set the engine speed to 650 rpm in FORWARD gear.
5. When idle speed is correct, shut the engine off. Hold the screw from moving and tighten the locknut.
6. Remove the tachometer and install the engine cover.

Idle Timing and Idle Speed Adjustment
1988-1990 Models

CAUTION
This procedure should be performed in a test tank with the proper test wheel installed or with the boat floating in the water (propeller installed) and unrestrained, not tied securely at the slip or dock. The use of the propeller in a test tank and/or a flushing device can result in an incorrect setting and possible engine damage.

Refer to **Figure 58**, **Figure 59**, **Figure 61** and **Figure 72** for this procedure.

1. Loosen lock ring and rotate lock ring and adjustment knob clockwise until completely seated (**Figure 58**).
2. Start the engine and warm to normal operating temperature.
3. Shift the engine into FORWARD gear and note the idle timing while making sure the throttle arm stop screw is firmly seated against crankcase (**Figure 72**).
4. The idle timing should be 6° ATDC.
5. If idle timing requires adjustment, shut the engine off as a safety precaution. Rotate idle timing screw clockwise (to advance) or counterclockwise (to retard) as required (**Figure 61**).
6. Repeat Steps 2-4. Repeat Step 5 if needed.
7. Start the engine and shift into FORWARD gear. Adjust throttle arm stop screw until engine idles at 950 rpm (**Figure 72**).
8. Stop the engine. With the throttle arm stop screw (D, **Figure 59**) firmly seated against the crankcase, turn adjustment knob (K, **Figure 58**) counterclockwise until the timer base just starts to move away from the idle timing screw (**Figure 61**).
9. Rotate lock ring against adjustment knob to secure setting (**Figure 58**).
10. With the throttle arm stop screw (**Figure 72**) lightly seated against the crankcase, rotate the stop screw counterclockwise unit the cam index line (I, **Figure 58**) intersects the center of the cam follower roller (B, **Figure 59**). Continue to rotate stop screw one complete turn counterclockwise from this point.

NOTE
*After performing the previous adjustments, the engine should idle between 575-700 rpm in gear and the cam follower roller should not touch the throttle cam when the engine is in NEUTRAL. Refer to **Throttle Cam Pickup Point Adjustment** in this chapter if incorrect adjustment is noted.*

ENGINE SYNCHRONIZATION AND LINKAGE ADJUSTMENTS

Table 1 TIMING SPECIFICATIONS

	Full Throttle Spark Advance degrees BTDC	Throttle Pickup degrees BTDC
48 hp		
1987-1988	19 ±1	2-4
1989-1990	19 ±1	—
50 hp		
1973-1975	19 ±1	3
1980-1981	19	2-4
1982-1988	19 ±1	2-4
1989-1990	19 ±1	—
55 hp		
1975-1976	19	4
1977-1979	19	3
1980-1981	19	2-4
1982-on	19 ±1	2-4
60 hp		
1980-1981	21	2-4
1982-1985	21 ±1	2-4
1986-on	19 ±1	1 ±1 ATDC
65 hp		
1973	22	8
70 hp		
1974	20	TDC
1975-1979	17	TDC
1980-1981	19	TDC
1982	19 ±1	TDC
1983-1985	See note 1	TDC ±1
1986-1987	19 ±1	4 ±1 ATDC
1988	19 ±1	1 ±1 ATDC
1989-1990	19 ±1	1 ±1 BTDC
75 hp		
1975-1979	16	TDC
1980-1981	19	TDC
1982	19 ±1	TDC
1983-1985	See note 1	TDC ±1
1986-1987	19 ±1	1 ±1 ATDC
85 hp		
1973	28	5
1974	28	4
1975-1976	26	4
1977-1979	26	3-5
1980	28	4-6
88 hp	28 ±1	4 ±1
90 hp		
1981	28	4-6
1982	28 ±1	4-6
1983-1985	See note 1	3-5
1986-on	28 ±1	4 ±1
100 hp	28	4-6
110 hp	28 ±1	4 ±1
115 hp		
1973	26	5
1974	26	4

(continued)

Table 1 TIMING SPECIFICATIONS (continued)

	Full Throttle Spark Advance degrees BTDC	Throttle Pick-up degrees BTDC
115 hp (continued)		
1975-1976	24	4
1977-1979	28	0-3
1980-1981	28	4-6
1982	28 ±1	4-6
1983-on	See note 1	3-5
120 hp		
1985	22	—
1986-on	18 ±1	—
135 hp		
1973	22	5
1974	22	4
1975-1976	20	4
140 hp		
1977-1979	28	0-3
1980-1981	28	4-6
1982	28 ±1	4-6
1983-1984	See note 1	3-5
1985	22	—
1986-on	18 ±1	—
150 hp		
1977-1979	26	6-8
1980-1981	28[2]	6-8
1982	28 ±1	6-8
1983-1985	See note 1	6-8
1986-on	32 ±1[3]	7 ±1
175 hp		
1977-1979	28	5
1980-1981	28[2]	6-8
1982	28 ±1	6-8
1983-1985	See note 1	6-8
1986-on	28 ±1	7 ±1
185 hp	See note 1	6-8
200 hp		
1975-1979	28	5
1980-1981	26[4]	6-8
1982	28 ±1	6-8
1983-1985	See note 1	6-8
1986-on	18 ±1	—
225 hp		
1986-on	18 ±1	—
235 hp		
1977-1979	28	6-8
1980-1981	30	6-8
1982	30 ±1	6-8
1983-1985	See note 1	6-8

1. Refer to timing decal on engine.
2. If timer base has red sleeves, 32° BTDC.
3. 1986-1987 150STL, 30 ±1° BTDC; 1988-1989 150STL, 28 ±1° BTDC.
4. 1981 CH model: timer base with red sleeves, 30° BTDC; 1981 CIA, CIB and CIM model: timer base with black sleeves, 28° BTDC.

Chapter Six

Fuel System

NOTE
Outboards equipped with Variable Ratio Oiling (VRO) use a self-contained mechanical pump mounted on the power head instead of the remote fuel pump. The VRO system is covered in Chapter Eleven.

This chapter contains removal, overhaul, installation and adjustment procedures for fuel pumps, carburetors, choke and primer solenoids, fuel tanks and connecting lines used with the Johnson and Evinrude outboards covered in this book. **Table 1** is at the end of the chapter.

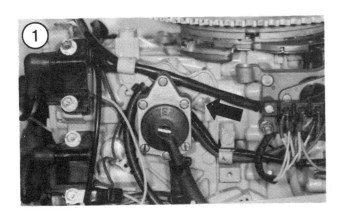

FUEL PUMP

Johnson and Evinrude outboards use a diaphragm-type fuel pump which operates by crankcase pressure (**Figure 1**). Since this type of fuel pump cannot create sufficient pressure to draw fuel from the tank during cranking, fuel is transferred to the carburetor for starting by operating the primer bulb installed in the fuel line.

Pressure pulsations created by movement of the pistons reach the fuel pump through a passageway between the crankcase and pump.

Upward piston motion creates a low pressure on the pump diaphragm. This low pressure opens the inlet check valve in the pump, drawing fuel from the line into the pump. At the same time, the low pressure draws the air-fuel mixture from the carburetor into the crankcase.

Downward piston motion creates a high pressure on the pump diaphragm. This pressure closes the inlet check valve and opens the outlet check valve, forcing the fuel into the carburetor and drawing the air-fuel mixture from the crankcase into the cylinder for combustion. **Figure 2** shows the operational sequence of a

CHAPTER SIX

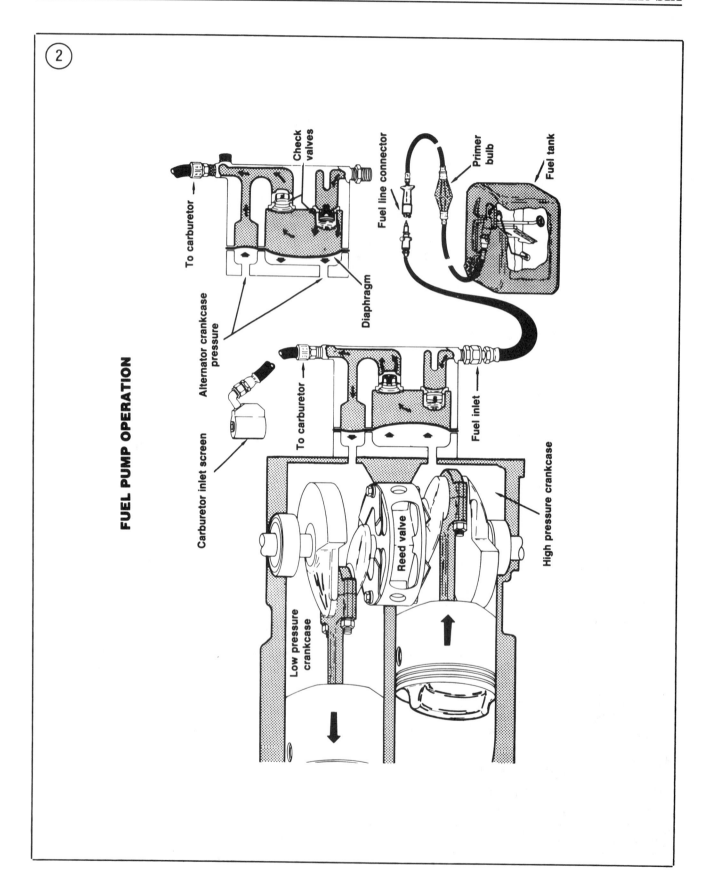

FUEL SYSTEM

typical Johnson and Evinrude outboard fuel pump.

Fuel Pump Problems

Johnson and Evinrude fuel pumps are self-contained, remote assemblies. Fuel pump design is extremely simple and reliable in operation. Diaphragm failures are the most common problem, although the use of dirty or improper fuel-oil mixtures can cause check valve problems. Repair kits are available to overhaul a defective pump.

V6 engines use dual fuel pumps mounted in series. Under certain operational conditions, a V6 engine may exhibit symptoms such as stalling when coming back from a full throttle position or surging with the throttle at a high setting. The symptoms point to a lack of fuel, yet the pumps may check out satisfactorily when pressure tested (Chapter Three).

When the motor is used on a boat containing permanent fuel tanks installed by the boat manufacturer, flow restrictions may be caused by the anti-siphon valve, primer, selector valves or other such devices that are part of the built-in tank system. Since low fuel flow can seriously damage an outboard, a built-in fuel tank system should be periodically checked for possible flow restrictions.

If the boat's fuel system functions properly and the engine has been correctly synchronized and the linkage adjusted as described in Chapter Five, Johnson and Evinrude recommend that 1976-1983 V6 fuel pumps be replaced with high performance pumps that will accomodate more vaporized fuel in the line.

NOTE
Some 1983 V6 engines are factory-equipped with the high performance pump. The standard and high performance pumps differ slightly in appearance. See **Figure 3**.

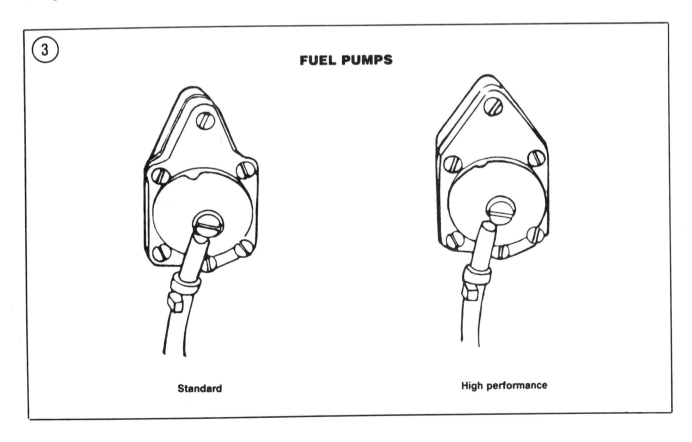

The high performance pump is supplied with a molded hose to be used on some applications but discarded on others. High performance pumps should be installed as follows:

a. 1976 V6—Upper pump, part No. 393868: discard hose. Lower pump, part No. 393868: modify and install hose.
b. 1977-1978 V6—Upper pump, part No. 393868: discard hose. Lower pump, part No. 393870: discard hose.
c. 1979-1983 V6—Upper pump, part No. 393870: discard hose. Lower pump, part No. 393868: install hose.

Removal/Installation

1. Unscrew and remove the filter cover and screen (**Figure 4**).
2. Remove the 2 screws at the bottom of the pump holding it to the engine (**Figure 5**).
3. Remove and discard any straps holding the fuel lines to the fuel pump. Disconnect the lines at the pump.
4. Remove the pump and gasket from the engine. Discard the gasket.
5. Clean all gasket residue from the engine mounting pad. Work carefully to avoid gouging or damaging the mounting surface.
6. Clean the filter screen in OMC Engine Cleaner and blow dry with compressed air. If extremely dirty or damaged, install a new screen in Step 7.
7. Installation is the reverse of removal. Use new mounting and filter screen gaskets. Install new straps on the fuel line connections.

Disassembly/Assembly

1. Remove the 4 screws holding the pump body together.
2. Separate the pump body. Remove and discard the diaphragm, check valves and gaskets.
3. Assembly is the reverse of disassembly. Install a new diaphragm, check valves and gaskets.

CARBURETORS

Many carburetors used on Johnson and Evinrude outboards have a fixed main jet orifice and require no high-speed adjustment.

Carburetors used on 1983 and earlier models have a white Delrin (plastic) needle valve retainer. Age and engine vibration can cause the retainer to lose its ability to prevent the low-speed needle from moving while the engine is running.

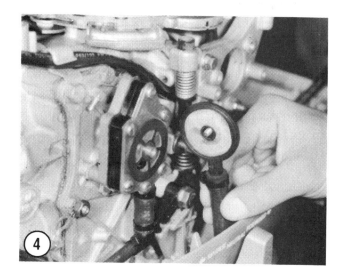

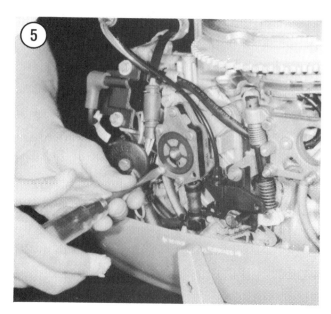

FUEL SYSTEM

An improved red retainer (part No. 315232) can be installed to correct the problem.

When removing and installing a carburetor, make sure that the mounting nuts are securely tightened. A loose carburetor will cause a lean-running condition.

High-elevation Modifications

Table 1 contains orifice recommendations suggested by Johnson and Evinrude when a 1973-1982 engine is used primarily at high elevations. Rejetting for high elevation operation will recover only that engine power lost due to the improper air-fuel ratio caused by the reduction in air density.

The propeller used must allow the engine to run within the recommended engine speed operating range. The correct propeller should place full throttle engine rpm in the middle of the recommended operating range. Changing the prop for high elevation operation will recover only that engine power lost by not operating within the proper rpm range.

Always rejet and prop the engine for the lowest elevation at which the boat will be operated to prevent the possibility of power head damage from a lean fuel mixture. If the boat is to be used extensively at both high and low elevations, you should have 2 sets of jets, 2 props and a fixed jet screwdriver (part No. 317002) for installation as required.

Your Johnson or Evinrude dealer can supply elevation modification stickers (part No. 393533) for application on the motor as a reminder of the original and elevation jet/prop sizes. Their use will assure that you always have the correct information readily available.

Cleaning and Inspection

Before removing and disassembling any carburetor, be sure you have the proper overhaul kit, the proper tools and a sufficient quantity of fresh cleaning solvent. Work slowly and carefully, follow the disassembly procedures, refer to the exploded drawing of your carburetor when necessary and do not apply excessive force at any time.

It is not necessary to disassemble the carburetor linkage or remove the throttle cam or other external components. Wipe the carburetor casting and linkage with a cloth moistened in solvent to remove any contamination and operating film. Clean the carburetor castings with an aerosol type solvent and a brush. Do not submerge them in a hot tank or carburetor cleaner. A sealing compound is used around the metering tubes and on the casting to eliminate porosity problems. A hot tank or submersion in carburetor cleaner will remove this sealing compound.

Spray the cleaner on the casting and scrub off any gum or varnish with a small bristle brush. Spray the cleaner through the casting metering passages. Never clean passages with a wire or drill as you may enlarge the passage and change the carburetor calibration.

Blow castings dry with low pressure (25 psi or less) compressed air. The use of higher pressures can damage the sealing compound.

Check the float for fuel absorption. Check the float arm for wear in the hinge pin and needle valve contact areas. Replace as required.

Check the needle valve tip for grooving, nicks or scratches. **Figure 6** shows a good valve tip

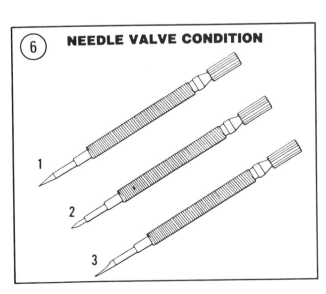

(1), a valve tip damaged from excessive pressure when seating (2) and one with wear on one side caused by vibration resulting from the use of a damaged propeller (3).

Check the throttle and choke shafts for excessive wear or play. The throttle and choke valves must move freely without binding. Replace carburetor if any of these defects are noted.

Clean all gasket residue from mating surfaces and remove any nicks, scratches or slight distortion with a surface plate and emery cloth.

Core Plugs and Lead Shot

Certain openings in the carburetor casting are covered with a core plug or have a lead shot installed. These usually require service only if the openings are leaking. **Figure 7** shows a carburetor with the core plug over the low-speed orifices removed (A) and a typical lead shot installed (B).

Core Plug Service

1. If leakage is noted, secure the carburetor in a vise with protective jaws.
2. Hold a flat end punch in the center of the core plug and tap sharply with a hammer to flatten the plug. Cover the plug area with OMC Adhesive M.

CAUTION
Do not drill more than 1/16 in. below the core plug in Step 3 or the casting will be damaged.

3. If this does not solve the leakage problem or if the low-speed orifices are completely plugged, carefully drill a 1/8 in. hole through the center of the plug and pry it from the casting with a punch.
4A. Clean all residue from the core plug hole in the casting. If the hole is out-of-round, replace the casting.
4B. If the low-speed orifices are plugged, clean with a brush and carburetor cleaner.
5. Coat the outer edge of a new core plug with OMC Adhesive M and position it in the casting opening with its convex side facing up.
6. Hold a flat end punch in the center of the core plug and tap sharply with a hammer to flatten the plug.
7. Coat the core plug with engine oil and blow compressed air (25 psi or less) through the casting passages to check for leakage.
8. Wipe the oil from the core plug and coat with OMC Adhesive M.

Lead Shot Service

1. If leakage is noted, secure the carburetor in a vise with protective jaws.
2. Tap the center of the lead shot sharply with a small hammer and appropriate size punch.
3. If leakage remains, carefully pry the lead shot from its opening with a suitable knife, awl or other sharp instrument.
4. Clean any residue from the lead shot opening in the casting.
5. Install a new lead shot in the opening and flatten out with a hammer and appropriate size punch.

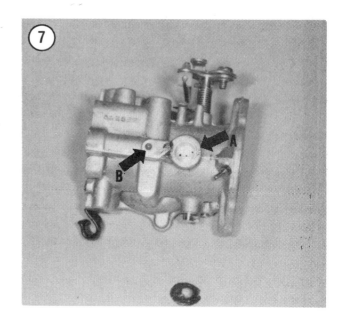

FUEL SYSTEM

6. Coat the core plug with engine oil and blow compressed air (25 psi or less) through the casting passages to check for leakage.

7. Clean the oil from the casting after pressure testing.

CARBURETOR (48-75 HP)

Removal/Installation

1. Remove the engine cover.
2. Disconnect the carburetor fuel line at the fuel pump. Plug the line to prevent leakage.
3. Remove the air silencer cover and gasket. Discard the gasket.
4. Remove and discard the air silencer base self-locking screws.
5. Disconnect the slow-speed needle valve link and carefully pull the slow-speed adjustment lever from the needle valve, if so equipped.
6. Disconnect the drain line and remove the air silencer base.
7. Disconnect the fuel lines between the carburetors.
8. Disconnect the throttle and choke linkage at the carburetors.
9. If trim/tilt junction box bracket on models so equipped interferes with upper carburetor removal, remove the box/bracket as an assembly and place to one side out of the way.
10. Remove the carburetor mounting nuts and lockwashers. Remove the carburetor. Remove and discard the gasket.
11. Disconnect the primer line from the carburetor nipple, if so equipped.
12. Clean all gasket residue from the manifold mounting surface.
13. Installation is the reverse of removal. Use a new gasket. Adjust choke linkage, throttle linkage and carburetor (Chapter Five) before reinstalling the air silencer. Install air silencer base with new self-locking screws.

Disassembly/Assembly

Refer to **Figure 8** or **Figure 9** for this procedure.

1. Remove the float chamber drain or fuel line plug. Drain any fuel remaining in the carburetor into a suitable container.
2. Remove the fixed high-speed orifice from the float bowl with screwdriver part No. 317002 or a suitable equivalent. **Figure 10** shows the plug and orifice removed.
3A. 1984 75 hp—Carefully mark position of low-speed adjustment needle slot. See **Figure 11**. Turn needle clockwise with screwdriver part No. 317002 (or equivalent) until it seats lightly and count the turns (approximately one), then back the needle out and remove. Remove the needle retainer.
3B. 1989 and 1990 48-60 hp—Carefully mark position of low-speed adjustment needle slot. See **Figure 12**. Turn needle clockwise with screwdriver until it seats lightly and count the turns (approximately 2-1/2—2-3/4), then back the needle out and remove. Remove the needle holder assembly.
3C. All others—Remove the low-speed needle valve (**Figure 13A**) or orifice plug from the carburetor:
   a. If equipped with a needle valve, insert a length of wire with a hooked end in the needle valve keyhole slot and remove the Delrin retainer.
   b. If equipped with an orifice plug, remove the orifice with screwdriver part No. 317002 or a suitable equivalent.
4. Remove the idle or low-speed orifice plug and orifice, if so equipped.
5. Remove the float chamber screws. Separate the float chamber from the main body (**Figure 13B**). Discard the float chamber gasket.
6. Remove the nylon float hinge pin. Remove the float and needle valve from the float chamber. See **Figure 14**.
7. Remove the needle valve seat (**Figure 15**) with a wide blade screwdriver. Discard the seat gasket.

CHAPTER SIX

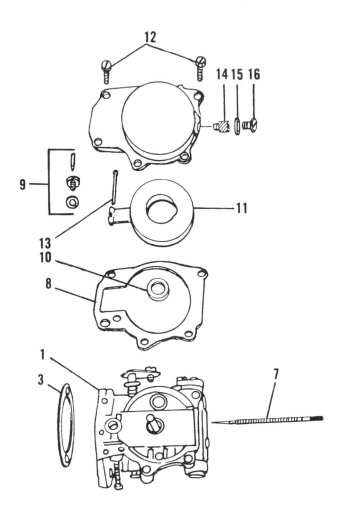

⑧ CARBURETOR (EARLY 50-70 HP)

1. Body
3. Mounting gasket
7. Low-speed needle valve
8. Fuel bowl gasket
9. Needle valve and seat assembly
10. Nozzle gasket
11. Float assembly
12. Float bowl screws
13. Float pin
14. High-speed orifice plug
15. Washer
16. Screw plug

FUEL SYSTEM

⑨

CARBURETOR (LATE 48-70 HP)

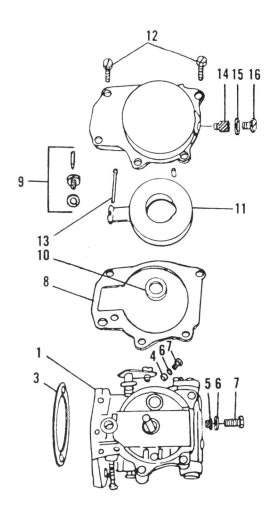

1. Body
3. Gasket
4. Idle orifice plug (55 and 60 hp)
5. Idle orifice plug (50 hp)
   Intermediate orifice plug
   (55 and 60 hp)
   Idle air bleed orifice plug
   (1989 and 1990 48-60 hp)
6. Washer
7. Screw plug
8. Gasket
9. Needle valve and seat assembly
10. Nozzle gasket
11. Float assembly
12. Float bowl screws
13. Float pin
14. High-speed orifice plug
15. Washer
16. Screw plug

8. Remove and discard all gaskets, O-rings or sealing washers in the carburetor body casting.

9. Assembly is the reverse of disassembly. Compare new gaskets to the old ones to make sure all holes are properly punched. Remove any loose gasket fibers or stamping crumbs adhering to the new gaskets. Adjust the float as described in this chapter. If equipped with a needle valve, lightly seat the needle and back it out the number of turns noted during disassembly. On 1984 75 hp and 1989-1990 48-60 hp carburetors, lightly seat low-speed adjustment needle and back out to align with mark scribed during disassembly. Reinstall on engine and adjust the choke linkage, throttle linkage and carburetor (Chapter Five).

Float Adjustment

1. Invert the carburetor body with its gasket surface horizontal.

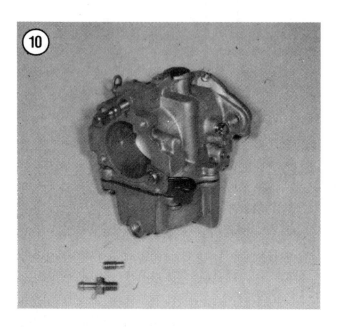

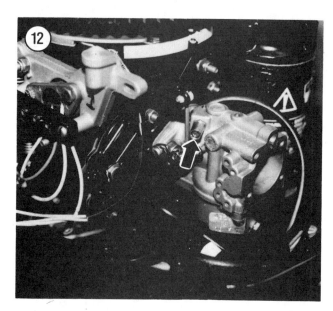

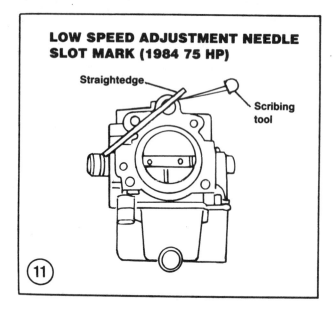

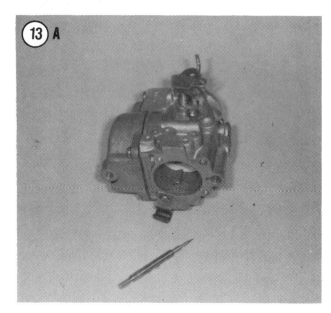

FUEL SYSTEM

2. Place float gauge (part No. 324891) on the gasket surface and hold it next to the float (**Figure 16**). Do not let gauge pressure hold float down.

3. If the top of the float is not between the gauge notches (**Figure 16**), bend the metal float arm carefully (to avoid forcing the needle valve into its seat) and bring the level within specifications.

4. Return the carburetor body to its normal running position and check float drop. The distance between the carburetor body and the float as shown in **Figure 17** should be 1 1/8-1 5/8 in.

5. If the float drop is incorrect, carefully bend the tang (**Figure 17**) until it comes within specifications.

V4 AND V6 CARBURETOR (EXCEPT 1985-ON 120-140 HP AND 1986-ON 200-225 HP)

Some early 1983 V4 engines are equipped with what appear to be slightly different carburetors. One is a 1982 design; the other is a 1983 design.

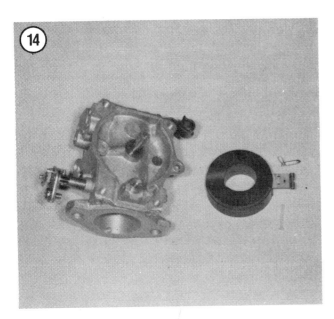

The 1982 carburetor has 2 orifice screw plugs vertically positioned on each side of the float bowl. The 1983 design uses a single plug. Both carburetors are calibrated the same and the difference in appearance will not affect performance.

The following Johnson and Evinrude 1983 150 hp engines may suffer from a lean condition and run poorly at or near wide-open throttle:

a. Johnson 150TLCTE—serial No. 5850458 and above.
b. Evinrude 150TRLCTE—serial No. 288678 and above.
c. Johnson 150TXCTE—serial No. 5852028 and above.
d. Evinrude 150TRXCTE—serial No. 290406 and above.

Some of these engines were equipped with a carburetor body containing venturis that are too large. If poor performance is encountered, remove the air silencer cover and base. The carburetors should have a "1 1/4" cast on their front. This is the correct venturi size. If "1 3/8" is cast on the carburetors, they are responsible for the poor performance and should be replaced with carburetors of the same part number which have 1 1/4 in. venturis.

Removal/Installation

1. Remove the air silencer cover and gasket. See **Figure 18** for V4; V6 is similar. Discard the gasket.
2. Remove the screws holding the air silencer base. See **Figure 19** for V4; V6 is similar. Discard the screws.
3. Disconnect drain line at rear of air silencer base. Remove the base and gasket. Discard the gasket.
4. Remove the VRO pump (V4) or VRO pump bracket (V6), if so equipped, and place out of the way without disconnecting the pump lines.
5. If equipped with a choke solenoid, disconnect solenoid link and retaining ring from upper choke arm.
6. If equipped with a choke primer, disconnect the primer lines to the carburetors on 1982 and later models.
7. Disconnect the throttle and choke lever linkage.
8. Remove the carburetor mounting nuts. Remove the carburetor(s) and gasket(s). Discard the gasket(s).

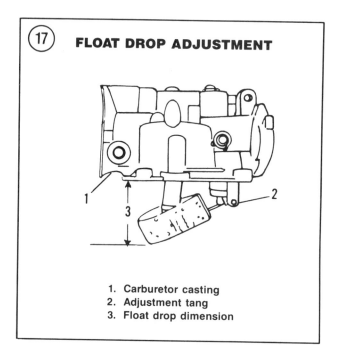

FLOAT DROP ADJUSTMENT

1. Carburetor casting
2. Adjustment tang
3. Float drop dimension

FUEL SYSTEM

9. Disconnect the fuel lines at the carburetor(s).
10. Clean all gasket residue from the manifold mounting surface.
11. Installation is the reverse of removal. Use new gasket(s). Adjust choke linkage, throttle linkage and carburetor(s) (Chapter Five) before reinstalling the air silencer. Install air silencer base with new screws.

Disassembly/Reassembly

Refer to **Figure 20** or **Figure 21** for this procedure.

1. Invert carburetor over a suitable container and tip the front down to drain any remaining fuel out the bowl vent.

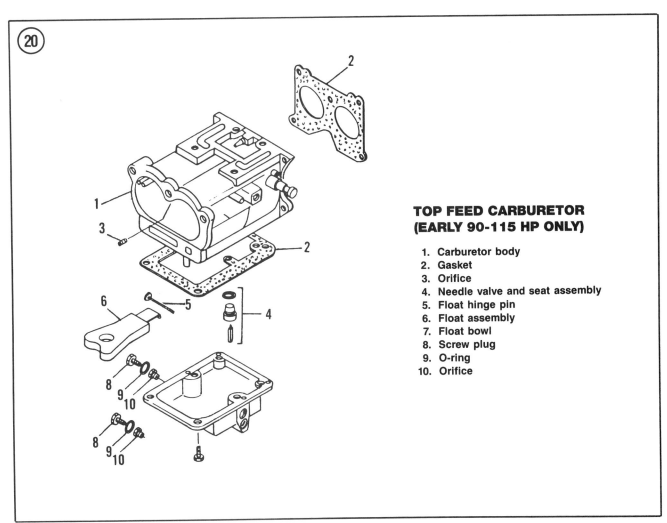

TOP FEED CARBURETOR (EARLY 90-115 HP ONLY)

1. Carburetor body
2. Gasket
3. Orifice
4. Needle valve and seat assembly
5. Float hinge pin
6. Float assembly
7. Float bowl
8. Screw plug
9. O-ring
10. Orifice

278 CHAPTER SIX

NOTE
If the orifice plugs have not been removed before, they may be difficult to loosen in Step 2. To prevent stripping the plug slot, support the float chamber on a solid surface and tap the plug gently with a small mallet. This should loosen the screw sufficiently for easy removal with a screwdriver.

2. Remove the orifice plugs and O-rings from the float chamber. Discard the O-rings.

3. Remove the high-speed orifice from each plug hole with screwdriver part No. 317002 or equivalent and record the number stamped on it for correct reassembly reference. **Figure 22** (dual plug design) shows the plugs and orifices on one side. **Figure 23** (single plug design) shows the plugs and orifices on both sides.

4. Remove the idle air bleeds from the carburetor casting (if so equipped). See **Figure 24**. Record the number stamped on each for correct reassembly reference.

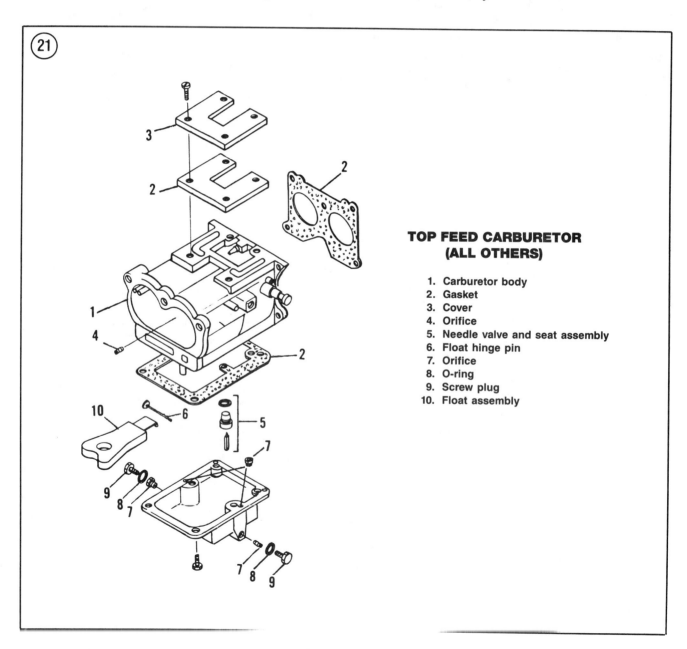

TOP FEED CARBURETOR (ALL OTHERS)

1. Carburetor body
2. Gasket
3. Cover
4. Orifice
5. Needle valve and seat assembly
6. Float hinge pin
7. Orifice
8. O-ring
9. Screw plug
10. Float assembly

FUEL SYSTEM

5. Remove the pull-over jet (**Figure 25**) from the carburetor casting, if so equipped.

6. Remove the cover plugs and intermediate orifices from the float chamber with screwdriver part No. 317002 or equivalent. See **Figure 26**.

7. Remove the float chamber screws. Separate the float chamber from the main body casting (**Figure 27**). Discard the gasket.

8. Remove the nylon float hinge pin. Remove the float (**Figure 28**).

9. Remove the inlet needle from the valve, then remove the valve seat with a wide blade screwdriver. See **Figure 29**. Discard the seat gasket.

10. Remove the idle chamber cover and gasket, if so equipped. See **Figure 30**. Discard the gasket.

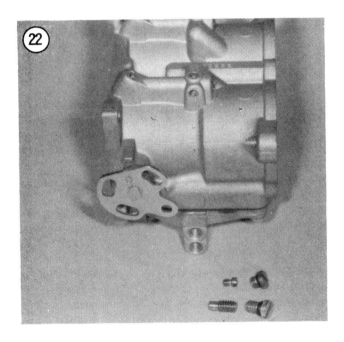

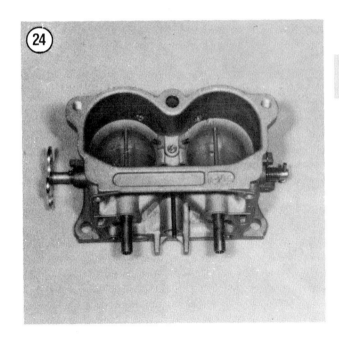

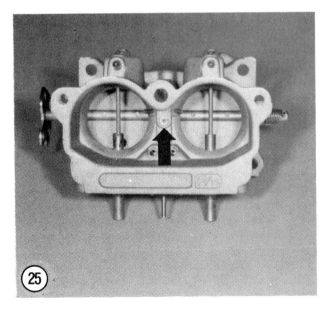

11. Assembly is the reverse of disassembly. Compare new gaskets to the old ones to make sure all holes are properly punched. Remove any loose gasket fibers or stamping crumbs adhering to the new gaskets. Adjust the float as described in this chapter. Reinstall on engine and adjust the choke linkage, throttle linkage and carburetor(s) (Chapter Five).

Float Adjustment

1. Invert the carburetor body with its gasket surface horizontal.

2. Place float gauge (part No. 324891) on the gasket surface and hold it next to the float (**Figure 31**). Do not let gauge pressure hold float down.

3. If the top of the float is not between the gauge notches (**Figure 31**), bend the metal float arm carefully (to avoid forcing the needle valve into its seat) and bring the level within specifications.

4. Return the carburetor body to its normal running position and check float drop. The distance between the carburetor body and the

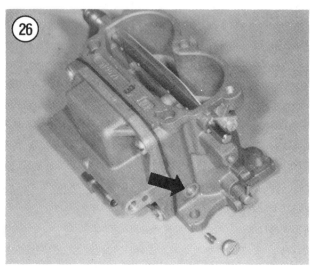

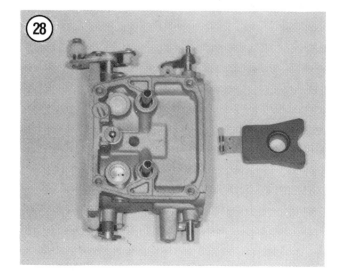

FUEL SYSTEM

float as shown in **Figure 32** should be 7/8-1 1/8 in.

5. If the float drop is incorrect, carefully bend the tang (**Figure 32**) until it comes within specifications.

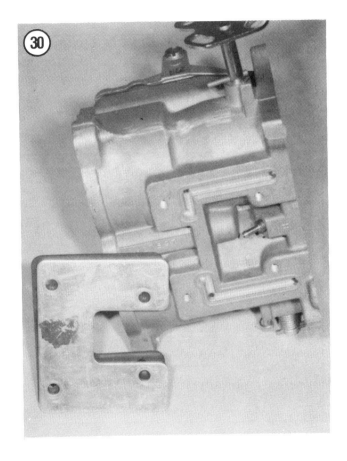

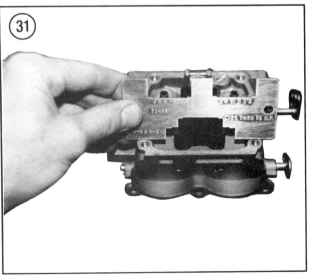

V4 AND V6 CARBURETOR (1985-ON 120-140 HP AND 1986-ON 200-225 HP)

These engines are equipped with a pair of 2-barrel carburetors containing fixed high- and low-speed jets. Each carburetor consists of a metal throttle body and 2 plastic main body/float chambers.

The main bodies on each carburetor can be removed individually for service or cleaning without disconnecting the throttle linkage or removing the throttle body. When overhauling the power head, the carburetors are removed as assemblies. Because of their special construction, observe the cleaning and inspection procedures, torque sequence and values specified to prevent carburetor damage.

Main Body
Removal/Installation

1. Remove the power steering hose support bracket, if so equipped.
2. Loosen the retaining screws holding the port and starboard lower engine covers. Remove the covers.
3. Remove the air silencer cover.

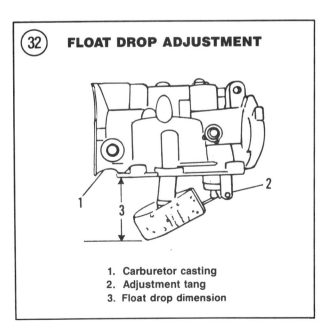

FLOAT DROP ADJUSTMENT

1. Carburetor casting
2. Adjustment tang
3. Float drop dimension

4. Remove the 4 mounting screws holding the main body to the throttle body.

5. Remove the main body/float chamber assembly. Remove and discard the carburetor body seals.

6. Cut the tie strap(s) and disconnect the fuel inlet hose(s) at the carburetor(s).

7. Installation is the reverse of removal. Use new carburetor body seal(s) and tie strap(s). Tighten mounting screws to 40-50 in.-lb. (4.5-5.6 N•m) in a crisscross pattern. Squeeze the primer bulb and check for fuel leaks.

Carburetor Assembly Removal/Installation

1. Remove the power steering hose support bracket, if so equipped.

2. Loosen the retaining screws holding the port and starboard lower engine covers. Remove the covers.

3. Remove the air silencer cover.

4. Cut the tie strap holding the main body fuel supply line. Carefully remove the line.

5. Loosen the throttle shaft link at one or both ends.

6. Cut the tie strap(s) and disconnect the fuel inlet hose(s) at the carburetor(s).

7. Cut the primer hose tie strap and disconnect the hose from the fuel manifold.

8. Cut the VRO pump outlet tie strap and separate the outlet hose.

9. Note the primer hose routing over the top of the carburetors, then remove the hoses from the intake manifold nipples.

10. Apply an even coat of Johnson or Evinrude 50:1 Lubricant to the fuel supply hose installed through the air silencer base.

11. Remove 2 nuts and 2 screws for each carburetor assembly to be removed.

12. Remove the carburetor assembly from the intake manifold, slowly drawing the fuel supply hose through the intake manifold grommet.

13. Remove and discard the carburetor flange gaskets.

14. Installation is the reverse of removal, plus the following:
  a. Use new gaskets without sealer.
  b. Lubricate fuel supply hose as in Step 10.
  c. Tighten mounting nuts/screws to 120-144 in.-lb. (14-16 N•m).
  d. Install new tie straps.
  e. Adjust throttle linkage and carburetor (Chapter Five) before installing the air silencer.
  f. Squeeze the primer bulb and check for fuel leaks.

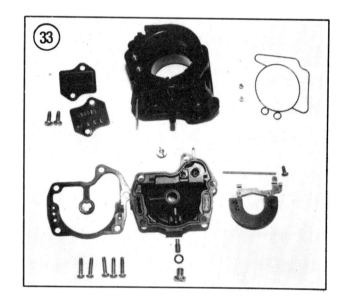

FUEL SYSTEM

Disassembly

Refer to **Figure 33** for this procedure.

1. Remove the main body attaching screws. Separate the main body/float chamber assembly from the throttle body. Remove and discard the O-ring seal.
2. Remove the float chamber screws. Separate the float chamber from the main body. Remove and discard the float chamber gasket.
3. Remove the float hinge pin anchor screw. See A, **Figure 34**. Remove the float and inlet needle from the float chamber.
4. Remove the inlet valve seat and gasket with a wide-blade screwdriver. Discard the gasket.
5. Remove the high-speed orifice plug from the float chamber with a wide-blade screwdriver.
6. Remove the high-speed orifice (B, **Figure 34**) with orifice plug screwdriver part No. 317002 or equivalent.
7. Remove the mixing chamber cover plate and discard the gasket.
8. Remove the idle air bleed and intermediate air bleed orifices with orifice plug screwdriver part No. 317002 or equivalent. See **Figure 35**.

Cleaning and Inspection

1. Clean all components using a mild aerosol solvent. Do not submerge components in hot tank or strong carburetor cleaner.
2. Flush all holes and passages with a syringe containing isopropyl alcohol. Blow holes and passages dry with low-pressure compressed air (under 25 psi).
3. Check inlet needle and valve seat for distortion or wear. Replace as required.
4. Check throttle body and shaft for excessive wear or play. Replace as required.

Assembly

Assembly is the reverse of disassembly, plus the following:
   a. Compare new gaskets to old ones to make sure all holes are properly punched.
   b. Remove any loose gasket fibers or stamping crumbs adhering to the new gaskets.
   c. Adjust the float as described in this chapter.
   d. Tighten mixing chamber cover and hinge pin anchor screws to 18-24 in.-lb. (2.2-2.7 N•m); high-speed orifice plug to 30-35 in.-lb. (3.3-3.9 N•m) and float chamber screws to 18-24 in.-lb. (2.2-2.7 N•m) in a criss-cross pattern.

Float Adjustment

Refer to **Figure 36** for this procedure.

1. Invert the carburetor float chamber with its gasket surface horizontal.

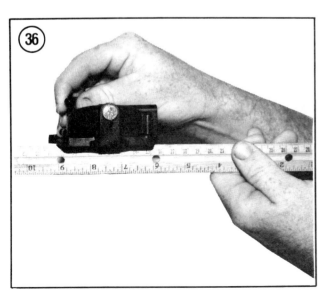

2. Check float height with a straightedge. Float should be level with gasket surface of float chamber ± 1/32 in.

3. If float requires adjustment, remove anchor screw and carefully bend tang on float lever. Reinstall float and anchor screw and repeat Step 1 and Step 2 to check adjustment.

CHOKE AND PRIMER SOLENOID SERVICE

Choke Solenoid Removal/Installation

1. Disconnect the solenoid link and retaining ring from the upper choke arm.
2. Disconnect the solenoid electrical lead(s) at the terminal board. Remove the leads from the wire routing clamp.

NOTE
Some solenoids are installed with a ground lead under one of the bracket screws. If not equipped with a ground lead, the solenoid ground is through the case.

3. Remove the solenoid bracket screws (**Figure 37**). Remove the solenoid.
4. Installation is the reverse of removal. Coat the terminal board (and ground lead) connections with OMC Black Neoprene Dip.

Choke Solenoid Test

1. Make sure the solenoid plunger is clean, dry and free of all corrosion.
2. 1973—Connect an ohmmeter between the solenoid case (ground) and the purple/yellow lead. The ohmmeter should read 7-8.5 ohms. Connect the ohmmeter between the solenoid case and the purple/white lead. The ohmmeter should read 3-4.5 ohms.
3. 1974-1976 (except 1976 V6)—Connect an ohmmeter between the solenoid case (ground) and the purple/yellow lead. The ohmmeter should read 4-6 ohms. Connect the ohmmeter between the solenoid case and the purple/white lead. The ohmmeter should read 3-5 ohms.
4. 1977-1983 50-60 hp, 1977-1981 V4, 1976-1981 V6—Connect an ohmmeter between the solenoid lead and case or between the 2 solenoid leads (if equipped with a ground lead). The ohmmeter should read 1.5-2.5 ohms.
5. If the ohmmeter reading is not as specified for the engine year and model, replace the choke solenoid.

Primer Solenoid Removal/Installation

Figure 38 shows the primer solenoid location on inline engines. On V4 and V6 models, it is mounted on the air silencer assembly.

FUEL SYSTEM

1. Disconnect the solenoid purple/white lead at the terminal board.
2. Remove the 2 screws and clamp holding the solenoid. Remove the solenoid.
3. Installation is the reverse of removal. Be sure to reinstall the ground lead under the lower clamp screw. Coat the screw with OMC Black Neoprene Dip.

Primer Solenoid Test

The solenoid plunger must be free of any dirt or corrosion that would prevent it from moving freely.

Connect an ohmmeter between the solenoid purple/white lead and the black ground lead. If the ohmmeter does not read 4-7 ohms, replace the solenoid.

Primer Solenoid Disassembly/Assembly

1. Remove the cover screws (**Figure 39**). Remove the cover and gasket. Discard the gasket.
2. Remove the valve seat, filter, valve, plunger and both springs. See **Figure 40**.
3. Clean or replace the filter as required.
4. Clean the plunger to remove any contamination or corrosion.

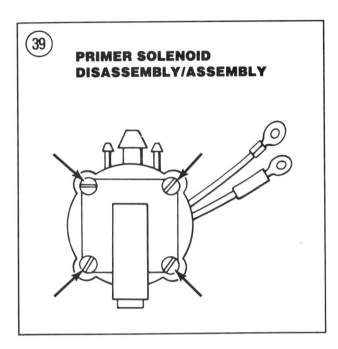

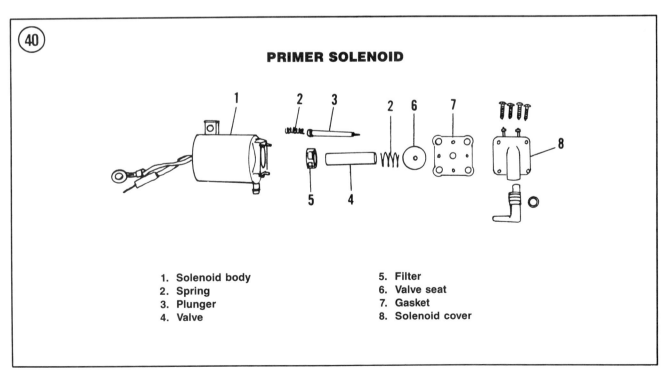

5. Assembly is the reverse of disassembly. Install a new valve seat and use a new cover gasket.

ANTI-SIPHON DEVICES

In accordance with industry safety standards, late-model boats equipped with a built-in fuel tank will have some form of anti-siphon device installed between the fuel tank outlet and the outboard fuel inlet. This device is designed to shut the fuel supply off in case the boat capsizes or is involved in an accident. Quite often, the malfunction of such devices leads the owner to replace a fuel pump in the belief that it is defective.

Anti-siphon devices can malfunction in one of the following ways:
    a. Anti-siphon valve: orifice in valve is too small or clogs easily; valve sticks in closed or partially closed position; valve fluctuates between open and closed position; thread sealer, metal filings or dirt/debris clogs orifice or lodges in the relief spring.
    b. Solenoid-operated fuel shut-off valve: solenoid fails with valve in closed position; solenoid malfunctions, leaving valve in partially closed position.
    c. Manually-operated fuel shut-off valve: valve is left in completely closed position; valve is not fully opened.

The easiest way to determine if an anti-siphon valve is defective is to bypass it by operating the engine with a remote fuel supply. If a fuel system problem is suspected, check the fuel filter first. If the filter is not clogged or dirty, bypass the anti-siphon device. If the engine runs properly with the anti-siphon device bypassed, contact the boat manufacturer for replacement of the anti-siphon device.

FUEL TANK

Figure 41 shows the components of the portable fuel tank, including the fuel gauge sender, in-tank filter and primer bulb assembly.

When some oils are mixed with gasoline and stored in a warm place, a bacterial substance will form. This substance is clear and covers the fuel pickup, restricting flow through the fuel system. Bacterial formation can be prevented by using OMC 2+4 Fuel Conditioner on a regular basis. If the bacteria is present, it can be removed with OMC Engine Cleaner.

To remove any dirt or water that may have entered the tank during refilling, clean the inside of the tank once each season by flushing with clean lead-free gasoline or kerosene.

Check the inside and outside of the tank for signs of rust, leakage or corrosion. Replace as required. Do not attempt to patch the tank with automotive fuel tank repair materials. Portable marine fuel tanks are subject to much greater pressure and vacuum conditions.

To check the fuel tank filter for possible restrictions, unscrew the fuel pickup nipple and withdraw the pickup tube and filter assembly from the tank. The filter on the end of the pickup tube can be cleaned with OMC Engine Cleaner.

Alcohol blended with gasoline may cause a gradual deterioration of the indicator lens in portable fuel tanks. The use of a tank with an alcohol-resistant indicator lens is recommended if blended fuels are used with any frequency.

FUEL LINE AND PRIMER BULB

When priming the engine, the primer bulb should gradually become firm. If it does not become firm or if it stays firm even when disconnected, a check valve inside the primer bulb is malfunctioning.

The line should be checked periodically for cracks, breaks, restrictions and chafing. The bulb should be checked periodically for proper operation. Make sure all fuel line connections are tight and securely clamped.

FUEL SYSTEM

FUEL TANK

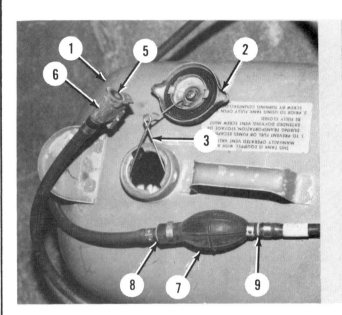
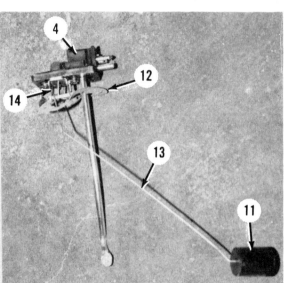

1. Fuel tank
2. Cap
3. Cap anchor
4. Upper housing assembly
5. O-ring
6. Connector assembly
7. Primer bulb
8. Outlet valve
9. Inlet valve
11. Indicator float
12. Housing gasket
13. Indicator arm
14. Indicator support

Table 1 is on the following pages.

Table 1 CARBURETOR ELEVATION ORIFICE CHART

Model	Sea level	3,000 to 6,000 ft.	6,000 to 10,000 ft.
1973-1979			
65 hp			
1973	59D	57D	55D
70 hp			
1974-1975	61D	58D	57D
1976-1978	54D	52D	50D
1979	52D	50D	45D
75 hp			
1975	61D	58D	57D
1976-1979	55D	53D	50D
85 hp			
1973	57C	54C	53C
1974-1976	54C	51C	50C
1977 (long shaft)	49C	47C	45C
1977-1978 (extra long shaft)	45C	43C	41C
1979	48C	46C	43C
100 hp	50C	48C	46C
115 hp			
1973-1975	67C	65C	63C
1976	63C	61C	59C
1977	59C	58C	54C
1978-1979	54C	51C	50C
135 hp			
1973	73C	70C	69C
1974	69C	66C	65C
1975-1976	61C	59C	58C
140 hp			
1977	65C	63C	61C
1978 (early)	61C	59C	58C
1978 (late)	65C	63C	61C
1979	65C	63C	61C
150 hp	51C	49C	47C
175 hp			
1977	58C	56C	54C
1978 (early)	58C	56C	54C
1978 (late)	56C	54C	52C
1979	58.5C	56C	54C
200 hp			
1976	58C	56C	54C
1977-1978	61C	59C	58C
1979	58.5C	56C	54C
235 hp			
1978 (early)	67C	65C	63C
1978 (late)	65C	63C	61C
1979	61C	59C	58C

(continued)

FUEL SYSTEM

Table 1 CARBURETOR ELEVATION ORIFICE CHART (continued)

	1980-1982			
	Sea level[1]		3,000-6,000 ft.[1]	
Model	Low speed	High speed	Low speed	High speed
50 hp				
1980[2]	0.034	52D	0.032	49D
1980[3]	0.046	55D top	0.044	53D top
		56D bottom		53D bottom
55-60 hp				
1980-1982	0.046	55D top	0.044	53D top
		56D bottom		53D bottom
70 hp[4]	0.055	52D	0.061	46D
70 hp[5]	0.068	52D	0.070	46D
75 hp[6]	0.061	55D	0.067	49D
75 hp[7]	0.059	56D	0.061	50D
85 hp (1980)	0.030 IAB	58C	0.034 IAB	56C
90 hp				
1981	0.035 IAB	65C	0.038 IAB	63C
1982	0.034 IAB	56C	0.041 IAB	54C
100 hp	0.034 IAB	54C	0.038 IAB	52C
115 hp				
1980	0.030 IAB	60C	0.034 IAB	58.5C
1981-1982	0.034 IAB	60C	0.038 IAB	58.5C
140 hp				
1980	0.025 IAB	67C	0.029 IAB	65C
1981	0.031 IAB	67C	0.035 IAB	65C
1982	0.036 IAB	67C	0.040 IAB	65C
150 hp	0.032 IAB	49C	0.034 IAB	46C
175 hp	0.033 IAB	60C	0.031 IAB	56C
	0.031 INT		0.035 INT	
200 hp				
1980	0.032 IAB	59C	0.028 IAB	See note 8
	0.033 INT		0.035 INT	56C
1981	0.035 IAB	62C	0.031 IAB	See note 8
	0.030 INT		0.032 INT	59C
1982	0.033 IAB	62C	0.029 IAB	58.5C
	0.033 INT		0.035 INT	
235 hp				
1980	0.030 IAB	62C	0.026 IAB	See note 8
	0.031 INT		0.033 INT	59C
1981-1982	0.030 IAB	62C	0.026 IAB	See note 8
	0.029 INT		0.031 INT	59C

(continued)

Table 1 CARBURETOR ELEVATION ORIFICE CHART (continued)

	1980-1982 (continued)	
	6,000-10,000 ft.[1]	
Model	Low speed	High speed
50 hp		
1980[2]	0.032	46D
1980[3]	0.044	50D top
		50D bottom
55-60 hp		
1980-1982	0.044	50D top
		50D bottom
70 hp[4]	0.67	40D
70 hp[5]	0.073	40D
75 hp[6]	0.073	43D
75 hp[7]	0.064	44D
85 hp (1980)	0.038 IAB	54C
90 hp		
1981	0.042 IAB	61C
1982	0.045 IAB	52C
100 hp	0.042 IAB	50C
115 hp		
1980	0.038 IAB	56C
1981-1982	0.042 IAB	56C
140 hp		
1980	0.033 IAB	63C
1981	0.038 IAB	63C
1982	0.043 IAB	63C
150 hp	0.036 IAB	43C
175 hp	0.029 IAB	52C
	0.031 INT	
200 hp		
1980	0.024 IAB	See note 8
	0.037 INT	53C
1981	0.027 IAB	See note 8
	0.034 INT	56C
1982	0.025 IAB	54C
	0.037 INT	
	(continued)	

FUEL SYSTEM

Table 1 CARBURETOR ELEVATION ORIFICE CHART (continued)

	1980-1982 (continued)	
	6,000-10,000 ft.[1]	
Model	Low speed	High speed
235 hp		
1980	0.023 IAB	See note 8
	0.035 INT	56C
1981-1982	0.026 IAB	See note 8
	0.033 INT	56C

1. IAB equals idle air bleed; INT equals intermediate orifice.
2. Die cast block.
3. Perm-mold block.
4. Carburetor part No. 391608 and part No. 391609.
5. Carburetor part No. 392574 and part No. 392575.
6. Carburetor part No. 391632 and part No. 391633.
7. Carburetor part No. 392576 and part No. 392577.
8. Remove high speed pull-over air bleed, solder it closed and reinstall.

Chapter Seven

Ignition and Electrical Systems

This chapter provides service procedures for the battery, starter motor, charging systems and ignition systems used on Johnson and Evinrude outboard motors during the years covered by this manual. Wiring diagrams are at the end of the book. **Tables 1-3** are at the end of the chapter.

BATTERY

Since batteries used in marine applications endure far more rigorous treatment than those used in an automotive charging system, they are constructed differently. Marine batteries have a thicker exterior case to cushion the plates inside during tight turns and rough weather. Thicker plates are also used, with each one individually fastened within the case to prevent premature failure. Spill-proof caps on the battery cells prevent electrolyte from spilling into the bilges. Automotive batteries are not designed to be run down and recharged repeatedly. For this reason, they should *only* be used in an emergency situation when a suitable marine battery is not available.

Johnson and Evinrude recommend that any 12-volt battery used to crank an outboard motor have a cold cranking amperage of 360 amps (50-140 hp) or 500 amps (V6) at 0° F (-18° C) and a reserve capacity of at least 115 minutes (50-140 hp) or 99 minutes (V6) at 80° F (27° C).

CAUTION
*Sealed or maintenance-free batteries are **not** recommended for use with the unregulated 4-6 amp charging systems used on Johnson and Evinrude 48-75 hp and some V4 outboards. Excessive charging during continued high-speed operation will cause the electrolyte to boil, resulting in its loss. Since water cannot be added to such batteries, such overcharging will ruin the battery.*

Separate batteries may be used to provide power for any accessories such as lighting, fish

IGNITION AND ELECTRICAL SYSTEMS

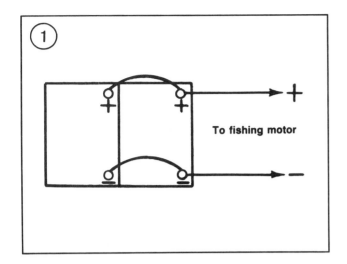

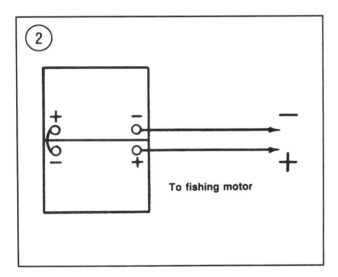

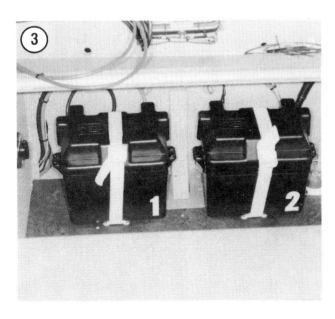

finders, depth finder, etc. To determine the required capacity of such batteries, calculate the average discharge rate of the accessories and refer to **Table 1**.

Batteries may be wired in parallel to double the ampere hour capacity while maintaining a 12-volt system. See **Figure 1**. For accessories which require 24 volts, batteries may be wired in series (**Figure 2**) but only accessories specifically requiring 24 volts should be connected into the system. Whether wired in parallel or in series, charge the batteries individually.

Battery Installation in Aluminum Boats

If a battery is not properly secured and grounded when installed in an aluminum boat, it may contact the hull and short to ground. This will burn out remote control cables, tiller handle cables or wiring harnesses.

Johnson and Evinrude recommend the following preventive steps be taken when installing a battery in a metal boat.

1. Choose a location as far as practical from the fuel tank while providing access for maintenance.
2. Install the battery in a plastic battery box with cover and tie-down strap (**Figure 3**).
3. If a covered container is not used, cover the positive battery terminal with a non-conductive shield or boot (**Figure 4**).
4. Make sure the battery is secured inside the battery box and that the box is fastened in position with the tie-down strap.

Care and Inspection

1. Remove the battery container cover (**Figure 3**) or hold-down strap (**Figure 4**).
2. Disconnect the negative battery cable. Disconnect the positive battery cable.

NOTE
*Some batteries have a built-in carry strap (**Figure 5**) for use in Step 3.*

3. Attach a battery carry strap to the terminal posts. Remove the battery from the battery tray or container.
4. Check the entire battery case for cracks.
5. Inspect the battery tray or container for corrosion and clean if necessary with a solution of baking soda and water.

NOTE
Keep cleaning solution out of the battery cells in Step 6 or the electrolyte will be seriously weakened.

6. Clean the top of the battery with a stiff bristle brush using the baking soda and water solution (**Figure 6**). Rinse the battery case with clear water and wipe dry with a clean cloth or paper towel.
7. Position the battery in the battery tray or container.
8. Clean the battery cable clamps with a stiff wire brush or one of the many tools made for this purpose (**Figure 7**). The same tool is used for cleaning the battery posts. See **Figure 8**.
9. Reconnect the positive battery cable, then the negative cable.

CAUTION
Be sure the battery cables are connected to their proper terminals. Connecting the

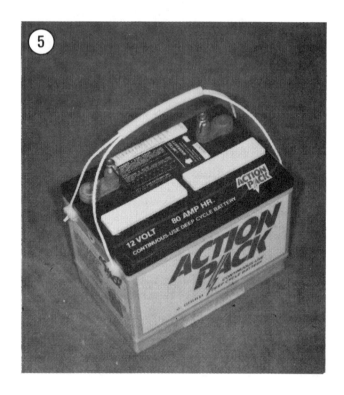

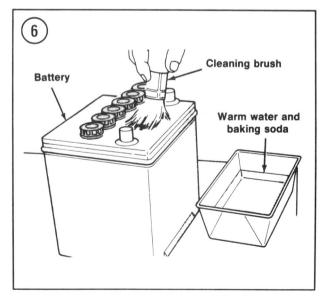

IGNITION AND ELECTRICAL SYSTEMS

battery backwards will reverse the polarity and damage the rectifier.

10. Tighten the battery connections and coat with a petroleum jelly such as Vaseline or a light mineral grease.

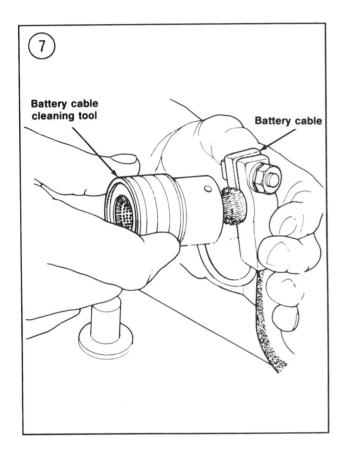

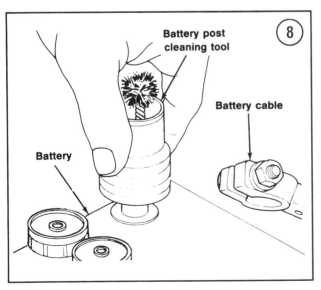

NOTE
Do not overfill the battery cells in Step 11. The electrolyte expands due to heat from charging and will overflow if the level is more than 3/16 in. above the battery plates.

11. Remove the filler caps and check the electrolyte level. Add distilled water, if necessary, to bring the level up to 3/16 in. above the plates in the battery case. See **Figure 9**.

Testing

Hydrometer testing is the best way to check battery condition. Use a hydrometer with numbered graduations from 1.100-1.300 rather than one with just color-coded bands. To use the hydrometer, squeeze the rubber ball, insert the tip in a cell and release the ball (**Figure 10**).

NOTE
Do not attempt to test a battery with a hydrometer immediately after adding water to the cells. Charge the battery for 15-20 minutes at a rate high enough to cause vigorous gassing and allow the water and electrolyte to mix thoroughly.

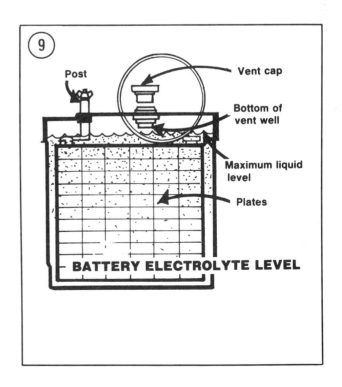

Draw enough electrolyte to float the weighted float inside the hydrometer. When using a temperature-compensated hydrometer, release the electrolyte and repeat this process several times to make sure the thermometer has adjusted to the electrolyte temperature before taking the reading.

Hold the hydrometer vertically and note the number in line with the surface of the electrolyte (**Figure 11**). This is the specific gravity for the cell. Return the electrolyte to the cell from which it came.

The specific gravity of the electrolyte in each battery cell is an excellent indicator of that cell's condition. A fully charged cell will read 1.260 or more at 80° F (25° C). A cell that is 75 percent charged will read from 1.220-1.230 while one with a 50 percent charge reads from 1.170-1.180. If the cell tests below 1.120, the battery must be recharged; one that reads 1.100 or below is dead. Charging is also necessary if the specific gravity varies more than 0.050 from cell to cell.

NOTE
If a temperature-compensated hydrometer is not used, add 0.004 to the specific gravity reading for every 10° above 80° F (25° C). For every 10° below 80° F (25° C), subtract 0.004.

Charging

A good state of charge should be maintained in batteries used for starting. Check the battery with a voltmeter as shown in **Figure 12**. Any

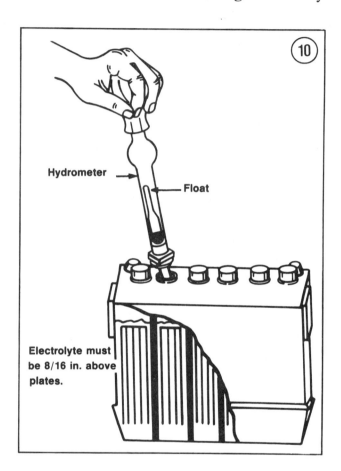

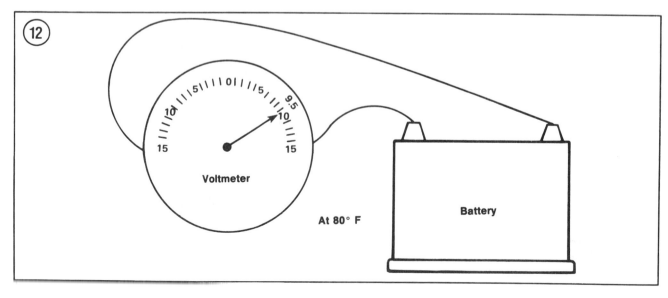

IGNITION AND ELECTRICAL SYSTEMS

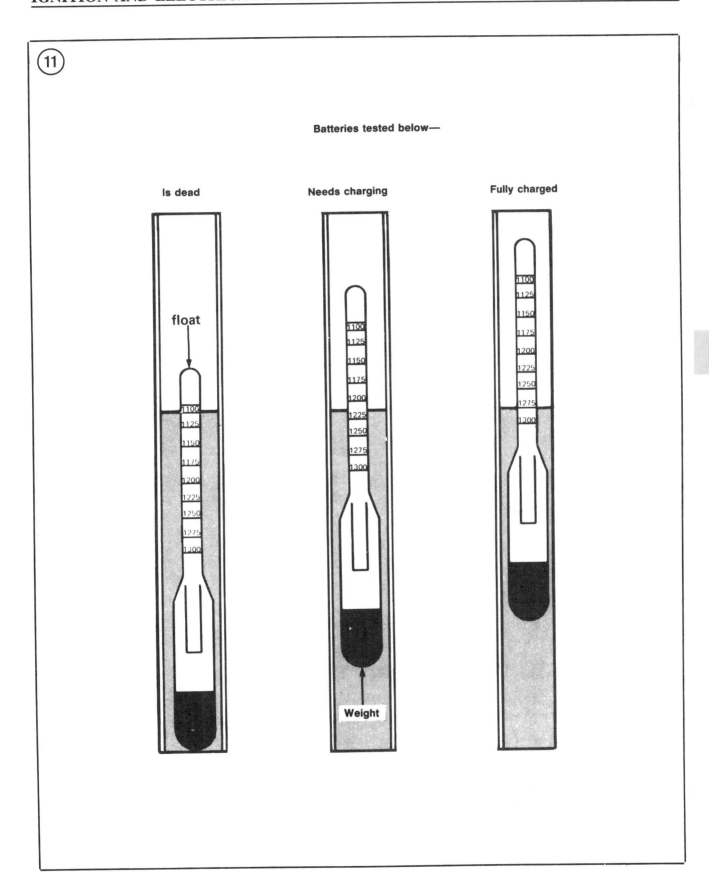

battery that cannot deliver at least 9.6 volts under a starting load should be recharged. If recharging does not bring it up to strength or if it does not hold the charge, replace the battery.

The battery does not have to be removed from the boat for charging, but it is a recommended safety procedure since a charging battery gives off highly explosive hydrogen gas. In many boats, the area around the battery is not well ventilated and the gas may remain in the area for hours after the charging process has been completed. Sparks or flames occuring near the battery can cause it to explode, spraying battery acid over a wide area.

For this reason, it is important that you observe the following precautions:

a. Do not smoke around batteries that are charging or have been recently charged.

b. Do not break a live circuit at the battery terminals and cause an electrical arc that can ignite the hydrogen gas.

Disconnect the negative battery cable first, then the positive cable. Make sure the electrolyte is fully topped up.

Connect the charger to the battery—negative to negative, positive to positive. If the charger output is variable, select a 5 amp setting. Set the voltage regulator to 12 volts and plug the charger in. If the battery is severely discharged, allow it to charge for at least 8 hours. Batteries that are not as badly discharged require less charging

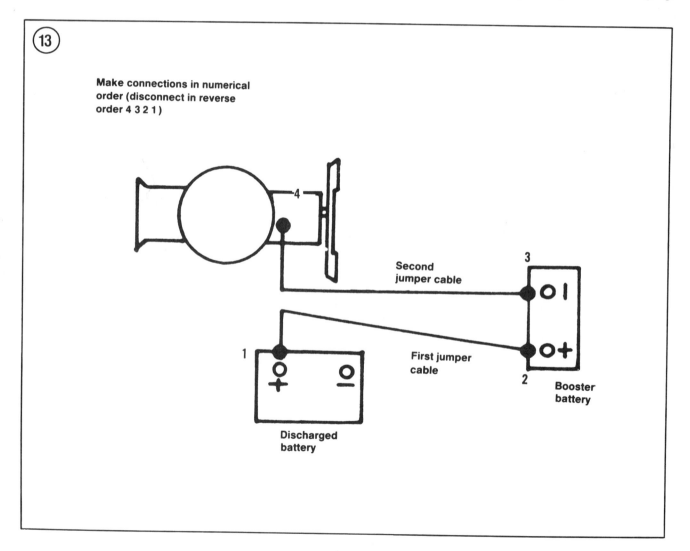

IGNITION AND ELECTRICAL SYSTEMS

time. **Table 2** gives approximate charge rates for batteries used primarily for cranking. Check the charging progress with the hydrometer.

Jump Starting

If the battery becomes severely discharged, it is possible to start and run an engine by jump starting it from another battery. If the proper procedure is not followed, however, jump starting can be dangerous. Check the electrolyte level before jump starting any battery. If it is not visible or if it appears to be frozen, do not attempt to jump start the battery.

WARNING
Use extreme caution when connecting a booster battery to one that is discharged to avoid personal injury or damage to the system.

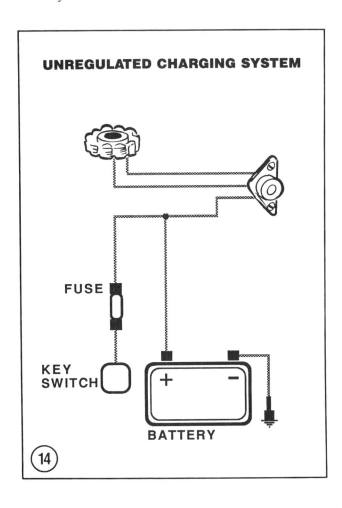

1. Connect the jumper cables in the order and sequence shown in **Figure 13**.

WARNING
An electrical arc may occur when the final connection is made. This could cause an explosion if it occurs near the battery. For this reason, the final connection should be made to a good ground away from the battery and not to the battery itself.

2. Check that all jumper cables are out of the way of moving engine parts.
3. Start the engine. Once it starts, run it at a moderate speed.

CAUTION
Running the engine at wide-open throttle may cause damage to the electrical system.

4. Remove the jumper cables in the exact reverse order shown in **Figure 13**. Remove the cables at point 4, then 3, 2 and 1.

BATTERY CHARGING SYSTEM

A battery charging system is standard on all electric start models.

The unregulated charging system on all 2- and 3-cylinder and some V4 engines consists of a flywheel with cast-in magnets, the stator assembly, a rectifier and the battery. See **Figure 14** (typical).

The regulated charging system on V4 and V6 engines equipped with a 10-35 amp system consists of a flywheel with cast-in magnets, the stator assembly, a rectifier, voltage regulator, fuse and the battery. See **Figure 15** (typical). Late model outboards have a combined rectifer/regulator unit.

Rotation of the flywheel stator magnets past the armature plate stator coils (**Figure 16**) creates alternating current. This current is sent to the rectifier (**Figure 17**) where it is converted into direct current to charge the battery or power accessories.

A malfunction in the unregulated charging system generally results in an undercharged battery. The regulated charging system may produce too little (undercharge) or too much (overcharge) current. Perform the following inspection to determine the cause of the problem. If the inspection proves satisfactory, test the stator coils and rectifier. With regulated systems, also check the voltage regulator. See Chapter Three.

1. Make sure the battery cables are connected properly. The red cable must be connected to the positive battery terminal. If polarity is reversed, check for a damaged rectifier.
2. Inspect the battery terminals for loose or corroded connections. Tighten or clean as required.
3. Inspect the physical condition of the battery. Look for bulges or cracks in the case, leaking electrolyte or corrosion build-up.
4. Carefully check the wiring between the stator coils and battery for signs of chafing, deterioration or other damage.
5. Check the circuit wiring for corroded, loose or disconnected connections. Clean, tighten or connect as required.
6. Determine if the electrical load on the battery from accessories is greater than the battery capacity.
7. Check the fuse on regulated charging systems.

Stator And Charge Coil Replacement

See *Armature Plate Removal/Installation* in this chapter for CD2 and CD2UL ignitions. See *Stator and Timer Base Removal/Installation* in this chapter for CD 3, 4 and 6 ignitions.

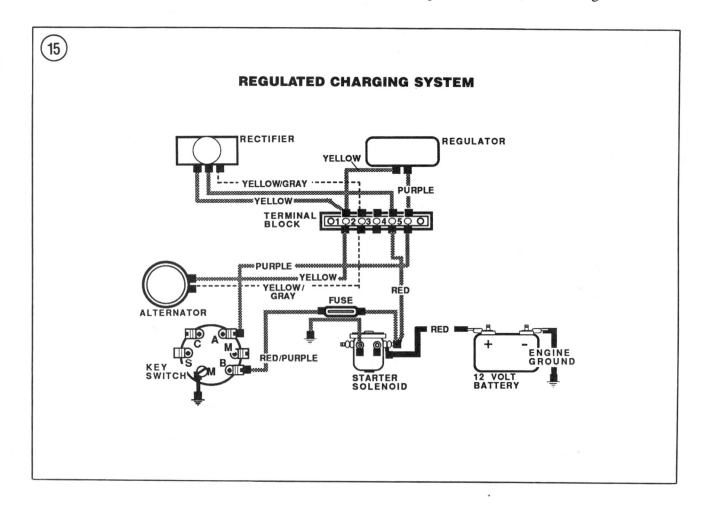

IGNITION AND ELECTRICAL SYSTEMS

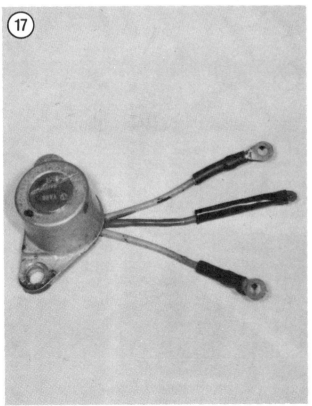

Rectifier Removal/Installation

1. Disconnect the red, yellow and yellow/gray rectifier leads at the terminal board. Some models will also use a yellow/blue wire that should be disconnected.
2. Remove the 2 rectifier mounting screws (**Figure 18**). Remove the rectifier.
3. Installation is the reverse of removal. Coat terminal board mounting screw threads with OMC Screw Lock. Coat rectifier lead connections with OMC Black Neoprene Dip.

Voltage Regulator Replacement

1. Disconnect the voltage regulator leads at the terminal board.
2. Remove the regulator mounting screws. Remove the regulator.
3. Installation is the reverse of removal.

Rectifier/Regulator Assembly Replacement

1. Remove the flywheel (Chapter Eight) and stator assembly as described in this chapter if necessary to provide access to the rectifier/regulator mounting screws.

2. Disconnect the rectifier/regulator leads at the terminal board and starter solenoid.

3. Remove the rectifier/regulator mounting screws. Remove the rectifier/regulator assembly. See **Figure 19**.

4. Installation is the reverse of removal.

ELECTRIC STARTING SYSTEM

All Johnson and Evinrude outboards covered in this manual (except 1973 50 hp and 1980-on 55 hp rope-start models) use an electric (starter motor) starting system. The electric starting system consists of the battery, starter solenoid, starter motor, neutral start switch, ignition switch and connecting wiring.

Starting system operation and troubleshooting is described in Chapter Three.

STARTER MOTOR

Marine starter motors are very similar in design, appearance and operation to those found on automotive engines, however, an automotive starter should never be used as a replacement.

The Bosch and Prestolite starters used on all Johnson and Evinrude outboards covered in this manual (except 1986-on 200-225 hp V6) have an inertia-type drive in which external spiral splines on the armature shaft mate with internal splines on the drive assembly. A Bosch gear-reduction starter is used on 1986-on 200-225 hp V6 motors.

The starter motor produces a very high torque but only for a brief period of time, due to heat buildup. Never operate the starter motor continuously for more than 30 seconds. Let the motor cool for at least 2 minutes before operating it again.

If the starter motor does not turn over, check the battery and all connecting wiring for loose or corroded connections. If this does not solve the problem, refer to Chapter Three. Except for brush replacement, amateur service to the starter motor is limited to replacement with a new or rebuilt unit.

Removal/Installation

1. Disconnect the negative battery cable.

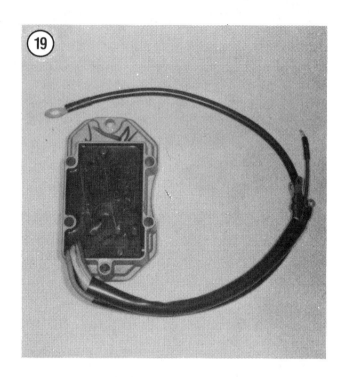

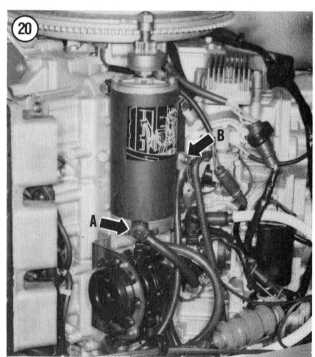

IGNITION AND ELECTRICAL SYSTEMS

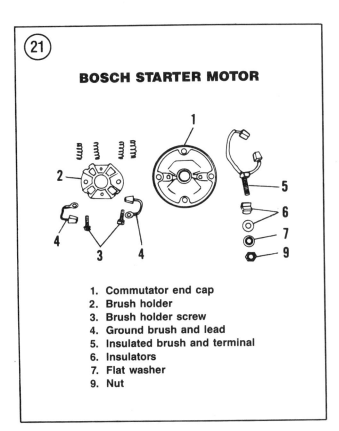

BOSCH STARTER MOTOR

1. Commutator end cap
2. Brush holder
3. Brush holder screw
4. Ground brush and lead
5. Insulated brush and terminal
6. Insulators
7. Flat washer
9. Nut

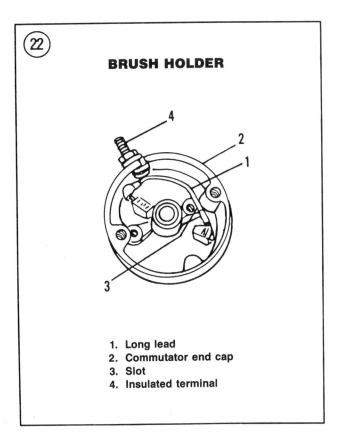

BRUSH HOLDER

1. Long lead
2. Commutator end cap
3. Slot
4. Insulated terminal

2. Remove the engine cover.
3. Disconnect the starter motor lead (A, **Figure 20**). On V6 models with a Bosch starter, disconnect the negative cable at the starter mounting flange (B, **Figure 20**).
4. Remove the air silencer assembly and/or starter solenoid on V4 and V6 models, if required to provide clearance for motor removal.
5. Remove the starter or starter bracket mounting bolts and lockwashers. Remove the starter or starter and bracket assembly.
6. If necessary, remove the through bolt nuts holding the bracket to the starter frame, if so equipped. Remove the bracket and washers.
7. Installation is the reverse of removal. Tighten the bolt behind the air silencer first (if removed), then tighten all bolts to specifications (**Table 3**). Coat the starter cable connection with OMC Black Neoprene Dip.

Brush Replacement (Bosch Direct Drive Starter)

Always replace brushes in complete sets. Refer to **Figure 21** for this procedure.
1. Remove the starter as described in this chapter.
2. Remove the 2 through-bolts and carefully tap the commutator end cap from the starter frame.
3. Inspect the brushes in the brush holder on the commutator end of the armature. Replace all brushes if any are pitted, oil-soaked or worn to 3/8 in. or less.
4. Remove brushes and springs from brush holder. Remove brush holder from commutator end.
5. Check brush holder straightness. If brushes do not show full-face contact with commutator, holder is probably bent.
6. Install the insulated brush and terminal set in the commutator end cap as shown in **Figure 22**.
7. Install the brush holder in the commutator end cap. Install the brush springs in the holder. Insert the brushes and tighten the brush lead screws to the holder.

8. Fit the ground brushes in the holder slots. **Figure 23** shows the reassembled brush holder and commutator end cap assembly.

9. Align the commutator and drive end cap marks. Hold the brushes in place and assemble the end cap to the frame. A putty knife with a 1 x ½ in. slot cut in its end makes a suitable tool for keeping the brushes in place during this step.

10. Wipe the through-bolt threads with engine oil. Install and tighten through-bolts to specifications (**Table 3**). Apply OMC Black Neoprene Dip around the end cap and frame joints.

Brush Replacement
(Bosch Gear Reduction Starter)

Always replace brushes in complete sets.

1. Remove the starter as described in this chapter.
2. Remove the through bolt nuts holding the bracket to the starter frame. Remove the bracket and washers.
3. Remove the 3 pinion housing screws. Remove the pinion housing. Remove the wave washer, spacer and pinion assembly (**Figure 24**).
4. Carefully pry the weather cover free. Remove the weather cover, driven gear and thrust washer (**Figure 25**).
5. Expand and remove the snap ring and drive gear from the armature shaft with snap ring pliers.
6. Mark the end cap and frame for reassembly reference. Remove the 2 through bolts.
7. Rap the starter frame with a soft-faced hammer to separate it from the gear housing.
8. Hold the armature and carefully slide the frame off the housing.
9. Slowly release pressure on the armature and allow it to separate from the housing. Do not lose the thrust washer. See **Figure 26**.
10. Inspect the brushes. Replace all brushes if any are pitted, oil-soaked or worn to 3/8 in. or less. If new brushes are required, use OMC

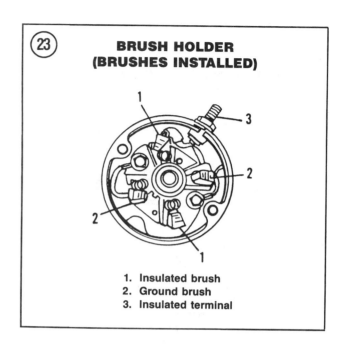

BRUSH HOLDER (BRUSHES INSTALLED)

1. Insulated brush
2. Ground brush
3. Insulated terminal

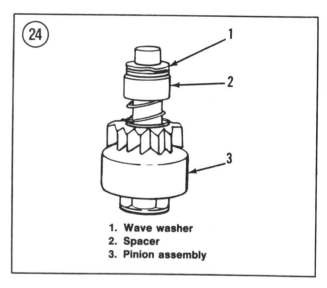

1. Wave washer
2. Spacer
3. Pinion assembly

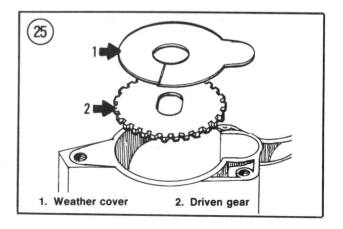

1. Weather cover
2. Driven gear

IGNITION AND ELECTRICAL SYSTEMS

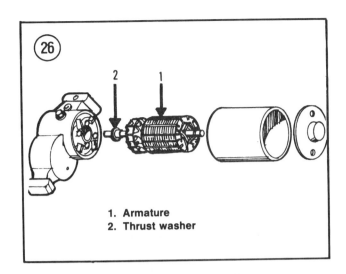

1. Armature
2. Thrust washer

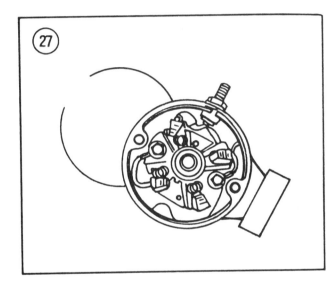

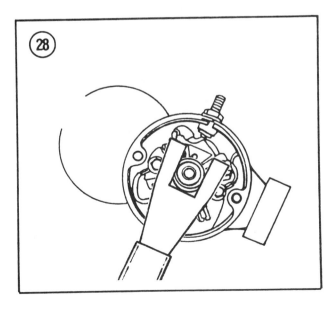

Locquic Primer and OMC Screw Lock on the brush card screws before installation.

11. Wipe the gear housing armature bearing with OMC Extreme Pressure grease.

12. Route the brush leads as shown in **Figure 27**. Install and compress the springs, then install the brushes.

13. Use a putty knife with a 1 x ½ in. slot cut in its end to hold the brushes and springs in place while the armature and thrust washer are reinstalled. See **Figure 28**.

14. Reverse Steps 1-8 to complete assembly. Install and tighten through bolts to specifications (**Table 3**). Apply OMC Black Neoprene Dip around the end cap and frame joints.

Brush Replacement (Prestolite Starter)

Replace all brushes if any are pitted, oil-soaked or worn to 3/8 in. or less. Always replace brushes in complete sets.

1. Remove the starter as described in this chapter.

2. Remove the 2 through-bolts and carefully tap the commutator end cap to separate it from the starter frame.

3. Remove the washers from the rear of the armature shaft, noting the quantity and sequence of installation for reassembly. Remove the armature and drive end cap from the front of the starter frame.

4. Remove the positive terminal and brushes from the starter frame. See **Figure 29**.

5. Cut the negative brush leads at the point where they are attached to the field coils.

6. File or grind off solder from ends of field coil leads where old brushes were connected.

7. Use rosin core soldering flux and solder the leads of new negative brushes to the field coils, making sure they are in the right position to reach the brush holders. See **Figure 29**.

NOTE
The leads should be soldered to the back sides of the coils so excessive solder will

not rub the armature. Do not overheat the leads, as the solder will run onto the lead and it will no longer be flexible.

8. Install new positive brushes and positive terminal assembly (**Figure 29**).

9. Install the brush plate with the notch facing the positive terminal. Route brush leads over plate as shown in **Figure 30**.

10. Install one brush spring and brush in its holder. Secure in place with a brush retaining tool or tie in place with a length of thread. Repeat this step for each remaining brush. See **Figure 30**.

11. Install armature in starter frame, aligning tab on drive end plate with starter frame slot.

12. Remove brush retaining tools or threads from the brushes.

13. Reinstall washers on armature shaft in the same order as removed.

14. Install end cap with raised lines facing positive terminal.

15. Wipe the through bolt threads with engine oil. Install and tighten through bolts to specifications (**Table 3**). Apply OMC Black Neoprene Dip around the end cap and frame joints.

IGNITION SYSTEM

All Johnson and Evinrude outboards covered in this manual use a CD breakerless ignition system. The ignition system is named for the number of cylinders in the engine. Thus, the 2-cylinder engine has a CD 2 ignition, the 3-cylinder engine has a CD 3 ignition and the V4 and V6 engines have a CD 4 or CD 6 ignition. The 1989-1990 48 and 50 hp models are equipped with a CD2USL ignition system. Refer to Chapter Three for troubleshooting and test procedures.

Operation
(All CD Ignition Systems Except CD2USL)

The major components of the CD ignition system on all models include the flywheel, charge coil(s), sensor coil(s), power pack(s), ignition

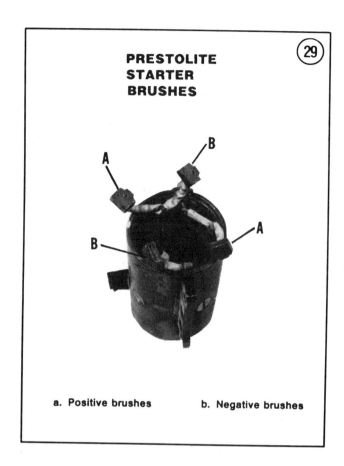

PRESTOLITE STARTER BRUSHES

a. Positive brushes   b. Negative brushes

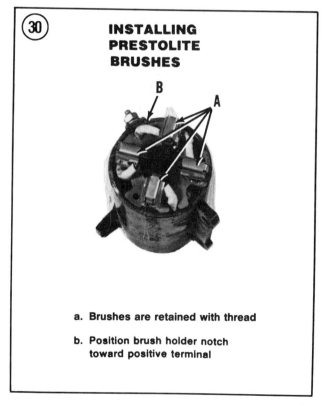

INSTALLING PRESTOLITE BRUSHES

a. Brushes are retained with thread

b. Position brush holder notch toward positive terminal

IGNITION AND ELECTRICAL SYSTEMS

coils, spark plugs and connecting wiring. The number of flywheel magnets, sensor and charge coils, ignition coils and power packs differ according to the number of cylinders in the engine.

The charge coil is located under the flywheel on the armature plate (CD 2) or stator assembly (CD 3, 4 and 6). See A, **Figure 31** (CD 2) or **Figure 32** (all others). The flywheel is fitted with permanent magnets inside its outer rim. As the crankshaft and flywheel rotate, the flywheel magnets pass the stationary charge coil(s). This creates an alternating current that is sent to the power pack(s) where it is stored in a capacitor for release.

The sensor coil(s) are also mounted under the flywheel on the armature plate (CD 2) or timer base (CD 3, 4 and 6). See B, **Figure 31** (CD 2) or **Figure 33** (all others). As the flywheel magnets pass the stationary sensor coil(s), an alternating current is created and sent to the No. 1 SCR (switch) in the power pack. The power pack contains electronic circuitry to produce ignition at the proper time. The No. 1 SCR discharges the capacitor into the No. 1 ignition coil at the proper time. While this is happening, the capacitor is recharging in preparation for discharge through the No. 2 SCR. The process continues until all cylinders have fired, then the sequence is repeated.

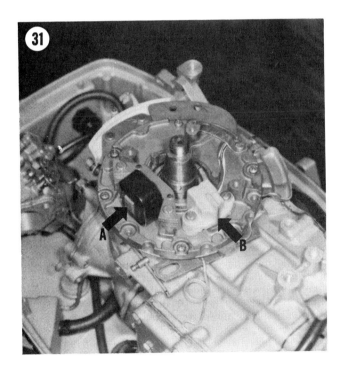

Operation (CD2USL)

The CD2USL ignition system is used on 1989-1990 48 and 50 hp models. The ignition system is completely contained under the flywheel with the exception of the ignition coils. The breakdown of the CD2USL model number is as follows: CD-capacitor discharge; 2-two cylinders; U-under the flywheel; S-contains S.L.O.W. (speed limiting overheat warning) mode; L-contains rpm rev. limiter. Ignition system model number is printed on top of ignition module located beneath engine flywheel. The ignition module incorporates the power pack and the sensor coil into one assembly instead of two assemblies as on CD 2 ignition systems. See A, **Figure 34**.

The major components of the CD2USL ignition system on all models include the flywheel, charge coil, ignition module, ignition coils, spark plugs and connecting wiring. The charge coil is located under the flywheel on the armature plate (B, **Figure 34**). The flywheel is fitted with permanent magnets inside its outer rim. As the crankshaft and flywheel rotate, the flywheel magnets pass the stationary charge coil. This creates an alternating current that is sent to the ignition module where it is stored in a capacitor for release. As the flywheel magnets pass the sensor coil (**Figure 35**) located in the ignition module, an alternating current is created and sent to the No. 1 SCR (switch) in the ignition module. The No. 1 SCR discharges the capacitor into the No. 1 ignition coil at the proper time. While this is happening, the capacitor is recharging in preparation for discharge through the No. 2 SCR.

Armature Plate Removal/Installation (CD 2)

1. Remove the rewind starter, if so equipped. See **Figure 36**.

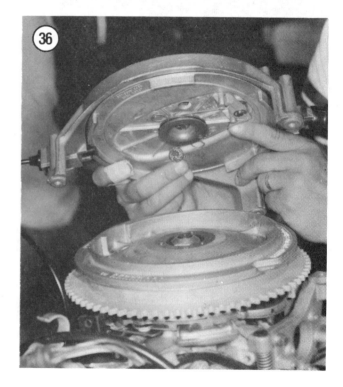

IGNITION AND ELECTRICAL SYSTEMS

2. Remove the flywheel. See Chapter Eight.

3. Disconnect the armature plate leads from their support clamps.

4. Disconnect the armature plate lead connectors and terminal board connections.

5. Remove the armature plate retaining screws. Remove armature plate and cable assembly from the power head (**Figure 37**).

6. If necessary to remove the retainer and support plates, disconnect the armature plate to throttle lever link and remove the support plate screws. Remove the retainer and support plates (**Figure 38**).

7. Installation is the reverse of removal. Apply OMC Nut Lock to support plate screw threads. Coat crankcase boss pilot with OMC Moly Lube. Lubricate armature plate bearing (Delrin ring) with Johnson or Evinrude Outboard Lubricant. Compress armature plate bearing with needlenose pliers (**Figure 39**) and guide armature plate into position. Tighten all fasteners to specifications (**Table 3**). Coat all terminal board connections with OMC Black Neoprene Dip. Check engine synchronization (Chapter Five).

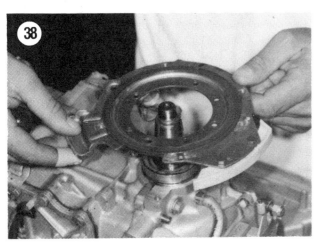

Armature Plate Removal/Installation (CD2USL)

1. Remove the flywheel. See Chapter Eight.
2. Disconnect the armature plate lead connectors.
3. Remove the 5 armature plate retaining screws. Remove armature plate assembly from the power head.
4. Installation is the reverse of removal. Apply OMC Nut Lock on threads of armature plate retaining screws. Tighten fasteners to specification (**Table 3**).

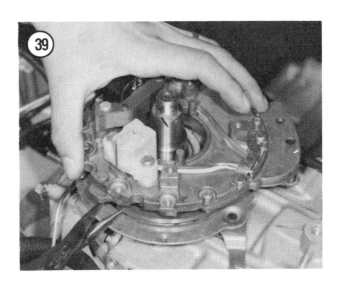

Charge Coil, Sensor Coil or Stator Coil Replacement (All CD Ignition Systems Except CD2USL)

The charge and sensor coils can be replaced individually on manual start models and all 1984 50 and 60 hp models. On all other electric start models, the stator and timer base assemblies containing the charge and sensor coils are potted and are replaced as an assembly.

1. Turn the armature plate over and remove the cover plate. See **Figure 40** (typical).
2. Remove the 2 screws holding the wiring lead strap in place (**Figure 41**).
3. Refer to **Figure 41** and remove the screws holding the defective component(s) to the armature plate.
4. On models using a 4-wire connector:

   a. Charge coil—Remove the A and D wire terminals from the 4-wire connector.
   b. Sensor coil—remove the B and C wire terminals from the 4-wire connector.
   c. See *Connector Terminal Removal/ Installation* in this chapter for proper wire terminal removal.

5. Pull the lead(s) of the defective component(s) from the insulation sleeve.

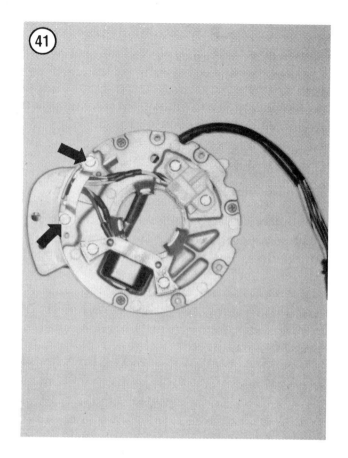

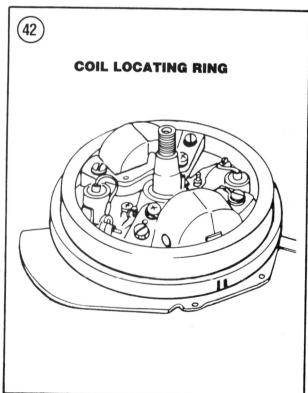

COIL LOCATING RING

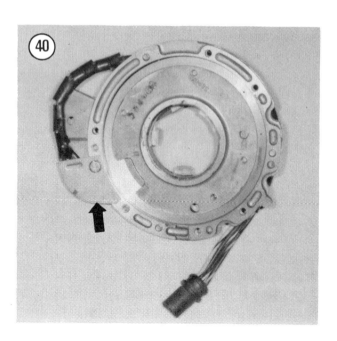

IGNITION AND ELECTRICAL SYSTEMS

A. Ignition module
B. Charge coil
C. Stator coil
D. Coil locating ring

6. Remove the defective component(s) and lead(s) from the armature plate.

7. Assembly is the reverse of disassembly. Each coil must be properly aligned when reinstalled to prevent contact with the flywheel magnets and to produce maximum output. Johnson and Evinrude recommend the use of a coil locating ring (part No. 317001) which is machined to fit over the armature plate bosses. See **Figure 42**. Push ring toward coil, pull coil toward ring and tighten coil mounting screws. Position all leads carefully so they will not rub against the flywheel. Coat all terminal board connections with OMC Black Neoprene Dip.

Charge Coil, Ignition Module or Stator Coil Replacement (CD2USL)

1. Remove the armature plate as previously outlined in this chapter.
2. Turn the armature plate over and remove the cover plate. See **Figure 40** (typical).
3. Remove the 2 screws holding the wiring lead strap in place (**Figure 43**).
4. Refer to **Figure 44** and remove the respective retaining screws to replace a defective ignition module (A), charge coil (B) or stator coil (C).
5. Pull the lead(s) of the defective component(s) from the insulation sleeve. See *Connector Terminal Removal/Installation* in this chapter.
6. Remove the defective component(s) and lead(s) from the armature plate.
7. Assembly is the reverse of disassembly. Each coil must be properly aligned when reinstalled to prevent contact with the flywheel magnets and to produce maximum output. Johnson and Evinrude recommend the use of a coil locating ring (part No. 334994) which is machined to fit over the armature plate bosses. See D, **Figure 44**. Position all leads carefully so they will not rub against the flywheel. Apply OMC Ultra Lock on threads of component retaining screws and install. Hold the component(s) against the

locating ring and tighten the retaining screws to specification (**Table 3**).

Stator and Timer Base Removal/Installation (CD 3, CD 4 and CD 6)

1. Disconnect the spark plug leads to prevent the engine from accidentally starting.
2. Remove the flywheel. See Chapter Eight.
3. Disconnect the stator leads at the terminal board. Disconnect the charge coil leads at the power pack(s).
4. Remove the 3 or 4 stator mounting screws. Remove the stator from the power head.
5. Remove the timer base retaining clips and screws. Note that the clips engage the timer base bearing (Delrin ring).
6. Installation is the reverse of removal, plus the following:
   a. Coat crankcase upper bearing and seal assembly with OMC Moly Lube.
   b. Lubricate the Delrin ring with Johnson or Evinrude Outboard Lubricant. Assemble Delrin ring to timer base, then install timer base.
   c. Wipe all retaining screw threads with OMC Locquic Primer, the OMC Nut Lock. Tighten fasteners to specifications (**Table 3**).
   d. Coat all terminal board connections with OMC Black Neoprene Dip.
   e. Check engine synchronization (Chapter Five).

Secondary Ignition Coil Replacement

1. Remove the engine cover.
2. Disconnect the spark plug lead of the coil to be replaced.
3A. Prior to 1985—Disconnect the coil lead from the power pack or separate the 2-wire

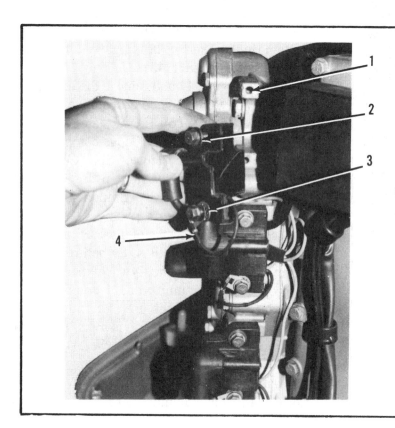

IGNITION COIL REPLACEMENT
1. Cylinder head
2. Washer
3. Lockwasher
4. Ground lead

IGNITION AND ELECTRICAL SYSTEMS

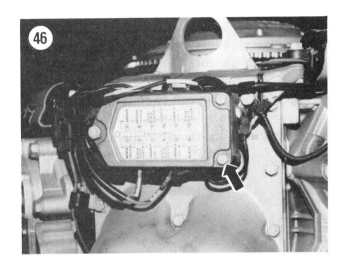

connector (CD 4), 3-wire connector (CD 2) or 4-wire connector (CD 3 and CD 6).

3B. *1985-on*—Disconnect primary wire from coil terminal.

4. On models with connectors, remove the coil lead terminal from the connector. See *Connector Terminal Removal/Installation* in this chapter.

5. If the coil uses a separate ground lead attachment point, disconnect the ground lead.

6. Remove the mounting bolts and washers (**Figure 45**). Remove the coil from the power head.

7. Installation is the reverse of removal. Attach ground lead with lockwasher between the flat washer or J-clamp and ground lead. On models so equipped, install stop button or key switch ground wire between lockwasher and ground wire connector. Fill spark plug boot with OMC Triple-Guard grease to prevent terminal corrosion and high tension arcing.

Power Pack Replacement

The power pack electrical circuits are encased in a potting material and can be serviced only by replacement.

1. Disconnect the negative battery cable, if so equipped.
2. Remove the engine cowling.
3. Separate the power pack connectors on later models or disconnect the stator and charge coil leads at the power pack on early models.

NOTE
On some engines, the ground lead is attached to one of the power pack mounting bolts and is removed in Step 5.

4. Remove the fastener holding the power pack ground lead.
5. Remove the power pack mounting bolts. See **Figure 47** (typical). Remove the power pack. See **Figures 46-49** for typical mounting locations.
6. Installation is the reverse of removal. Position the lockwasher on the engine side of the ground

wire connector to assure a good ground. Make sure there is sufficient slack in the ground wire; if too tight, it can cause an ignition failure. Tighten the mounting bolts to specifications (**Table 3**).

Connector Terminal Removal/Installation

Waterproof plug-in connectors are used with 1978-on 50-60 hp and V4 engines and 1979-on 70-75 hp and V6 engines. The connector halves are secured by a retaining clamp which must be removed before they can be separated. See **Figure 50** (typical).

Whenever an ignition component is replaced on these models, the component wires and their plug terminals must be removed from the connector. A set of 3 special tools is available

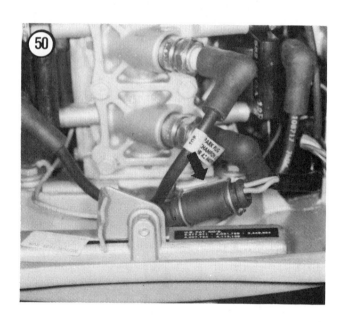

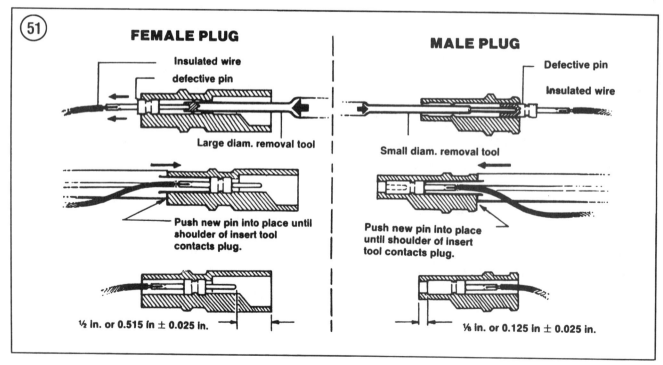

IGNITION AND ELECTRICAL SYSTEMS

for quick and easy terminal removal/installation: insert tool part No. 322697, pin remover tool part No. 322698 and socket remover tool part No. 322699. Each tool has the appropriate tip for its intended use.

Connector terminals should be removed and installed according to this procedure. Use of tools or lubricant other than specified can result in high resistance connections, short circuits between terminals or damage to the connector material.

1. Lubricate the terminal pin or socket to be removed with rubbing alcohol at both ends of the connector cavity.
2. Hold the connector against the edge of a flat surface, allowing sufficient clearance for terminal/socket removal.
3. Insert the proper removal tool in the connector end of the plug and carefully push the terminal or socket from the plug. See **Figure 51**.
4. If the pin or socket requires replacement, install a new one on the end of the wire. Make sure the insulation is stripped back far enough to allow the new pin or socket to make complete contact with the wire.
5. Crimp the new terminal onto the wire with crimping pliers (part No. 322696) or equivalent. If crimping pliers are not available, solder the wire in the pin or socket.
6. Lubricate the connector cavity with rubbing alcohol.
7. Place the insert tool against the pin or socket shoulder. Carefully guide the pin or socket into the rear of the connector plug cavity and press it in place until the insert tool shoulder rests against the connector plug. Withdraw the insert tool.
8. Reconnect the connector plug halves and install the retaining clamp.

Table 1 BATTERY CAPACITY (HOURS)

Accessory draw	80 Amp-hour battery provides continuous power for:	Approximate recharge time
5 amps	13.5 hours	16 hours
15 amps	3.5 hours	13 hours
25 amps	1.8 hours	12 hours
Accessory draw	105 Amp-hour battery provides continuous power for:	Approximate recharge time
5 amps	15.8 hours	16 hours
15 amps	4.2 hours	13 hours
25 amps	2.4 hours	12 hours

Table 2 APPROXIMATE STATE OF CHARGE

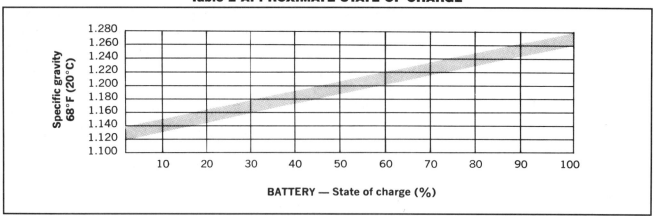

Table 3 TIGHTENING TORQUES

Fastener	in.-lb.	ft.-lb.
Armature support plate	60-80	
Armature-to-support plate	25-35[1]	
Charge coil screws	30-40[2]	
Ignition coil screws		
1973-1984	48-60	
1985-on	60-84	
Ignition module screws	30-40[2]	
Power pack screws		
1973-1984	48-60	
1985-on	60-84	
1988-on 120, 140, 200 and 225 hp	48-60	
Starter motor		
Mounting bolts		14-16
Through bolts		
Bosch direct drive	90-105	
Bosch gear reduction	55-60	
Prestolite	110-122	
Stator screws		
2-cylinder		
1973-1984	60-80	
1985-on	120-144	
3-cylinder		
1973-1984	48-60	
1985-on	120-144	
V4 and V6		10-12
Stator coil screws	30-40[2]	
Standard bolts and nuts		
No. 6	7-10	
No. 8	15-22	
No. 10	25-35	
No. 12	35-40	
1/4 in.		5-7
5/16 in.		10-12
3/8 in.		18-20
7/16 in.		28-30

1. Apply OMC Nut Lock on fastener threads prior to installation.
2. Apply OMC Ultra Lock on fastener threads prior to installation.

Chapter Eight

Power Head

Basic repair of Johnson and Evinrude outboard power heads is similar from model to model, with minor differences. Some procedures require the use of special tools, which can be purchased from a dealer. Certain tools may be fabricated by a machinist, often at substantial savings. Power head stands are available from specialty shops such as Bob Kerr's Marine Tool Co. (P.O. Box 1135, Winter Garden, FL 32787).

Work on the power head requires considerable mechanical ability. You should carefully consider your own capabilities before attempting any operation involving major disassembly of the engine.

Much of the labor charge for dealer repairs involves the removal and disassembly of other parts to reach the defective component. Even if you decide not to tackle the entire power head overhaul after studying the text and illustrations in this chapter, it can be cheaper to perform the preliminary operations yourself and then take the power head to your dealer. Since many marine dealers have lengthy waiting lists for service (especially during the spring and summer season), this practice can reduce the time your unit is in the shop. If you have done much of the preliminary work, your repairs can be scheduled and performed much quicker.

Repairs go much faster and easier if your motor is clean before you begin work. There are special cleaners for washing the motor and related parts. Just spray or brush on the cleaning solution, let it stand, then rinse it away with a garden hose. Clean all oily or greasy parts with fresh solvent as you remove them.

WARNING
Never use gasoline as a cleaning agent. It presents an extreme fire hazard. Be sure to work in a well-ventilated area when using cleaning solvents. Keep a fire extinguisher rated for gasoline and oil fires nearby in case of emergency.

Once you have decided to do the job yourself, read this chapter thoroughly until you have a good idea of what is involved in completing the overhaul satisfactorily. Make arrangements to buy or rent any special tools necessary and obtain replacement parts before you start. It is frustrating and time-consuming to start an overhaul and then be unable to complete it because the necessary tools or parts are not at hand.

NOTE
A limited number of new motors are assembled every year with cylinders that have been bored larger than standard. Such motors are identified by the letters "OS" stamped on the power head serial number welch plug. Cylinder reboring on these motors is not possible; a new cylinder and crankcase is required. If you have one of these engines, contact your Johnson or Evinrude dealer if cylinder repair becomes necessary.

Before beginning the job, re-read Chapter Two of this manual. You will do a better job with this information fresh in your mind.

Since this chapter covers a large range of models over a lengthy time period, the procedures are somewhat generalized to accommodate all models. Where individual differences occur, they are specifically pointed out. The power heads shown in the accompanying pictures are current designs. While it is possible that the components shown in the pictures may not be identical with those being serviced, the step-by-step procedures may be used with all models covered in this manual.

Tables 1-4 are at the end of the chapter.

CAUTION
Whenever a power head is rebuilt, it should be treated as a new engine. On models prior to 1986, use a 50:1 mixture of gasoline and Johnson or Evinrude Outboard Lubricant on both non-VRO and VRO models during the 10 hour break-in period. On 1986-on models, use a 50:1 mixture (non-VRO models) or a 100:1 mixture (VRO models) of gasoline and Johnson or Evinrude Outboard Lubricant during the 10 hour break-in period. Keep maximum engine speed under ¾ throttle during this time. Before switching to straight fuel on VRO equipped models, make sure oil level has dropped in holding tank to ensure fuel and oil is being mixed.

ENGINE SERIAL NUMBER

Johnson and Evinrude outboards are identified by engine serial number. This number is stamped on a plate riveted to the transom clamp (**Figure 1**). It is also stamped on a welch plug installed on the power head (**Figure 2**). Exact location of the transom clamp plate and welch plug varies according to model.

This information identifies the outboard and indicates if there are unique parts or if internal

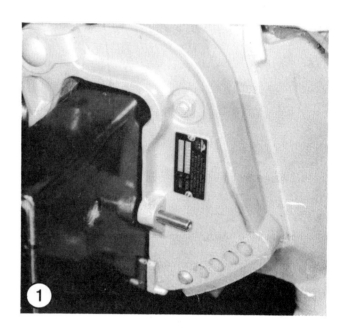

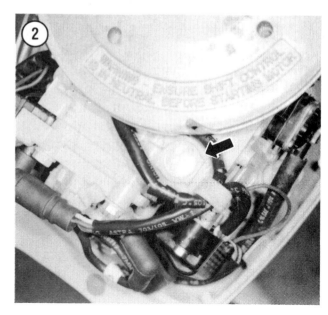

changes have been made during the model run. The serial number should be used when ordering any replacement parts for your outboard.

Starting with the 1980 model year, the model year designation is coded. The last 2 letters of the model code indicate the model year. To determine the year of a given model (1980-on), write the word "INTRODUCES." Below the word, number the letters 1-0. Match these numbers to the last 2 letters of the model code. As an example, J25TELCT is a 1983 model. The C represents 8 and the T represents 3.

FASTENERS AND TORQUE

Always replace a worn or damaged fastener with one of the same size, type and torque requirement.

Power head tightening torques are given in **Table 1**. Where a specification is not provided for a given bolt, use the standard bolt and nut torque according to fastener size.

Where specified, clean fastener threads with OMC Locquic Primer and then apply OMC Nut Lock or Screw Lock as required.

Power head fasteners should be tightened in 2 steps. Tighten to 50 percent of the torque value in the first step, then to 100 percent in the second step.

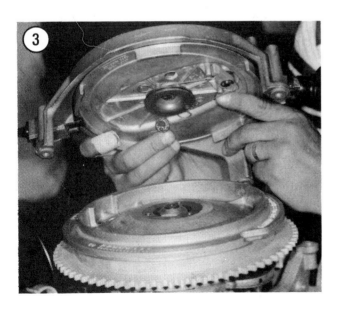

Retighten the cylinder head screws after the engine has been run and warmed up.

To retighten the power head mounting fasteners properly, back them out one turn and then tighten to specifications.

When spark plugs are reinstalled after an overhaul, tighten to the specified torque. Warm the engine to normal operating temperature, let it cool down and retorque the plugs.

GASKETS AND SEALANTS

Three types of sealant materials are recommended: OMC Gasket Sealing Compound, Adhesive M and OMC Gel Seal II. Unless otherwise specified, OMC Gasket Sealing Compound is used with gaskets on older engines. Adhesive M is used primarily with crankcase spaghetti seals (gasket strips). Gel Seal II replaces the original Gel Seal and is used instead of a crankcase gasket with newer engines. Be sure to use the appropriate sealant specified for your engine.

FLYWHEEL

Removal/Installation (All Engines)

A strap wrench can be used to hold the flywheel on manual start models. Use a flywheel holding tool on electric start models. The OMC universal puller kit (part No. 378103) or equivalent is recommended for flywheel removal.

If oil is found under the flywheel, the upper main bearing seal is leaking. The power head should be removed, disassembled and a new upper main bearing seal installed.

1. Disconnect stator or armature plate-to-power pack connections or disconnect the spark plug leads to disable the ignition and prevent accidental starting of the motor.

2. Remove the rewind starter assembly on manual start engines (**Figure 3**).

3. Remove the flywheel nut with an appropriate size socket and flywheel holding tool (**Figure 4**).

4. Install puller on flywheel with its flat side facing up.

5. Hold puller body with puller handle and tighten center screw. See **Figure 5**. If flywheel does not pop from the crankshaft taper, pry up on the rim of the flywheel with a large screwdriver while tapping the puller center screw with a brass hammer.

6. Remove puller from flywheel. Remove flywheel from crankshaft (**Figure 6**).

7. Remove the Woodruff key from the crankshaft key slot.

8. Clean the crankshaft and flywheel tapers with OMC Cleaning Solvent.

9. Install Woodruff key with its outer edge parallel to the crankshaft centerline. On models with a single mark on key, the key side with the mark must face downward. See **Figure 7**.

10. Install the flywheel and flywheel nut (**Figure 8**). Hold flywheel with the holding tool and tighten flywheel nut to specification (**Table 2**).

11. Install rewind starter on manual models.

12. Reconnect the wiring disconnected in Step 1.

POWER HEAD

When removing any power head, it is a good idea to make a sketch or take an instant picture of the location, routing and positioning of wires and J-clamps for reassembly reference. Take notes as you remove wires, washers and engine grounds so they may be reinstalled in their correct position. Unless specified otherwise, install lockwashers on the engine side of the electrical lead to assure a good ground.

Removal/Installation
(1973-1986 50-55 hp and
1973-1985 60-75 hp)

1. Disconnect the negative battery cable.

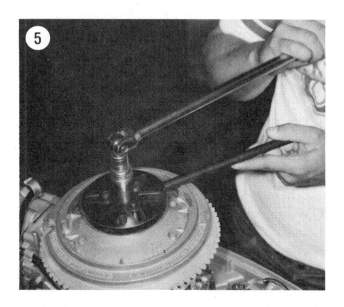

POWER HEAD

2. Remove the engine cover.

3. Disconnect the connections between the armature plate and power pack.

4. Remove the flywheel as described in this chapter.

5. Remove the armature plate or stator. See Chapter Seven.

6. Remove the fuel pump and carburetor. See Chapter Six.

7. Remove the starter motor, if so equipped. See Chapter Seven.

8. Disconnect all wiring connectors and ground leads. Remove wires from J-clamps.

9. Remove the throttle and shift levers. Remove the throttle cam. See **Figure 9**.

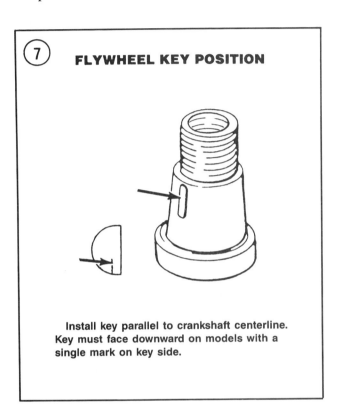

⑦ **FLYWHEEL KEY POSITION**

Install key parallel to crankshaft centerline. Key must face downward on models with a single mark on key side.

⑧

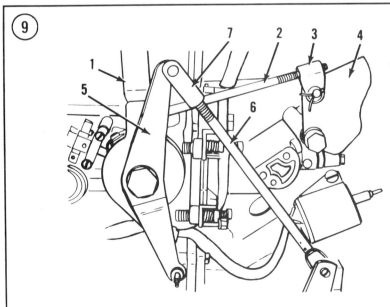

⑨ **THROTTLE AND SHIFT LINKAGE**

1. Throttle lever
2. Throttle link
3. Throttle link yoke
4. Cam
5. Bellcrank
6. Shift link
7. Shift link yoke

10. Remove the screw holding the shift clevis to the shift rod (**Figure 10**).
11. Remove the screws holding the crankcase front bracket to the lower motor cover.
12. Remove the front and rear exhaust cover screws.
13. Remove the nut and washer from the stud at the rear of the power head.
14. Remove the 4 screws on each side holding the power head to the exhaust housing and adapter assembly.

NOTE
At this point, there should be no hoses, wires or linkage connecting the power head to the exhaust housing. Recheck this to be sure nothing will hamper removal.

15. Carefully lift power head from exhaust housing and adapter assembly. Place on a clean workbench for disassembly.
16. Clean all gasket residue from the adapter assembly mating surfaces.
17. Installation is the reverse of removal. Lubricate drive shaft splines with OMC Moly Lube. Coat both sides of a new power head-to-adapter assembly gasket with OMC Gasket Sealing Compound. Rotate power head slightly in a clockwise direction as it is lowered onto the exhaust housing to engage the gearcase drive shaft splines. Align power head stud with hole in adapter and seat power head. Coat fasteners with OMC Screw Lock and tighten to specifications (**Table 1**). Complete engine synchronization and linkage adjustments. See Chapter Five.

Removal/Installation (1987-1988 48-55 hp)

1. Disconnect the negative battery cable.
2. Remove the engine cover.
3. Disconnect the connections between the armature plate and power pack.
4. Remove positive battery cable from starter solenoid and negative battery cable from power head ground.
5. Remove yellow/red wire from starter solenoid.
6. Disconnect fuel supply hose from fuel pump or VRO pump.
7. Tiller models—Remove stop switch ground wire from power head connection. Remove stop switch black wire from connector leading to armature plate by using OMC Amphenol connector removal tools.
8. Remove purple/white wire from primer solenoid.
9. Remove red wires from terminal board. See **Figure 11**.
10. Remove 2 fasteners retaining throttle cable and bracket. See **Figure 11**.
11. Remove hairpin clip from shift rod through-shaft at base of power head, then push through-shaft in and disconnect shift rod from through-shaft lever.
12. Remove 4 bolts retaining power head to lower engine cover.
13. Remove the front and rear exhaust cover.
14. Remove the nut and washer from the stud at the rear of the power head.
15. Remove the 4 bolts on each side holding the power head to the exhaust housing and adapter assembly.

NOTE
At this point, there should be no hoses, wires or linkage connecting the power head to the exhaust housing. Recheck this to be sure nothing will hamper removal.

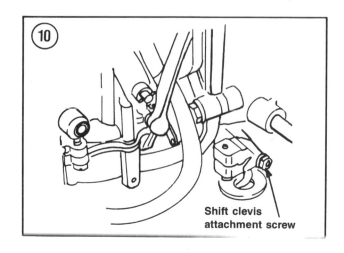

Shift clevis attachment screw

POWER HEAD

16. Carefully lift power head from exhaust housing and adapter assembly. Place on a clean workbench for disassembly.
17. Clean all gasket residue from the adapter assembly mating surfaces.
18. Installation is the reverse of removal. Lubricate drive shaft splines with OMC Moly Lube. Install a new power head-to-adapter assembly gasket. Install gasket dry (no sealer). Rotate power head slightly in a clockwise direction as it is lowered onto the exhaust housing to engage the gearcase drive shaft splines. Align power head stud with hole in adapter and seat power head. Coat fasteners with OMC Gasket Sealing Compound and tighten to specification (**Table 1**). Complete engine synchronization and adjustments. See Chapter Five.

Removal/Installation (1989-1990 48-55 hp)

1. Disconnect the negative battery cable.
2. Remove the engine cover.
3. Remove positive battery cable from starter solenoid and negative battery cable from power head ground.
4. Remove yellow/red wire from starter solenoid.
5. Disconnect fuel supply hose from fuel pump or VRO pump.
6. Tiller models—Remove stop switch ground wire from power head connection. Disconnect stop switch black wire from single wire connector.
7. Remove purple/white wire from primer solenoid.
8. Remove red wires from terminal board.
9. Remove fasteners retaining throttle cable and bracket.
10. Remove the screw holding the shift clevis to the shift rod.
11. Remove the 4 screws to split the lower engine cover halves.
12. Remove the nut and washer from the stud at the rear of the power head.
13. Remove the 4 bolts on each side holding the power head to the exhaust housing and adapter assembly.

NOTE
At this point, there should be no hoses, wires or linkage connecting the power head to the exhaust housing. Recheck this to be sure nothing will hamper removal.

14. Carefully lift power head from exhaust housing and adapter assembly. Place on a clean workbench for disassembly.
15. Clean all gasket residue from the adapter assembly mating surfaces.
16. Installation is the reverse of removal. Lubricate drive shaft splines with OMC Moly Lube. Install a new power head-to-adapter assembly gasket. Install gasket dry (no sealer). Rotate power head slightly in a clockwise direction as it is lowered onto the exhaust housing to engage the gearcase drive shaft splines. Align power head stud with hole in adapter and seat power head. Coat fasteners with OMC Gasket Sealing Compound and tighten to specification (**Table 1**). Complete engine synchronization and adjustments. See Chapter Five.

Removal/Installation (1986-on 60-75 hp)

1. Disconnect the negative battery cable.
2. Remove the engine cover.

3. Remove positive battery cable from starter solenoid and negative battery cable from power head ground.

4. Remove yellow/red wire from starter solenoid.

5. Disconnect fuel supply hose from fuel pump or VRO pump.

6. Tiller models—Remove stop switch ground wire from power head connection. Remove stop switch black wire from connector leading to stator by using OMC Amphenol connector removal tools.

7. Remove primer hose from manual primer valve.

8. Remove red wires from terminal board.

9. Disconnect starter lockout cable from shift lever.

10. Remove throttle cable and shift cable from engine mounting linkage.

11A. Non-VRO models—Remove the screw holding the shift clevis to the shift rod.

11B. VRO models—Remove hairpin clip from shift rod through-shaft at base of power head, then push through-shaft in and disconnect shift rod from through-shaft lever.

12. Remove the lower engine cover assembly.

13. Remove the port and starboard lower pan support screws. Detach the tell-tale hose.

14. Remove the nut and washer from the stud at the rear of the power head.

15. Remove the 3 bolts on each side holding the power head to the exhaust housing and adapter assembly.

NOTE
At this point, there should be no hoses, wires or linkage connecting the power head to the exhaust housing. Recheck this to be sure nothing will hamper removal.

16. Carefully lift power head from exhaust housing and adapter assembly. Place on a clean workbench for disassembly.

17. Clean all gasket residue from the adapter assembly mating surfaces.

18. Installation is the reverse of removal. Lubricate drive shaft splines with OMC Moly Lube. On 1986-1988 models, coat both sides of a new power head-to-adapter assembly gasket with OMC Gasket Sealing Compound. On 1989-1990 models, install a new power head-to-adapter assembly gasket dry (no sealer). Rotate power head slightly in a clockwise direction as it is lowered onto the exhaust housing to engage the gearcase drive shaft splines. Align power head stud with hole in adapter and seat power head. Coat fasteners with OMC Gel Seal II and tighten to specification (**Table 1**). Complete engine synchronization and adjustments. See Chapter Five.

Removal/Installation (All 1974-1990 V4 and V6 except 1985-on 120 and 140 hp and 1986-on 200 and 225 hp)

Carefully identify wiring leads as components are removed and make notes of lead routing for reinstallation reference.

POWER HEAD

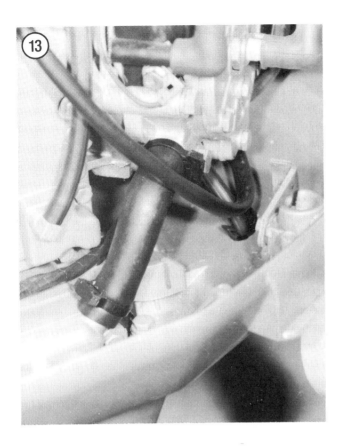

1. Disconnect the negative battery cable.
2. Remove the engine cover.
3. Disconnect the stator and timer base-to-power pack connections.
4. Remove the starter motor (**Figure 12**). See Chapter Seven.
5. Remove the air silencer cover and base assembly.
6. Remove the fuel pumps and carburetors. See Chapter Six.
7. Remove the flywheel as described in this chapter.
8. Remove the stator and timer base assemblies. See Chapter Seven.
9. V4—Disconnect the thermostat hoses at each cylinder head. See **Figure 13**.
10. Disconnect the spark plug leads and remove the spark plugs.
11. Remove the ignition coils (**Figure 14**) and power pack (**Figure 15**) on each side of the engine. See Chapter Seven.
12. Disconnect all ground leads from the power head.
13. Remove the front rubber mount screws on each side of the power head.
14. Remove the shift rod connector screw (**Figure 16**).
15. Remove power head fasteners as follows:
    a. 2 exhaust cover screws on each side of the power head.
    b. 2 lower motor cover to rear exhaust cover screws. Remove rear exhaust cover.
    c. Front exhaust cover-to-lower motor cover screw on each side.
    d. Power head-to-adapter housing nuts at rear of power head.
    e. Screw and nut at front of power head on each side.
    f. 3 long screws on each side holding power head to adapter housing.

NOTE
At this point, there should be no hoses, wires or linkage connecting the power head to the exhaust housing. Recheck this to be sure nothing will hamper removal.

16A. V4—Attach a hoist to the lift bracket and lift power head from exhaust housing adapter.

16B. V6—Temporarily reinstall flywheel. Secure body of OMC puller part No. 378103 to flywheel and thread lifting eye (part No. 321537) into puller until its point touches the crankshaft taper. Tighten lifting eye slightly and attach a hoist to the lifting eye. Remove power head from exhaust housing adapter with hoist.

17. With power head on hoist, reinstall nuts on studs at power head base until flush with stud ends to protect threads.

18. Lower power head onto a clean workbench with crankcase side facing up, then remove the hoist from the lifting eye or bracket.

19. V6—Remove lifting eye and puller body. Remove flywheel.

20. Remove remaining components from outside of power head to provide access to power head assembly fasteners.

21. Installation is the reverse of removal. Lubricate drive shaft splines with OMC Moly Lube. Remove nuts from power head studs. Install a new power head-to-adapter gasket (no sealer). Install a new drive shaft O-ring. Lower power head in position with hoist. Apply OMC Gel Seal II on threads of each power head-to-exhaust housing screw. Complete engine synchronization and linkage adjustments. See Chapter Five.

Removal/Installation
(1985 120-140 hp V4)

1. Disconnect the negative battery cable.
2. Disconnect the charge coil connector between the stator and power pack.
3. Remove the lower engine cover as follows:
   a. Disconnect the water hose at the fitting inside the starboard lower engine cover.
   b. Remove the 2 front and 2 rear lower engine cover screws.
   c. Remove the wire retaining clip from the covers, pull the covers apart and remove from the engine.

4. Remove the VRO pump from the power head:
   a. Cut the tie strap holding the fuel outlet hose to the pump and carefully work the hose off the fitting. Do *not* remove the inlet hose from the pump.
   b. Remove the 2 screws holding the VRO pump bracket to the power head.
   c. Cut the tie strap holding the crankcase pulse hose to the pump and carefully work the hose off the fitting.
   d. Remove the VRO pump and place to one side out of the way.

5. Disconnect the water control valve hose from the adaptor housing fitting.
6. Remove the 2 screws and 1 nut at the rear of the adaptor housing holding the power head to the housing.
7. From underneath the exhaust housing, remove the 6 attaching screws.
8. Remove the screw and nut from the front of the power head on each side.

POWER HEAD

9. Remove the engine cover mount grommets and washers.
10. Remove the shift rod connector screw.
11. Install the body of OMC universal puller kit (part No. 378103) to the flywheel (flat side facing upward) with 3 screws (part No. 309492) and 3 washers (part No. 307640).
12. Thread OMC lifting eye part No. 321537 or equivalent into the puller body until its taper touches the crankshaft taper. Tighten the lifting eye in place.

NOTE
At this point, there should be no hoses, fasteners, wires or linkage connecting the power head to the exhaust housing. Recheck this to make sure nothing will hamper removal.

13. Attach a hoist to the lifting eye. Remove the power head from the exhaust housing adapter with the hoist.
14. With power head on hoist, reinstall nuts on studs at power head base until flush with stud ends to protect threads.
15. Lower the power head onto a clean workbench with the crankcase side facing up, then remove the hoist from the lifting eye.
16. Remove the lifting eye and puller body.
17. Remove flywheel, fuel system, power trim/tilt junction box and electrical components from outside of power head to provide access to power head assembly fasteners.
18. Installation is the reverse of removal, plus the following:

   a. Lubricate drive shaft splines with OMC Moly Lube.
   b. Remove nuts from power head studs.
   c. Install a new power head-to-adapter gasket with OMC Gasket Sealing Compound on both sides. Silicon bead side of gasket should face the power head.
   d. Install a new drive shaft O-ring.
   e. Align the crankshaft and drive shaft splines.
   f. Coat power head attaching screw and stud threads with OMC Gel Seal II.

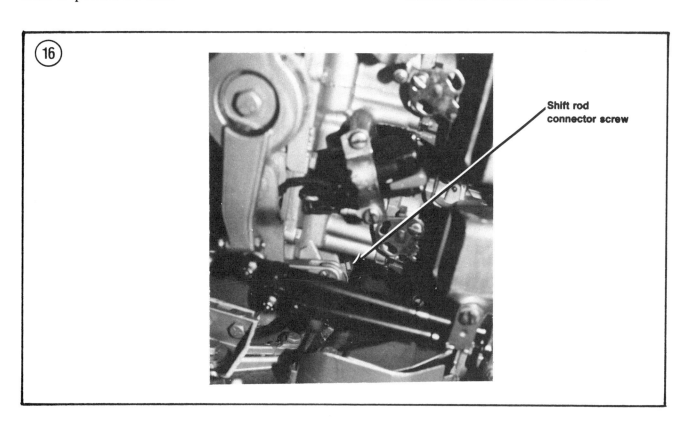

g. Tighten attaching screws and nuts to specifications (**Table 3**).
h. Purge air from VRO pump system. See Chapter Eleven.
i. Complete engine synchronization and linkage adjustments (Chapter Five).

Removal/Installation (1986-on 120-140 hp V4 and 200-225 hp V6)

NOTE
If your engine is not equipped with power steering, omit those steps in the procedure which refer to it and proceed with the next step in sequence.

1. Disconnect the negative battery cable.
2. Remove the bracket holding the power steering hoses.
3. Disconnect the water hose at the fitting inside the starboard lower engine cover.
4. Loosen the 4 lower engine cover screws and remove one lower cover half with spring clip at a time.
5. Loosen the nut and lockwasher to release tension on the power steering belt idler pulley. Remove the pulley from the flywheel assembly.
6. Remove the flywheel as described in this chapter with OMC universal puller part No. 378103.
7. Remove the screws holding the stator assembly to the engine block. Tilt the stator and pulley to remove the screws holding the power steering reservoir to the engine block. Remove the power steering assembly from the power head without disconnecting any hydraulic lines and place to one side out of the way.
8. Disconnect the power trim wiring harnesses at the power head.
9. Remove the shift rod pin.
10. Remove 4 screws and 1 nut from each side holding the exhaust housing to the power head.
11. Remove 2 screws and 1 nut holding the rear of the power head to the exhaust adapter.
12. Thread OMC lifting eye part No. 396748 or equivalent to the crankshaft.

NOTE
At this point, there should be no hoses, fasteners, wires or linkage connecting the power head to the exhaust housing. Recheck this to make sure nothing will hamper removal.

13. Attach a hoist to the lifting eye. Remove the power head from the exhaust housing adapter with the hoist.
14. Lower the power head onto a clean workbench with the crankcase side facing up, then remove the hoist from the lifting eye.
15. Remove the lifting eye. Remove the VRO pump, fuel system and electrical components from outside of power head to provide access to power head assembly fasteners.
16. Installation is the reverse of removal, plus the following:
   a. Lubricate drive shaft splines with OMC Moly Lube.
   b. Install a new power head base gasket (rib side up) with no sealer.
   c. Install a new drive shaft O-ring.
   d. Align the crankshaft and drive shaft splines.
   e. Coat power head-to-exhaust housing threads with OMC Locquick Primer and Loctite No. 515 Gasket Eliminator.
   f. Tighten attaching screws and nuts to specifications (**Table 3**). Use a new flywheel nut.
   g. Complete engine synchronization and linkage adjustments (Chapter Five).

Disassembly (Inline Engines)

The following procedure can be adapted to disassembly and 2- or 3-cylinder power head covered in this manual. The 1986-on 70 hp engine uses a permanent-mold cylinder block containing shrunk-in-position cylinder sleeves. The exhaust system is an integral part of the cylinder casting and the cylinder head does not require the separate cover used with other 2- and 3-cylinder engines. This new design results in

POWER HEAD

fewer power head joints and gaskets, which means easier and faster disassembly/reassembly.

Refer to **Figure 17** (2-cylinder power head), **Figure 18** (3-cylinder power head) or **Figure 19** (crankshaft assembly) for this procedure.

1. Remove the crankcase head assembly (**Figure 20**). Remove and discard the seal housing O-rings and the O-ring inside the crankshaft. Drive the seal from the housing with the appropriate size punch.

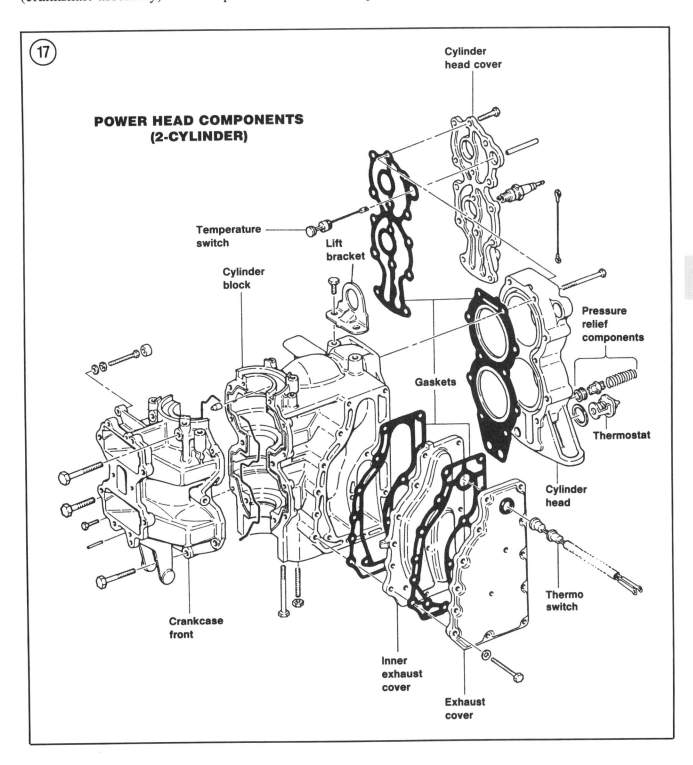

17 POWER HEAD COMPONENTS (2-CYLINDER)

CHAPTER EIGHT

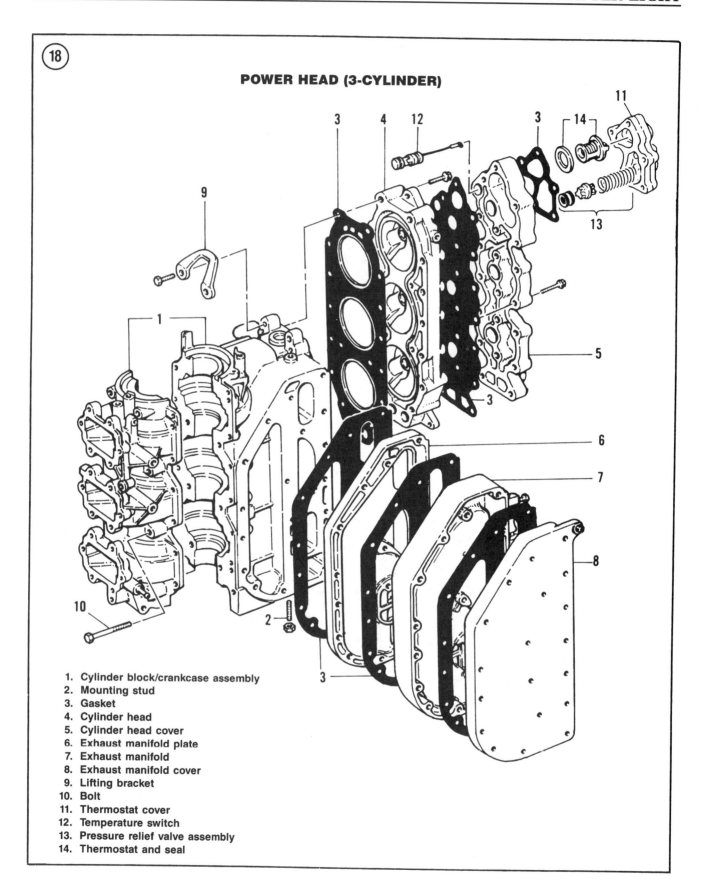

Power Head

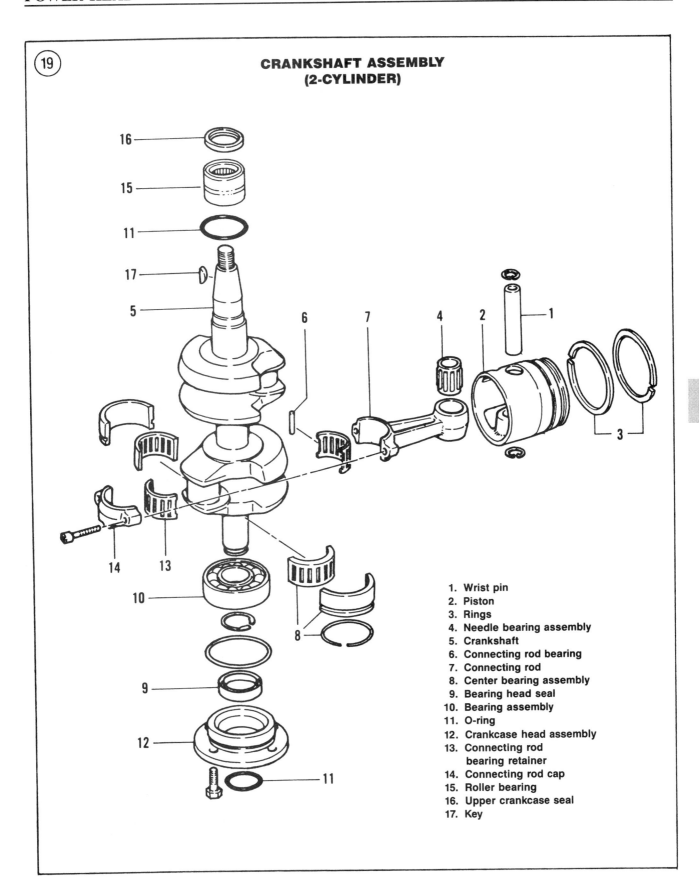

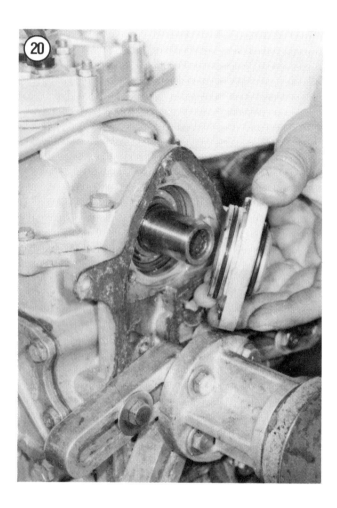

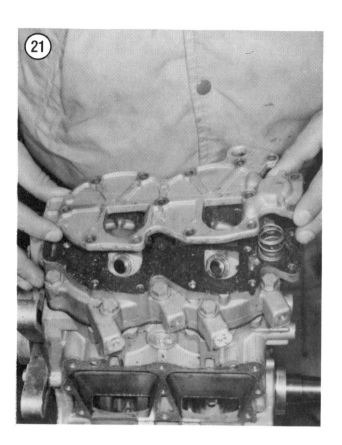

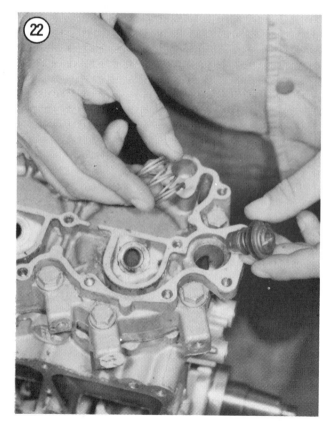

2. Remove the intake manifold and reed valve assembly.

3. Remove the cylinder head cover and gasket (**Figure 21**). Discard the gasket.

4. Remove the thermostat spring, thermostat and seal (**Figure 22**). Discard the seal.

5. Remove the cylinder head and gasket (**Figure 23**). Discard the gasket.

6. Remove the exhaust cover screws. Tap cover if necessary to break the seal. Remove the outer cover and gasket (**Figure 24**). Discard the gasket.

7. Remove the inner exhaust cover and gasket (**Figure 25**). Discard the gasket.

8. 70-75 hp—Remove the exhaust manifold plate and gasket. Discard the gasket. Check plate and discard if pitted.

9. Remove the crankcase taper pins with a punch (**Figure 26**). Drive pins from front to back of crankcase.

POWER HEAD

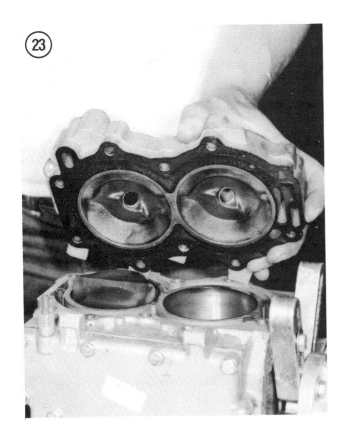

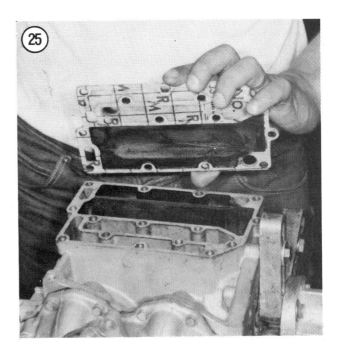

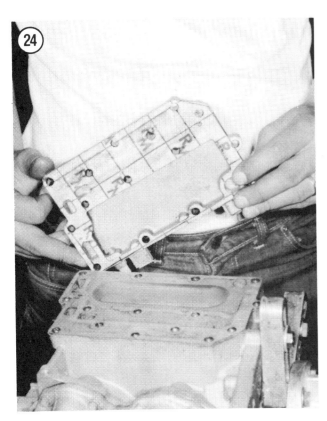

10. Remove the crankcase-to-cylinder block screws and nuts. Tap the side of the crankshaft lightly with a rubber mallet to break the gasket seal, then remove the cylinder half from the crankcase. See **Figure 27**.

NOTE
The connecting rod caps must be removed before the crankshaft can be removed from the cylinder block. Note that one side of the rod and cap has raised dots and that the corners are chamfered for proper rod/cap alignment. Caps are not interchangable and cannot be turned.

11. Check connecting rod and cap alignment with a pencil point or dental pick (**Figure 28**). Alignment should be correct at 3 of the 4 corners. Record alignment points of each cap for reassembly reference.

12. Remove each connecting rod cap and bearing retainer. See **Figure 29**. Place in a clean container.

13. Remove the crankshaft from the cylinder block (**Figure 30**).

14. Remove the remaining connecting rod bearing retainers and roller bearings. Place

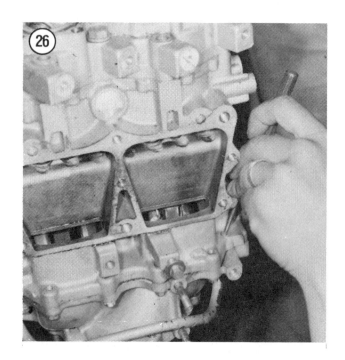

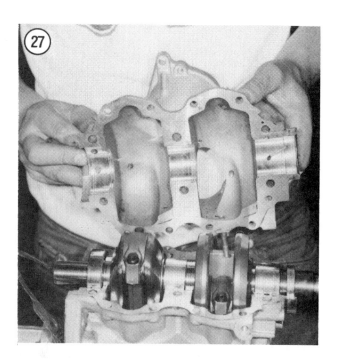

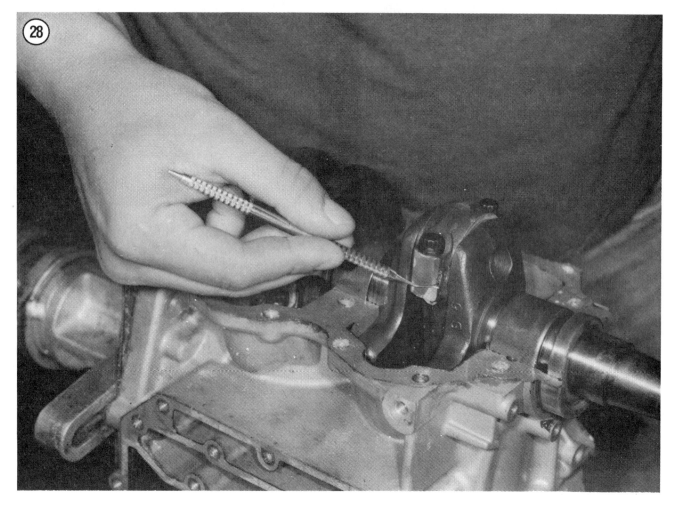

POWER HEAD

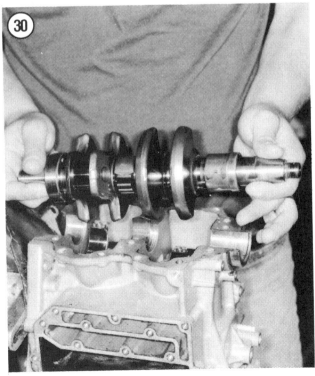

bearings and retainers in their respective containers.

15. Orient each rod cap to its respective connecting rod and reinstall with cap screws finger-tight. See **Figure 31**.

16. Remove the piston and rod assemblies from their cylinders. Mark the cylinder number on the top of each piston with a felt-tipped pen.

17. Remove the flywheel Woodruff key, if not removed when the flywheel was removed. Slide the upper seal and main bearing assembly off the crankshaft.

18. If lower main bearing requires removal, remove snap ring and then remove bearing with an appropriate puller.

19. Slide the crankshaft center main bearing retaining ring from its groove. Remove the

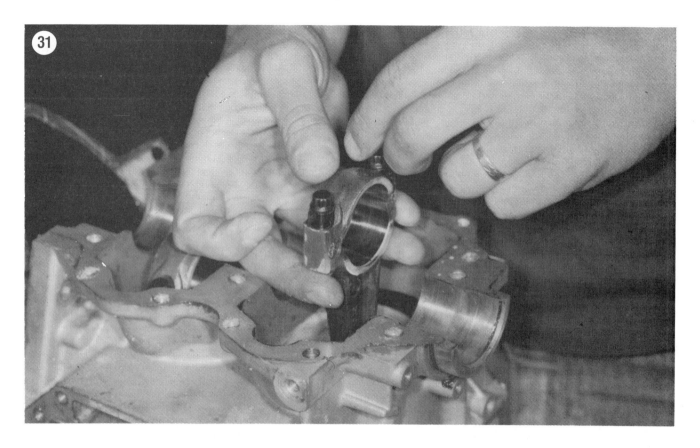

bearing split sleeves. Remove the bearing cages and roller bearings. Place cages and bearings in a clean container.

20. Pry each ring far enough from the piston to grip it with pliers, then break the rings off the piston and discard.

NOTE
If piston pin bore has a small cutout facing the top of the piston, the retaining rings can be removed in Step 21 by inserting a small screwdriver blade or awl in the cutout and prying the ring free.

21. If the piston is to be removed from the connecting rod, remove the piston pin retaining rings with tool part No. 325937. See **Figure 32**.

NOTE
Pistons used on early engines and some later engines are not marked with an

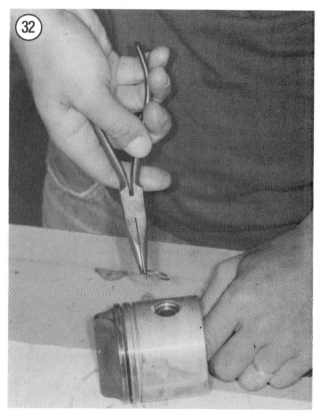

POWER HEAD

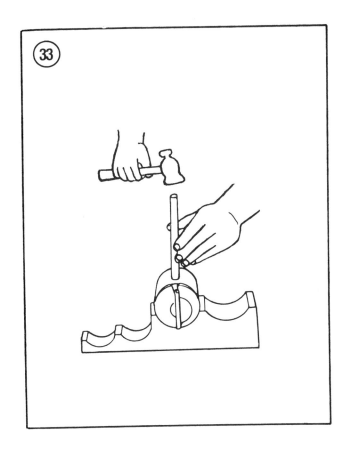

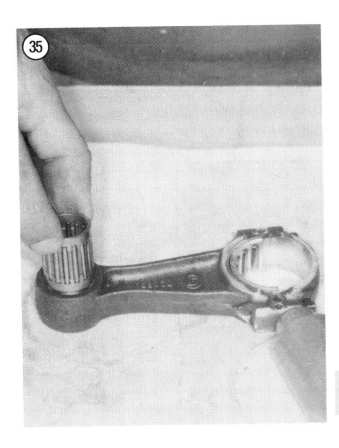

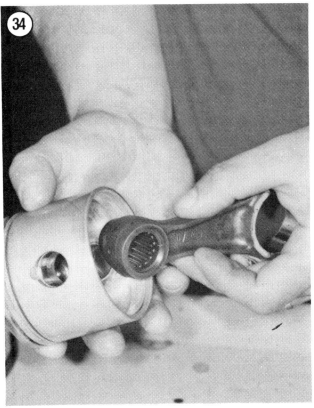

"L" as described in Step 22. The piston pin can be removed from either side of such pistons. Both sides are a loose fit. No heat is required for removal.

22. Place piston in cradle part No. 326573 with the "L" mark on the inside of the piston boss facing upward. This positions the driver on the loose end of the piston pin.

23. Heat the piston to 200-400° F with a heat lamp, then press the piston pin through the piston using an appropriate size driver. See **Figure 33**.

24. Let the piston cool, then remove the connecting rod and caged bearing retainer (**Figure 34**).

25. On models prior to 1985, slide bearing retainer from connecting rod (**Figure 35**) and place in the container with the other bearings from that piston/cylinder assembly. On 1985-on models, 28 loose needle roller bearings and 2 thrust washers are used. Be careful not to lose needle roller bearings.

Disassembly
(V4 and V6 Cross-Flow Engines)

Figure 36 shows the V4 crankshaft assembly. **Figure 37** shows the V4 power head components. The V6 crankshaft assembly and power head are similar.

1. Note location of intake bypass cover fasteners, primer and recirculation hoses, clamps and hose connectors. The cover on which the starter solenoid is mounted (V4 and V6) and the V6 cover with the terminal block must be reinstalled in the same location as removed; the other covers

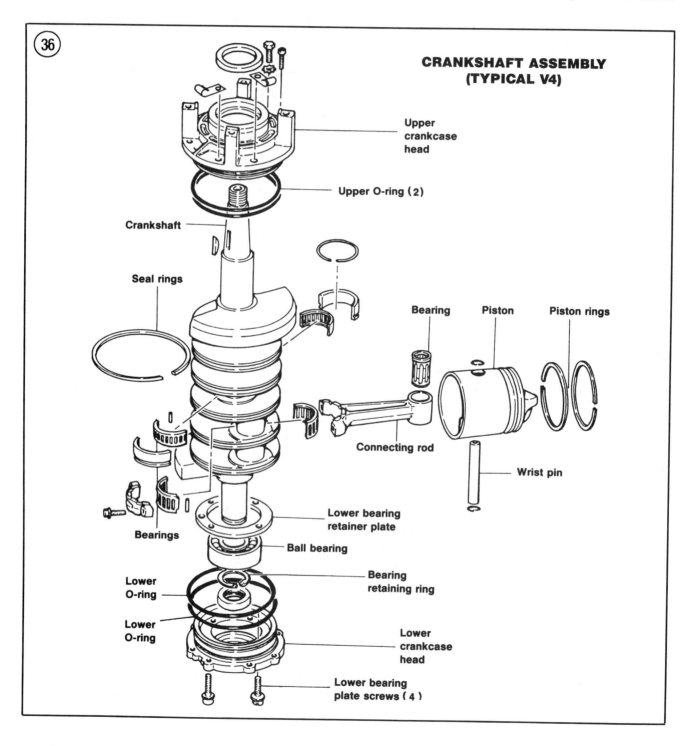

Power Head

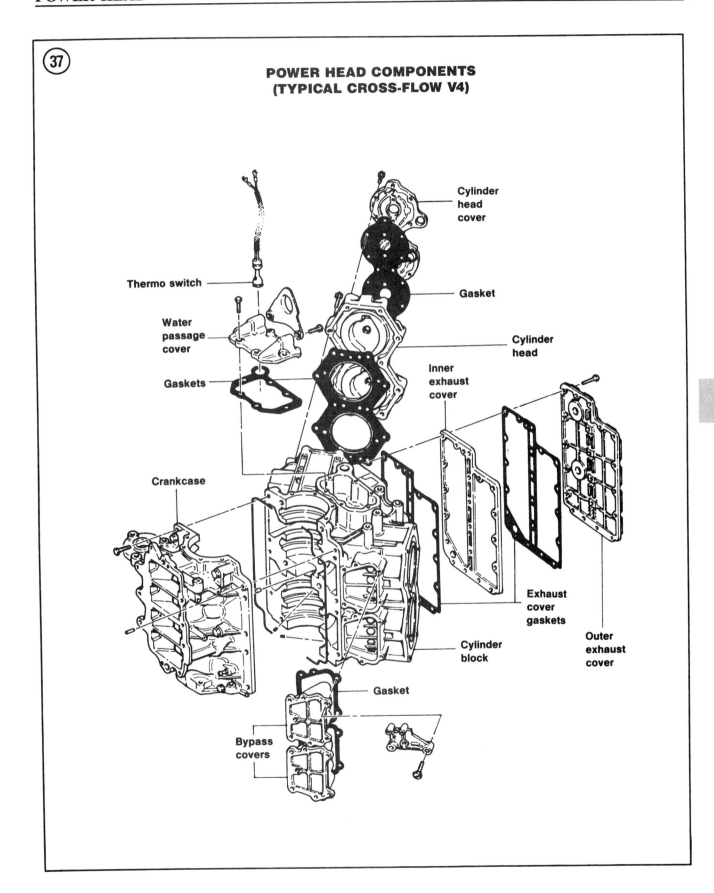

are interchangeable. Remove each cylinder bypass cover.

2. Remove the upper (**Figure 38**) and lower (**Figure 39**) crankcase head screws.

3. Locate the taper pins at opposite ends of the crankcase and cylinder block assembly. Drive the pins from the back to the front of the crankcase with a suitable drift and hammer.

4. Loosen but do not remove the 4 screws holding the lower bearing retainer plate to the lower crankcase head (**Figure 39**).

5A. V4—Tap the side of the crankshaft with a soft mallet to break the gasket seal, then remove the crankcase.

5B. V6—Temporarily reinstall one bypass cover to provide a pry point. Use a flat blade screwdriver to pry the crankcase halves apart, using the bypass cover as a fulcrum. Remove the crankcase.

NOTE
The connecting rod caps must be removed before the crankshaft can be removed from the cylinder block. Note that one side of the rod and cap has raised dots and that the corners are chamfered for proper rod/cap alignment. Caps are not interchangable and cannot be turned.

6. Rotate the crankshaft to position the No. 1 and No. 2 connecting rods as shown in **Figure 40**. Remove the connecting rod cap screws with a 5/16 in. 12-point deep socket. Remove the connecting rod caps, retainers and bearing assemblies. Mark each rod cap according to cylinder number. Place retainers and bearings from each rod cap in individual containers marked with the cylinder number.

7. Repeat Step 6 to remove the remaining rod caps, retainers and bearings.

8. Grasp the upper and lower crankcase heads and lift the crankshaft from the cylinder block. Place crankshaft on a clean workbench.

9. Remove the bearings and retainers from the connecting rods. Place in the containers with the rod cap bearings and retainers.

10. Remove the cylinder head covers and gaskets. Discard the gaskets.

11. Remove the cylinder heads and gaskets. Discard the gaskets.

12. Reinstall rod caps to their respective rods, then remove each piston from its cylinder.

13. Remove the flywheel Woodruff key, if not removed when the flywheel was removed. Slide the upper crankcase head from the crankshaft. Remove and discard the O-rings. Carefully drive seal from crankcase head with a punch inserted through the bottom of the bearing to engage the seal lip.

14. Remove and discard the screws loosened in Step 4, then remove the lower crankcase head from the crankshaft. Remove and discard the O-rings. Carefully drive seal from crankcase head with a punch inserted through the bottom of the bearing to engage the seal lip.

15. Remove the center main bearing sleeve retaining rings. Note that the retaining ring

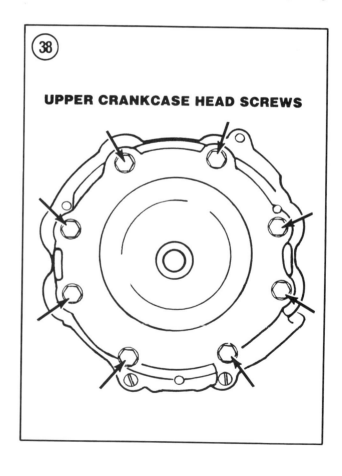

UPPER CRANKCASE HEAD SCREWS

POWER HEAD

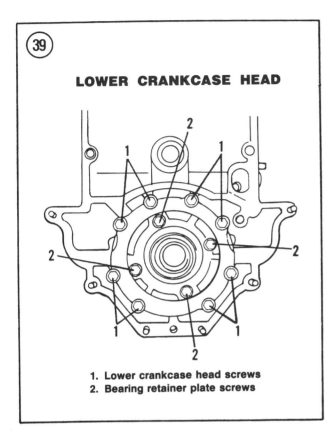

39

LOWER CRANKCASE HEAD

1. Lower crankcase head screws
2. Bearing retainer plate screws

40

groove in the bearing sleeve faces toward the bottom of the crankshaft, then separate the bearing sleeve halves and remove from the crankshaft.

16. If lower main bearing requires removal, remove snap ring and then remove bearing with an appropriate puller.

17. Pry each ring far enough from the piston to grip it with pliers, then break the rings off the piston and discard.

NOTE
If piston pin bore has a small cutout facing the top of the piston, the retaining rings can be removed in Step 18 by inserting a small screwdriver blade or awl in the cutout and prying the ring free.

18. If the piston is to be removed from the connecting rod, remove the piston pin retaining rings with tool part No. 325937. See **Figure 32**.

NOTE
Pistons used on engines after 1978 and some engines 1978 and prior are not marked with an "L" as described in Step 19. The piston pin can be removed from either side of such pistons. Both sides are a loose fit. No heat is required for removal.

19. Place piston in cradle part No. 326572 with the "L" mark on the inside of the piston boss facing upward. This positions the driver on the loose end of the piston pin.

20. Heat the piston to 200-400° F with a heat lamp, then press the piston pin through the piston using an appropriate size driver. See **Figure 33**.

21. Let the piston cool, then remove the connecting rod and caged bearing retainer (**Figure 34**).

22. On models prior to 1985, slide bearing retainer from connecting rod (**Figure 35**) and place in the container with the other bearings from that piston/cylinder assembly. On 1985-on models, 28 loose needle roller bearings and 2 thrust washers are used. Be careful not to lose needle roller bearings.

Disassembly
(V4 and V6 Loop-Charged Engines)

Figure 36 shows the V4 crankshaft assembly used on a cross-flow engine. The crankshaft assembly used on a loop-charged engine is similar except piston domes are near flat and intake windows are used in piston skirt. **Figure 37** shows the V4 power head assembly used on a cross-flow engine. On a loop-charged engine, the cylinder head and cylinder head cover are cast as one piece, the exhaust covers and cylinder block are cast as one piece and bypass covers are not used.

1. Remove stator yellow and yellow/gray wires from terminal block.
2. Remove rectifier/regulator red wire from starter solenoid.
3. Remove bracket retaining stator wires at upper crankcase head.
4. Label and remove all wiring and electrical components mounted on top of power head.
5. Remove electric starter motor.
6. Remove the air silencer cover and baffle.
7. Remove the 2 nuts and 2 screws retaining each carburetor assembly. Disconnect fuel primer hoses and rotate each carburetor assembly to allow access to air silencer base screws. Remove all screws retaining air silencer base. Disconnect VRO pulse hose from crankcase fitting and unbolt VRO mounting bracket. Lift complete fuel system as an assembly from the intake manifold.
8. Remove the throttle cam from the intake manifold.
9A. V4 models—Remove the fasteners retaining the intake manifold and lift the manifold from the lower crankcase half.
9B. V6 models—Remove the fasteners retaining the upper and lower intake manifolds. Lift the lower intake manifold first and then the upper intake manifold from the lower crankcase half.
10. Label location of recirculation hoses to lower crankcase half fittings then disconnect.
11. Remove cooling hose from starboard cylinder head.
12. Remove the upper (**Figure 38**) and lower (**Figure 39**) crankcase head screws.
13. Remove the crankshaft main bearing screws in small increments. The upper port screw requires a 3/8 in. Allen head socket for removal.
14. Locate the taper pins at opposite ends of the crankcase and cylinder block assembly. Drive the pins from the back to the front of the crankcase with a suitable drift and hammer.
15. Tap the end of the crankshaft toward the lower crankcase half with a soft-faced mallet to break the gasket seal, then remove the crankcase.

NOTE
The connecting rod caps must be removed before the crankshaft can be removed from the cylinder block. Note that one side of the rod and cap has raised dots and that the corners are chamfered for proper rod/cap alignment. Caps are not interchangeable and cannot be turned.

16. Rotate the crankshaft to position the No. 1 and No. 2 connecting rods as shown in **Figure 40**. Remove the connecting rod cap screws with a 5/16 in. 12-point deep socket. Remove the connecting rod caps, retainers and bearing assemblies. Mark each rod cap according to cylinder number. Place retainers and bearings from each rod cap in individual containers marked with the cylinder number.
17. Repeat Step 16 to remove the remaining rod caps, retainers and bearings.
18. Grasp the upper and lower crankcase heads and lift the crankshaft from the cylinder block. Place crankshaft on a clean workbench.
19. Remove the bearings and retainers from the connecting rods. Place in the containers with the rod cap bearings and retainers.
20. Disconnect the water hose between cylinder heads. Remove the cylinder head screws in small increments, then remove cylinder heads.
21. Reinstall rod caps to their respective rods, then remove each piston from its cylinder.

POWER HEAD

22. Remove the flywheel Woodruff key, if not removed when the flywheel was removed. Slide the upper crankcase head from the crankshaft. Remove and discard the O-rings. Carefully drive seal from crankcase head with a punch inserted through the bottom of the bearing to engage the seal lip.

23. Remove the 4 screws holding the lower bearing retainer plate to the lower crankcase head (**Figure 39**). Remove the lower crankcase head from the crankshaft. Remove and discard the O-rings. Carefully drive seals from crankcase head with a punch placed from the outside of the crankcase head onto the seals and driving the seals into the inside.

24. Remove the center main bearing sleeve retaining rings. Note that the retaining ring groove in the bearing sleeve faces toward the bottom of the crankshaft, then separate the bearing sleeve halves and remove from the crankshaft.

25. If lower main bearing requires removal, remove snap ring and then remove bearing with an appropriate puller.

26. Remove each ring from the piston with a pair of ring expander pliers and discard.

NOTE
If piston pin bore has a small cutout facing the top of the piston, the retaining rings can be removed in Step 27 by inserting a small screwdriver blade or awl in the cutout and prying the ring free.

27. If the piston is to be removed from the connecting rod, remove the piston pin retaining rings with tool part No. 325937. See **Figure 32**.

28. Place piston in cradle part No. 326572 and remove piston pin using driver assembly part No. 396747.

29. Remove the connecting rod and caged bearing retainer (**Figure 34**).

30. Slide bearing retainer from connecting rod (**Figure 35**) and place in the container with the other bearings from that piston/cylinder assembly.

Cylinder Block and Crankcase Cleaning and Inspection (All Engines)

Johnson and Evinrude outboard cylinder blocks and crankcase covers are matched and line-bored assemblies. For this reason, you should not attempt to assemble an engine with parts salvaged from other blocks. If inspection indicates that either the block or cover requires replacement, replace both as an assembly.

WARNING
Remove your wrist watch and other jewelry and wear hand and eye protection when working with OMC Gel Seal and Gasket remover. The substance is powerful enough to etch holes in a watch crystal or badly irritate any cut in the skin.

Carefully remove all gasket and sealant residue from the cylinder block and crankcase cover mating surfaces. Gel Seal or Gel Seal II should be cleaned from mating parts with OMC Gel Seal and Gasket Remover. After spraying the area to be cleaned, allow the solvent to stand for 5-10 minutes before cleaning off the old Gel Seal and the remover. Since Gel Seal is bright red in color, a visual inspection will quickly tell you when it has been completely removed.

Clean the aluminum surfaces carefully to avoid nicking them. A dull putty knife can be used, but a piece of Lucite with one edge ground to a 45 degree angle is more efficient and will also reduce the possibility of damage to the surfaces. See **Figure 41**. Once the area is clean, apply OMC Cleaning Solvent to remove all traces of the Gel Seal and Gasket Remover. When sealing the crankcase cover and cylinder block, both mating surfaces must be free of all sealant residue or dirt and oil or leaks will develop.

Once the gasket surfaces are cleaned, place the mating surface of each component on a large pane of glass. Apply uniform downward pressure on the component and check for warpage.

Replace each component if more than a slight degree of warpage exists. In cases where there is a slight amount of warpage, it can often be eliminated by placing the mating surface of each component on a large sheet of 120 emery cloth. Apply a slight amount of pressure and move the component in a figure-8 pattern. See **Figure 42**. Remove the component and emery cloth and recheck surface flatness on the pane of glass.

If warpage exists, the high spots will be dull while low areas will remain unchanged in appearance. It may be necessary to repeat this procedure 2-3 times until the entire mating surface has been polished to a dull luster. Do not remove more than a total 0.010 in. from the cylinder block and head. Finish the resurfacing with 180 emery cloth.

1. Clean the cylinder block and crankcase cover thoroughly with solvent and a brush.
2. Carefully remove all gasket and sealant residue from the cylinder block and crankcase cover mating surfaces.
3. Check the cylinder heads and exhaust ports for excessive carbon deposits or varnish. Remove with a scraper or other blunt instrument.
4. Check the block, cylinder head and cover for cracks, fractures, stripped bolt or spark plug holes or other defects.
5. Check the gasket mating surfaces for nicks, grooves, cracks or excessive distortion. Any of these defects will cause a compression leakage. Replace as required.
6. Check all oil and water passages in the block and cover for obstructions. Make sure any plugs installed are properly tightened.

NOTE
1980 Johnson and Evinrude 150, 175 and 200 hp V6 engines with an overheating problem may have been built with a faulty core. Such engines have mislocated water restrictor locating ribs which must be modified. To determine if this is the problem, check the bottom of the starboard water jacket on the exhaust port side between cylinders No. 3 and No. 5 for the letter designation. If it is an "E," take the motor to your dealer for recommended OMC modification.

7. Make sure all water passage restrictors are in good condition and properly installed (**Figure 43** shows typical water passage restrictors and **Figure 44** shows V4 water circulation and water passage restrictor location). Damaged, loose or missing restrictors will interfere with cooling water circulation and result in possible engine overheating.

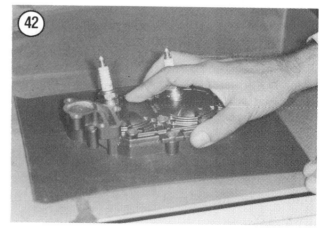

POWER HEAD

8. Check the inner and outer exhaust covers for signs of overheating or warpage. Replace as required.

9. Check crankcase recirculation orifice, if so equipped, and clean with tool part No. 326623.

NOTE
With older engines, it is a good idea to have the cylinder walls lightly honed with a medium stone even if they are in good condition. This will break up any glaze that might reduce compression.

10. Check each cylinder bore for signs of aluminum transfer from the pistons to the cylinder walls. If scoring is present but not excessive, have the cylinders honed by a dealer or qualified machine shop.

11. Check each cylinder bore for size and taper in the port area with an inside micrometer or bore gauge. See **Figure 45**. On inline engines, if bore

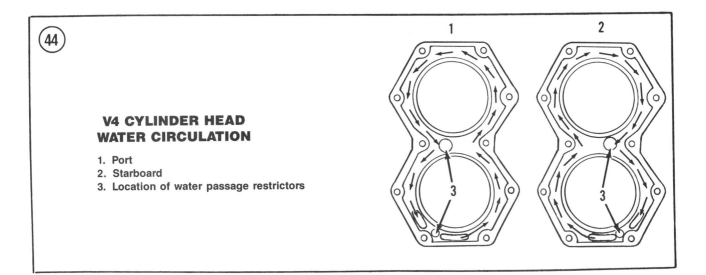

V4 CYLINDER HEAD WATER CIRCULATION

1. Port
2. Starboard
3. Location of water passage restrictors

is tapered by 0.002 in. or more or out-of-round or worn by 0.003 in. or more, have the cylinder(s) honed or rebored by a dealer or qualified machine shop. On V4 and V6 engines, if bore is tapered by 0.002 in. or more or out-of-round or worn by 0.004 in. or more, have the cylinder(s) honed or rebored by a dealer or qualified machine shop.

CAUTION
If cylinders are rebored for oversize pistons, be sure to allow 0.003-0.005 in. clearance between the new bore diameter and the oversize piston on 1973-1984 models. For 1985-on models, no manufacturer recommended piston-to-cylinder clearance is provided. When an Evinrude or Johnson replacement piston (provided by OMC) is used, a built-in clearance is maintained by boring the cylinder to the oversize diameter determined by adding the oversize dimension of the piston to the standard bore.

Crankshaft and Connecting Rod Bearings Cleaning and Inspection (All Engines)

Bearings can be reused if they are in good condition. To be on the safe side, however, it is a good idea to discard all bearings and install new ones whenever the engine is disassembled. New bearings are inexpensive compared to the cost of another overhaul caused by the use of marginal bearings.

1. Place ball bearings in a wire basket and submerge in a suitable container of fresh solvent. The bottom of the basket should not touch the bottom of the container.
2. Agitate basket containing bearings to loosen all grease, sludge and other contamination.
3. Dry ball bearings with dry filtered compressed air. Be careful not to spin the bearings.
4. Lubricate the dry bearings with a light coat of Johnson or Evinrude Outboard Lubricant and

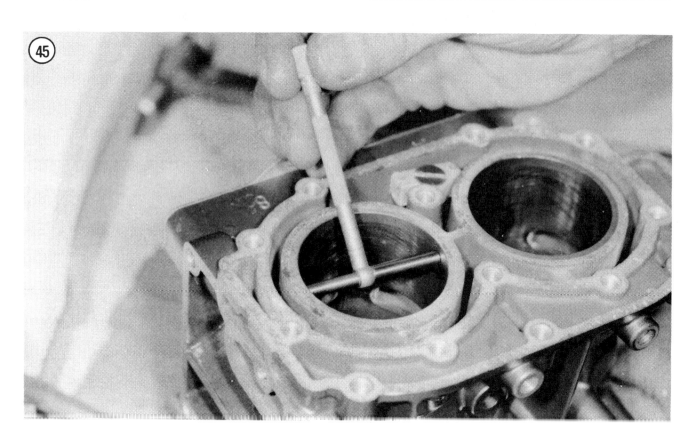

POWER HEAD

inspect for rust, wear, scuffed surfaces, heat discoloration or other defects. Replace as required.

5. If roller bearings are to be reused, repeat Steps 1-4, cleaning one set at a time to prevent any possible mixup. Check bearings for flat spots. If one roller bearing is defective, replace all in the set with new bearings and liners.

6. Repeat Step 5 to check caged piston pin bearings. If bearing is defective, replace the bearing and its corresponding piston pin.

Piston Cleaning and Inspection (All Engines)

1. Check the pistons for signs of scoring, cracking, cracked or worn piston pin bosses or metal damage. Replace piston and pin if any of these defects are noted.

2. Check piston ring grooves for distortion, loose ring locating pins or excessive wear. If the flexing action of the rings has not kept the lower surface of the ring grooves free of carbon, clean with a bristle brush and solvent.

NOTE
Do not use an automotive ring groove cleaning tool in Step 3 as it can damage the piston ring locating pin.

3. Clean the piston skirt, ring grooves and dome with the recessed end of a broken ring to remove any carbon deposits.

4. Immerse pistons in a carbon removal solution to remove any carbon deposits not removed in Step 3. If the solution does not remove all of the carbon, carefully use a fine wire brush and avoid burring or rounding of the machined edges. Clean the piston skirt with crocus cloth.

5A. Check 1981-1984 50-60 hp pistons for size and roundness:

   a. Measure the piston at a point 1/8 in. above the bottom edge of the skirt (A, **Figure 46**) and at a 90° angle to the piston pin hole (B, **Figure 46**).

   b. Repeat the measurement at a point parallel to the piston pin hole. See **Figure 47**.

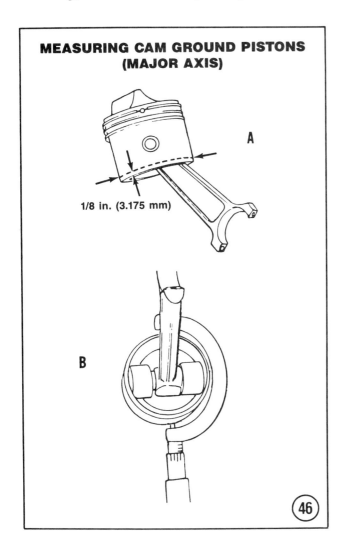

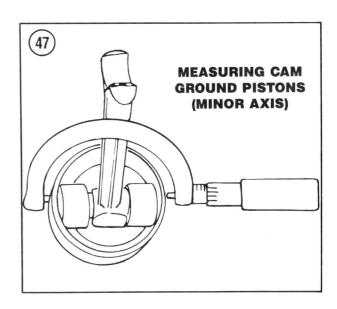

c. Subtract the second measurement from the first one. If the difference is not between 0.0025-0.0055 in., replace the piston.

5B. Check 1985-on 48-50 hp pistons for size and roundness:
   a. Measure the piston at a point 1/8 in. above the bottom edge of the skirt (A, **Figure 46**) and at a 90° angle to the piston pin hole (B, **Figure 46**).
   b. Repeat the measurement at a point parallel to the piston pin hole. See **Figure 47**.
   c. Subtract the second measurement from the first one. If the difference is more than 0.004 in., replace the piston.
   d. If piston skirt grooves are worn smooth in area 90° to piston pin hole, replace the piston.

5C. Check all other pistons with a micrometer at the top and piston skirt. See **Figure 48**. Measure at 90° angle to the piston pin and at a point parallel with the piston pin bosses. Compare the 2 top and skirt measurements to determine if piston is out-of-round. Replace piston and pin if piston is more than 0.004 in. out-of-round.

Crankshaft Cleaning and Inspection (All Engines)

1. Clean the crankshaft thoroughly with solvent and a brush. Blow dry with dry filtered compressed air, if available, and lubricate with a light coat of Johnson or Evinrude Outboard Lubricant.

2. Check the crankshaft journals and crankpins for scratches, heat discoloration or other defects. See **Figure 49** (2-cyl.) or **Figure 50** (V4). The 3-cylinder and V6 crankshafts are similar.

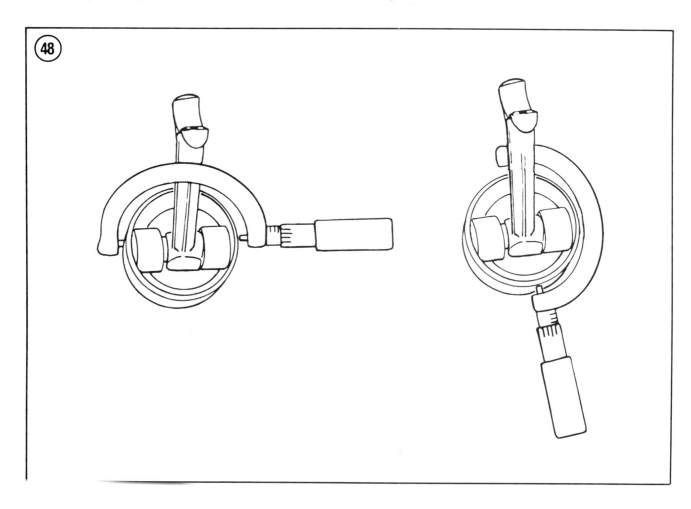

POWER HEAD

3. Measure the journals and crankpins with a micrometer and compare to **Table 3**. If journals or crankpins are not within specifications, replace the crankshaft.

4. Check drive shaft splines and flywheel taper threads for wear or damage. Replace crankshaft as required.

5. Check V4 and V6 crankshaft seal rings for excessive wear or damage and replace if they do not seal tightly around the crankshaft web. On 1973-1984 models, measure seal ring side clearance with a feeler gauge. If not between 0.0015-0.0025 in., replace the seal ring. See **Table 4** for available ring thicknesses. Do not remove seal rings unless replacement is required.

6. If V4 or V6 crankshaft seal ring breaks during crankshaft removal, measure the ring thickness near its outer edge with a micrometer. Refer to **Table 4** and select a new seal ring of the same thickness. Carefully spread the end gap just enough to slip the ring over the crankshaft journal and complete installation with a piston ring expander.

7. If the lower crankshaft ball bearing has not been removed, grasp its inner race and try to work it back and forth. Replace the bearing if excessive play is noted.

8. Lubricate the ball bearing with Johnson or Evinrude Outboard Lubricant and rotate its outer race. Replace the bearing if it sounds or feels rough or if it does not rotate smoothly.

Piston and Connecting Rod Assembly (All Engines)

If the pistons were removed from the connecting rods, they must be correctly oriented when reassembling. On cross-flow models, the exhaust or slanted side of the piston dome must face the exhaust ports when installed. See **Figure 51**. On V4 and V6 engines, remember to consider the difference between port and starboard cylinders. See **Figure 52**. On inline loop-charged models, the side of the piston dome stamped "UP" must be installed toward the flywheel end of the engine. On V4 and V6 loop-charged models, the piston domes are stamped "STBD" or "PORT". The "STBD" piston must be installed on the starboard side with the piston

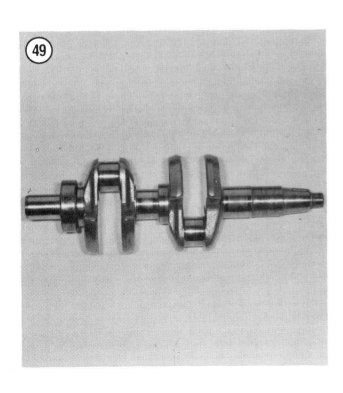

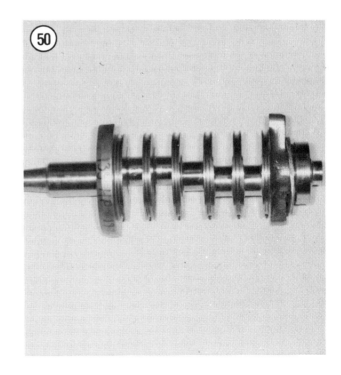

CORRECT PISTON INSTALLATION (INLINE CROSS-FLOW ENGINE)

1. Intake side of piston
2. Intake side of cylinder
3. Exhaust side of cylinders
4. Exhaust side of piston

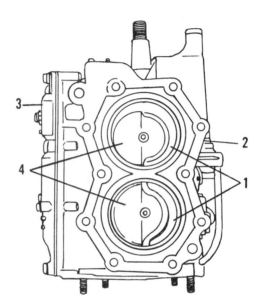

CORRECT PISTON INSTALLATION (V4 AND V6 CROSS-FLOW ENGINE)

1. Intake side of piston
2. Exhaust side of piston
3. Cylinder exhaust ports

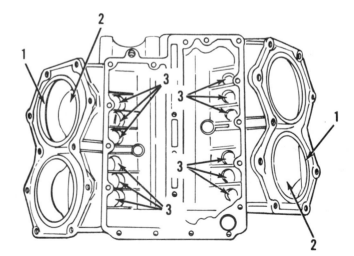

POWER HEAD

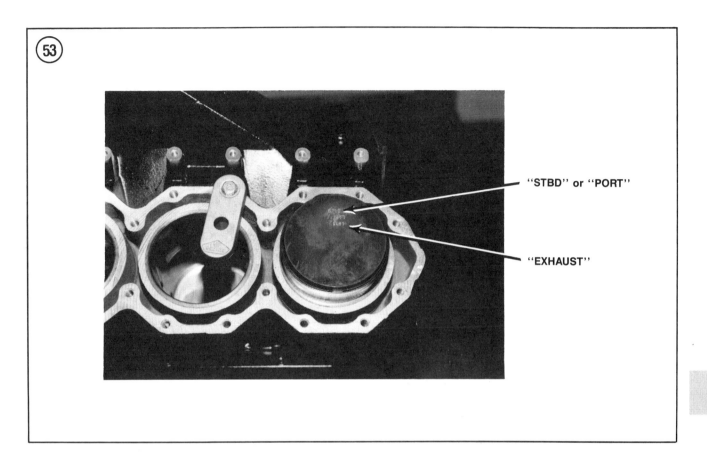

53 "STBD" or "PORT" / "EXHAUST"

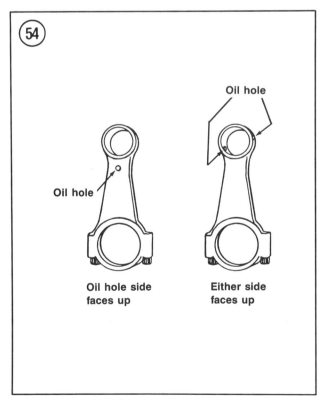

54 Oil hole / Oil hole / Oil hole side faces up / Either side faces up

dome side stamped "EXHAUST" toward the cylinder exhaust ports. The "PORT" piston must be installed on the port side with the piston dome side stamped "EXHAUST" toward the cylinder exhaust ports. See **Figure 53**. The connecting rod oil hole on models with hole in center of connecting rod must face toward the flywheel end of the engine. The connecting rod on models with 2 oil holes in side of connecting rod small end can be installed with either side toward the flywheel end of the engine. See **Figure 54**. Double-check rod and piston orientation before installing the piston pin.

1A. If a caged bearing was removed from the connecting rod piston pin end:
   a. Reinstall the bearing with an arbor press.
   b. Fit piston over connecting rod piston pin end to provide correct piston-to-cylinder orientation during reassembly as previously outlined. On models with one oil hole in center of connecting rod, the

oil hole must face toward the flywheel end of the engine.

1B. If needle bearings were removed from the connecting rod piston pin end:
   a. Position connecting rod with oil hole facing up on single hole models. See **Figure 54**.
   b. Wipe inside of piston pin bore with OMC Needle Bearing grease.
   c. Install a suitable bushing to act as a spacer and insert needle bearings individually. See **Figure 55**.
   d. When all bearings are in place, fit washers at top and bottom of bearing assembly.
   e. Fit piston over connecting rod piston pin end to provide correct piston-to-cylinder orientation during reassembly as previously outlined. On models with one oil hole in center of connecting rod, the oil hole must face toward the flywheel end of the engine.

2. Lightly coat piston pin and lubricate each piston pin hole with a drop or two of Evinrude or Johnson Outboard Lubricant.

NOTE
Position piston in Step 3 to press piston pin in place from the "L" or "LOOSE" side of the piston. See piston pin boss inside piston skirt for marking. If there is no marking, the pin can be installed from either side.

3. Insert piston pin through piston pin hole and engage connecting rod. Make sure piston dome and connecting rod oil hole (single hole models) are properly oriented. Position piston on the same cradle used to disassemble it and press piston pin in place with the same piston pin tool used to remove it. See **Figure 56**.

4. Install new piston pin retaining rings on each side of the piston. Insert the retaining ring in the tapered end of cone part No. 396747 for V4 and V6 loop-charged models or part No. 318600 for all other models. Position cone over piston pin hole and insert driver part No. 331913 for V4 and V6 loop-charged models or part No. 318599 for all other models. See **Figure 57**. Press driver into cone as far as possible to install the retaining ring. Turn piston over and install the other ring. Make sure both rings fit into their grooves in the piston pin bore and ring openings face downward.

5. On pistons with one piston pin boss marked "L" or "LOOSE", measure the bottom of piston skirt at a point parallel with the piston pin and 90° from the piston pin. Compare the measurements. If measurements vary more than 0.003 in. (inline) or 0.004 in. (V4 and V6), the piston was distorted during assembly and should be replaced.

6. Check end gap of new rings before installing on piston. Place ring in cylinder bore, then square it up by inserting the bottom of an old piston. Do not push ring into bore more than 3/8-1/2 in. Measure the gap with a feeler gauge (**Figure 58**) and compare to specifications (**Table 3**).

7. If ring gap is excessive in Step 6, repeat the step with the ring in another cylinder. If gap is also excessive in that cylinder, discard and replace with a new ring.

8. If ring gap is insufficient in Step 6, the ends of the ring can be filed slightly. Clean ring thoroughly and recheck gap as in Step 6.

NOTE
The upper ring is a pressure-back type and has a tapered groove which cannot

POWER HEAD

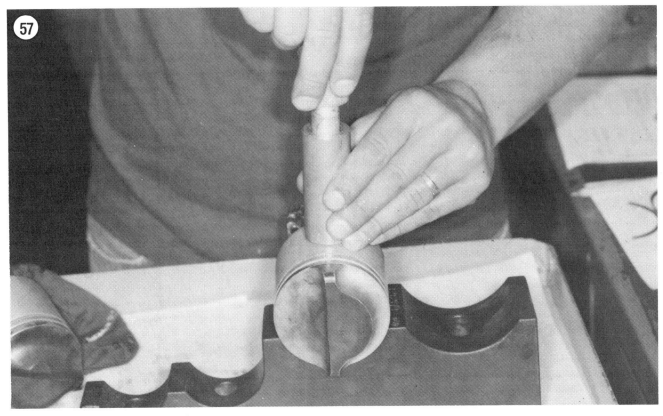

354 CHAPTER EIGHT

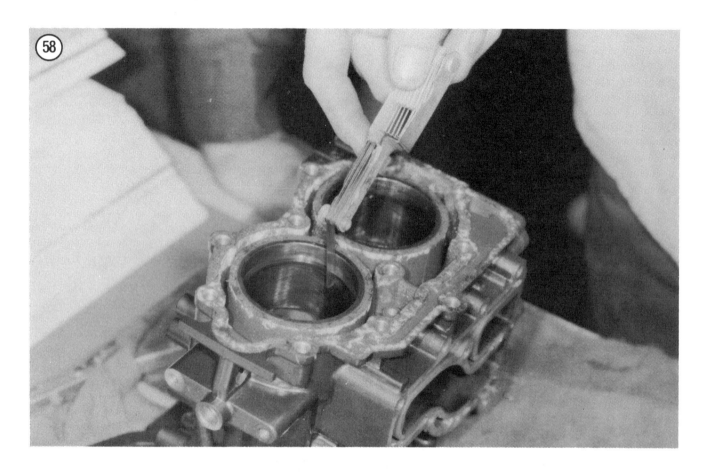

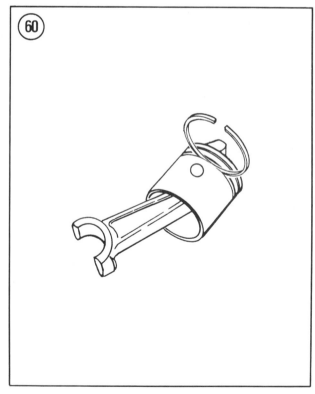

POWER HEAD

be checked following procedure in Step 9. To check pressure-back ring-to-piston groove fit, install ring on piston. Place a straightedge across the ring groove. See **Figure 59**. *The straightedge should touch on both sides of the ring groove and not touch the ring face. If straightedge does not touch on both sides or touches the ring face, remove ring and clean ring groove.*

9. Once the ring gaps are correctly established, roll the lower ring around the piston ring groove to check for binding or tightness. See **Figure 60**.

10. Install the lower ring on the piston with a ring expander. Spread the ring just enough to fit it over the piston head and into position. See **Figure 61**.

11. Repeat Step 10 to install the upper ring.

12. Position each ring so the piston groove locating pin fits in the ring gap. See **Figure 62**. Proper ring positioning is necessary to minimize compression loss and prevent the ring ends from catching on the cylinder ports.

13. Coat the piston and cylinder bore with Johnson or Evinrude Outboard Lubricant, then proceed as follows:

   a. Check the piston dome number made during disassembly and match piston with its correct cylinder.

   b. Orient exhaust side of piston to exhaust port side of cylinder block or "UP" side toward flywheel end of crankshaft.

   c. Make sure connecting rod oil hole (single hole models) faces flywheel end of engine.

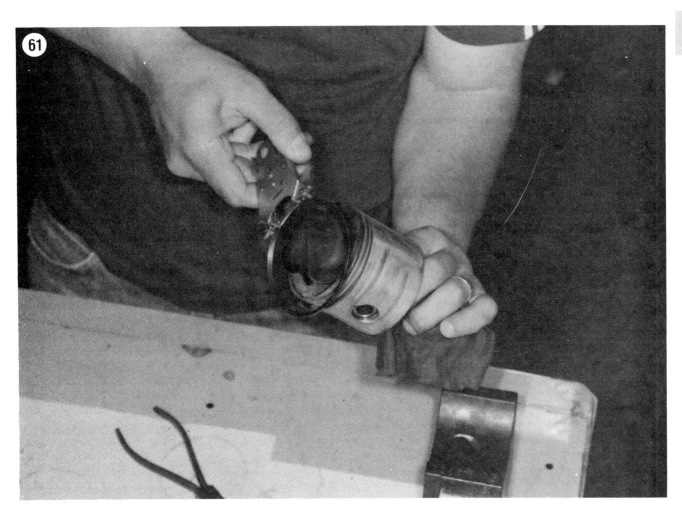

d. Insert piston into cylinder bore. Recheck ring gap and groove locating pin alignment.
e. Install ring compressor over piston dome and rings. Use part No. 326592 (48, 50, 55, 60 [1980-1985] and 70 [1986-1990] hp), part No. 326591 (60 [1986-1990], 70 [1974-1985] and 75 hp), part No. 327018 (V4 [Except 1988-1990 120 and 140 hp]), part No. 327018 (150, 175 [1977-1985], 200 [1975-1987] and 225 [1986-1987] hp V6), part No. 326734 (175 [1986-1990] and 235 hp V6), or part No. 334140 (1988-1990 120, 140, 200 and 225 hp) as appropriate.
f. Hold connecting rod end with one hand to prevent it from scraping or scratching the cylinder bore and slowly push piston into cylinder. See **Figure 63**.

14. Remove the ring installer tool and repeat Step 13 to install the remaining pistons.

15. On cross-flow models (uses transfer covers), reach through exhaust port and lightly depress each ring with a pencil point or small screwdriver blade. See **Figure 64**. The ring should snap back when pressure is released. If it does not, the ring was broken during piston installation and will have to be replaced.

16. Temporarily reinstall the cylinder head(s) with 2 screws to prevent the pistons from slipping out while the crankshaft is being installed.

Connecting Rod and Crankshaft Assembly (All Engines)

1. V4 and V6—If lower crankshaft bearing was removed, install retainer plate and press bearing on crankshaft. Install a new snap ring and lubricate the bearing with Johnson or Evinrude Outboard Lubricant. See **Figure 65**.

2. V4 and V6—Install lower crankcase head with new O-rings. Align holes in retainer plate and crankcase head. Apply Gel Seal II on threads of new retainer plate screws. Install screws and securely tighten, then back off each screw 2 turns to ease crankshaft installation.

3. V4 and V6—Install new upper crankcase head seal with tool part No. 325453. Seal lip should face inward and be flush with top of crankcase

POWER HEAD

head. Install upper crankcase head and bearing assembly on crankshaft.

4. Inline engine—Install new upper main bearing seal with tool part No. 326567. Seal lip should face inward.

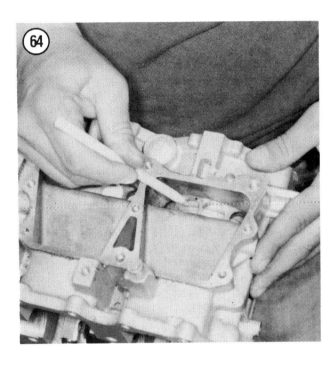

5. Install center main bearing(s) and sleeve(s) with retaining ring groove toward the bottom of the crankshaft. Slip retainer ring(s) into sleeve groove(s).

6. Inline engine—Install a new O-ring on the upper bearing.

7. Remove the connecting rod caps from the rods. Discard the cap screws.

8. Coat connecting rod bearing surface with OMC Needle Bearing Grease and install bearing retainer half. Insert roller bearings in each retainer slot. Repeat this step for each connecting rod.

NOTE
Location of bearing hole and dowel pin varies according to model. Determine which is in the block and align the bearings accordingly so that pin and hole will mate when crankshaft is installed in Step 9.

9. Install crankshaft in cylinder block. Align upper main bearing dowel and pin hole. See

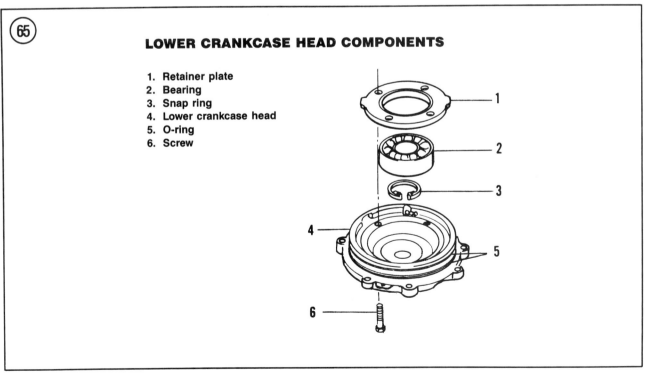

LOWER CRANKCASE HEAD COMPONENTS

1. Retainer plate
2. Bearing
3. Snap ring
4. Lower crankcase head
5. O-ring
6. Screw

358   CHAPTER EIGHT

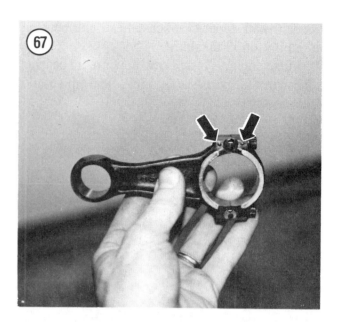

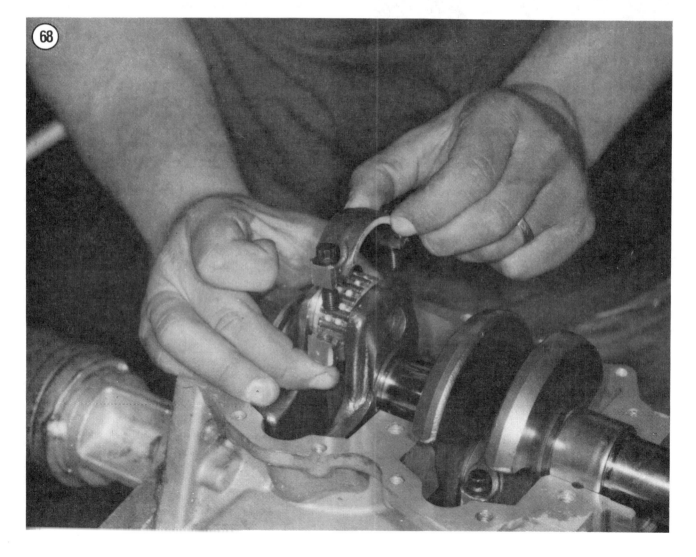

POWER HEAD

Figure 66 (typical). The center main bearing sleeve hole should align with the dowel pin in th eblock. Make sure the end gap on all crankshaft seal rings are facing UP.

10. Draw connecting rods up around crankshaft crankpin journal. Coat crankpins with OMC Needle Bearing grease and install remaining retainer halves and roller bearings on each crankpin.

11. Orient the connecting rod cap according to the embossed dots and the marks made during disassembly. See **Figure 67**.

12. Install the caps with new screws (**Figure 68**). Tighten screws finger-tight.

CAUTION
Steps 13A and 13B are very important to proper engine operation as they affect bearing action. If not performed correctly, major engine damage can result. This involves a time-consuming and frustrating process. Work slowly and with patience. If alignment was satisfactory when checked during disassembly, it should be possible to achieve a similar alignment on reassembly.

NOTE
*Connecting rods on models prior to 1985 and some 1985 models are hand sanded. Connecting rods on all other 1985 models and all 1986-on models are precision ground. See **Figure 69**.*

13A. Proceed as follows for models equipped with hand-sanded connecting rods:
  a. Run a pencil point or dental pick over the rod and cap chamfers to check alignment.

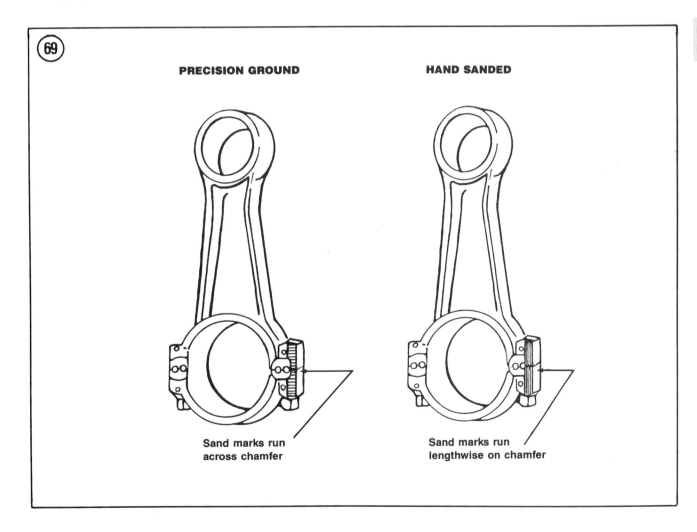

See **Figure 67** and **Figure 70**. Refer to notes made during disassembly. Rod and cap must be aligned so that the pencil point or dental pick will pass smoothly across the break line on at least 3 of the 4 corners.

b. If rod and cap alignment is not satisfactory, gently tap cap in direction required with a soft mallet and recheck alignment. Repeat this procedure as many times as necessary to achieve alignment of at least 3 corners.

c. If satisfactory alignment cannot be achieved, the connecting rod and cap should be replaced.

d. Once rod and cap alignment is correct, carefully torque rod caps to specifications (**Table 1**).

13B. Proceed as follows for models equipped with precision-ground connecting rods:

a. Assemble OMC Rod Cap Alignment Fixture part No. 396749 on connecting rod (**Figure 71**, typical) following directions provided with special tool.

b. Once rod and cap alignment is correct, use special torque socket part No. 331638 and a suitable torque wrench and extension and carefully torque rod cap screws to specification (**Table 1**).

14. Rotate the crankshaft to check for binding. If the crankshaft does not float freely over the full length of the crankpins, loosen the rod caps and repeat Steps 13A or 13B.

Cylinder Block and Crankcase Assembly (All Engines)

The cylinder head gasket on some models is impregnated with sealant during manufacture and requires no additional sealer when installed. Check the gasket package to determine if it is this type.

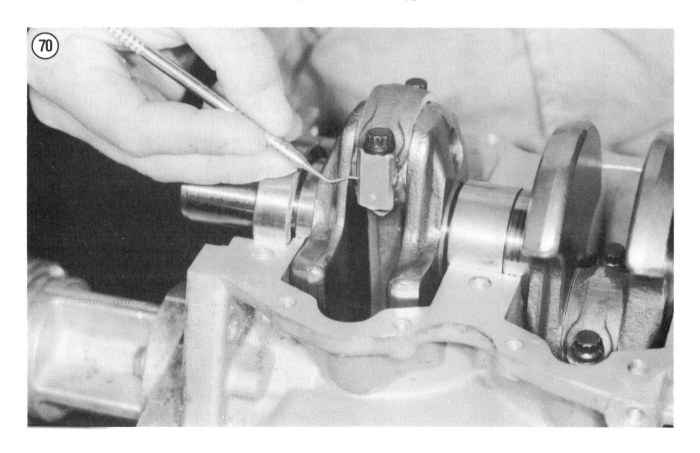

POWER HEAD

All gaskets which have not been impregnated with sealant (except 1978-on 235 V6 inner exhaust gaskets) should be lightly coated on both sides with OMC Gasket Sealing Compound. The outer diameter of the crankshaft upper seal should also be coated with OMC Gasket Sealing Compound.

The crankcase face on some models is grooved for the use of a spaghetti or rubber seal. See **Figure 72**. The new seal should be fully seated in the grooves and then cut 1/2 in. longer at each end to assure a good butt seal against both crankcase bearings. Run a thin bead of OMC Adhesive M in the groove before installing the seal. Force the seal into the groove and let it set 15-20 minutes. Trim the ends of the seal with a sharp knife, leaving about 1/32 in. of the seal end to butt against the bearings. Apply another thin bead of OMC Adhesive M to the crankcase face. Use care not to apply an excessive amount, as it can squeeze over when the parts are mated and may block oil or water passages.

A motor reassembled with Gel Seal II should sit overnight before it is started and run. If the motor must be run the same day, spray the surface opposite the coated one with OMC Locquic Primer and allow several hours for the Gel Seal II to cure.

Gel Seal II starts to cure as soon as the parts are mated. The process is shortened when OMC Locquic Primer has also been used. For this reason, it is important that you install and torque all fasteners as soon as possible. If Gel Seal II starts to set up before the fasteners are torqued, it can act as a shim and result in bearing misalignment or mislocation or tight armature or stator bearings.

Inline engines

1. Smooth crankcase flange—Squeeze a 1/4 in. (50-60 hp) or 3/8 in. (70-75 hp) ball of Gel Seal II on each crankcase flange. Spread sealant along flange and inside all bolt holes but keep it at least 1/4 in. from seals. Use additional Gel Seal II at points where crankcase flanges are narrow.
2. Grooved crankcase flange—Install a new spaghetti seal with OMC Adhesive M as described in this section. Apply a very thin bead of Adhesive M to the crankcase flange.
3. Install the crankcase to the cylinder block. Install the screws finger-tight.
4. Install the crankcase taper pin(s) with a mallet and punch.
5. Tap bottom of crankshaft with a mallet to seat the bearings, then temporarily install flywheel and rotate the crankshaft to check for binding. If crankshaft does not turn easily, disassemble and correct the interference.
6. Tighten all crankcase screws to specifications (**Table 1**).
7. Install the intake manifold and reed block assembly with a new gasket.
8. Remove the cylinder head temporarily installed to hold the pistons in place during crankshaft installation. Install a new head gasket (lightly coated with OMC Gasket Sealing Compound, if so required).
9. Install the thermostat with a new seal. Install the thermostat spring.
10. Install the cylinder head and tighten head bolts to specifications following the sequence shown in **Figure 73** (48, 50, 55 and 60 [1980-1985] hp) or **Figure 74** (60 [1986-1990], 70 and 75 hp).
11. Install the cylinder head cover with a new gasket.
12. Install the exhaust covers with new gaskets. Tighten fasteners to specifications (**Table 1**).
13. Install a new seal (lip facing downward) in lower main bearing housing.
14. Install the power head as described in this chapter.

V4 and V6 engines

1. Smooth crankcase flange—Squeeze a 3/8 in. ball of Gel Seal II on each crankcase flange. Spread sealant along flange and inside all bolt holes but keep it at least 1/4 in. from seals. Use additional Gel Seal II at points where crankcase flanges are narrow.

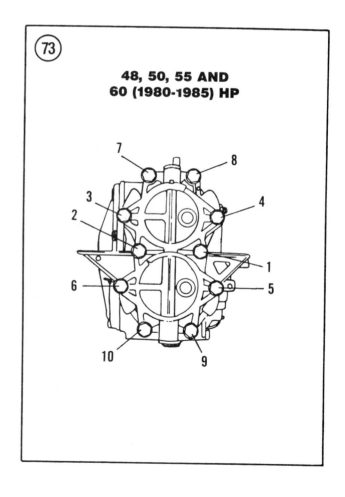

73

48, 50, 55 AND 60 (1980-1985) HP

POWER HEAD

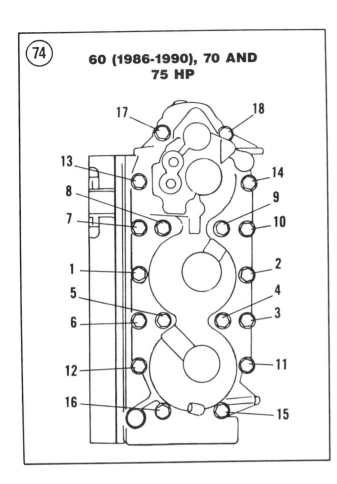

74 60 (1986-1990), 70 AND 75 HP

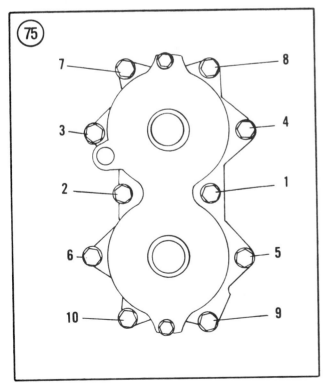

75

2. Grooved crankcase flange—Install a new spaghetti seal with OMC Adhesive M as described in this section. Apply a very thin bead of Adhesive M to the crankcase flange.

3. Install the crankcase to the cylinder block carefully to avoid damage to the crankcase head seals. Rotate the crankcase heads to align the screw holes.

4. Install the upper and lower crankcase head screws finger-tight.

5. Install the crankcase taper pins with a mallet and punch.

6. Install the main bearing and crankcase screws finger-tight.

7. Temporarily install the flywheel and rotate the crankshaft to check for binding. If crankshaft does not turn easily, disassemble and correct the interference.

8. Tighten the crankcase-to-block screws to specifications (**Table 1**).

9. Tighten the upper and lower crankcase head screws to specifications (**Table 1**).

10. Remove the cylinder heads (temporarily installed to hold the pistons in place during crankshaft installation).

11. Cross-Flow Engines—Install the inner and outer exhaust covers with new gaskets. Apply OMC Gasket Sealing Compound on both sides of new gaskets prior to assembly. Make sure alignment is correct between inner exhaust cover and cylinder block when tightening the screws to prevent leakage.

12. Install thermostats (models so equipped) as described in this chapter.

13. Install the cylinder heads with new gaskets. On cross-flow engines, tighten cylinder head bolts to specifications (**Table 1**) following the sequence shown in **Figure 75** (V4) or **Figure 76** (V6). On loop-charged engines, tighten cylinder head bolts to specifications (**Table 1**) following the sequence marked on the cylinder head.

14. Cross-Flow Engines—Install cylinder head covers with new gaskets. Tighten fasteners to specifications (**Table 1**).

15. Cross-Flow Engines—Install water cover, exhaust and intake bypass covers with new gaskets, as needed.

16. Install the power head as described in this chapter.

REED BLOCK SERVICE

The reed block or leaf valve assemblies are located behind the intake manifold. **Figure 77** shows a typical 60 (1986-1990), 70 and 75 hp intake manifold with reed block assemblies.

The reeds remain in contact with the leaf plate until crankcase pressure is exerted on them. A reed stop limits the amount of travel from the plate. Once the crankcase pressure is relieved, the reed returns to contact the plate.

Do not attempt to bend or flex the reeds if they are distorted or do not contact the plate as designed. If the reeds or reed stops are defective, replace the entire assembly.

Whenever the intake manifold is removed, inspect the reed blocks for signs of gum or varnish and broken, chipped or distorted reeds. If gum or varnish is present, remove the screws holding the reed blocks to the intake manifold. Carefully clean all components in OMC Engine Cleaner and reassemble with a new intake manifold gasket.

Some reed blocks contain a recirculation valve hole and reed; others may use a recirculation valve and screen assembly. These require no service beyond an occasional cleaning in OMC Engine Cleaner to remove any gum or varnish.

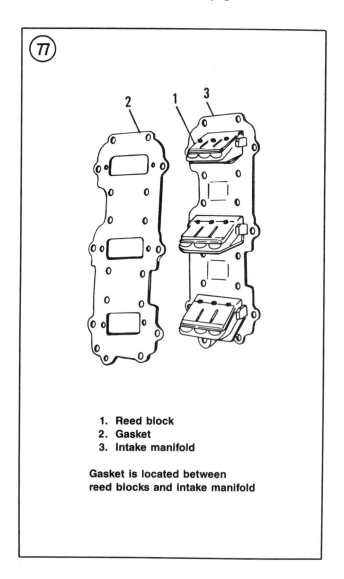

1. Reed block
2. Gasket
3. Intake manifold

Gasket is located between reed blocks and intake manifold

POWER HEAD

THERMOSTAT SERVICE

On cross-flow V4 engines, thermostats and pressure valves are a part of the exhaust housing adapter. Inline engines have a thermostat in the cylinder head; cross-flow V6 engines have a thermostat and pressure valve installed in each cylinder head; loop-charged V4 and V6 engines have a single unit thermostat and pressure valve installed in each cylinder head. A temperature switch in each cylinder head is connected to a warning horn in the remote control box. Whenever water jacket temperatures exceed the temperature switch rating, the warning horn will sound to alert the user to potential engine damage from overheating. The thermostats and pressure valves should be serviced whenever the power head is removed or when engine overheating is indicated.

NOTE
*Inline engine thermostat removal/installation is described under **Power Head Disassembly** and **Assembly** in this chapter.*

Removal/Installation (Cross-Flow V4 Engines)

Refer to **Figure 78** for this procedure.
1. Remove the thermostat cover screws.
2. Tap the cover with a soft hammer to break the gasket seal, then remove the cover.

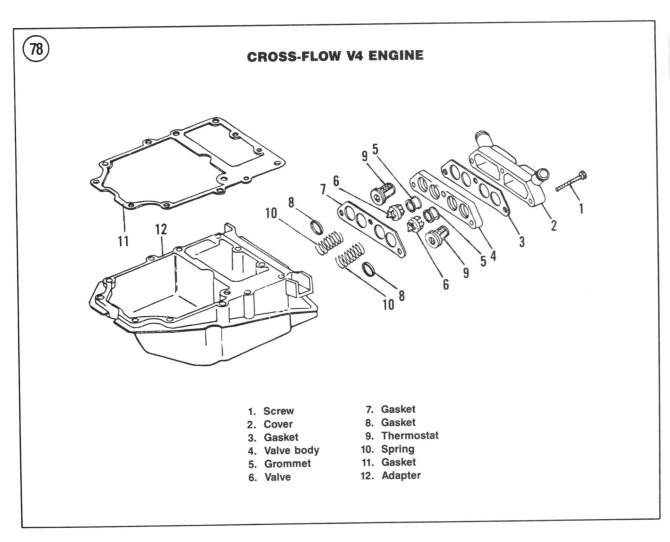

CROSS-FLOW V4 ENGINE

1. Screw
2. Cover
3. Gasket
4. Valve body
5. Grommet
6. Valve
7. Gasket
8. Gasket
9. Thermostat
10. Spring
11. Gasket
12. Adapter

3. Remove the gaskets, thermostats, grommets, springs and pressure valves. Discard the gaskets and grommets.

4. Clean all gasket residue from the exhaust housing adapter, valve body and thermostat cover mating surfaces.

5. Coat both sides of a new valve body-to-adapter gasket with OMC Gasket Sealing Compound. Install gasket to adapter.

6. Install the 2 relief valve springs.

7. Coat both sides of a new valve body-to-cover gasket with OMC Gasket Sealing Compound. Install gasket to thermostat cover.

8. Install the relief valves and grommets in the valve body.

9. Install the thermostats in the valve body. Install new gaskets on the thermostats.

10. Install the thermostat cover to the valve body.

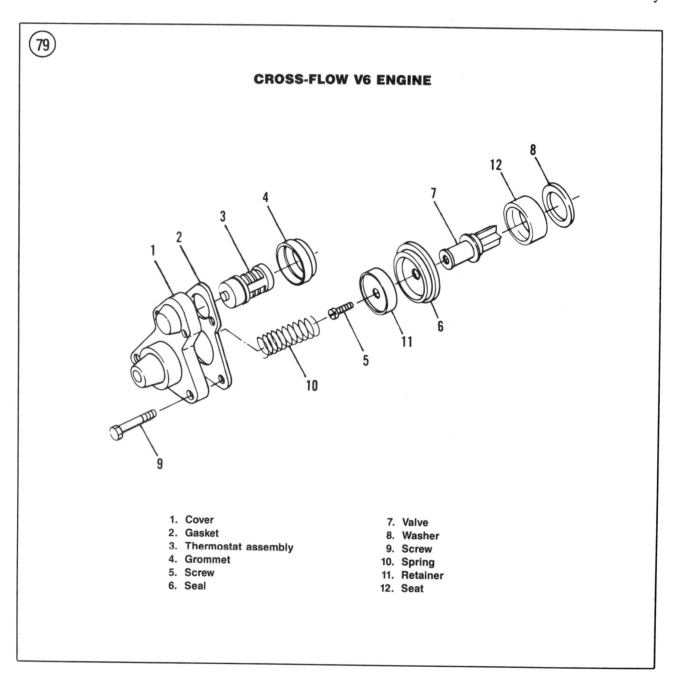

CROSS-FLOW V6 ENGINE

1. Cover
2. Gasket
3. Thermostat assembly
4. Grommet
5. Screw
6. Seal
7. Valve
8. Washer
9. Screw
10. Spring
11. Retainer
12. Seat

POWER HEAD

11. Align the relief valves in the valve body with the springs in the adapter. Install the valve body and cover to the adapter.
12. Install cover screws and tighten to 60-80 in.-lb.
13. Start and run the engine in a test tank or in the water and check for leakage.

Removal/Installation
(Cross-Flow V6 Engines)

Refer to **Figure 79** for this procedure.
1. Remove the 4 screws holding the thermostat housing to its cylinder head (**Figure 80**).
2. Tap the cover with a soft hammer to break the gasket seal, then remove the cover.
3. Remove the gasket, thermostat, O-ring, spring and pressure valve. Discard the gasket and O-ring.
4. Check grommet under cylinder head cover. If deteriorated, remove cylinder head cover and gasket. Discard the grommet and gasket. Install a new grommet in the cylinder head and reinstall cover with a new gasket.
5. Clean all gasket and sealer residue from the thermostat cover and cylinder head cover mating surfaces.

6. Install the thermostat and pressure valve in the cylinder head with new O-rings.
7. Coat both sides of a new thermostat cover gasket with OMC Gasket Sealing Compound and install gasket to cover.
8. Fit spring in thermostat cover and install cover to cylinder head. Tighten screws to 60-80 in.-lb.
9. Start and run the engine in a test tank or in the water and check for leakage.

Cleaning and Inspection
(All Cross-Flow Engines)

1. Remove thermostat (A, **Figure 81**) as described in this chapter.
2. Discard gasket or seal (B, **Figure 81**).
3. Wash thermostat in clean water.
4. Test thermostat if suspected of malfunctioning. See Chapter Three.
5. Check thermostat spring (C, **Figure 81**) for loss of tension, distortion or corrosion. Replace as required.

Removal/Installation
(Loop-Charged V4 and V6 Engines)

1. Remove the 4 screws holding the thermostat cover to its cylinder head (**Figure 82**).

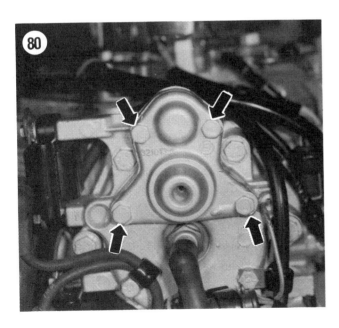

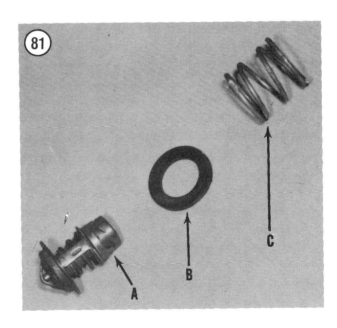

2. Tap the cover with a soft mallet to break the cover loose, then remove the cover with the diaphragm and thermostat assembly.

3. Separate cover from diaphragm and thermostat assembly from diaphragm.

4. Inspect thermostat seal in cylinder head passage. If seal is damaged, replace cup and seal. The lip on the cup must face toward the thermostat assembly.

5. Inspect diaphragm for damage. Make sure vent hole is unrestricted. Replace diaphragm if damage is noted or diaphragm's resilience is no longer present.

6. Clean cylinder head and cover sealing surfaces.

7. Install the thermostat assembly in the diaphragm. Position the diaphragm and thermostat assembly on the cylinder head. Install the cover and tighten the retaining screws to 60-80 in.-lb.

8. Start and run the engine in a test tank or in the water and check for leakage.

Cleaning and Inspection (Loop-Charged V4 and V6 Engines)

NOTE
The pressure-relief valve is contained within the thermostat assembly.

1. Remove the thermostat assembly as described in this chapter.

2. Unscrew the 2 halves and remove the vernatherm and spring.

3. Inspect the thermostat housings for physical or heat damage. Replace complete thermostat assembly if damage is noted.

4. Inspect the vernatherm and spring for corrosion, pitting or heat damage. Replace component if damage is noted.

5. Install vernatherm and spring between thermostat halves. Make sure rounded end of vernatherm pin protrudes from vernatherm fore end.

6. Screw the 2 thermostat halves together making sure the vernatherm and spring properly seat within housings.

Power Head

Table 1 POWER HEAD TIGHTENING TORQUES

Fastener	in.-lb.	ft.-lb.
Adapter-to-power head screws		
3-cylinder	60-84	
V4 and V6	144-168	
Armature plate assembly		
All except 1986-on 2-cylinder	60-80	
1986-on		
To support plate	25-35	
Support plate screws	48-60	
Bearing retainer plate screws	96-120	
Bypass cover screws	60-84	
Connecting rod screws		
1985-on 120-140 hp		42-44
1986-on 200-225 hp		42-44
All others		30-32
Crankcase main bearing screws/nuts		
1985-on 120-140 hp		26-30
1986-on 200-225 hp		26-30
All others		18-20
Crankcase flange screws	60-84	
Cylinder head screws		
All 235 hp and 1986-on 175 hp		20-22
1987-on 120, 140, 200 and 225 hp		20-22
All others		18-20
Cylinder head cover screws	60-84	
Exhaust cover screws	60-84	
Intake manifold screws	60-84	
Crankcase head screws		
60-75 hp	60-84	
1973-1985 V4		
Upper		10-12
Lower		8-10
1986 V4		
Upper		
90-110 hp	120-144	
120-140 hp	72-96	
Lower		
90-110 hp	96-120	
120-140 hp	72-96	
1987-on 88-110 hp V4	96-120	
1987-on 120-140 hp V4	72-96	
1976-1985 V6		6-8
1986 V6	72-96	
1987-on 150-175 hp V6	96-120	
1987-on 200-225 hp V6	72-96	

(continued)

Table 1 POWER HEAD TIGHTENING TORQUES (continued)

Fastener	in.-lb.	ft.-lb.
Power head-to-exhaust housing		
Inline engines		18-20
1986-on 88-110 hp V4		16-18
1986 120-140 hp V4		18-20
All other V4		12-14
1986 200-225 hp V6		18-20
1986-on 150-175 hp V6		18-20
All other V6		12-14
Spark plugs		18-21
Stator screws		
3-cylinder		
All except 1986	48-60	
1986-on	120-144	
V4 and V6	120-144	
Standard bolts and nuts		
No. 6	7-10	
No. 8	15-22	
No. 10	25-35	2-3
No. 12	35-40	3-4
1/4 in.	60-80	5-7
5/16 in.	120-140	10-12
3/8 in.	220-240	18-20

Table 2 FLYWHEEL NUT TORQUE

Model	ft.-lb.
2-cylinder	
1973-1977, 1985-on	100-105
1978-1984	80-85
3-cylinder and V4	
1985-1986 120-140 hp	140-145
1987-on 120-140 hp	140-150
All others	100-105
V6	140-145

Table 3 POWER HEAD SPECIFICATIONS

2-CYLINDER INLINE ENGINES	
Standard bore size	3.1870-3.1880 in.
Bore	3.1875 in.
Stroke	2.820 in.
Piston ring	
Width (1973-1984)	
Upper	0.0895-0.0900 in.
Lower	0.0615-0.0625 in.
Gap	0.007-0.017 in.
Side clearance	0.0015-0.0040 in.

(continued)

Power Head

Table 3 POWER HEAD SPECIFICATIONS (continued)

2-CYLINDER INLINE ENGINES (continued)

Crankshaft bearings	
Upper and center	Roller bearing
Lower	Ball bearing
Crankshaft journal diameter	
Top	1.4974-1.4979 in.
Center	1.3748-1.3752 in.
Bottom	1.1810-1.1815 in.
Crankpin diameter	1.1823-1.1828 in.

3-CYLINDER INLINE ENGINES

Standard bore size	
1989-1990 60 and 1986-on 70 hp	3.1870-3.1880 in.
All others	2.9985-3.0005 in.
Bore	
1989-1990 60 and 1986-on 70 hp	3.1880 in.
All others	3.0000 in.
Stroke	2.3437 in.
Piston ring	
Width (1973-1984)	
Upper	0.0895-0.0900 in.
Lower	0.0615-0.0625 in.
Gap	0.007-0.017 in.
Side clearance	0.0015-0.0040 in.
Crankshaft bearings	
Upper and center	Roller bearing
Lower	Ball bearing
Crankshaft journal diameter	
Top	1.4974-1.4979 in.
Center	1.3748-1.3752 in.
Bottom	1.1810-1.1815 in.
Crankpin diameter	1.1823-1.1828 in.

V4 ENGINES (EXCEPT 1985-ON 120-140 HP)

Standard bore size	
85 hp	3.375 in.
All others	3.4995-3.5005 in.
Bore	
85 hp	3.375 in.
All others	3.500 in.
Stroke	2.588 in.
Piston ring	
Width (1973-1984)	
Upper	0.0895-0.0900 in.
Lower	0.0615-0.0625 in.

(continued)

Table 3 POWER HEAD SPECIFICATIONS (continued)

V4 ENGINES (EXCEPT 1985-ON 120-140 HP) (continued)	
Gap	
1973-1986	0.007-0.017 in.
1987-on	0.019-0.031 in.
Side clearance	0.0020-0.0040 in.
Crankshaft bearings	
Upper and center	Roller bearing
Lower	Ball bearing
Crankshaft journal diameter	
1973-1978	
Top	1.6199-1.6204 in.
Center	1.3748-1.3752 in.
Bottom	1.3779-1.3784 in.
1979-on	
Top	1.6199-1.6204 in.
Center	2.1870-2.1875 in.
Bottom	1.3779-1.3784 in.
Crankpin diameter	
1973-1978	1.1823-1.1828 in.
1979-on	1.3757-1.3762 in.

V4 ENGINES (1985-ON 120-140 HP)	
Standard bore size	
1985-1987	3.4955-3.5055 in.
1988-1990	3.6845-3.6855 in.
Bore	
1985-1987	3.500 in.
1988-1990	3.685 in.
Stroke	2.860 in.
Piston ring	
Gap	
1985-1986	0.007-0.017 in.
1987-on	0.019-0.031 in.
Side clearance	0.0020-0.0040 in.
Crankshaft bearings	
Upper and center	Roller bearing
Lower	Ball bearing
Crankshaft journal diameter	
Top	1.6199-1.6204 in.
Center	2.1870-2.1875 in.
Bottom	1.5747-1.5752 in.
Crankpin diameter	1.4995-1.5000 in.

(continued)

Power Head

Table 3 POWER HEAD SPECIFICATIONS (continued)

V6 ENGINES (EXCEPT 1986-ON 200-225 HP)

Standard bore size	
235 hp	3.620-3.630 in.
1986-on 175 hp	3.6245-3.6255 in.
All others	3.4995-3.5005 in.
Bore	
235 hp and 1986-on 175 hp	3.625 in.
All others	3.500 in.
Stroke	2.588 in.
Piston ring	
Width (1973-1984)	
Upper	0.0895-0.0900 in.
Lower	0.0615-0.0625 in.
Gap	
1973-1985	0.007-0.017 in.
1986 150 hp	0.007-0.017 in.
1987-on 150 hp	0.019-0.031 in.
1986-on 175 hp	0.020-0.033 in.
Side clearance	0.0020-0.0040 in.
Crankshaft bearings	
Upper and center	Roller bearing
Lower	Ball bearing
Crankshaft journal diameter	
Top	1.6199-1.6205 in.
Center	2.1870-2.1875 in.
Bottom	1.3779-1.3784 in.
Crankpin diameter	1.3757-1.3762 in.

V6 ENGINES (1986-ON 200-225 HP)

Standard bore size	
1986-1987	3.4995-3.5005 in.
1988-1990	3.6845-3.6855 in.
Bore	
1986-1987	3.500 in.
1988-1990	3.685 in.
Stroke	2.860 in.
Piston ring	
Gap	
1986	0.007-0.017 in.
1987-1990	0.019-0.031 in.
Side clearance	0.004 in. max.
Crankshaft bearings	
Upper and center	Roller bearing
Lower	Ball bearing
Crankshaft journal diameter	
Top	1.6199-1.6204 in.
Center	2.1870-2.1875 in.
Bottom	1.5747-1.5752 in.
Crankpin diameter	1.4995-1.5000 in.

Table 4 V4 AND V6 CRANKSHAFT SEAL RINGS

Identification markings	Width in.	mm
1973-1984		
One stripe	0.1565-0.1570	3.975-3.988
Two stripes	0.1575-0.1580	4.000-4.013
Three stripes	0.1585-0.1590	4.026-4.039
Four stripes	0.1605-0.1610	4.077-4.089
1985-on	0.154 minimum	3.91 minimum

Chapter Nine

Gearcase

Torque is transferred from the engine crankshaft to the gearcase by a drive shaft. A pinion gear on the drive shaft meshes with a drive gear in the gearcase to change the vertical power flow into a horizontal flow through the propeller shaft. On Johnson and Evinrude outboards covered in this manual, a sliding clutch engages a forward or reverse gear in the gearcase. This creates a direct coupling that transfers the power flow from the pinion to the propeller shaft.

Gearcase design differs primarily in the shift mechanisms used. Hydro-mechanical gearcases have a hydraulic assist shift cylinder and oil pump. Mechanical gearcases have a forward bearing housing with shift shaft and cradle connected to the shift rod.

The gearcase can be removed without removing the entire outboard from the boat. This chapter contains removal, overhaul and installation procedures for the propeller, hydro-mechanical gearcase, mechanical gearcase and water pump. **Table 1** and **Table 2** are at the end of the chapter.

The gearcases covered in this chapter differ somewhat in design and construction and thus require slightly different service procedures. The chapter is arranged in a normal disassembly/assembly sequence. When only a partial repair is required, follow the procedure(s) for your gearcase to the point where the faulty parts can be replaced, then reassemble the unit.

Since this chapter covers a wide range of models, the gearcases shown in the accompanying pictures are the most common ones. While it is possible that the components shown in the pictures may not be identical with those being serviced, the step-by-step procedures may be used with all models covered in this manual.

PROPELLER

The propeller rides on thrust bushings and is retained by a castellated nut and cotter pin (**Figure 1**). Any impact caused by hitting an underwater object is absorbed by the propeller hub.

Removal/Installation

1. Remove and discard the cotter pin.
2. Remove the castellated nut.
3. Remove the thrust spacer, propeller and thrust bushing assembly from the propeller shaft.
4. Wipe propeller shaft splines with OMC Gasket Sealing Compound.
5. Install the thrust bushing in the propeller hub and slide propeller on drive shaft.
6. Install thrust spacer on drive shaft splines.
7. Install castellated nut finger-tight, then tighten with a wrench to align the cotter pin hole in the nut and propeller shaft.
8. Install a new cotter pin and bend the ends over the nut.

WATER PUMP

Johnson and Evinrude outboards use a pressure-type water pump. An optional chrome water pump is available and recommended for use in areas where the water contains considerable sand or silt.

The water pump impeller is secured to the drive shaft by a pin that fits between a flat area on the drive shaft and a similar cutout in the impeller hub. As the drive shaft rotates, the impeller rotates with it. Water between the impeller blades and pump housing is pumped up to the power head through the water tube.

All seals and gaskets should be replaced whenever the water pump is removed. Since proper water pump operation is critical to outboard operation, it is also a good idea to install a new impeller at the same time.

Do not turn a used impeller over and reuse it. The impeller rotates in a clockwise direction with the drive shaft and the vanes gradually take a "set" in one direction. Turning the impeller over will cause the vanes to move in a direction opposite to that which caused the "set." This will result in premature impeller failure and can damage a power head extensively.

Removal and Disassembly

1. Remove the gearcase as described in this chapter and secure it in a holding fixture or a vise with protective jaws. If protective jaws are

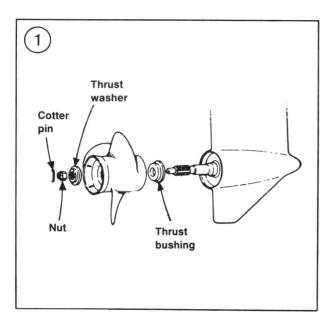

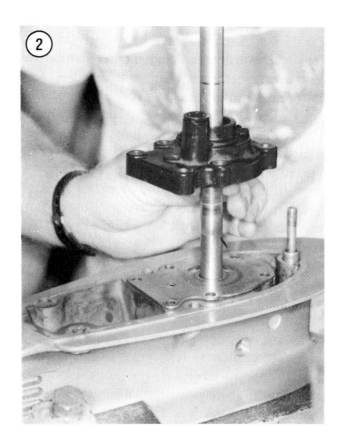

GEARCASE

not available, position the gearcase upright in the vise with the skeg between wooden blocks.

2. Remove the impeller housing screws. Insert screwdrivers at the fore and aft ends of the impeller housing and pry it loose.

3. Slide the impeller housing and impeller up and off the drive shaft. See **Figure 2** (typical).

NOTE
In extreme cases, the impeller hub may have to be split with a hammer and chisel to remove it in Step 4.

4. If the impeller did not come off with the impeller housing, carefully pry the impeller up and off the shaft. Remove the drive pin.

5. Carefully pry the impeller plate loose with a screwdriver, then slide the plate and gasket (if used) up and off the drive shaft. Discard the gasket.

6. Remove the guide bracket from the impeller housing, if used.

7. Remove and discard the impeller housing water tube grommet. See A, **Figure 3** (typical).

8. Remove and discard the impeller housing seal or grommet and O-ring. See B, **Figure 3** (typical).

9. Remove the impeller from the housing. Remove the impeller cup and insert. See **Figure 4** (typical).

Cleaning and Inspection

1. Check the housing for cracks, distortion or melting. Replace as required.

2. Clean all metal parts in solvent and blow dry with compressed air, if available.

3. Carefully remove all gasket residue from the mating surfaces.

4. Check impeller plate and cup for grooving or rough surfaces. Replace if any defects are found.

5. If original impeller is to be reused, check bonding to hub. Check side seal surfaces and vane ends for cracks, tears, wear or a glazed or melted appearance. If any of these defects are noted, do *not* reuse impeller.

Assembly and Installation

1A. Inline engines—Install a new drive shaft grommet on impeller housing.

1B. V4 and V6—Wipe the outer diameter of a new drive shaft seal with OMC Gasket Sealing Compound and install in impeller housing (lip facing gearcase) with installer part No. 326546 or equivalent. After installation, wipe the seal lips with OMC Triple-Guard grease.

2. Lubricate a new water tube grommet with OMC HI-VIS Gearcase Lubricant and install in impeller housing.

3. If impeller housing uses a guide bracket, install bracket to housing with a new grommet.

4. Install impeller plate on gearcase. If a gasket is used, coat both sides with OMC Gasket Sealing Compound and install on gearcase under the impeller plate.

5. Wipe the drive shaft flat with OMC Needle Bearing Grease and install impeller drive pin on shaft flat.

CAUTION
If the original impeller is to be reused, install it in the same rotational direction as removed to avoid premature failure.

6. Lubricate the impeller blade tips with OMC HI-VIS Gearcase Lubricant. Install the impeller on the drive shaft and engage its hub cutout over the drive pin. Seat impeller on the impeller plate.

NOTE
Always install a new impeller liner and cup when the impeller is replaced.

7. Install the impeller housing liner, then align the impeller cup tabs with the housing cutouts and install the cup.

8. Install a new O-ring in bottom of impeller housing with OMC Adhesive M.

9. Carefully slide impeller housing on drive shaft to prevent seal or grommet damage, then push housing over impeller while rotating the drive shaft clockwise to correctly position impeller blades in housing.

CAUTION
Correct housing fastener torque is important in Step 10. Excessive torque can

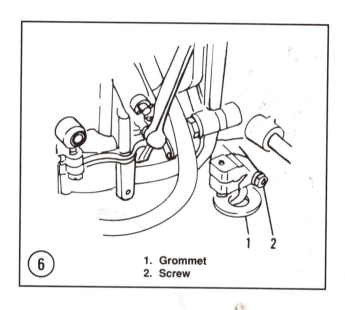

1. Grommet
2. Screw

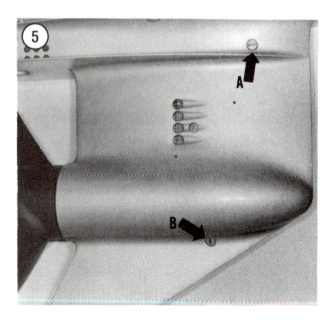

Shift rod retaining screw

GEARCASE

cause the pump to crack during operation; insufficient torque may result in leakage and exhaust induction which will cause overheating.

10. Wipe housing screw threads with OMC Gasket Sealing Compound and tighten screws to specifications (**Table 1**).

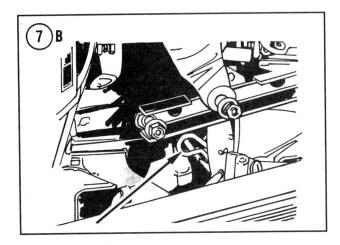

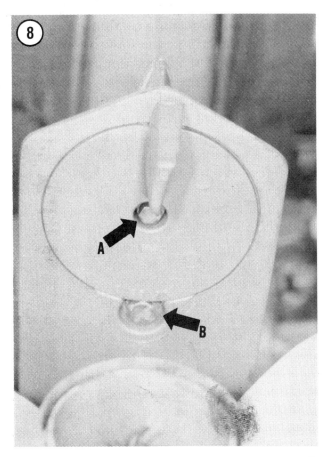

GEARCASE REMOVAL/INSTALLATION (ALL MODELS)

1. Disconnect the spark plug leads as a safety precaution to prevent any accidental starting of the engine during lower unit removal.

2. Place a container under the gearcase. Remove the oil level plug (A, **Figure 5**), then remove the drain/fill plug (B, **Figure 5**). Drain the lubricant from the unit.

> *NOTE*
> *If the lubricant is creamy in color or metallic particles are found in Step 3, the gearcase must be completely disassembled to determine and correct the cause of the problem.*

3. Wipe a small amount of lubricant on a finger and rub the finger and thumb together. Check for the presence of metallic particles in the lubricant. Note the color of the lubricant. A white or creamy color indicates water in the lubricant. Check the drain container for signs of water separation from the lubricant.

4. Remove the propeller as described in this chapter.

5. Move the shift lever into REVERSE. If necessary, rotate the propeller shaft slightly to help unit engage.

6. Remove the shift rod retaining screw. See **Figure 6** (inline) or **Figure 7A** (V6). On V4 models, remove the shift rod retaining clip (**Figure 7B**).

7. On inline electric start models, remove the starter motor. See Chapter Seven.

8. Mark the trim tab position relative to the gearcase for reinstallation, then remove the trim tab screw and trim tab (A, **Figure 8**).

9. Remove the countersunk 5/8 in. screw between the trim tab cavity and gearcase housing with a narrow wall socket (B, **Figure 8**).

10. Remove the screw from inside the trim tab cavity (if used) with a 1/2 in. socket and extension. See **Figure 9**.

11. Remove the 2 gearcase-to-exhaust housing screws on each side of the gearcase. See A, **Figure 10**.

12. Separate the gearcase from the exhaust housing. Work carefully to avoid damage to the shift rod and remove from the unit.

13A. On inline models, remove and discard the inner exhaust housing seal and shift rod grommet.

13B. On V4 and V6 models, remove lower inner exhaust housing and shift rod seals.

14. Mount the gearcase in a suitable holding fixture.

15. Remove and discard the drive shaft O-ring.

16. To reinstall the gearcase, lubricate the exhaust housing water pump grommet with liquid soap.

17. Install the shift rod spacer between the shift rod cover and grommet, if used.

18. Install a new drive shaft O-ring.

19. Make sure the shift lever is in REVERSE gear.

CAUTION
Do not grease the top of the drive shaft in Step 20. This may excessively preload the drive shaft and crankshaft when the mounting bolts are tightened and cause a premature failure of the power head or gearcase.

20. Lightly lubricate the drive shaft splines with OMC Moly Lube.

21A. Inline engines—Coat outside of a new gearcase-to-exhaust housing seal with OMC Adhesive M. Install seal in gearcase.

21B. V4 and V6—Install new seals on lower inner exhaust housing. Coat seals with OMC Triple-Guard grease.

22. Apply OMC Adhesive M to gearcase and exhaust housing mating surfaces.

CAUTION
Do not rotate the flywheel counterclockwise in Step 23. This can damage the water pump impeller.

23. Position gearcase under exhaust housing. Align water tube with grommet, drive shaft with crankshaft splines and the shift rod with the shift rod bushing. Push the gearcase into place, rotating the flywheel clockwise as required to let the drive shaft and crankshaft engage.

24. Coat gearcase screw threads with OMC Gasket Sealing Compound. Install and tighten gearcase screws to specifications (**Table 1**).

25. Install countersunk screw in anti-ventilation plate and screw into trim tab cavity, if used.

26. Align trim tab and gearcase marks made during disassembly. Install trim tab and tighten screw securely.

27A. Inline engines—Position new shift rod grommet as shown in **Figure 6**. Engage shift cap with shift rod connector and tighten screw.

27B. V4 and V6—Install new shift rod seal and tighten shift rod connector screw.

28. Install the propeller as described in this chapter.

29. Rotate propeller shaft while moving shift lever into FORWARD to make sure the clutch dog engages fully with the forward gear.

30. Reconnect the spark plug leads and refill the gearcase with proper type and quantity of lubricant. See Chapter Four.

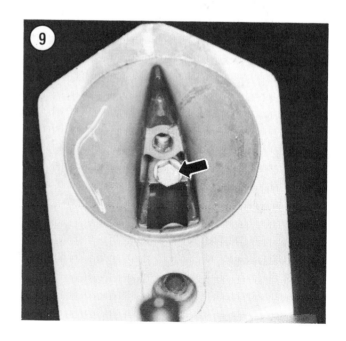

GEARCASE

HYDRO-MECHANICAL GEARCASE

Disassembly

Refer to **Figure 11** for this procedure.

1. Remove the gearcase as described in this chapter.
2. Secure the gearcase in a holding fixture or a vise with protective jaws. If protective jaws are not available, position the gearcase upright with the skeg between wooden blocks.
3. Remove the water pump as described in this chapter.
4. Remove the propeller bearing housing screws with a thin wall deep socket. Discard the O-rings on the screws.
5A. V6—Assemble flywheel puller (part No. 378103) with 2 puller legs (part No. 321631). Fit puller legs around bearing housing flange legs. Tighten puller nut until the bearing housing comes loose.
5B. All others—Install flywheel puller (part No. 378103) and two 8 in. 5/16 × 18 bolts (part No. 316982) in bearing housing screw holes. Tighten puller nut until the bearing housing comes loose. See **Figure 12**.
6. Remove the puller assembly from the bearing housing. Remove the bearing housing from the gearcase (**Figure 13**). Remove and discard the bearing housing O-ring.
7. Reach into the gearcase propeller bore and remove the thrust washer and thrust bearing, if used.
8. Use snap ring pliers (part No. 311879) and carefully remove the 2 snap rings in the gearcase propeller bore (**Figure 14**).
9. Remove the shift rod cover boot, if used. Remove the cover screws. Tap cover with a soft mallet to break the gasket seal and slide it upward on the shift rod away from the gearcase. See **Figure 15**.
10. Grasp shift rod with one hand. Grasp propeller shaft with the other hand. Pull on shift rod and propeller shaft at the same time sufficiently to disengage them.
11. Disengage the shift rod from the assist cylinder and remove with the shift cover (**Figure 16**). Remove the power assist cylinder (**Figure 17**).
12. Remove the propeller shaft with reverse gear and retainer plate from the gearcase (**Figure 18**).
13. Install drive shaft holding socket on drive shaft splines and connect a breaker bar. Use socket part No. 316612 (2-cylinder), part No. 312752 (3-cylinder) or part No. 311875 (V4 and V6).
14. Hold pinion locknut with a socket and flex handle. Pad the gearcase where the flex handle will hit with shop cloths to prevent housing damage.
15. Hold the pinion nut from moving and turn the drive shaft to break the pinion locknut loose. See **Figure 19**. Remove the pinion locknut and drive shaft holding tool.
16. Remove the pinion locknut from the gearcase. Remove the pinion gear (**Figure 20**) from the gearcase.
17. Remove the upper drive shaft bearing housing screws. Pull the drive shaft and bearing housing from the gearcase. See **Figure 21**. With V6 models, the 18 rollers in the pinion bearing are now loose and may fall out. Remove the

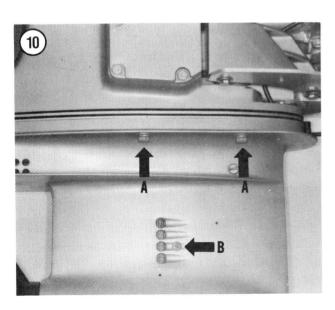

CHAPTER NINE

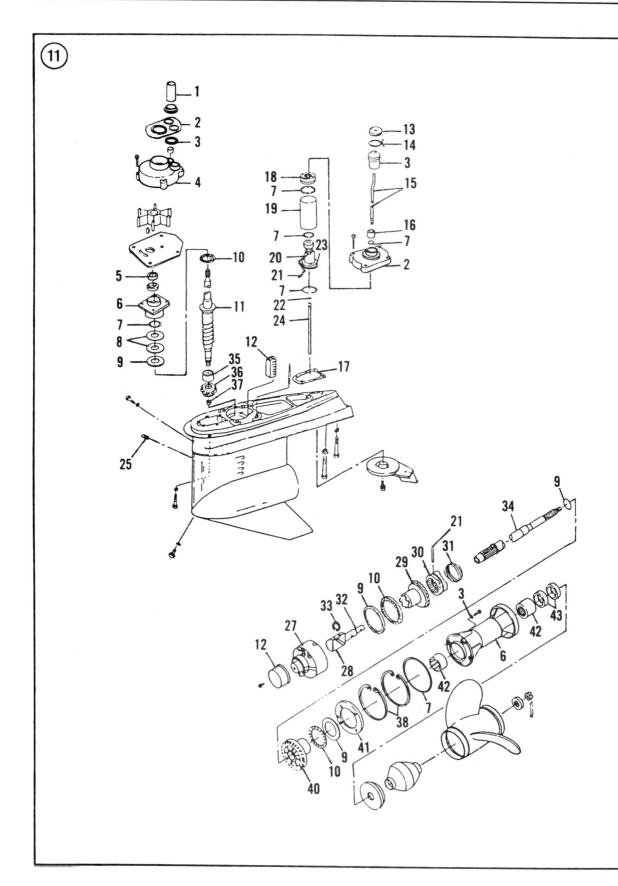

GEARCASE

HYDRO-MECHANICAL GEARCASE

1. Guide
2. Cover
3. Seal
4. Housing
5. Drive shaft oil retainer
6. Bearing housing
7. O-ring
8. Shim
9. Thrust washer
10. Thrust bearing
11. Drive shaft
12. Screen
13. Grommet
14. Clamp
15. Shift rod
16. Bushing
17. Gasket
18. Cap
19. Cylinder
20. Valve
21. Pin
22. Screw
23. Piston
24. Pushrod
25. Lower bearing setscrew
27. Oil pump
28. Plunger
29. Forward gear
30. Clutch dog
31. Spring
32. Shift rod and bearing
33. Plunger snap ring
34. Propeller shaft
35. Bearing
36. Pinion gear
37. Nut
38. Retaining ring
40. Reverse gear
41. Retainer plate
42. Needle bearing
43. Oil retainer

CHAPTER NINE

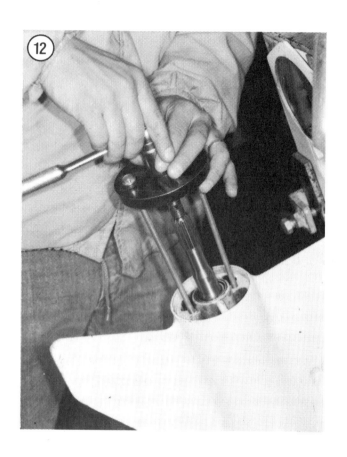

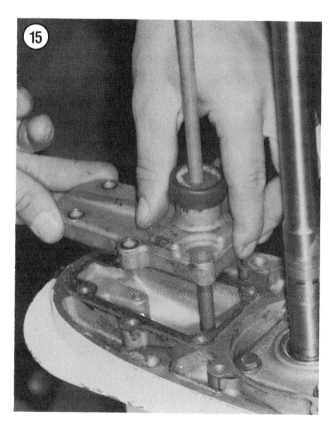

GEARCASE

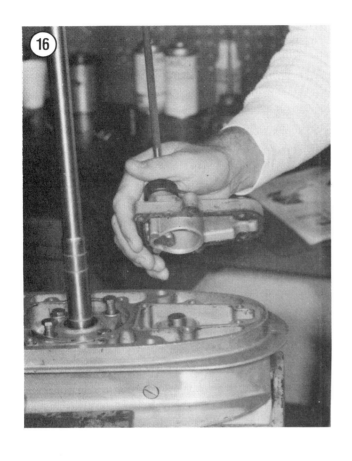

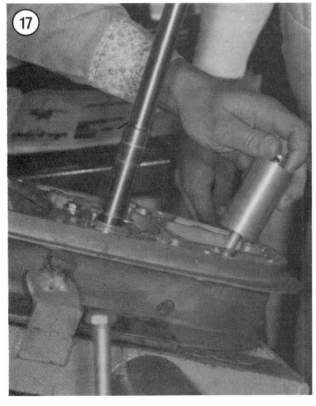

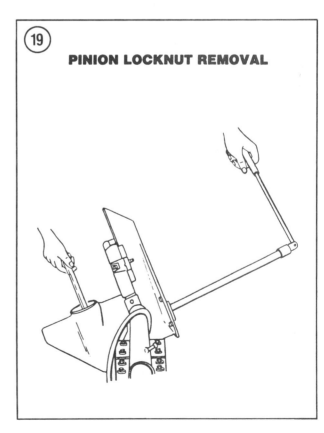

PINION LOCKNUT REMOVAL

thrust washer, bearing and shims from the drive shaft.

NOTE
The drain plug must be removed before the forward gear can be removed in Step 18.

18. Remove drain plug (if reinstalled after draining lubricant) and then remove the forward gear (**Figure 22**).

NOTE
On V6 models, the oil pump may come free without using the slide hammers in Step 19.

19. Thread two 16 in. 1/4 × 20 threaded rods in the oil pump screw holes. Attach a slide hammer to each threaded rod and pull oil pump free of gearcase housing.

20. If lower drive shaft bearing requires replacement, remove the setscrew from the starboard side of the water intake grill on the exterior of the gearcase. See 6, **Figure 23**.

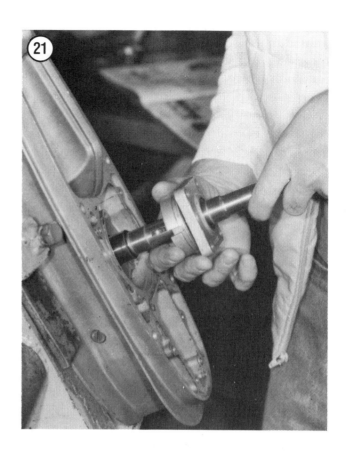

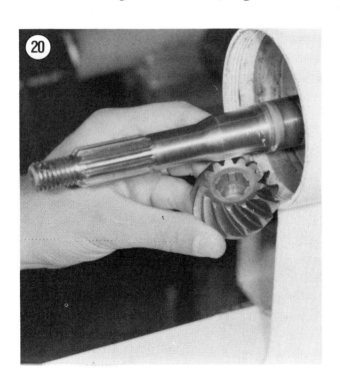

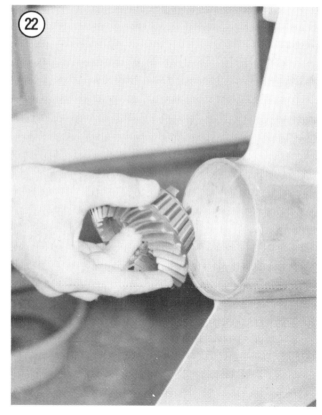

GEARCASE

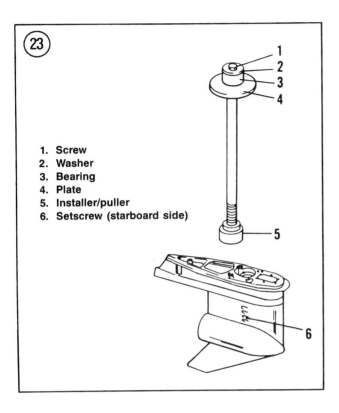

1. Screw
2. Washer
3. Bearing
4. Plate
5. Installer/puller
6. Setscrew (starboard side)

21A. Inline engines—Assemble bearing remover-installer tool (part No. 385546) as shown in **Figure 23** and tighten screw to pull the bearing from the gearcase.

21B. V4—Install tool part No. 384096 (1973) or part No. 318117 (1974-on) to drive handle part No. 311880. Insert assembly in gearcase and drive the bearing out.

21C. V6—Assemble bearing remover-installer tool (part No. 385546) with remover (part No. 321521) and tighten screw to pull the bearing from the gearcase.

22. Drive oil seals and bearings from each end of the bearing housing (**Figure 24**) with a punch and mallet. Remove and discard the bearing housing O-ring.

23. Remove the 4 screws and lockwashers from the oil pump cover. Remove the cover from the housing. See **Figure 25**.

24. Remove the rubber oil seal and band from the pump cover. Remove the screw holding the

screen to the cover. See **Figure 26**. Carefully remove screen (sealed with OMC Adhesive M) from the cover.

25. If pump disassembly is necessary, remove the snap ring, then remove the plug, spring guide and ball valve from the pump body.

26. Clamp one-half of spanner wrench part No. 386112 in a vise. Place the power assist cylinder on the tool, pushing the rod down to engage the tool pins with the piston holes.

27. Install other half of spanner wrench part No. 386112 on top of cylinder and engage the tool pins in the piston cap holes.

28. Turn top half of spanner wrench counterclockwise with a socket wrench and loosen the piston cap from the cylinder. Remove tool. Refer to **Figure 27** and remove piston cap (A) from cylinder and cylinder (B) from assist cylinder (C). If further disassembly is required, refer to **Figure 28**.

29. Carefully lift one end of the clutch dog retaining spring and insert a screwdriver blade under it as shown in **Figure 29**. Holding the screwdriver stationary, rotate the propeller shaft to unwind the spring.

30. Remove the retainer pin with an appropriate size punch. See **Figure 30**.

NOTE
Some later model shift shafts may have 3 detent balls which will come free when the shift rod assembly is removed.

31. Remove the plunger, shift rod assembly and spring from the propeller shaft. See **Figure 31**. If necessary, separate the plunger from the shift rod by removing the snap ring.

Cleaning and Inspection

1. Clean all parts in fresh solvent. Blow dry with compressed air, if available.
2. Clean all nut and screw threads thoroughly if OMC Screw Lock or OMC Nut Lock has been used. Soak nuts and screws in solvent and use a fine wire brush to remove residue.

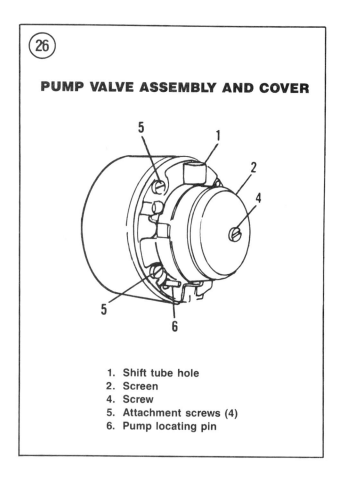

PUMP VALVE ASSEMBLY AND COVER

1. Shift tube hole
2. Screen
4. Screw
5. Attachment screws (4)
6. Pump locating pin

GEARCASE

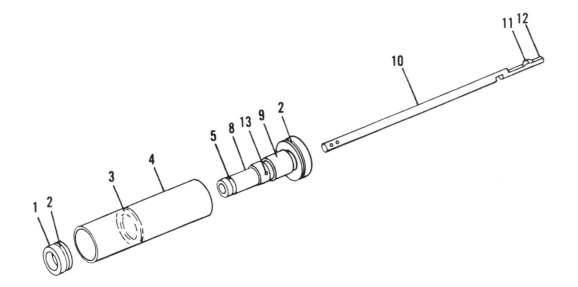

SHIFT ASSIST VALVE COMPONENTS

1. Piston cap
2. O-ring
3. O-ring
4. Cylinder
5. Slot
8. Valve
9. Piston
10. Pushrod
11. Key
12. Flat surface
13. Retaining pin

3. Remove and discard all O-rings, gaskets and seals. Clean all residue from gasket mating surfaces.

4. Check drive shaft splines for wear or damage. If gearcase has struck a submerged object, the drive shaft and propeller shaft may suffer severe damage. Replace drive shaft as required and check crankshaft splines for similar wear or damage.

5. Check propeller shaft splines and threads for wear, rust or corrosion damage. Replace shaft as necessary.

6. Install V-blocks under the drive shaft bearing surfaces at each end of the shaft. Slowly rotate the shaft while watching the crankshaft end. Replace the shaft if any signs of wobble are noted.

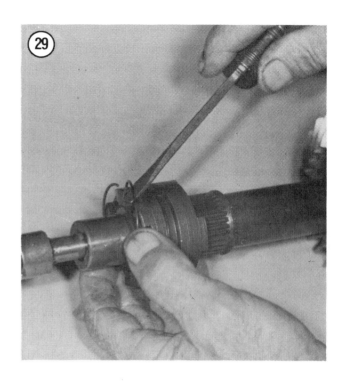

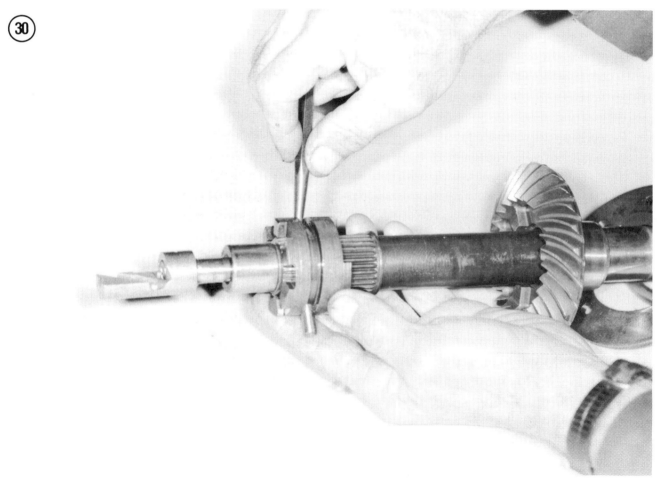

GEARCASE

7. Repeat Step 6 with the propeller shaft. Also check the shaft surfaces where oil seal lips make contact. Replace the shaft as required.

8. Check bearing housing and needle bearing for wear or damage. See **Figure 32**. Replace as required.

9. Check bearing housing contact points on the propeller shaft. If shaft shows signs of pitting, grooving, scoring, heat discoloration or embedded metallic particles, replace shaft and bearings.

10. Inspect water pump as described in this chapter.

11. Check all shift components for wear or damage. Look for excessive wear on the clutch dog and forward/reverse gear engagement surfaces. See **Figure 33**. If clutch dogs are pitted, chipped, broken or excessively worn, replace the gear(s) and clutch dog.

12. Clean all roller bearings with solvent and lubricate with OMC HI-VIS Gearcase Lubricant to prevent rusting. Check all bearings for rust, corrosion, flat spots or excessive wear. Replace as required.

13. Check the forward, reverse and pinion gears for wear or damage. If teeth are pitted, chipped, broken or excessively worn, replace the gear. Check pinion gear splines for excessive wear or damage. Replace as required. See **Figure 34** (typical).

14. Check the propeller for nicks, cracks or damaged blades. Minor nicks can be removed with a file, taking care to retain the shape of the propeller. Replace any propeller with bent, cracked or badly chipped blades.

15. Check oil pump pressure relief valve. If grooved or worn, replace the ball and seat.
16. Check shift rod cover bushing. If worn, remove bushing and O-ring and install a new set with an appropriate driver.

Assembly

Refer to **Figure 11** for this procedure.
1. If oil pump was disassembled, reinstall valve, ball guide, spring and plug. Rounded side of plug should face out. Install snap ring with rounded side of ring facing rounded side of plug.

NOTE
Pump gears can be installed without regard to aligning any match marks in Step 2.

2. Lubricate pump gears with OMC HI-VIS Gearcase Lubricant. Install pump gears (**Figure 35**). Install cover and tighten fasteners to specifications (**Table 1**).
3. If pump cover screen was removed, coat mating edge of screen with OMC Adhesive M. Reinstall and tighten screw securely. Install band and oil seal, if used.

NOTE
On some models, it may be necessary to thread the ¼ × 20 slide hammer rods into the oil pump to properly install it in Step 4.

4. Install oil pump in gearcase, aligning pin on front of pump with hole in gearcase casting. Make sure pump is properly seated to prevent misalignment of assist cylinder pushrod when installed.
5. Lubricate all assist cylinder components with OMC HI-VIS Gearcase Lubricant. Install new O-rings. Slip cylinder over piston and position piston cap.
6. Clamp one-half of spanner wrench part No. 386112 in a vise. Place the power assist cylinder on the tool, pushing the rod down to engage the tool pins with the piston holes.
7. Install other half of spanner wrench part No. 386112 on top of cylinder and engage the tool pins in the piston cap holes.
8. Turn top half of spanner wrench clockwise with a socket wrench and tighten the piston cap on the cylinder.
9A. If shift shaft has no detent balls, reassemble as follows:

  a. Install the spring and shift rod/plunger assembly in the propeller shaft.

  b. Slide clutch dog on shaft with grooved end facing propeller end of shaft (some clutch dogs may be marked with the words "PROP END" to indicate correct placement).

  c. Align holes in clutch dog and shaft, then install the retainer pin.

GEARCASE

9B. If shift shaft uses detent balls, reassemble as follows:

   a. Slide clutch dog on shaft with grooved end facing propeller end of shaft.
   b. Insert 2 detent balls in shift rod detent holes, then install the other ball and detent spring in shift rod end.
   c. Holding balls in shaft, align shift rod and clutch dog holes with shaft slot.
   d. Insert shift rod into shaft until detent balls slip into the shaft grooves.
   e. Use an awl to depress the detent spring and install the retainer pin in the clutch dog and shaft holes.

10. Reinstall one end of the clutch dog retaining spring over the clutch dog, then rotate the propeller shaft to wind the spring back in place.

11. If drive shaft pinion bearing was removed, install a new bearing as follows:

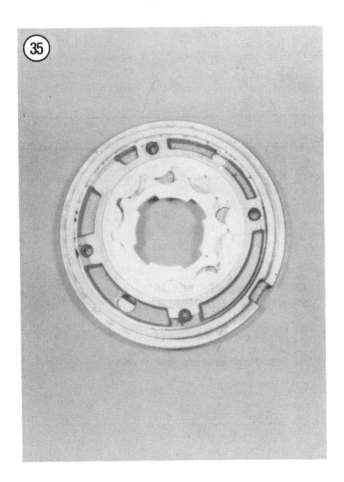

   a. Inline engines—Assemble bearing remover-installer tool (part No. 385546). Position new bearing on tool (lettered side facing tool shoulder). Insert assembly into gearcase and drive the bearing in until the tool plate touches the top of the gearcase.
   b. 1973 V4—Position new bearing (lettered side up) on upper end of tool part No. 384096. Fit stop ring over bearing and attach a suitable driver handle. Insert assembly into gearcase and drive the bearing in until the tool shoulder seats against the gearcase flange.
   c. 1974-on V4—Position new bearing (lettered side up) on tool part No. 318117. Insert tool and bearing assembly in propeller shaft housing and position under bearing bore. Assemble tool part No. 385547 components as shown in **Figure 36**. Hold installer tool with a one-inch wrench and tighten the tool bolt on top of gearcase to draw bearing into place. Remove the tool.
   d. V6—Assemble bearing remover-installer tool (part No. 385546) with installer (part No. 321518). Position new bearing (lettered side up) on installer with OMC Needle Bearing Grease. Insert assembly into gearcase and drive the bearing in until the tool plate touches the top of the gearcase.

12. Wipe bearing setscrew threads with OMC Screw Lock and install setscrew (6, **Figure 23**).

13. Install pinion gear to drive shaft and tighten pinion nut to specifications (**Table 1**).

14. Place the shims removed during disassembly on the drive shaft shoulder and install shim gauge part No. 321520 (V6) or part No. 315767 (all others) on drive shaft. See **Figure 37**.

15. Hold gauge against shims and measure the clearance between the bottom of the gauge and the pinion nut. If the clearance is not 0.000-0.002 in., remove or add shims as required.

16. Remove the shims, pinion gear nut and pinion gear from the drive shaft.

17. Install the thrust bearing and thrust washer in that order on the forward gear (**Figure 38**). Install forward gear and bearing assembly in the gearcase.

18. Install pinion gear in gearcase propeller shaft bore and mesh with forward gear teeth.

19. Install drive shaft to engage pinion gear and install locknut.

20. Install drive shaft holding socket on drive shaft splines and connect a breaker bar. Use socket part No. 316612 (50 hp), part No. 312752 (65 hp) or part No. 311875 (V4 and V6).

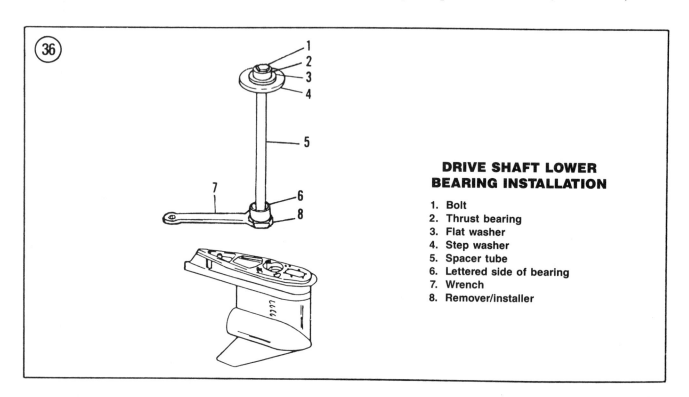

DRIVE SHAFT LOWER BEARING INSTALLATION

1. Bolt
2. Thrust bearing
3. Flat washer
4. Step washer
5. Spacer tube
6. Lettered side of bearing
7. Wrench
8. Remover/installer

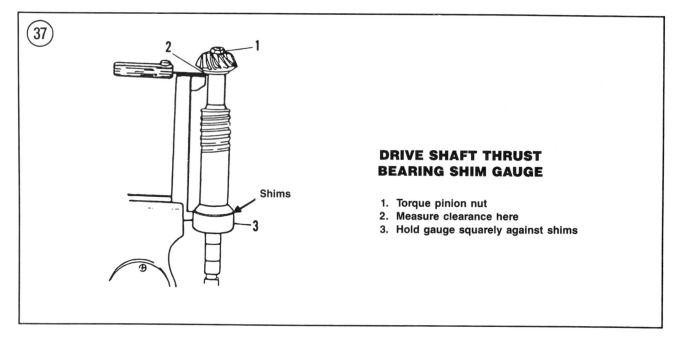

DRIVE SHAFT THRUST BEARING SHIM GAUGE

1. Torque pinion nut
2. Measure clearance here
3. Hold gauge squarely against shims

GEARCASE

21. Hold pinion locknut with a socket and flex handle. Pad the gearcase where the flex handle will hit with shop cloths to prevent housing damage.

22. Hold the pinion nut from moving and turn the drive shaft to tighten the pinion locknut to specifications (**Table 1**). See **Figure 39**. Remove the pinion locknut and drive shaft holding tool.

23. Install the drive shaft thrust bearing, thrust washer and shims selected in Step 15.

24. Install a new bearing housing seal, then install a new O-ring on the bearing housing. Slide the housing over the drive shaft and seat in the gearcase. Tighten screws to specifications (**Table 1**).

25. Remove gearcase and place on a clean workbench port side up. Insert propeller shaft in gearcase with the flat side of the shift rod plunger facing up (**Figure 40**) and engage the oil pump.

26. Position the flat side of the assist cylinder rod down and install in gearcase until it contacts the shift rod plunger. Apply gentle downward pressure on the assist cylinder rod while easing

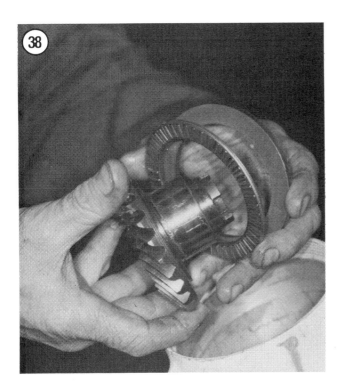

PINION LOCKNUT REMOVAL

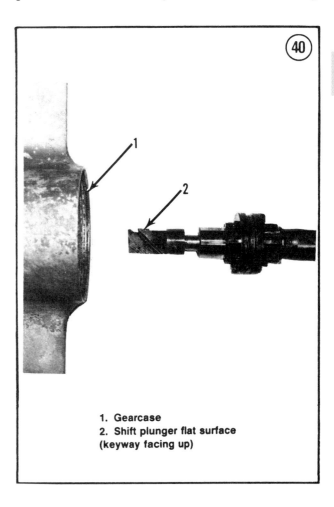

1. Gearcase
2. Shift plunger flat surface (keyway facing up)

the propeller shaft to the rear until the assist cylinder rod key slips into the shift rod plunger keyway. **Figure 41** shows the relative position when properly installed.

27. Once the key and keyway align, push the propeller shaft and assist cylinder rod into complete engagement. When properly engaged, the assist cylinder rod cannot be rotated.

28. Reinstall gearcase in the holding fixture. Coat both sides of a new shift cover gasket with OMC Gasket Sealing Compound and install on gearcase.

29. Insert shift rod through shift rod cover bushing. If rod is threaded, screw into assist cylinder valve. If rod has a toe, engage it with slot in assist cylinder valve.

30. Install shift rod cover with a new O-ring and tighten screws to specifications (**Table 1**).

31. If shift rod uses a boot, lubricate about 3 in. of the rod from the gearcase up with OMC Triple-Guard grease and install the boot.

32. Install the thrust washer, reverse gear and retainer plate in the gearcase. See **Figure 42**.

33. Slip one snap ring (flat side facing out) over the propeller shaft (**Figure 43**) and install with snap ring pliers part No. 311879. Repeat this step to install the other snap ring.

34. If the bearing was removed from the propeller shaft bearing housing, install a new one with a suitable mandrel.

35. Wipe the outer diameter of 2 new propeller bearing housing seals with OMC Gasket Sealing Compound. Install the seals back to back with a suitable seal installer. Pack the cavity between the seals with OMC Triple-Guard grease.

36. Install a new bearing housing O-ring and lubricate with OMC Triple-Guard grease.

37. Thread two 10 in. guide pins (part No. 383175) into retainer plate about 2-3 turns. Position bearing housing with drain slot facing down or "UP" mark facing up and slide it over the guide pins and into gearcase.

38. Install new O-rings on bearing housing screws. Coat screw threads with OMC Gasket Sealing Compound. Install 2 screws with a long screwdriver or a screwdriver tip socket and extension. Remove the guide pins and install the other 2 screws.

39. Install the water pump as described in this chapter.

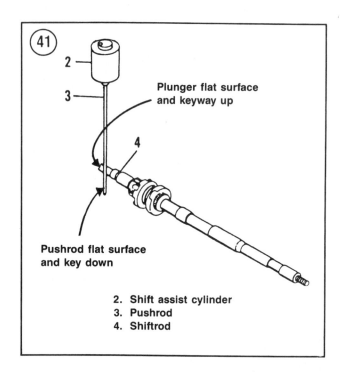

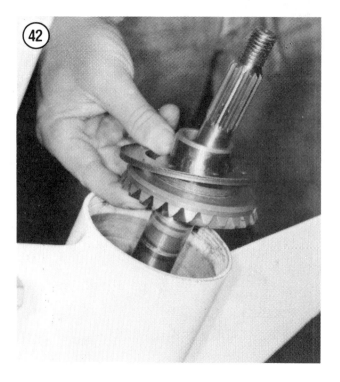

GEARCASE

40. Pressure and vacuum test the gearcase as described in this chapter.
41. Install the gearcase as described in this chapter. Fill with the recommended type and quantity of lubricant. See Chapter Four.
42. Check the gearcase lubricant level after the engine has been run. Change the lubricant after 10 hours of operation (break-in period). See Chapter Four.

MECHANICAL GEARCASE (EXCEPT 1989-1990 48 AND 50 HP)

Disassembly

Refer to **Figure 44** for this procedure.
1. Remove the gearcase as described in this chapter.

2. Secure the gearcase in a holding fixture or a vise with protective jaws. If protective jaws are not available, position the gearcase upright with the skeg between wooden blocks.
3. Remove the water pump as described in this chapter.
4. Remove the propeller bearing housing screws with a thin wall deep socket. Discard the O-rings on the screws.
5A. Inline engines:

   a. 1973-1985—Assemble flywheel puller (part No. 378103) with 3 puller legs (part No. 320737). Fit puller legs around bearing housing flange legs. Tighten puller screw until the bearing housing comes loose.
   b. 1986-on—Assemble flywheel puller (part No. 378103) with 2 puller legs (part No. 330278). Fit puller legs around bearing housing flange legs. Tighten puller screw until the bearing housing comes loose.

5B. V4—Install flywheel puller (part No. 378103) and two puller bolts (part No. 316982) in bearing housing screw holes. Tighten puller screw until the bearing housing comes loose. See **Figure 12**.
5C. V6 engines:

   a. 1979-1983—Assemble flywheel puller (part No. 378103) with 2 puller legs (part No. 321631). Fit puller legs around bearing housing flange legs. Tighten puller screw until the bearing housing comes loose.
   b. 1984-on—Assemble flywheel puller (part No. 378103) with 2 puller legs (part No. 330278). Fit puller legs around bearing housing flange legs. Tighten puller screw until the bearing housing comes loose.

6. Remove the puller assembly from the bearing housing. Remove the bearing housing from the gearcase (**Figure 45**). Remove and discard the bearing housing O-ring.
7A. 1973-1984—Use snap ring pliers part No. 311879 (**Figure 46**) and carefully remove the 2 snap rings in the gearcase propeller bore.

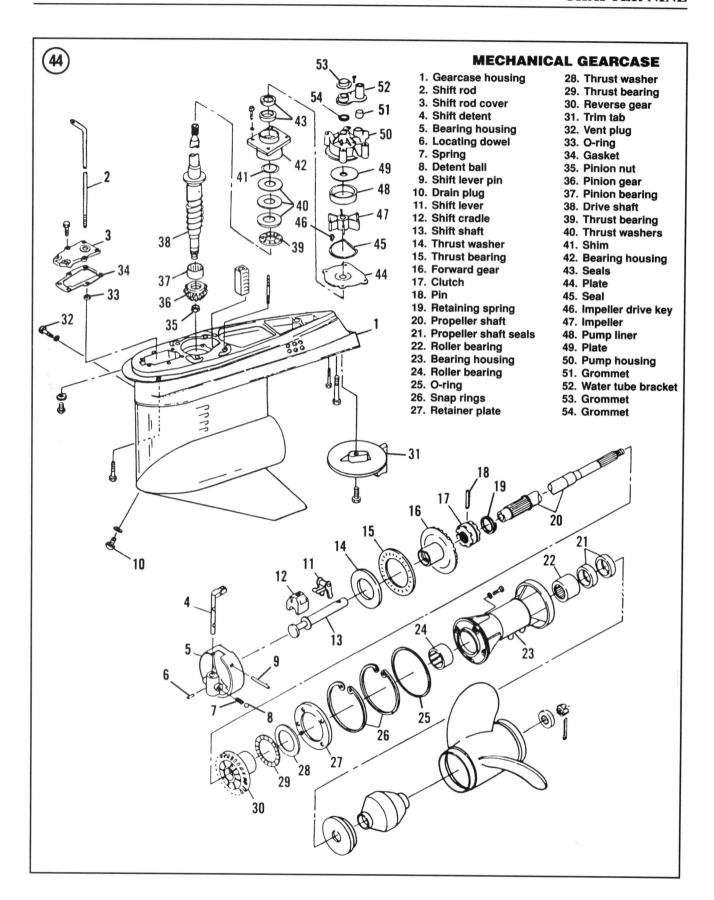

GEARCASE

7B. *1985-on*—Use snap ring pliers part No. 331045 and carefully remove the 2 snap rings in the gearcase propeller bore.

8. Remove the retainer plate, thrust washer, thrust bearing (if used) and reverse gear from the propeller shaft. See **Figure 47**.

9. Pull up on shift rod to shift the clutch dog into FORWARD.

10. Install drive shaft holding socket on drive shaft splines and connect a breaker bar. Use socket part No. 316612 (2-cylinder), part No. 312752 (3-cylinder [1973-1988]), part No. 334995 (3-cylinder [1989]) or part No. 311875 (V4 and V6).

11. Hold pinion locknut with a socket and flex handle (**Figure 48** shows locknut position with propeller shaft removed). Pad the gearcase where the flex handle will hit with shop cloths to prevent housing damage.

12. Hold the pinion nut from moving and turn the drive shaft to break the pinion locknut loose.

See **Figure 49**. Remove the pinion locknut and drive shaft holding tool.

13. Remove the pinion locknut from the gearcase. Remove the pinion gear (**Figure 50**) from the gearcase.
14. Remove the upper drive shaft bearing housing screws. Pull the drive shaft and bearing housing from the gearcase. Remove the thrust washer, bearing and shims from the drive shaft.
15. Push the shift rod down to engage REVERSE. Remove the shift rod cover boot, if used. Remove the cover screws. Tap cover with a soft mallet to break the gasket seal and slide it upward on the shift rod away from the gearcase.
16. Disengage the shift rod and remove from the gearcase with the shift rod cover.

NOTE
The shift detent must be in REVERSE to provide necessary clearance for propeller shaft removal in Step 17.

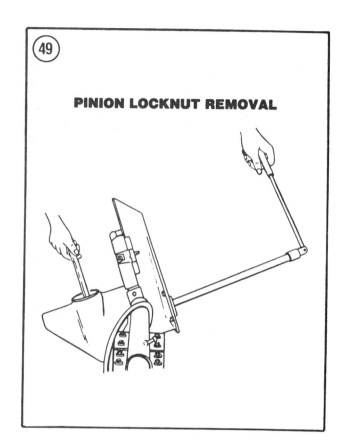

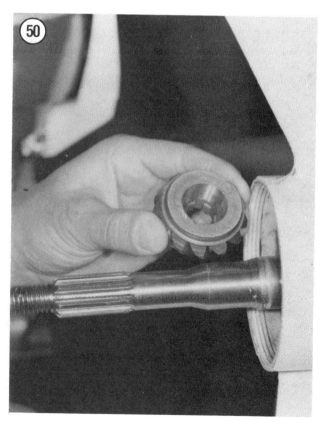

GEARCASE

17. Remove the propeller shaft, forward gear and bearing housing as an assembly from the gearcase. See **Figure 51**. On V6 models, the forward gear needle bearings are loose and may fall out.

18. If lower drive shaft bearing requires replacement, remove the setscrew from the starboard side of the water intake grill on the exterior of the gearcase. See 6, **Figure 23**.

19A. 48-60 (2-cylinder) hp—Assemble components (part No. 326586, nut; part No. 391260, plate; part No. 326582, rod; part No. 326581, remover) of bearing remover-installer tool (part No. 391257) as shown in **Figure 52** and tighten nut to pull the bearing from the gearcase.

19B. 70-75 (1974-1985) hp—Assemble components (part No. 326586, nut; part No. 391260, plate; part No. 326582, rod; part No. 326581, remover) of bearing remover-installer tool (part No. 391257) as shown in **Figure 52** and tighten nut to pull the bearing from the gearcase.

19C. 60 (3-cylinder)-70 (1986-1989) hp—Assemble components (part No. 326586, nut; part No. 391260, plate; part No. 326582, rod; part No. 326580, remover) of bearing remover-installer tool (part No. 391257) as shown in **Figure 52** and tighten nut to pull the bearing from the gearcase.

19D. V4—Assemble components (1/4-20 x 1/2 in. hex head screw; 1 in. O.D. flat washer; part No. 326582, rod; part No. 391260, plate; part No. 326574, remover; 1/4-20 x 1-1/4 in. hex head screw) of bearing remover-installer tool (part No. 391257). Insert assembly in gearcase and drive the bearing out.

19E. V6—Assemble components (part No. 326586, nut; part No. 391260, plate; part No. 326582, rod; part No. 326579, remover) of bearing remover-installer tool (part No. 391257) as shown in **Figure 52** and tighten nut to pull the bearing from the gearcase.

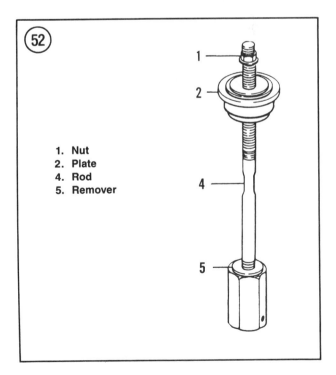

52
1. Nut
2. Plate
4. Rod
5. Remover

20. Drive oil seals and bearings from each end of the bearing housing (**Figure 53**) with a punch and mallet. Remove and discard the bearing housing O-ring.

21. Carefully lift one end of the clutch dog retaining spring and insert a screwdriver blade or the tip of an awl under it. Holding the screwdriver or awl in a stationary position, rotate the propeller shaft to unwind the spring.

22. Remove the clutch dog retainer pin (**Figure 54**) and separate the bearing housing, forward gear and clutch dog from the propeller shaft.

23. If equipped with a standard bearing housing, disassemble as follows:
   a. Remove the shift lever pin from the bearing housing with a suitable punch.
   b. Disconnect the shift lever from the shifter shaft cradle, then remove shaft, cradle and lever. See **Figure 55**.
   c. Depress the detent ball and spring by rotating the shifter detent 90-180°.
   d. Catch ball and spring as detent is removed from bearing housing. See **Figure 56**.

24. If equipped with a setscrew bearing housing, disassemble as follows:
   a. Remove the setscrew, spring and detent ball on each side of the housing.
   b. Remove the shift lever pin from the bearing housing with a suitable punch.
   c. Remove the shift lever, shifter detent and shift dog shaft from the housing.

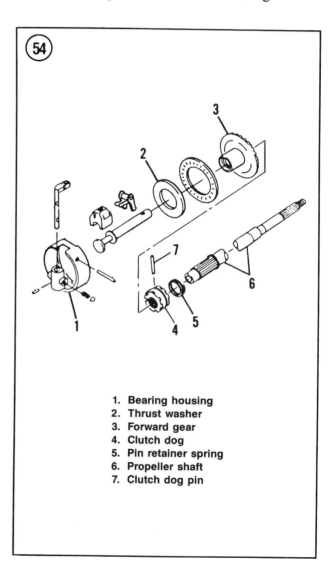

1. Bearing housing
2. Thrust washer
3. Forward gear
4. Clutch dog
5. Pin retainer spring
6. Propeller shaft
7. Clutch dog pin

GEARCASE

Cleaning and Inspection

See *Hydro-Mechanical Gearcase Cleaning and Inspection* in this chapter.

Assembly

Refer to **Figure 44** for this procedure.

1. Reassemble standard forward bearing housing as follows:
   a. Coat end of forward bearing housing detent spring and ball with OMC Needle Bearing Grease. Insert spring and ball in forward bearing housing.
   b. Install shifter detent in forward bearing housing. Depress ball and spring with a suitable punch and push detent into housing.
   c. Coat shifter cradle with OMC HI-VIS Gearcase Lubricant and install on shift shaft. Rotate shifter detent 90-180°, then insert cradle and shaft in housing.
   d. Insert the shift lever to engage the cradle and shifter detent, then install retaining pin.
2. Reassemble setscrew forward bearing housing as follows:
   a. Install shift lever and shifter detent in forward bearing housing.
   b. Install shift dog shaft and shift lever pin.
   c. Coat setscrew threads with OMC Screw Lock.
   d. Insert a detent ball and spring on one side, then install the setscrew and tighten until flush with housing. Repeat this step to install the other detent ball and spring.
3. V6—Coat forward gear needle bearings with OMC Needle Bearing Grease and position in bearing race.
4. Place thrust bearing on forward gear shoulder. Place shims and thrust washer on bearing housing shoulder. Install forward gear to bearing housing.

NOTE
The clutch dog on 48, 50, 55, 60 (2-cylinder), 65, 70 (1974-1985) and 75

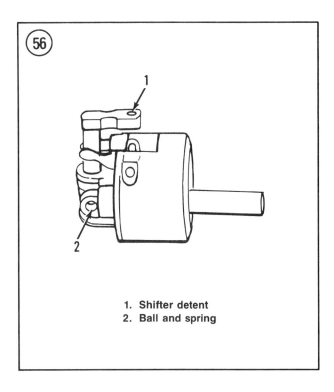

55
1. Shifter shaft
2. Shift lever
3. Shifter detent
4. Pin

56
1. Shifter detent
2. Ball and spring

hp models is marked "PROP END" to indicate correct placement. With 60 (3-cylinder) and 70 (1986-on) hp models, the grooved end of the clutch dog should face toward the forward end of the propeller shaft. With V4 models, the clutch dog lugs with no splines should face the rear of the propeller shaft. On 1979-1986 V6 models, the 6-lug side of the clutch dog should face forward. The clutch dog on 1987-on V6 models is marked "PROP END" to indicate correct placement.

5. Align the clutch dog holes and propeller shaft slot, then install clutch dog on shaft.

6. Install the propeller shaft over the shift shaft and align the clutch dog and shift shaft holes. Insert the retaining pin.

7. Reinstall one end of a new clutch dog retaining spring over the clutch dog, then rotate the propeller shaft to wind the spring back in place.

8. Push shifter detent into REVERSE, then install bearing housing, forward gear and propeller shaft assembly in the gearcase. The locating pin on the bearing housing must engage the locating hole in the gearcase.

9. If pinion bearing was removed, lubricate a new bearing with OMC Needle Bearing Grease and install as follows:

  a. 48-60 (2-cylinder) hp—Assemble components (1/4-20 × 1/2 in. hex head screw; part No. 326584, spacer; 1 in. O.D. flat washer; part No. 326582, rod; part No. 326576, installer; 1/4-20 × 1-1/4 in. hex head screw; part No. 391260, plate) of bearing remover-installer tool (part No. 391257). Position new bearing on tool (lettered side facing tool shoulder). Insert assembly into gearcase and drive the bearing in until the tool washer touches the tool spacer.

  b. 70-75 (1974-1980) hp—Assemble bearing remover-installer tool (part No. 385546 and part No. 320673). Position new bearing on tool (lettered side facing tool shoulder). Insert assembly into gearcase and drive the bearing in until the tool touches the gearcase.

  c. 70-75 (1981-on) hp—Assemble components (1/4-20 × 1/2 in. hex head screw; part No. 326584, spacer; 1 in. O.D. flat washer; part No. 326582, rod; part No. 326575, installer; 1/4-20 × 1-1/4 in. hex head screw; part No. 391260, plate) of bearing

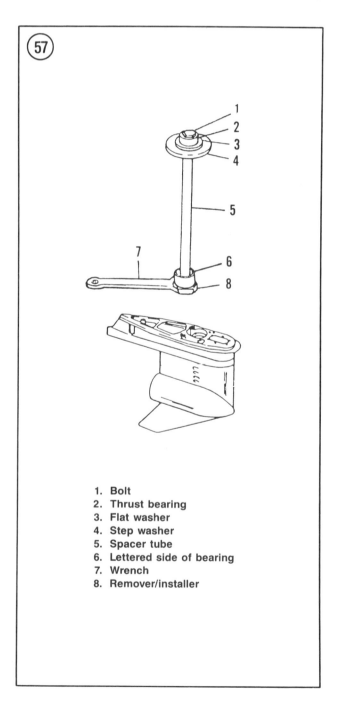

1. Bolt
2. Thrust bearing
3. Flat washer
4. Step washer
5. Spacer tube
6. Lettered side of bearing
7. Wrench
8. Remover/installer

GEARCASE

remover-installer tool (part No. 391257). Position new bearing on tool (lettered side facing tool shoulder). Insert assembly into gearcase and drive the bearing in until the tool washer touches the tool spacer.

d. V4 (1973-1979)—Position new bearing (lettered side up) on tool part No. 318117. Insert tool and bearing assembly in propeller shaft housing and position under bearing bore. Assemble tool part No. 385547 components as shown in **Figure 57**. Hold installer tool with a one-inch wrench and tighten the tool bolt on top of gearcase to draw bearing into place. Remove the tool.

e. V4 (1980-on)—Assemble components (1/4-20 × 1/2 in. hex head screw; part No. 326585, spacer; 1 in. O.D. flat washer; part No. 326582, rod; part No. 326574, installer; 1/4-20 × 1-1/4 in. hex head screw; part No. 391260, plate) of bearing remover-installer tool (part No. 391257). Position new bearing on tool (lettered side facing tool shoulder). Insert assembly into gearcase and drive the bearing in until the tool washer touches the tool spacer.

f. V6 (1979)—Assemble bearing remover-installer tool (part No. 385546) with installer (part No. 321518). Position new bearing (lettered side up) on installer with OMC Needle Bearing Grease. Insert assembly into gearcase and drive the bearing in until the tool plate touches the top of the gearcase.

g. V6 (1980-on)—Assemble components (1/4-20 × 1/2 in. hex head screw; part No. 326584, spacer; 1 in. O.D. flat washer; part No. 326582, rod; part No. 326574, installer; part No. 326587, washer; 1/4-20 × 1-1/4 in. hex head screw; part No. 391260, plate) of bearing remover-installer tool (part No. 391257). Position new bearing on tool (lettered side facing tool shoulder). Insert assembly into gearcase and drive the bearing in until the tool washer touches the tool spacer.

10. Wipe bearing setscrew threads with OMC Screw Lock and install setscrew (6, **Figure 23**).

11. If drive shaft tapered roller bearing requires replacement, remove and install the bearing with an arbor press and universal bearing plate.

12. Install pinion gear on drive shaft and tighten locknut to specifications (**Table 1**).

13. Place the shims removed during disassembly on drive shaft shoulder, then install the drive shaft in the proper shim gauge. Use part No. 320739 (48, 50 and 60 [1980-1985] hp), part No. 315767 (55 and 60 [1986-on] hp), part No. 320739 (70 and 75 hp [1974-1985]), part No. 315767 (70 and 75 hp [1986-on]), part No. 315767 (V4) or part No. 321520 (V6). See **Figure 58A** (typical).

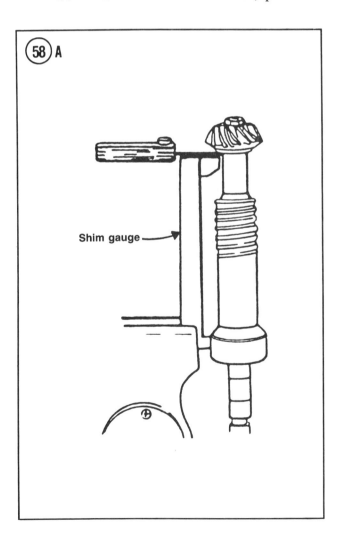

14. Measure the clearance between the shim gauge and pinion gear. If clearance is not zero, add or subtract shims as required.

NOTE
*A universal pinion thrust bearing shim gauge (part No. 393185) was made available in 1983. See **Figure 58B**. If universal shim gauge is used to determine pinion shimming, follow directions provided with tool for correct setup and measuring procedures.*

15. Remove the drive shaft from the shim gauge. Remove the shim(s). Remove the pinion gear from the drive shaft.
16. Coat the outside diameter of 2 new drive shaft bearing housing seals with OMC Gasket Sealing Compound. Install the seals (**Figure 59**) back to back in the housing (one lip facing in, the other facing out). Pack the cavity between the seals with OMC Triple-Guard grease.
17. Install a new bearing housing O-ring (**Figure 60**) and lubricate with OMC Triple-Guard grease.
18. Install drive shaft thrust bearing, thrust washer and shim pack on drive shaft.
19. Coat both sides of a new bearing housing gasket with OMC Gasket Sealing Compound. Install gasket to bearing housing.
20. Wrap drive shaft splines with one thickness of masking tape and carefully slide bearing housing and gasket over drive shaft.
21. Wipe the bearing housing screw threads with OMC Gasket Sealing Compound and install the screws. Tighten to specifications (**Table 1**).
22. Remove the masking tape from the drive shaft splines and clean splines as required.
23. Coat both sides of a new shift cover gasket with OMC Gasket Sealing Compound and install on gearcase.
24. Insert shift rod through shift rod cover bushing and thread it into the shifter detent until it stops, then back it out enough to position offset on top of rod with the port side of the gearcase. Tighten shift rod cover screws to specifications (**Table 1**).

NOTE
After adjustment in Step 25, shift rod offset should face drive shaft.

GEARCASE

25. Place shift rod in REVERSE (1973-1974 50 hp electric start and 1975 70 hp with "B" model suffix) or NEUTRAL (all others). Position a universal shift rod gauge (part No. 389997) on the gearcase beside the vertical shift rod and align gauge and shift rod holes. Insert gauge pin in gauge hole. See **Figure 61**. Screw the shift rod in or out of the shifter detent to obtain the dimension specified in **Table 2**, then pull up on shift rod to engage clutch dog with forward gear.

26. If shift rod uses a boot, lubricate about 3 in. of the rod from the gearcase up with OMC Triple-Guard grease and install the boot.

27. Install pinion gear in gearcase propeller shaft bore and mesh with forward gear teeth.

28. Install drive shaft to engage pinion gear and install locknut.

29. Install drive shaft holding socket on drive shaft splines and connect a breaker bar. Use socket part No. 316612 (2-cylinder), part No. 312752 (3-cylinder [1973-1988]), part No. 334995 (3-cylinder [1989]) or part No. 311875 (V4 and V6).

30. Hold pinion locknut with a socket and flex handle. Pad the gearcase where the flex handle will hit with shop cloths to prevent housing damage.

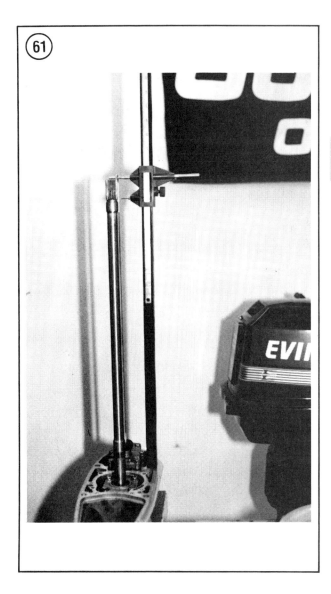

31. Hold the pinion nut from moving and turn the drive shaft to tighten the pinion locknut to specifications (**Table 1**). See **Figure 62**. Remove the pinion locknut and drive shaft holding tools.

32. Install the thrust bearing and thrust washer on reverse gear. Install reverse gear assembly with retainer plate on propeller shaft.

33. Slip one snap ring (flat side facing out) over the propeller shaft and install with snap ring pliers part No. 311879 on 1973-1984 models and with snap ring pliers part No. 331045 on 1985-on models.

34. If the bearing was removed from the propeller shaft bearing housing, install a new one with a suitable mandrel.

35. Wipe the outer diameter of 2 new propeller bearing housing seals with OMC Gasket Sealing Compound. Install the seals back to back (one lip facing in, the other facing out) with a suitable seal installer. Pack the cavity between the seals with OMC Triple-Guard grease.

36. Install a new O-ring on the bearing housing and lubricate with OMC Triple-Guard grease. Install new O-rings on the housing screws and wipe the screw threads with OMC Gasket Sealing Compound.

37. Install the guide pins used for bearing housing removal. Position the housing on the guide pins with drain slot facing downward or "UP" mark facing upward and slide the housing into the gearcase. Tighten screws to specifications (**Table 1**).

38. Install the water pump as described in this chapter.

39. Pressure and vacuum test the gearcase as described in this chapter.

40. Install the gearcase as described in this chapter. Fill with the recommended type and quantity of lubricant. See Chapter Four.

41. Check the gearcase lubricant level after the engine has been run. Change the lubricant after 10 hours of operation (break-in period). See Chapter Four.

MECHANICAL GEARCASE (1989-1990 48 AND 50 HP)

Disassembly

Refer to **Figure 63** for this procedure.

1. Secure the gearcase in a holding fixture or a vise with protective jaws. If protective jaws are not available, position the gearcase upright with the skeg between wooden blocks.

2. Remove the propeller, if not previously removed, and inspect for damage.

3. Remove the propeller shaft bearing housing anode retaining screws and remove the anode.

4. Remove the propeller shaft bearing housing retaining screws and retainers. See **Figure 64**.

5. Assemble flywheel puller (part No. 378103) with two 1/4-20 × 8 in. cap screws and flat washers or a suitable length of 1/4-20 threaded rod, nuts and flat washers. Screw the cap screws or threaded rod into the threaded holes for the retaining screws in the propeller shaft bearing housing. Tighten the puller jackscrew until the bearing housing comes loose.

6. Remove the puller assembly from the bearing housing. Remove the bearing housing from the

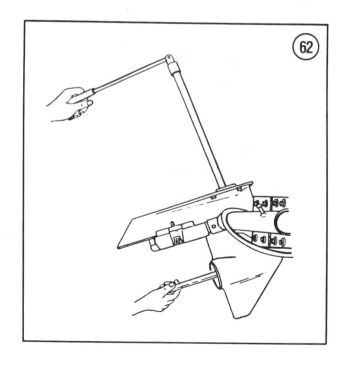

GEARCASE

gearcase. Remove and discard the bearing housing O-ring. See **Figure 65**.

7. Remove thrust washer (A, **Figure 66**) from the bearing housing end.

8. Tilt the gearcase housing with the propeller shaft angled down and slide reverse gear (C, **Figure 66**) and thrust bearing (B, **Figure 66**) off the propeller shaft. Reposition gearcase housing with anti-ventilation plate level.

9. Remove shift lever pivot pin. See **Figure 67**. Remove and discard the screw O-ring.

10. Remove 2 shift rod cover retaining screws. See **Figure 68**.

11. Grasp shift rod and lift assembly from gearcase.

12. Disassemble shift rod and discard 2 O-rings. See **Figure 69**.

13. Remove the propeller shaft assembly (B, **Figure 70**) and cradle (A, **Figure 70**) from the gearcase.

14. Carefully lift one end of the clutch dog retaining spring and insert a screwdriver blade or the tip of an awl under it. Holding the screwdriver or awl in a stationary position, rotate the propeller shaft to unwind the spring.

15. Remove the clutch dog retainer pin, then slide the clutch dog off of the propeller shaft.

WARNING
To prevent possible personal injury, eye protection is recommended prior to disassembly of shift detent.

16. Remove the shift shaft, 3 detent balls and spring from the propeller shaft. See **Figure 71**.

17. Remove the 3 cap screws holding the drive shaft bearing housing.

18. Install drive shaft holding socket, part No. 316612, on drive shaft splines and connect a breaker bar.

19. Hold pinion nut with a 11/16 in. wrench. Pad the gearcase, where the wrench will hit, with shop cloths to prevent housing damage.

20. Hold the pinion nut from moving and turn the drive shaft to break the nut loose. Remove the pinion nut and drive shaft holding tool.

21. Remove the pinion nut and gear from the gearcase.

22. Pull the drive shaft and bearing housing from the gearcase.

23. Separate the bearing housing, shim(s), thrust washers and thrust bearing from the drive shaft. See **Figure 72**.

24. Remove the forward gear, thrust washer and thrust bearing from the gearcase.

25. If lower drive shaft (pinion) bearing requires replacement, remove the bearing retaining screw (**Figure 73**) from the gearcase. Remove and discard the O-ring on the screw.

26. Assemble rod (part No. 326582 from tool part No. 391257), remover/installer (part No. 326575 from tool part No. 391257), guide plate (part No. 334987 from tool part No. 433033), 1/4-20 × 1-1/4 in. hex head screw, 1 inch O.D. flat washer and 1/4-20 × 1/2 in. hex head screw. Install assembled tool into top of gearcase and drive the lower drive shaft bearing into the propeller shaft cavity using a suitable mallet.

27. If the two forward gear bearings are to be replaced, assemble forward bearing service kit (part No. 433034) and extract the bearings from the gearcase bore.

28. Use bearing puller (part No. 432130) and a suitable slide hammer to extract the bearings and oil seals from the propeller shaft bearing housing if replacement is required.

29. Drive oil seals from drive shaft bearing housing using a suitable punch and mallet. Remove and discard the bearing housing outer O-ring.

NOTE
Bearing in drive shaft bearing housing is only replaceable as a complete assembly with the drive shaft bearing housing.

410 CHAPTER NINE

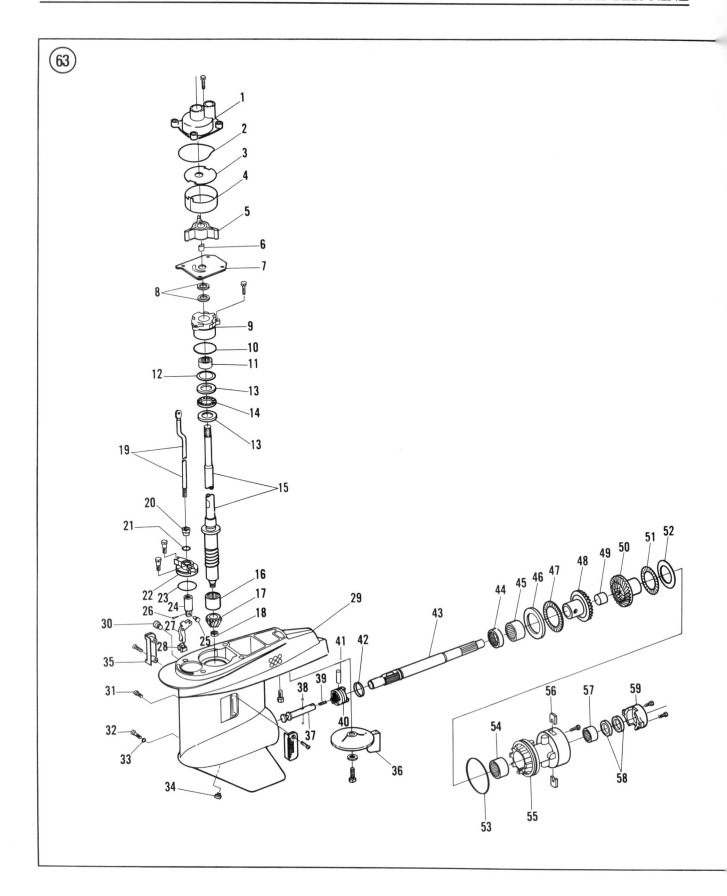

1989-1990 48-50 HP GEARCASE

1. Impeller housing
2. Housing plate seal
3. Impeller plate
4. Impeller liner
5. Impeller
6. Impeller drive key
7. Impeller housing plate
8. Seals
9. Drive shaft bearing housing
10. O-ring
11. Bearing
12. Shim(s)
13. Thrust washers
14. Thrust bearing
15. Drive shaft
16. Bearing
17. Pinion gear
18. Nut
19. Shift rod
20. Bushing
21. O-ring
22. Shift rod cover
23. O-ring
24. Connector
25. Pin
26. Cotter pin
27. Shift lever
28. Cradle
29. Gearcase
30. Oil level plug
31. Bearing retaining screw
32. Shift lever pivot pin
33. O-ring
34. Drain/fill plug
35. Water intake screen
36. Trim tab
37. Shift shaft
38. Detent ball (3)
39. Spring
40. Clutch dog
41. Pin
42. Spring
43. Propeller shaft
44. Bearing (short)
45. Bearing (long)
46. Thrust washer
47. Thrust bearing
48. Forward gear
49. Bushing
50. Reverse gear
51. Thrust bearing
52. Thrust washer
53. O-ring
54. Bearing
55. Propeller shaft bearing housing
56. Retainer (2)
57. Bearing
58. Seals
59. Anode

CHAPTER NINE

Cleaning and Inspection

See *Hydro-Mechanical Gearcase Cleaning and Inspection* in this chapter.

Assembly

CAUTION
Bearings must always be replaced after removal. Never reuse a bearing.

Refer to **Figure 63** for this procedure.
1. Reassemble the propeller shaft bearing housing as follows:
   a. Use bearing installer (part No. 334997) and drive *new* fore and aft bearings into the bearing housing, with the lettered side of the bearing facing outward, until tool contacts bearing housing.
   b. Position the oil seals back-to-back (open sides facing away from each other). Apply OMC Gasket Sealing Compound onto the metal outer surfaces of the seals. Using seal installer (part No. 910585 or part No. 326556), drive the seals into the aft end of the bearing housing. Apply OMC Triple-Guard Grease onto the seal lips after installation.
   c. Apply OMC Triple-Guard Grease onto the bearing housing outer O-ring and install onto housing. See **Figure 65**.

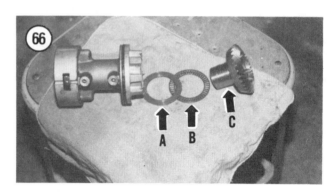

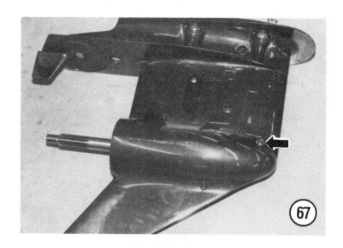

GEARCASE

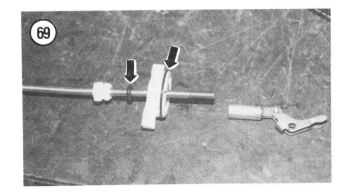

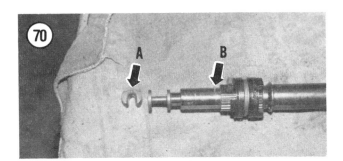

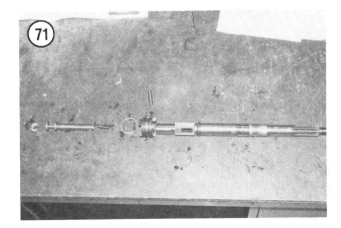

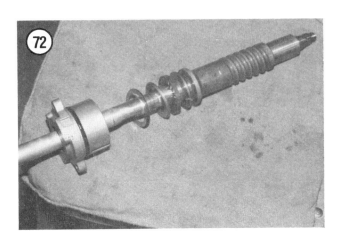

2. Reassemble the drive shaft bearing housing as follows:

NOTE
Bearing in drive shaft bearing housing is only replaceable as a complete assembly with the drive shaft bearing housing.

a. Position the oil seals back-to-back (open sides facing away from each other). Apply OMC Gasket Sealing Compound onto the metal outer surfaces of the seals. Using seal installer (part No. 335823), drive the seals into the bearing housing. Apply OMC Triple-Guard Grease onto the seal lips after installation. See **Figure 74**.

b. Apply OMC Triple-Guard Grease onto the bearing housing outer O-ring and install onto housing.

3. If lower drive shaft (pinion) bearing was removed, install as follows:

a. Assemble rod (part No. 326582 from tool part No. 391257), remover/installer (part No. 326575 from tool part No. 391257), guide plate (part No. 334987 from tool part No. 433033), 1/4-20 × 1-1/4 in. hex head screw, 1 inch O.D. flat washer, spacer (part No. 334986 from tool part No. 433033) and 1/4-20 × 1/2 in. hex head

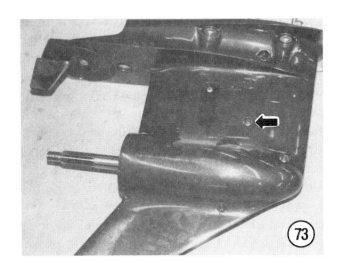

screw. Apply OMC Needle Bearing Assembly Grease onto *new* bearing to retain bearing on installer (part No. 326575). Position bearing on installer with lettered side of bearing facing toward tool assembly (facing top of gearcase after installation). Install assembled tool into top of gearcase and drive the lower drive shaft bearing into bearing bore, using a suitable mallet, until correctly positioned (tool's flat washer contacts top of spacer).

  b. Install a new O-ring on the bearing retaining screw. Apply OMC Nut Lock on the screw threads, then install screw (**Figure 73**) and tighten to specification (**Table 2**).

4. If the two forward gear bearings were removed, install as follows:
  a. Assemble forward bearing service kit (part No. 433034) and drive handle (part No. 311880).
  b. Support the nose of the gearcase with a suitable thickness of wood and position the propeller shaft cavity in a vertical direction on a solid surface.
  c. First, install the short bearing into the gearcase with the lettered side of the bearing facing toward the installer tool. Drive the bearing into the gearcase, using a suitable mallet, until the installer tool contacts the gearcase. Then install the long bearing following same procedure. Note that one side of the installer tool is used to install the short bearing and the other side is used to install the long bearing.

5. Use universal pinion thrust bearing shim gauge (part No. 393185) and service kit part No. 433032 to determine proper drive shaft shim thickness. See **Figure 75**. Follow directions provided with tool for correct setup and measuring procedures in determining correct shim thickness. Use fewest number of shims possible to obtain desired thickness.

6. Position the clutch dog with the end stamped "PROP END" facing the propeller end of the shaft. Align the clutch dog holes and propeller shaft slot, then install clutch dog on shaft.

WARNING
To prevent possible personal injury, eye protection is recommended prior to assembly of shift detent.

7. Install the three detent balls and spring into the shift shaft. While preventing the detent balls from falling free, align the shift shaft holes and propeller shaft slot then carefully slide the shift shaft into the propeller shaft.

8. Using a tool similar to that shown in **Figure 76** to align components, slide clutch dog retainer pin into position while pushing aligning tool from components.

GEARCASE

9. Install one end of a new clutch dog retaining spring over the clutch dog, then rotate the propeller shaft to wind the spring back in place. Make sure no coils overlap.
10. Install forward gear thrust washer into the gearcase with the chamfered side facing toward the gearcase nose.
11. Apply OMC Needle Bearing Assembly Grease onto the forward gear thrust bearing and assemble the bearing onto the forward gear. Install the forward gear and thrust bearing assembly into the gearcase.
12. Assemble thrust washers and thrust bearing onto drive shaft. See **Figure 72**. Install the bottom thrust washer with the chamfered side facing down and the top thrust washer with the chamfered side facing up.

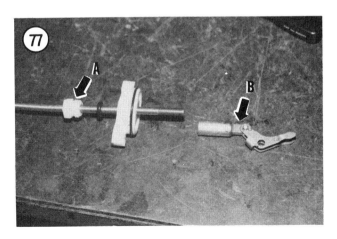

13. Lightly coat the drive shaft shim(s) with OMC Needle Bearing Assembly Grease and position them in the drive shaft bearing housing.
14. Slide drive shaft bearing housing onto drive shaft.
15. Apply a light coat of OMC Gasket Sealing Compound onto the gearcase area where the drive shaft bearing housing mounting flange contacts.
16. Install drive shaft and bearing housing assembly into gearcase.
17. Install the pinion gear and nut onto the drive shaft.
18. Install drive shaft holding socket, part No. 316612, onto drive shaft splines and connect a torque wrench.
19. Hold pinion nut with a 11/16 in. wrench. Pad the gearcase, where the wrench will hit, with shop cloths to prevent housing damage.
20. Hold the pinion nut from moving and turn the drive shaft until the specified torque is obtained. See **Table 1**.
21. Apply OMC Gasket Sealing Compound onto the threads of the drive shaft bearing housing retaining screws and tighten to specification. See **Table 1**.
22. Apply OMC Triple-Guard Grease onto the 2 shift rod cover O-rings. See **Figure 69**. Install the larger, thinner O-ring onto the cover and the smaller, thicker O-ring onto the shift rod.
23. Apply OMC Adhesive M onto the threads of shift rod cover bushing. See A, **Figure 77**. Screw bushing into cover and tighten to 48-60 in.-lb.
24. Screw the shift lever and connector assembly (B, **Figure 77**) *nine* turns onto the shift rod.
25. Apply OMC Needle Bearing Assemble Grease onto shift shaft cradle and position cradle onto shift shaft with part number facing upward. See **Figure 78**.
26. Tilt gearcase nose down and slide propeller shaft assembly into gearcase with shift shaft cradle facing upward.
27. Install shift rod assembly into gearcase. Shift lever tangs must engage slots in shift shaft cradle.

28. Install a new O-ring on the shift lever pivot pin. Apply OMC Nut Lock on the pin threads, then install pin (**Figure 67**) and tighten to 60-84 in.-lb.

29. Apply OMC Gasket Sealing Compound onto the threads of the shift rod cover retaining screws, then install and tighten the screws (**Figure 68**) to specification. See **Table 1**.

30. Apply OMC Needle Bearing Assembly Grease onto the reverse gear thrust bearing and assemble the bearing onto the reverse gear. Install the reverse gear and thrust bearing assembly into the gearcase.

31. Apply a light coat of OMC Gasket Sealing Compound onto the O-ring flange and rear support flange on the propeller shaft bearing housing.

32. Apply OMC Needle Bearing Assembly Grease onto the reverse gear thrust washer and position the washer onto the propeller shaft bearing housing. See **Figure 79**.

33. Position the propeller shaft bearing housing with mounting screw holes in a vertical position and the drain slot on the bottom. Then slide the propeller shaft bearing housing into the gearcase.

34. Install propeller shaft bearing housing retainers and screws (**Figure 64**), then tighten to specification. See **Table 1**.

35. Install propeller shaft bearing housing anode and secure with retaining screws. Tighten to specification shown in **Table 1**.

36. Place shift rod in NEUTRAL. Position a universal shift rod gauge (part No. 389997) on the gearcase beside the vertical shift rod and align gauge and shift rod holes. Insert gauge pin in gauge hole. See **Figure 61**, typical. Screw the shift rod in or out of the connector to obtain a dimension of 16-15/16 ± 1/32 in. (standard shaft) or 21-15/16 ± 1/32 in. (long shaft).

37. Install the water pump as described in this chapter.

38. Pressure and vacuum test the gearcase as described in this chapter.

39. Install the gearcase as described in this chapter. Fill with 16.4 oz. of OMC HI-VIS Gearcase Lube as outlined in Chapter Four. Tighten oil level plug and drain/fill plug to specification. See **Table 1**.

40. Check gearcase lubricant level after engine has been run. See Chapter Four.

PRESSURE AND VACUUM TEST

Whenever a gearcase is overhauled, it should be pressure and vacuum tested before refilling it with lubricant. If the gearcase fails either the pressure or vacuum test, it must be disassembled and the source of the problem located and corrected. Failure to perform a pressure and vacuum test or ignoring the results and running

GEARCASE

a gearcase which failed one or both portions of the test will result in major gearcase damage.

1. Install a new seal on the oil level plug.
2. Thread a pressure test gauge into the fill/drain plug hole. See **Figure 80** (typical).
3. Pump the pressure to 3-6 psi. If pressure holds, increase it to 16-18 psi. If it does not hold, submerge the gearcase in water and check for the presence of air bubbles to indicate the source of the leak.
4. If pressure holds at 16-18 psi, release the pressure and remove the pressure tester. If pressure does not hold at this level, submerge the gearcase in water and check for the presence of air bubbles to indicate the source of the leak.
5. Thread a vacuum test gauge into the fill/drain plug hole. See **Figure 80** (typical).
6. Draw a 3-5 in. Hg vacuum. If vacuum holds, increase it to 15 in. Hg. If vacuum does not hold at this level, coat the suspected seal with lubricant to see if the leak stops or the lubricant is sucked in.
7. If vacuum holds at 15 in. Hg, release the vacuum and remove the tester. If vacuum does not hold, coat the suspected seal with lubricant to see if the leak stops or the lubricant is sucked in.
8. If the source of a pressure of vacuum leak cannot be determined visually, disassemble the gearcase and locate it.
9. If the gearcase passes the pressure and vacuum test, fill it with the required type and quantity of lubricant. See Chapter Four.

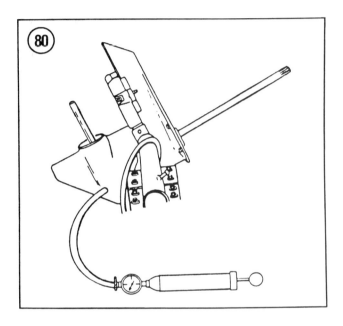

Table 1 GEARCASE TIGHTENING TORQUES

Fastener	in.-lb.	ft.-lb.
Bearing housing anode screws	108-132	
Drain/fill/oil level plugs	60-84	
Drive shaft bearing housing screws		
Inline engines		
1973-1984	60-84	
1985-1986	96-120	
1987-on	120-144	
V4 and V6		
1985-on	168-192	
All others		10-12
(continued)		

Table 1 GEARCASE TIGHTENING TORQUES (continued)

Fastener	in.-lb.	ft.-lb.
Gearcase mounting screws		
Inline engines		
1973-1985		18-20
1986-on		
3/8 in.	216-240	
7/16 in.		28-30
V4 and V6 (1976-1986)		
3/8 in.		22-24
7/16		30-32
V6 (1987-on)		
3/8 in.		37-40
7/16 in.		65-70
Pinion bearing retaining screw	48-80	
Pinion gear locknut		
Inline engines		40-45
V4 and V6		
1985 120-140 hp V4		40-45
All 1986-on		70-75
All others		60-65
Propeller shaft bearing housing		
Inline engines		
1973-1984	60-84	
1985-1986		8-10
1987-on		10-12
V4 and V6	120-144	
Shift rod cover screws	60-84	
Water pump screws	60-84	
Standard screws and nuts		
No. 6	7-10	
No. 8	15-22	
No. 10	25-35	2-3
No. 12	35-40	3-4
1/4 in.	60-80	5-7
5/16 in.	120-140	10-12
3/8 in.	220-240	18-20
7/16 in.		28-30

Table 2 UNIVERSAL SHIFT ROD GAUGE DIMENSIONS

Model	Shift rod dimension gearcase to center of shift rod hole*
48-50 hp (1989-1990)	
Short shaft	16 15/16 ± 1/32 in.
Long shaft	21 15/16 ± 1/32 in.
(continued)	

GEARCASE

Table 2 UNIVERSAL SHIFT ROD GAUGE DIMENSIONS (continued)

Model	Shift rod dimension gearcase to center of shift rod hole*
All other 48-60 hp (2-cyl. electric start)	
Short shaft	
1973-1974	16 7/32 ± 1/32 in.
1975-on	15 29/32 ± 1/32 in.
Long shaft	
1973-1974	21 7/32 ± 1/32 in.
1975-1981, 1985-on	20 29/32 ± 1/32 in.
1982-1984	21 23/32 ± 1/32 in.
All other 50-55 hp (2-cyl. manual start)	
Short shaft	16 23/32 ± 1/32 in.
Long shaft	21 23/32 ± 1/32 in.
65 hp	
Short shaft	16 11/32 ± 1/32 in.
Long shaft	21 11/32 ± 1/32 in.
Extra long shaft	24 7/32 ± 1/32 in.
60-75 hp (3-cyl.)	
Standard shaft	16 13/32 ± 1/32 in.
Short shaft	
1973-1975	16 11/32 ± 1/32 in.
1976-1981	16 13/32 ± 1/32 in.
1982-on	15 29/32 ± 1/32 in.
Long shaft	
1973-1975	21 11/32 ± 1/32 in.
1976-1977	21 25/64 ± 1/32 in.
1978-on	21 23/32 ± 1/32 in.
V4 (except 1985-on 120-140 hp)	
Long shaft	
1973-1977	21 13/32 ± 1/32 in.
1978	21 53/64 ± 1/32 in.
1979-on	21 27/32 ± 1/32 in.
Extra-long shaft	
1973-1977	26 13/32 ± 1/32 in.
1978	26 53/64 ± 1/32 in.
1979-on	26 27/32 ± 1/32 in.
V4 (1985-on 120-140 hp)	
Long shaft (20 in.)	21 15/16 ± 1/32 in.
Extra-long shaft (25 in.)	26 15/16 ± 1/32 in.
V6	
Long shaft	
1976	21 13/32 ± 1/32 in.
1977-1978	21 19/32 ± 1/32 in.
1979-1984	21 1/16 ± 1/32 in.
1985-on 150-175 hp	22 1/16 ± 1/32 in.
1986-on 200-225 hp	21 15/16 ± 1/32 in.
Extra-long shaft	
1976	26 13/32 ± 1/32 in.
1977-1978	26 19/32 ± 1/32 in.
1979-on 150-175 hp	27 1/16 ± 1/32 in.
1986-on 200-225 hp	26 15/16 ± 1/32 in.

* 1973-1974 50 hp electric start and 1975 70 hp with "B" model suffix in REVERSE; all others in NEUTRAL.

Chapter Ten

Power Trim and Tilt Systems

The usual method of raising and lowering the outboard gearcase is a mechanical one, consisting of a series of holes in the transom mounting bracket. To trim the engine, an adjustment stud is removed from the bracket, the outboard is repositioned and the stud reinserted in the proper set of holes to retain the unit in place.

A power trim and tilt system is available on all models covered in this manual and can be retrofitted on most models with accessory kits provided by Johnson and Evinrude. Power trim provides low-effort control when the boat is underway or at rest.

All 2-cylinder, some 3-cylinder and early V4 engines are fitted with an external trim and tilt unit with the pump and cylinders mounted outside the stern brackets. Some 3-cylinder, later V4 and all V6 engines use an integral trim and tilt system with the pump and cylinders mounted inside the stern brackets.

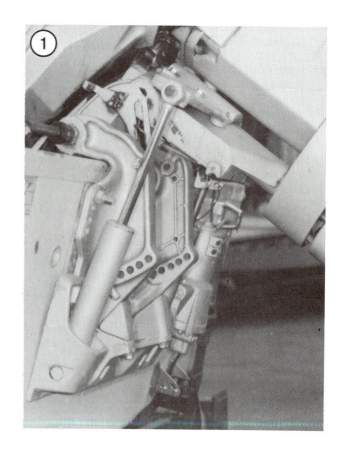

POWER TRIM AND TILT SYSTEMS

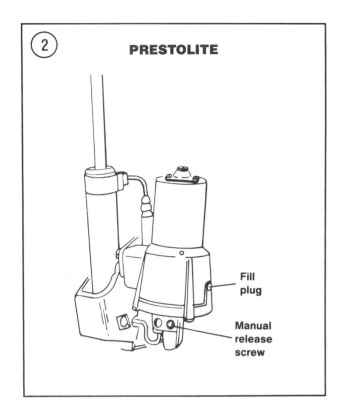

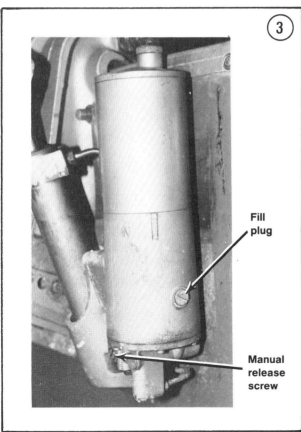

This chapter includes maintenance, troubleshooting procedures and trim/tilt cylinder replacement for both power trim/tilt designs. **Table 1** and **Table 2** are at the end of the chapter.

EXTERNAL POWER TRIM AND TILT SYSTEM

Components

The external system consists of a hydraulic pump (containing an electric motor, oil reservoir, oil pump and valve body), hydraulic trim (starboard) and tilt (port) cylinders, a trim/tilt switch, indicator gauge and the necessary hydraulic and electrical lines. See **Figure 1**.

The external power trim and tilt system differs slightly according to the pump type used. The pump may be a Prestolite (**Figure 2**) or a Calco (**Figure 3**). System operation is basically the same.

Operation

Moving the trim switch to the UP position closes the pump motor circuit. The motor drives the oil pump, forcing oil into the UP side of the trim and tilt cylinders. The engine will trim upward from 0° to 15°. At this point, an internal piston stop in the trim cylinder forces the hydraulic flow into the tilt cylinder, which now functions at a faster rate of travel until it reaches a maximum of 68-70°.

Moving the trim switch to the DOWN position also closes the pump motor circuit. The reversible motor runs in the opposite direction, driving the oil pump to force oil into the DOWN side of the trim and tilt cylinders and bringing the engine back to the desired position. If the switch is not released when the engine reaches the limit of its downward travel, an overload cutout switch opens to shut the pump motor off and prevent system damage.

The power tilt will temporarily maintain the engine at any angle within its range to allow shallow water operation at slow speed, launching, beaching or trailering.

A manual release valve with a slotted head permits manual raising and lowering of the engine if the electrical system fails.

Hydraulic Pump Fluid Check

1. Tilt the outboard to its full UP position.
2. Clean area around pump fill plug. See **Figure 2** or **Figure 3**. Remove the plug and visually check the fluid level in the pump reservoir. It should be at the bottom of the fill hole threads.
3. Top up if necessary with OMC Power Trim/Tilt fluid.

Hydraulic Pump Fluid Refill

Follow this procedure when a large amount of fluid has been lost due to an overhaul of the system or leakage that has been corrected. See **Figure 2** or **Figure 3**.

1. Remove the fill plug and top off the reservoir with OMC Power Trim/Tilt Fluid. Make sure the pistons are in the down position.
2. Open the manual release valve. Depress the tilt switch in the UP position and hold for 10 seconds (the engine will not move), then top off the reservoir again.
3. Close the manual release valve. Move the tilt switch to the UP position while an assistant helps to lift the engine to a full tilt position. Top off the reservoir as the pistons start to move.
4. Install the reservoir plug just enough to hold it in place (about 1-2 threads). Run the engine all the way down, then all the way up. Top off the reservoir.
5. Continue running the engine first up, then down, topping off the reservoir until it will accept no more fluid with the engine raised. This should take about 4-6 cycles. Tighten reservoir plug securely.
6. Recharge the battery. See Chapter Seven.

Troubleshooting

Whenever a problem develops in the power trim system, the initial step is to determine whether the problem is in the electrical or hydraulic system. Electrical tests are given in this chapter. If the problem appears to be in the hydraulic system, refer it to a dealer or qualified specialist for necessary service.

To determine whether the problem is in the electric or hydraulic system, proceed as follows:

1. Make sure the plug-in connectors are properly engaged and that all terminals and wires are free of corrosion. Tighten and clean as required.
2. Make sure the battery is fully charged. Charge or replace as required.
3. Check the system fuse, if so equipped.
4. Check the fluid level as described in this chapter. Top off if necessary.
5. Make sure the manual release valve is fully closed.
6. Run the system pump in the UP direction and clock the system operation. It should take a maximum of 9 seconds for the engine to travel through the trim range and a maximum of 6 seconds through the tilt range.
7. Perform an amperage test of the system as described in this chapter and compare results to **Table 1**.
8. If the system does not function within the time limits specified in Step 6 or amperage values specified in Step 7, check the motor pivot for corrosion and clean as required, then check the voltage drop at the motor with a voltmeter. With a fully charged battery, the voltage drop should not exceed 3 volts in an upward direction or 1.5 volts downward. If voltage drop is excessive, clean or replace the cables as required.
9A. Calco pump:
   a. Remove the pump fill port plug and connect a pressure gauge. Trim the engine from full in to full out. Note the time required and pressure reading and compare to **Table 1**.

POWER TRIM AND TILT SYSTEMS

b. Run the engine from the full trim to the full tilt position. Note the time required and pressure reading and compare to **Table 1**.

c. Move the pressure gauge to the high pressure line fitting on the pump and reinstall the fill port plug. Run the unit in the tilt position for about 5 seconds. The pressure gauge should read a minimum of 1,450 psi.

9B. Prestolite pump:

a. Run the unit to the full down position and loosen the manual release valve one full turn.

b. Quickly cycle the switch up, down and up again.

c. Insert a screwdriver in manual release valve slot, wrap end of screwdriver with a shop cloth and slowly remove the valve. Check valve port for contamination and clean if necessary.

d. Install the A gauge of pressure tester part No. 395415 in manual release valve port. Cycle the engine up and down several times and top off the reservoir.

e. Run the engine to the full up position and note the pressure reading. It should not drop more than 100-200 psi after the engine is stopped.

f. Carefully remove the A gauge and install the B gauge in its place. Cycle the engine up and down several times and top off the reservoir, if necessary.

g. Run the engine to the full down position and note the pressure reading. It should not drop more than 100-200 psi after the engine is stopped.

10. If the system fails any of Steps 6-8, the problem is electrical. If it fails Step 9, the problem is hydraulic.

Trim and Tilt Switch Circuit Test (2-cylinder)

Refer to **Figure 4** for this procedure.

1. Disconnect the 3-wire connector between the pump motor and battery at point 1.
2. Connect the voltmeter black test lead to the connector plug terminal (point 2) on the battery side of the wiring harness.
3. Connect the red test lead to the terminal at point 3. Move the tilt switch to the UP position. If battery voltage is not shown, check the fuse, key switch, battery connections and harness wires for an open circuit.
4. Move the red test lead to point 4. Move the tilt switch to the DOWN position. If battery voltage is not shown, check the fuse, key switch, battery connections and harness wires for an open circuit.
5. Disconnect the negative battery cable and reconnect the 3-wire connector.
6. Connect the ohmmeter black test lead to point 5. Connect the red test lead to point 6, then to point 7, operating the switch each time. If the ohmmeter does not show high resistance with the switch in one position and low resistance in the other position, replace the switch.
7. If the system passes Steps 1-6 but the pump motor does not run, replace the pump motor.

Trim and Tilt Switch Circuit Test (3-cylinder and V4)

Refer to **Figure 5** for this procedure.

1. With the key switch ON, connect the voltmeter red test lead to the inside terminal at point 1. Connect the black test lead to the solenoid G terminal. Move the trim switch to the DOWN position. If the motor does not run, test it as described in this chapter.
2. If battery voltage is not shown in Step 1, move the red test lead to point 2 and the black meter lead to the indicator gauge G terminal. Move the trim switch to the DOWN position. Check for an open between points 1 and 2 if voltage is shown; check for an open between the indicator gauge ground terminal and the remote control box if no voltage is shown.

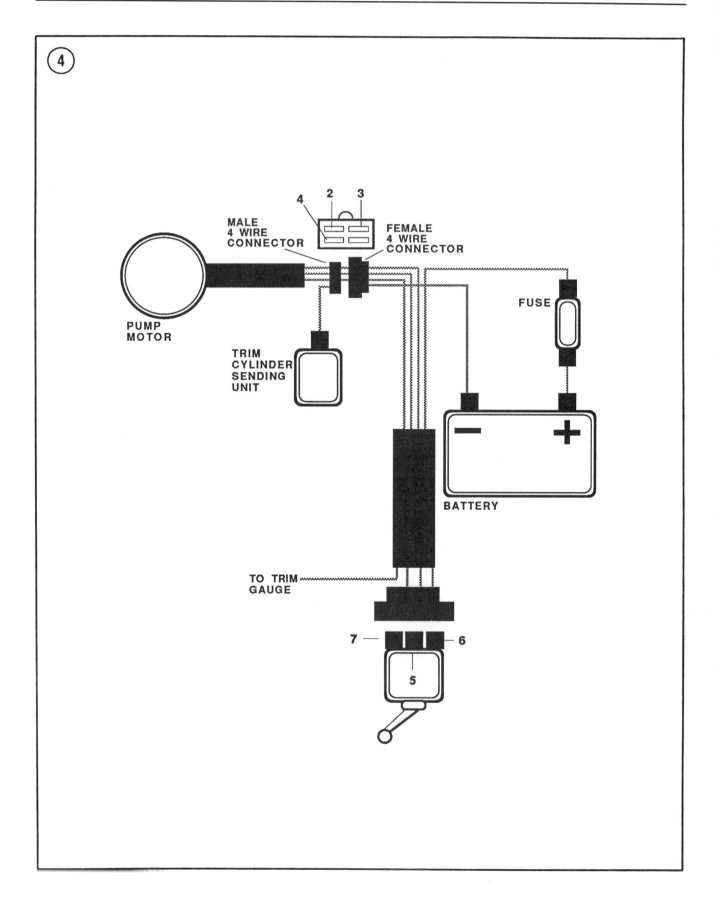

POWER TRIM AND TILT SYSTEMS

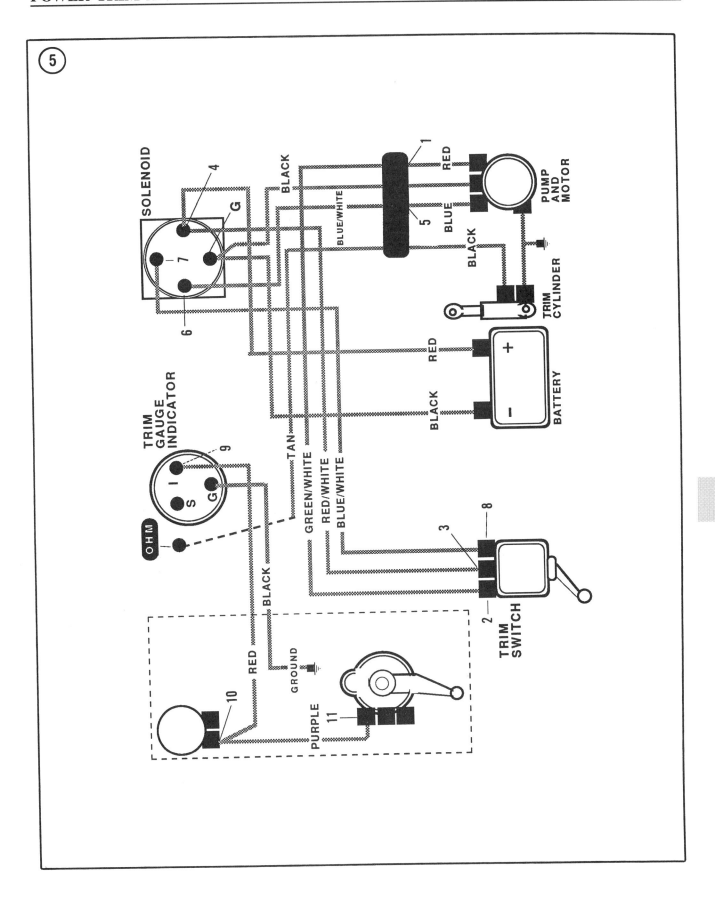

3. If no voltage is shown in Step 2 and the lead between the indicator gauge ground terminal and remote control box is good, move the red test lead to point 3. Replace the trim switch if voltage is shown.

4. Move the red test lead to point 4 and the black test lead to the solenoid G terminal. If voltage is shown, look for an open in the lead between points 3 and 4. If no voltage is shown, check both battery-to-solenoid leads for a poor connection or open circuit and correct as required.

5. Connect the voltmeter red test lead to the inside terminal at point 5. Connect the black test lead to the solenoid G terminal. Move the trim switch to the UP position. If the motor does not run, test it as described in this chapter.

6. If battery voltage is not shown in Step 5, move the red test lead to point 6. Move the trim switch to the UP position. Check for an open between points 5 and 6 if voltage is shown.

7. If no voltage is shown in Step 6, move the red test lead to point 7. Move the trim switch to the UP position. If voltage is shown and the motor does not run, move the red test lead to point 4 and repeat the step. If voltage is still shown, replace the solenoid.

8. If no voltage is shown in Step 7, move the red test lead to point 8 and the black test lead to the indicator gauge G terminal. Move the trim switch to the UP position. Check for an open between points 7 and 8 if voltage is shown; check for an open between the indicator gauge ground terminal and the remote control box if no voltage is shown.

9. If no voltage is shown in Step 8 and the lead between the indicator gauge ground terminal and remote control box is good, move the red test lead to point 3. Replace the trim switch if voltage is shown.

10. Move the red test lead to point 4 and the black test lead to the solenoid G terminal. If voltage is shown, look for an open in the lead between points 3 and 4. If no voltage is shown, check both battery-to-solenoid leads for a poor connection or open circuit and correct as required.

Trim Indicator Circuit Test

Refer to **Figure 5** for this procedure.

1. Turn the key switch ON.
2. Connect the voltmeter red test lead to point 9 on the indicator gauge. Connect the black test lead to the indicator gauge G terminal. If battery voltage is shown, proceed with Step 5.
3. If battery voltage is not shown in Step 2, check for an open in the lead between the solenoid G terminal and the remote control box. If the circuit is good, move the red test lead to point 10 (remote control warning horn). If voltage is shown, there is an open in the lead between points 9 and 10.
4. If no voltage is shown in Step 3, move the red test lead to point 11. If no voltage is shown at point 11, look for a bad fuse or open in the key switch wiring. If the fuse and wiring are good, replace the switch.
5. Turn the key switch OFF.
6. Disconnect the tan wire at the indicator gauge S terminal. With an ohmmeter on the low scale, connect the red test lead to the disconnected tan wire and the black test lead to the indicator gauge G terminal. Move the trim switch to the UP position and run the engine to the full UP position. The ohmmeter should read 0-6 ohms. Move the trim switch to the DOWN position and return the engine to the full DOWN position. The ohmmeter should read 84-93 ohms.

NOTE
The tan wire changes to black at the connector in Step 7.

7. If the ohmmeter does not read resistance as described in Step 6, look for an open in the tan wire to the trim cylinder. If the wire is good, replace the trim indicator sending unit.

Pump Motor Circuit Test

Refer to **Figure 5** for this procedure.
1. Turn the key switch ON.

POWER TRIM AND TILT SYSTEMS

2. Connect the voltmeter black test lead to the indicator gauge G terminal. Connect the red test lead to the red/white wire at point 3. Move the red test lead to the blue/white wire at point 8 and place the switch in the UP position. Move the red test lead to the green/white wire at point 2 and place the switch in the DOWN position. If voltage is not shown at each point, replace the switch.

3. Disconnect the motor harness connector. With the switch in the UP position, connect the red test lead to the connector blue/white wire terminal and the black test lead to the connector black wire terminal. Battery voltage should be shown.

4. Repeat Step 3 with the switch in the DOWN position and the red test lead connected to the connector green/white wire terminal. Battery voltage should be shown.

5. Plug the 2 halves of the motor harness connector back together and probe the blue and black, then the red and black leads on the motor side of the connector. Both should show voltage.

6. If no voltage is shown in Steps 3-5, the problem is in the wiring between the connector and trim switch, solenoid or battery. If there is voltage in Steps 3-5 and the motor still does not run, turn the key switch OFF and proceed with Step 7.

7. Scribe alignment marks on the motor frame and end cap. Loosen the cable harness nut at the end cap and carefully remove the end cap, feeding the cable through the end cap as it is lifted. Make sure all three spade terminals are clean and properly connected inside the motor. If the results of Steps 1-6 are satisfactory and the connections inside the motor are good, remove the motor for further testing by a dealer or qualified specialist.

Trim Switch Test

A corroded trim switch or one that has an accumulation of water will often self-energize, draining battery power. While this condition can often (but not always) be determined by visual inspection, a resistance check is recommended.

1. Disconnect the negative battery cable.
2. With the trim switch off, connect the ohmmeter black test lead to its center terminal. Connect the red test lead to the UP terminal and then to the DOWN terminal. If the ohmmeter reads less than 50,000 ohms at each terminal, replace the switch.
3. Disconnect the red/white lead at the switch. Connect the red test lead to the center terminal of the switch and the black test lead to a good ground. If the ohmmeter reads less than 50,000 ohms, replace the switch.

Trim Gauge Test

If the trim gauge constantly reads FULL UP, the lead between the gauge and the trim cylinder is shorted. Look for a pinched wire between the stern bracket and boat.

If the trim gauge constantly reads FULL DOWN, there is an open in the wiring between the indicator gauge S terminal and the trim cylinder. Connect the black sending unit wire to a good ground. If the gauge does not read FULL UP, the trim cylinder sending unit is corroded or defective or there is an open circuit in the trim cylinder bobbin.

INTERNAL POWER TRIM AND TILT SYSTEM

Components

The internal system consists of a manifold containing an electric motor, oil reservoir, oil pump, all valving and 2 hydraulic trim cylinders. A combination hydraulic tilt cylinder and shock absorber is attached to the manifold. See **Figure 6**. A tilt/trim switch, indicator gauge and the necessary hydraulic and electrical lines complete the system.

The internal power trim and tilt system differs slightly in design and troubleshooting. One system is used on 1976-1979 models, with a variation used on 1980 and 1981 CIH engines. Another system is used on 1981 CIM, CIA and CIB and 1982-1983 engines. A third variation is used on 1984 models. The pump motor may be a Prestolite, Bosch or a Showa. The tilt cylinder may be a Prestolite or Showa. System operation is basically the same, although troubleshooting procedures differ slightly.

Operation

Moving the trim switch to the UP position closes the pump motor circuit. The motor drives the oil pump, forcing oil into the UP side of the trim cylinders. The trim cylinder pistons push on the swivel bracket thrust pads to trim the engine upward from 0° to 15°. At 15°, the trim cylinders are fully extended and the hydraulic fluid is diverted into the tilt cylinder, which moves the engine through the final 50° of travel.

Moving the trim switch to the DOWN position also closes the pump motor circuit. The reversible motor runs in the opposite direction, forcing oil into the tilt cylinder and bringing the engine back to the 15° position, where the swivel brackets rest on the trim rods. The trim rods then lower the engine the remainder of the way.

The power tilt will temporarily maintain the engine at any angle within its range to allow shallow water operation at slow speed, launching, beaching or trailering. At about 1,500 rpm, however, an overload switch will automatically lower the engine to its fully trimmed out (15 degree) position.

A manual release valve with a slotted head permits manual raising and lowering of the engine if the electrical system fails. See **Figure 7**.

A trim gauge sending unit is located on the inside of the port stern bracket (**Figure 8**). Access to the sending unit requires the engine to be fully tilted.

Hydraulic Pump Fluid Check

1. Tilt the outboard to its full UP position.
2. Clean area around pump fill plug. See **Figure 9**. Remove the plug and visually check the fluid level in the pump reservoir. It should be at the bottom of the fill hole threads.
3. Top up if necessary with OMC Power Trim/Tilt fluid.

Hydraulic Pump Fluid Refill

Follow this procedure when a large amount of fluid has been lost due to an overhaul of the system or a leakage that has been corrected.

1. Remove the fill plug and top off the reservoir with OMC Power Trim/Tilt Fluid. Make sure the pistons are fully extended.
2. Run the pump motor while rechecking the oil level. Top up if necessary, then cycle the motor full up and full down at least 5 times, adding oil as required between cycles when the trim and tilt cylinders are fully extended.
3. Install and tighten the reservoir plug securely.
4. Recharge the battery. See Chapter Seven.

POWER TRIM AND TILT SYSTEMS

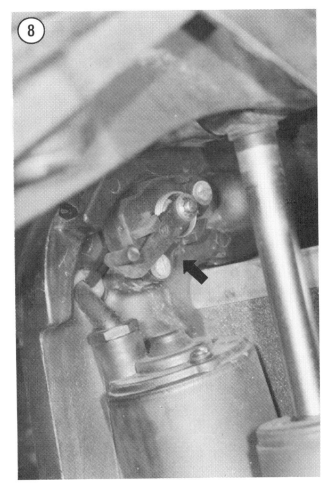

Troubleshooting

Whenever a problem develops in the power trim system, the initial step is to determine whether the problem is in the electrical or hydraulic system. Electrical tests are given in this chapter. If the problem appears to be in the hydraulic system, refer it to a dealer or qualified specialist for necessary service.

To determine whether the problem is in the electric or hydraulic system, proceed as follows:

1. Make sure the plug-in connectors are properly engaged and all terminals and wires are free of corrosion. Tighten and clean as required.
2. Make sure the battery is fully charged. Charge or replace as required.
3. Check the system fuse, if so equipped.
4. Check the fluid level as described in this chapter. Top off if necessary.
5. 1976-1978—Make sure the manual release valve is fully closed.

NOTE
The power trim and tilt mechanism must be removed from the engine for the remainder of this test in order to provide access to the test points.

6A. 1976-1978—Locate the 1/16 in. dry seal pipe plugs shown in **Figure 10**. Clean the area around the 3 plugs, then remove the one on each end (Nos. 1 and 3) and install a pressure gauge in each plug hole. Seal the gauge fittings with Teflon tape.

6B. 1979-on—Proceed as follows to install OMC Trim and Tilt Pressure Tester part No. 390010 (**Figure 11**):

   a. Remove manual release valve retaining ring.
   b. Operate trim/tilt motor to completely retract all cylinders.
   c. Rotate manual release valve 1 full turn counterclockwise.
   d. Momentarily operate trim/tilt motor in the "UP" direction and then in the "DOWN" direction.

CAUTION
Residual hydraulic pressure may be present when manual release valve is removed. To prevent personal injury, make sure protective eye wear is worn and cover manual release valve with a cloth during removal.

   e. Slowly rotate manual release valve counterclockwise to remove.
   f. Securely screw pressure gauge and adapter together. Adapter "A" is used to test the UP circuit and adapter "B" is used to test the DOWN circuit.
   g. Install pressure gauge and adapter into manual release valve passage and tighten to 5-10 in.-lb. See **Figure 12**. Do not overtighten as damage to O-ring in end of adapter may result.
   h. Operate unit UP and DOWN several cycles, then recheck and add fluid to reservoir as needed.

7A. 1976-1978—Perform the following operations and note the pressure gauge readings:

   a. Trim out—Plug No. 1 reading of 0-200 psi.
   b. Tilt out—Plug No. 1 reading of 0-400 psi.
   c. Cylinders bottomed out in UP direction—Plug No. 1 reading of 1,300-1,600 psi.
   d. Tilting and trimming down—Plug No. 1 reading of 0-600 psi; Plug No. 3 reading of 500-1,000 psi.
   e. Cylinders bottomed out in DOWN direction—Plug No. 3 reading of 500-1,000 psi.

7B. 1979-on—Perform the following operations and note the pressure gauge readings:

   a. Install pressure gauge and adapter "A" into manual release valve passage.
   b. Operate trim/tilt motor to extend cylinders (UP direction). Pressure gauge should read approximately 200 psi as cylinders are extending. Pressure gauge should read approximately 1500 psi as the unit stalls. Pressure gauge should not drop by more than 200 psi from the stall reading when the trim/tilt motor is stopped.
   c. Install pressure gauge and adapter "B" into manual release valve passage.
   d. Operate trim/tilt motor to retract cylinders (DOWN direction). Pressure gauge should read approximately 800 psi as cylinders are retracting. Pressure gauge should read approximately 900 psi as the unit stalls. Pressure gauge should not drop by more than 200 psi from the stall reading when the trim/tilt motor is stopped.

POWER TRIM AND TILT SYSTEMS

8. If the system fails any part of Step 7, the problem is hydraulic. If the hydraulic system checks out satisfactorily, the problem is in the electric system.

Trim and Tilt Switch Circuit Test (1976-1979)

Refer to **Figure 13** for this procedure.

1. With the key switch ON, connect the voltmeter red test lead to point A in the junction box. Connect the black test lead to point G in the junction box. The meter should read battery voltage.
2. If battery voltage is not shown in Step 1, move the red test lead to point B. If battery voltage is shown, check the junction box fuse. If no voltage is shown, look for poor connections or an open circuit in the wiring between the junction box and the battery.
3. Move the red test lead to point C in the junction box. Move the trim switch to the DOWN position. If battery voltage is shown and the trim motor does not run, test the motor as described in this chapter.
4. If there is no voltage shown in Step 3, move the red test lead to point D and the black test lead to the G1 terminal on the indicator gauge. Move the trim switch to the DOWN position. If voltage is shown, there is an open in the circuit between points C and D.
5. If no voltage is shown in Step 4 and the lead between points C and D is good, move the red test lead to point E. Replace the trim switch if voltage is shown. If no voltage is shown, look for an open in the trim gauge to junction box ground lead.
6. Move the red test lead to point A and the black test lead to the G1 terminal. If voltage is shown, look for an open in the lead between points A and E.
7. Connect the voltmeter red test lead to the junction box at point F. Connect the black test lead to the junction box G terminal. Move the trim switch to the UP position. If the motor does not run, test it as described in this chapter.
8. If battery voltage is not shown in Step 7, move the red test lead to point G2 on the solenoid. Move the trim switch to the UP position. Check for an open in the wiring between points F and G2 if voltage is shown.
9. If no voltage is shown in Step 8, move the red test lead to point H. Move the trim switch to the UP position. If voltage is shown and the motor does not run, move the red test lead to point B and repeat the step. If voltage is still shown, replace the solenoid.

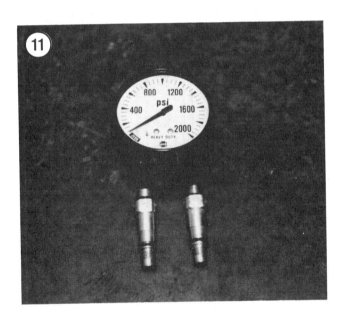

CHAPTER TEN

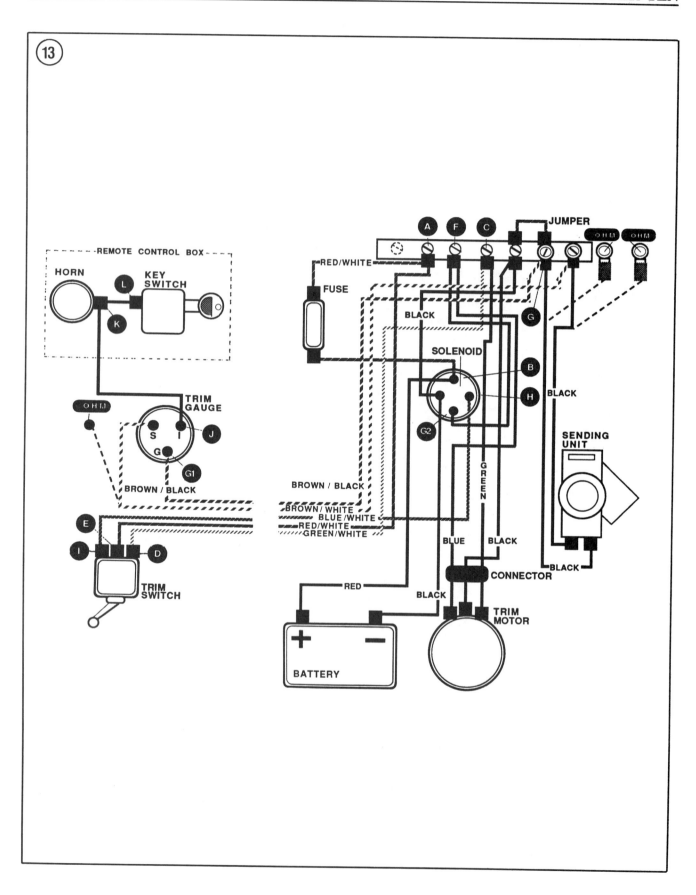

POWER TRIM AND TILT SYSTEMS

10. If no voltage is shown in Step 9, move the red test lead to point I and the black test lead to the indicator gauge G1 terminal. Move the trim switch to the UP position. If voltage is shown, check for an open between points H and I.

11. If no voltage is shown in Step 10, move the red test lead to point E. Replace the trim switch if voltage is shown; check for an open in the ground lead between the junction box and indicator gauge if no voltage is shown.

12. If no voltage is shown in Step 11 and the lead between the indicator gauge and junction box ground terminals is good, move the red test lead to point A. If voltage is shown, there is an open in the lead between points A and E.

Trim Indicator Circuit Test
(1976-1979)

Refer to **Figure 13** for this procedure.
1. Turn the key switch ON.
2. Connect the voltmeter red test lead to point J on the indicator gauge. Connect the black test lead to the indicator gauge G1 terminal. If battery voltage is shown, proceed with Step 5.
3. If battery voltage is not shown in Step 2, check for an open in the lead between the solenoid G2 terminal and the remote control box. If the circuit is good, move the red test lead to point K (remote control warning horn). If voltage is shown, there is an open in the lead between points J and K.
4. If no voltage is shown in Step 3, move the red test lead to point L. If no voltage is shown at point L, look for a bad fuse or open in the key switch wiring. If the fuse and wiring are good, replace the switch.
5. Turn the key switch OFF.
6. Disconnect the tan/white wire at the indicator gauge S terminal. With an ohmmeter on the low scale, connect the red test lead to the disconnected tan/white wire and the black test lead to the indicator gauge G1 terminal. Move the trim switch to the UP position and run the engine to the full trim in (down) and then the full tilt (up) positions. The ohmmeter should read 82-88 ohms (full trim) and 0-10 ohms (full tilt).
7. If the resistance readings are not as specified in Step 6, check for an open circuit in the ground wire between the indicator gauge and the junction box or in the tan/white wire disconnected from the indicator gauge S terminal.
8. If the wiring is good, disconnect the black sending leads at the terminal strip. Connect an ohmmeter between the disconnected leads and check the resistance of the sending unit. If not within specifications in Step 6, replace the sending unit.

Trim and Tilt Switch Circuit Test
(All 1980 and 1981 CIH)

Refer to **Figure 14** for this procedure.
1. With the key switch ON, connect the voltmeter red test lead to point E1 in the junction box. Connect the black test lead to point G in the junction box. The meter should read battery voltage.
2. If battery voltage is not shown in Step 1, move the red test lead to point E2. If battery voltage is shown, check the junction box fuse. If no voltage is shown, look for poor connections or an open circuit in the wiring between the junction box and the battery.
3. Move the red test lead to point E3 in the junction box. Move the trim switch to the DOWN position. If battery voltage is shown and the trim motor does not run, test the motor as described in this chapter. If no voltage is shown, the problem is in the trim switch or connector leads.
4. Connect the voltmeter red test lead to the junction box at point E4. Connect the black test lead to the junction box G terminal. Move the trim switch to the UP position. If the motor does not run, test it as described in this chapter.
5. If battery voltage is not shown in Step 4, move the red test lead to point E5 on the solenoid. Move the trim switch to the UP position. There should be battery voltage.

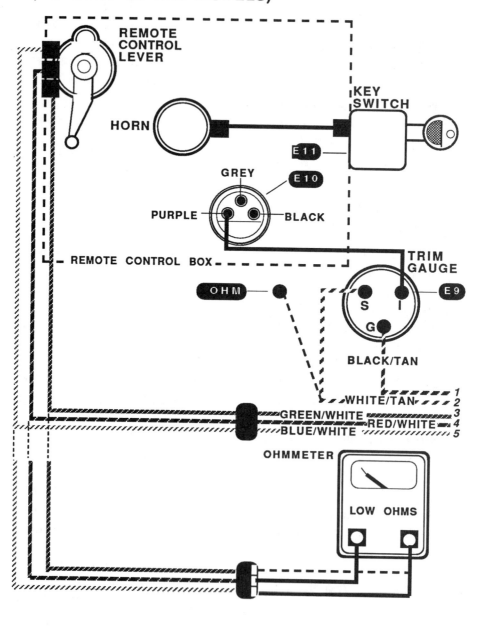

POWER TRIM AND TILT SYSTEMS

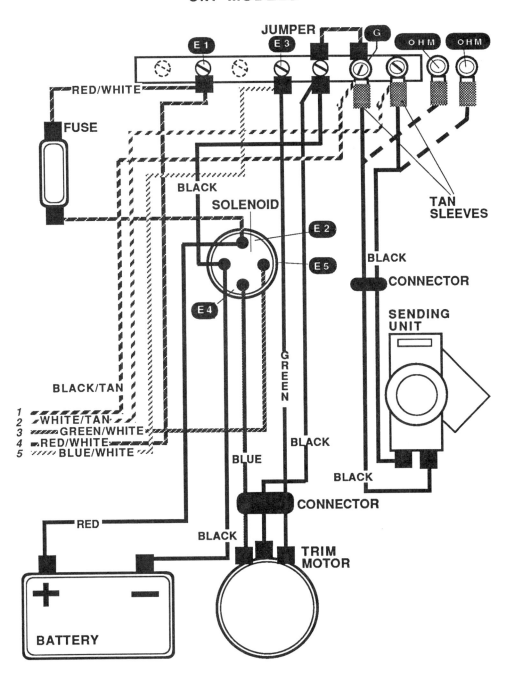

6. If voltage is shown in Step 5 and the motor still does not run, move the red test lead to point E2 on the solenoid. If voltage is shown, replace the solenoid.

7. If no voltage is shown in Step 6, the problem is in the trim switch or the connector leads.

Trim Indicator Circuit Test (All 1980 and 1981 CIH)

Refer to **Figure 14** for this procedure.
1. Turn the key switch ON.
2. Connect the voltmeter red test lead to point E9 on the indicator gauge. Connect the black test lead to the indicator gauge G terminal. If battery voltage is shown, proceed with Step 5.
3. If battery voltage is not shown in Step 2, move the red test lead to point E10 (accessory plug). If voltage is shown, there is an open in the circuit between the accessory plug and the trim indicator gauge.
4. If there is no voltage in Step 3, move the red test lead to point E11 (key switch accessory terminal). If no voltage is shown at point E11, look for a bad fuse or open in the key switch wiring. If the fuse and wiring are good, replace the switch.
5. Turn the key switch OFF.
6. Disconnect the tan/white wire at the indicator gauge S terminal. With an ohmmeter on the low scale, connect the red test lead to the disconnected tan/white wire and the black test lead to the indicator gauge G terminal. Move the trim switch to the UP position and run the engine to the full trim and then the full tilt positions. The ohmmeter should read 82-88 ohms (full trim) and 0-10 ohms (full tilt).
7. If the resistance readings are not as specified in Step 6, check for an open circuit in the ground wire between the indicator gauge and the junction box or in the tan/white wire disconnected from the indicator gauge S terminal.
8. If the wiring is good, disconnect the black sending leads at the terminal strip. Connect an ohmmeter between the disconnected leads and check the resistance of the sending unit. If not within specifications in Step 6, replace the sending unit.

Trim and Tilt Switch Circuit Test (1981 CIM, CIA and CIB Models; All 1982-1983)

Refer to **Figure 15** for this procedure.
1. With the key switch ON, connect the voltmeter red test lead to point E1 in the junction box. Connect the black test lead to point G in the junction box. The meter should read battery voltage.
2. If battery voltage is not shown in Step 1, move the red test lead to point E2 on the terminal strip. If battery voltage is shown, check the junction box fuse. If no voltage is shown, look for poor connections or an open circuit in the wiring between the junction box and the battery.
3. Move the red test lead to point E3 in the junction box. Move the trim switch to the DOWN position. If battery voltage is shown, move the red test lead to points E4 and E5. With the trim switch in the DOWN position, there should be voltage at each point. If voltage is shown and the trim motor does not run, probe connector C2 at the trim motor. If voltage is shown at the connector and the motor does not run, test the motor as described in this chapter.
4. If no voltage is shown at point E3 in Step 3, check connector C1 between the junction box and remote control box. If the connector is good, check for an open circuit in the following:
   a. Green/white wire between trim switch and junction box.
   b. Trim switch between the green/white and red/white leads with the switch in the DOWN position.
   c. Red/white wire between trim switch and point E1 on the terminal strip.
5. Move the voltmeter red test lead to the junction box at point E6. Move the trim switch to the UP position. If voltage is shown, move

POWER TRIM AND TILT SYSTEMS

the red test lead to point E7 (trim switch DOWN) and then point E8 (trim switch UP). If voltage is shown at each point, check connector C2 at the trim motor. If voltage is shown at the connector and the motor still does not run, test it as described in this chapter.

6. If there is no voltage at point E6, check connector C1 between the junction box and remote control box. If the connector is good, check for an open circuit in the following:
 a. Blue/white wire between the junction box and remote control box.
 b. Trim switch between the blue/white and red/white leads with the switch in the up position.
 c. Red/white wire between trim switch and point E1 on the terminal strip.

Trim Indicator Circuit Test
(1981 CIM, CIA and CIB Models; All 1982-1983)

Refer to **Figure 15** for this procedure.
1. Turn the key switch ON.
2. Connect the voltmeter red test lead to point E9 on the indicator gauge. Connect the black test lead to the indicator gauge G terminal. If battery voltage is shown, proceed with Step 5.
3. If battery voltage is not shown in Step 2, move the red test lead to point E10 (accessory plug). If voltage is shown, there is an open in the circuit between the accessory plug and the trim indicator gauge.
4. If there is no voltage in Step 3, move the red test lead to point E11 (key switch accessory terminal). If no voltage is shown at point E11, look for a bad fuse or open in the key switch wiring. If the fuse and wiring are good, replace the switch.
5. Turn the key switch OFF.
6. Disconnect the white/tan wire at the indicator gauge S terminal. With an ohmmeter on the low scale, connect the red test lead to the disconnected tan/white wire and the black test lead to the indicator gauge G terminal. Move the trim switch to the UP position and run the engine to the full trim in (down) and then the full tilt (up) positions. The ohmmeter should read 82-88 ohms (full trim) and 0-10 ohms (full tilt).
7. If the resistance readings are not as specified in Step 6, check for an open circuit in the ground wire between the indicator gauge and the junction box or in the white/tan wire disconnected from the indicator gauge S terminal.
8. If the wiring is good, disconnect the black sending leads at the terminal strip. Connect an ohmmeter between the disconnected sender leads and check the resistance of the sending unit. If not within specifications in Step 6, replace the sending unit.

Trim and Tilt Switch
Circuit Test (1984-on)

Refer to **Figure 16** (fused) or **Figure 17** (non-fused) for this procedure.
1. With the key switch ON, connect the voltmeter red test lead to point E1 in the junction box. Connect the black test lead to point G in the junction box. The meter should read battery voltage.
2. If battery voltage is not shown in Step 1, move the red test lead to point E2 on the starter solenoid. If battery voltage is shown, check the red lead and connections to the junction box (non-fused) or the junction box fuse. If no voltage is shown, look for poor connections or an open circuit in the wiring between the junction box and the battery.
3. Move the red test lead to point E3 in the junction box. Move the trim switch to the DOWN position. If battery voltage is shown, move the red test lead to points E4 and E5. With the trim switch in the DOWN position, there should be voltage at each point. If voltage is shown and the trim motor does not run, probe connector C2 at the trim motor. If voltage is shown at the connector and the motor does not run, test the motor as described in this chapter.

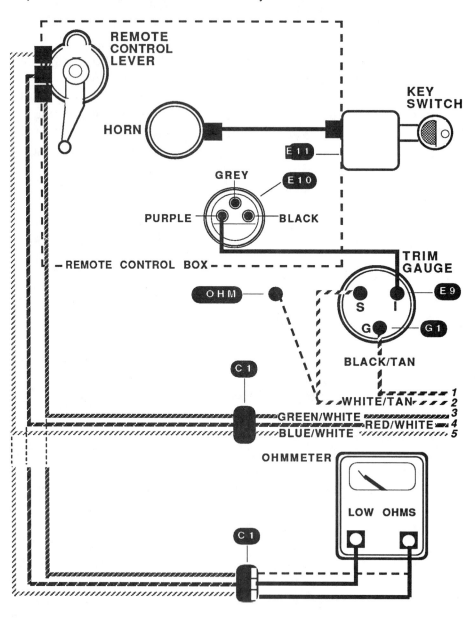

POWER TRIM AND TILT SYSTEMS

439

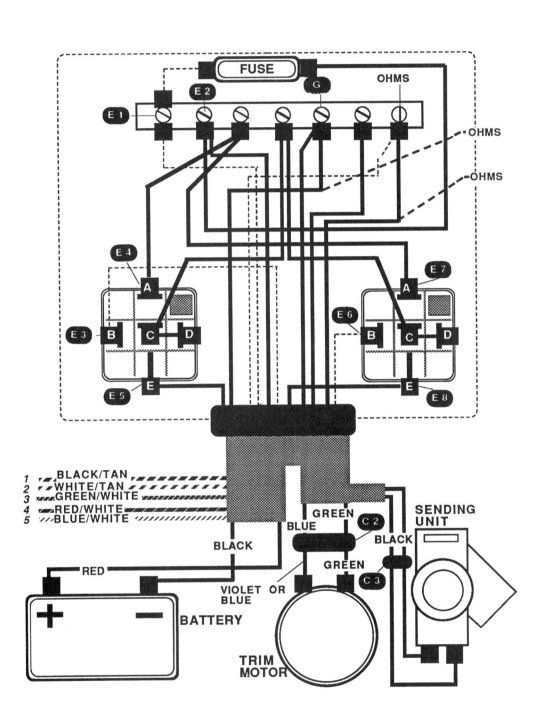

CHAPTER TEN

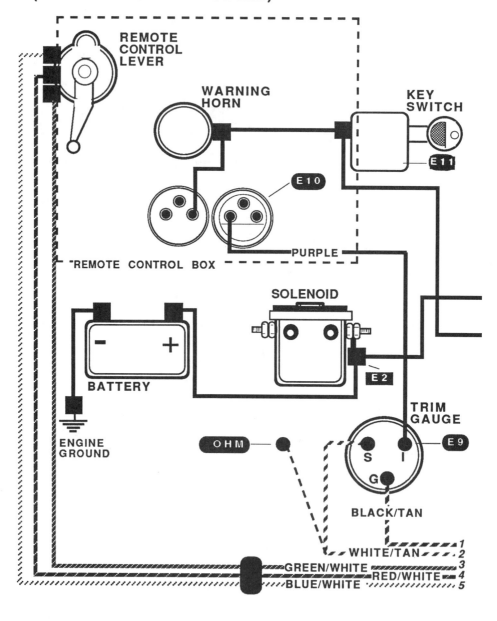

POWER TRIM AND TILT SYSTEMS

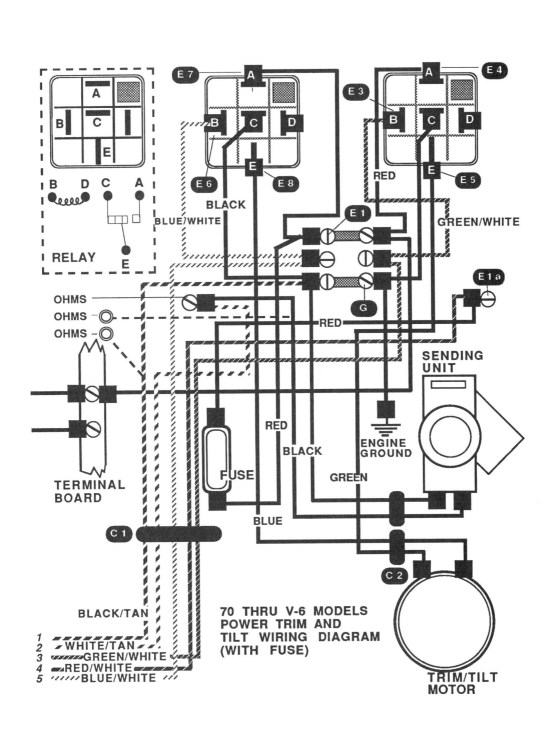

70 THRU V-6 MODELS POWER TRIM AND TILT WIRING DIAGRAM (WITH FUSE)

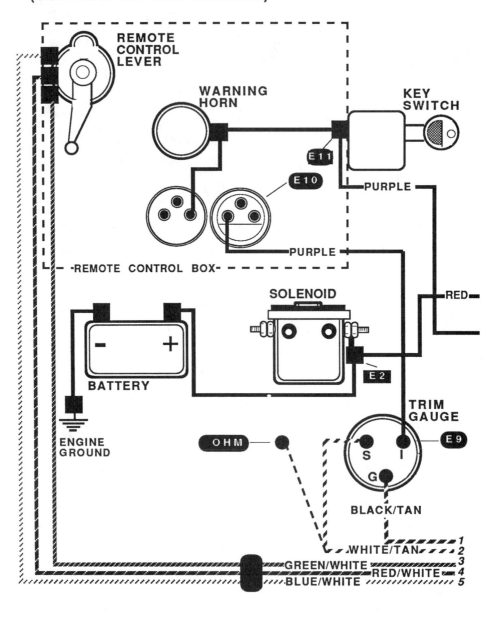

POWER TRIM AND TILT SYSTEMS

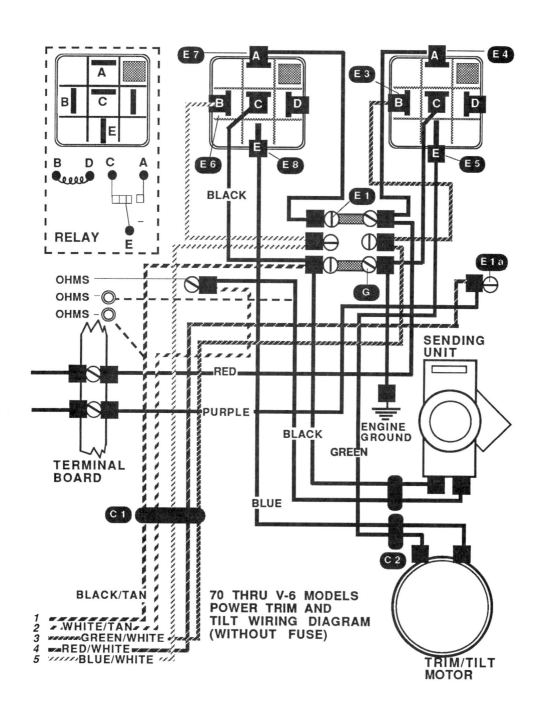

4. If no voltage is shown at point E3 in Step 3, check connector C1 between the junction box and remote control box. If the connector is good, check for an open circuit in the following:

   a. Green/white wire between trim switch and junction box.
   b. Trim switch between the green/white and red/white leads with the switch in the DOWN position.
   c. Red/white wire between trim switch and point E1 on the terminal strip (disconnect connector C1).

5. On non-fused models, check for an open in the the purple lead between the key switch and point E1A with an ohmmeter:

   a. If the wire checks out good, test key switch with a voltmeter.
   b. If there is no voltage at the key switch, check the switch continuity with an ohmmeter.
   c. If the switch is good, check the engine fuse and lead to the key switch.

6. Move the voltmeter red test lead to the junction box at point E6. Move the trim switch to the UP position. If voltage is shown, move the red test lead to point E7 (trim switch DOWN) and then point E8 (trim switch UP). If voltage is shown at each point, check connector C2 at the trim motor. If voltage is shown at the connector and the motor still does not run, test it as described in this chapter.

7. If there is no voltage at point E6, check connector C1 between the junction box and remote control box. If the connector is good, check for an open circuit in the following:

   a. Blue/white wire between the junction box and remote control box.
   b. Trim switch between the blue/white and red/white leads with the switch in the UP position.
   c. Red/white wire between trim switch and point E1 on the terminal strip (connector C1 disconnected).

8. On non-fused models, repeat Step 5.

**Trim Indicator
Circuit Test (1984-on)**

Refer to **Figure 16** (non-fused) or **Figure 17** (fused) for this procedure.

1. Turn the key switch ON.
2. Connect the voltmeter red test lead to point E9 on the indicator gauge. Connect the black test lead to the indicator gauge G terminal. If battery voltage is shown, proceed with Step 5.
3. If battery voltage is not shown in Step 2, move the red test lead to point E10 (accessory plug). If voltage is shown, there is an open in the circuit between the accessory plug and the trim indicator gauge.
4. If there is no voltage in Step 3, move the red test lead to point E11 (key switch accessory terminal). If no voltage is shown at point E11, look for a bad fuse or open in the key switch wiring. If the fuse and wiring are good, replace the switch.
5. Turn the key switch OFF.
6. Disconnect the white/tan and black/tan wires from gauge terminals S and G, respectively. Calibrate an ohmmeter on the low-ohm scale and connect the meter between the white/tan and black/tan wires.
7. Tilt the outboard to the fully UP position and note the meter reading. Then, trim the outboard to the fully DOWN/IN position and note the meter.
8. With the outboard tilted fully UP, the ohmmeter should indicate approximately 1 ohm. With the outboard fully DOWN/IN, the meter should indicate approximately 88 ohms.
9. If resistance is as specified in Step 8, replace the trim gauge. If resistance is not as specified (Step 8), disconnect the trim indicator sending unit 2-pin connector. Connect the ohmmeter between the terminals in the sending unit connector half and repeat Step 7.

POWER TRIM AND TILT SYSTEMS

10. If the resistance is now as specified in Step 8, repair or replace the white/tan and/or black/tan wires between the sending unit and gauge. If resistance is still not as specified in Step 8, replace the sending unit.

TRIM AND TILT MOTOR TESTING

Amperage Test

This test evaluates the pump output pressure by determining the current requirements of the pump motor. Readings must be taken with the boat at rest.

1. Connect an ammeter between the positive battery terminal and the heavy red cable at the trim/tilt solenoid.
2. Move the trim switch to the UP position and note the meter reading, then move the trim switch to the DOWN position and note that reading.
3. Compare the readings obtained in Step 2 to **Table 1**:
   a. If the current draw is within specifications, the motor, hydraulic pump and relief valves are working properly.
   b. A low current draw indicates an internal leak in the pump.
   c. A high current draw in either the UP or DOWN position indicates a problem in the motor or pump. Remove the motor and perform a no-load current draw test as described in this chapter.

No-load Current Draw Test

This procedure can be used with any trim and tilt motor. It requires the use of a volt-ammeter with a variable resistance.

1. Remove the motor from the trim/tilt system as described in this chapter.
2. Connect a volt-ammeter to the motor and a fully charged battery according to manufacturer's instructions.
3. With the variable resistance at a maximum, note the voltage reading.

4. Disconnect the motor from the battery and reduce the resistance of the rheostat until the voltmeter reading in Step 3 is obtained.
5. Read the ammeter scale and compare to **Table 2**. If the current draw is not within specifications, repair or replace the motor.

SYSTEM REMOVAL/ INSTALLATION

The entire system must be removed on both external and internal power trim and tilt models in order to service the motor or the trim/tilt cylinders.

External System

1. Disconnect the trim motor electrical connector.
2. Remove the nuts and bolts holding the trim and tilt cylinder piston eyes to the motor.
3. Remove the bolts holding the power trim and tilt unit to the motor and transom.
4. Remove the trim and tilt unit.
5. Installation is the reverse of removal. Tighten all fasteners securely.

Internal System (1976-1980)

1. Remove the motor from the boat.
2. Hold the tilt tube bolt head with one wrench and remove the tilt tube nut.
3. Mark the thrust rod location and remove the rod.

NOTE
Corrosion affects the stern bracket screws on these models when the boat is used in salt water and the motor or boat has a sacrificial anode. Corrosion-resistant screws part No. 318612 and part No. 552968 should be installed on such applications during reinstallation.

4. Remove the screws on each side holding the power trim manifold to the stern brackets. Remove the stern brackets.

5. Remove the tilt cylinder pin spring clip (**Figure 18**). Support the power trim manifold weight and remove the tilt cylinder pin with a punch and hammer.

6. Remove the power trim manifold.

7. Installation is the reverse of removal. Lubricate the tilt cylinder pin with OMC Triple-Guard grease. Tighten the tilt tube nut to 24-26 ft.-lb.

Internal System (1981-on)

1. Mark the angle adjusting rod location. Remove the rod.

2. Lift the engine by hand and engage the tilt trail lock to hold the engine up.

3. Remove the tilt cylinder pin spring clip (**Figure 18**). Remove the tilt cylinder pin with a punch and hammer and retract the cylinder manually.

NOTE
Corrosion affects the stern bracket screws on these models when the boat is used in salt water and the motor or boat has a sacrificial anode. Corrosion-resistant screws part No. 318612 and part No. 552968 should be installed on such applications during reinstallation.

4. Remove the screws on each side holding the power trim manifold to the stern brackets.

5. Angle the power trim manifold down and out of the stern brackets, pulling the control cable through the bracket.

6. Installation is the reverse of removal. Lubricate the tilt cylinder pin with OMC Triple-Guard grease. Tighten all fasteners securely.

POWER TRIM AND TILT SYSTEMS

Table 1 CURRENT DRAW (AMPS) AND DURATION (UNDER LOAD CONDITIONS)

Unit movement	Prestolite Part No. 393259[1]	Prestolite Part No. 393988[2] & 394176	Bosch Part No. 393988 & 394176	Showa (3 in. Dia.) Part No. 394176	Showa (2.4 in. Dia.) Part No. 394176
Trimming up	20-25	11-15	7-10	5-8	7-9
Tilting up	20-25	11-15	7-10	8-10	9-12
Full tilt up (stall)	40-44	30-35	30-35	19-23	25-29
Tilting down	18-22	12-16	10-13	—	—
Trimming down	18-22	12-16	10-13	—	—
Full trim down (stall)	28-32	21-25	18-32	14-18	12-17
Time in Seconds					
Trimming up	5-7	7-9	7-9	8-10	8-10
Tilting up	5-7	7-9	6-9	7-9	6-7
Tilting down	5-6	8-11	8-10	—	—
Trimming down	6-8	7-10	6-8	—	—
Trim in to full tilt up	16-17	15-20	14-16	16-18	14-16
Full tilt up to trim in	16-17	15-20	15-20	16-18	15-17

1. Code dated 8-R or before.
2. Code dated 9-R or after.

Table 2 TRIM AND TILT MOTOR NO-LOAD CURRENT DRAW

External system	
Calco pump	20 amps
Prestolite pump	18 amps
Internal system	
Prestolite	
Part No. 393259	7-9 amps
Part No. 393988 and 394176 (Prior to 1988)	6-8 amps
1988-1989	7 amps
Bosch	
Prior to 1988	6-8 amps
1988-1989	4.5 amps
Showa	4.5 amps

Chapter Eleven

Oil Injection Systems

The fuel-oil ratio required by outboard motors depends upon engine demand. Without oil injection, oil must be hand-mixed at a 50:1 ratio to assure that sufficient lubrication is provided at all operating speeds and engine load conditions. With oil injection, however, the ratio of oil provided with the fuel sent to the engine cylinders can be varied instantly and accurately to provide the optimum ratio for proper lubrication at any operating speed or engine load condition.

Late model Johnson and Evinrude outboards may be equipped with one of two oil injection systems. The OMC Economixer is an electronically controlled oil injection system that can be installed as an option on all 1980-1983 V4 and V6 models. Variable ratio oil (VRO) is a factory-installed automatic oil injection system on 1984 and later V4 and V6 models and 1985-on 2- and 3-cylinder models except 48 and 88 SPL models.

This chapter covers the operation and troubleshooting of the Economixer and VRO systems. **Table 1** is at the end of the chapter.

OMC ECONOMIXER

The OMC Economixer injection system consists of a remote mounted oil tank containing a pump and microprocessor unit, a throttle position sender, rpm sender and a warning gauge. **Figure 1** shows the major components of the system. **Figure 2** shows the power supply and charging circuits. **Figure 3** shows the ignition circuit.

Operation

The microprocessor in the oil tank receives voltage signals from a throttle position sender at the carburetors and a lead wire in the alternator system. The microprocessor compares the signals to its program and determines the operating rate of the tank-mounted oil pump. The pump feeds Johnson or Evinrude Outboard Lubricant into the fuel system at a ratio of approximately 150:1 during idle, approximately 80:1 during cruising and 50:1 at high speed.

When the key switch is turned ON, an alarm horn in the instrument panel gauge sounds for

OIL INJECTION SYSTEMS

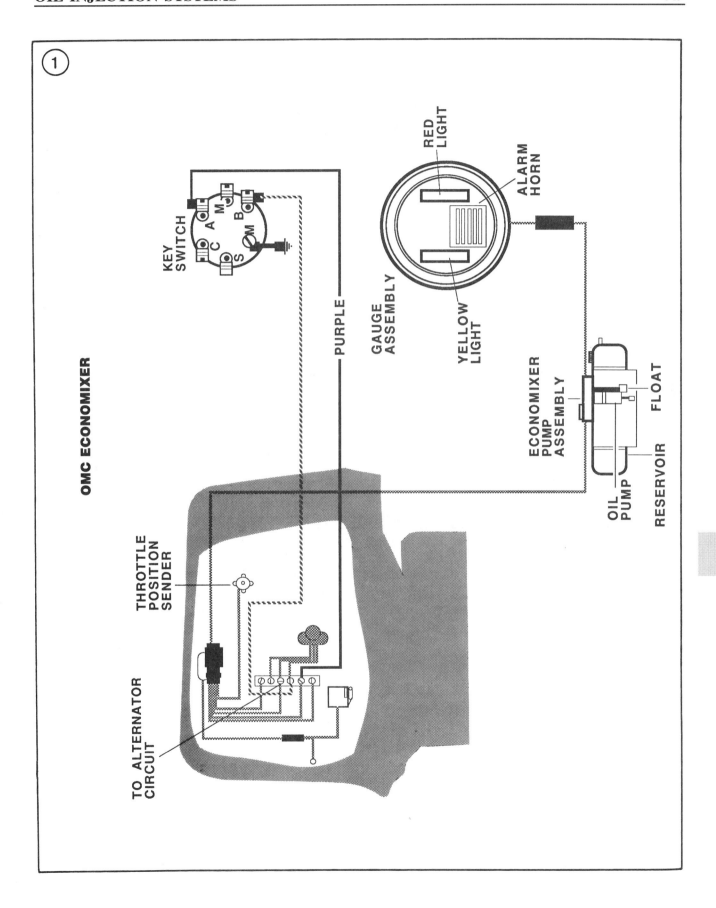

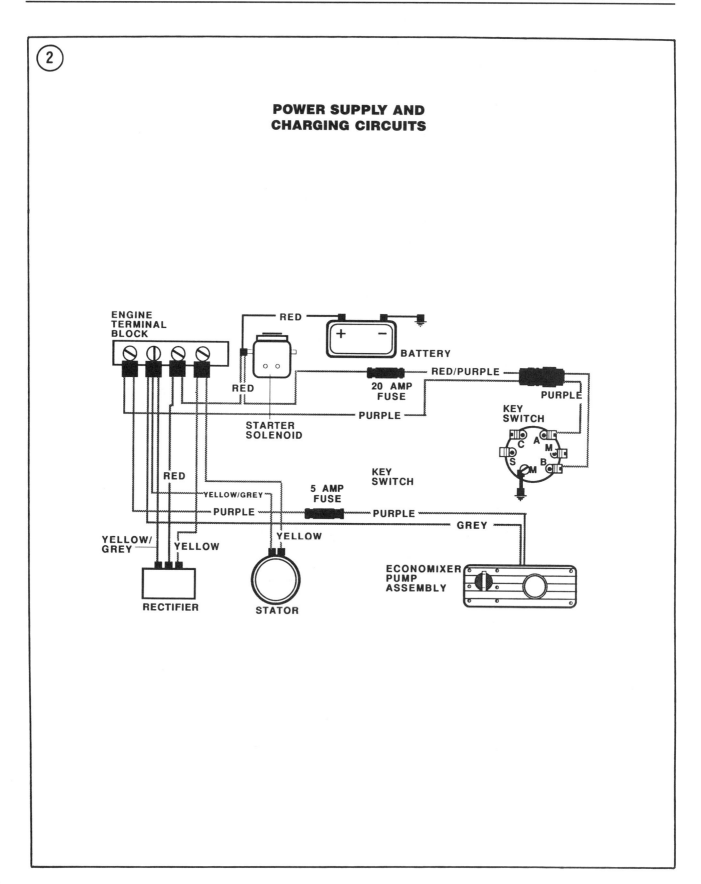

OIL INJECTION SYSTEMS

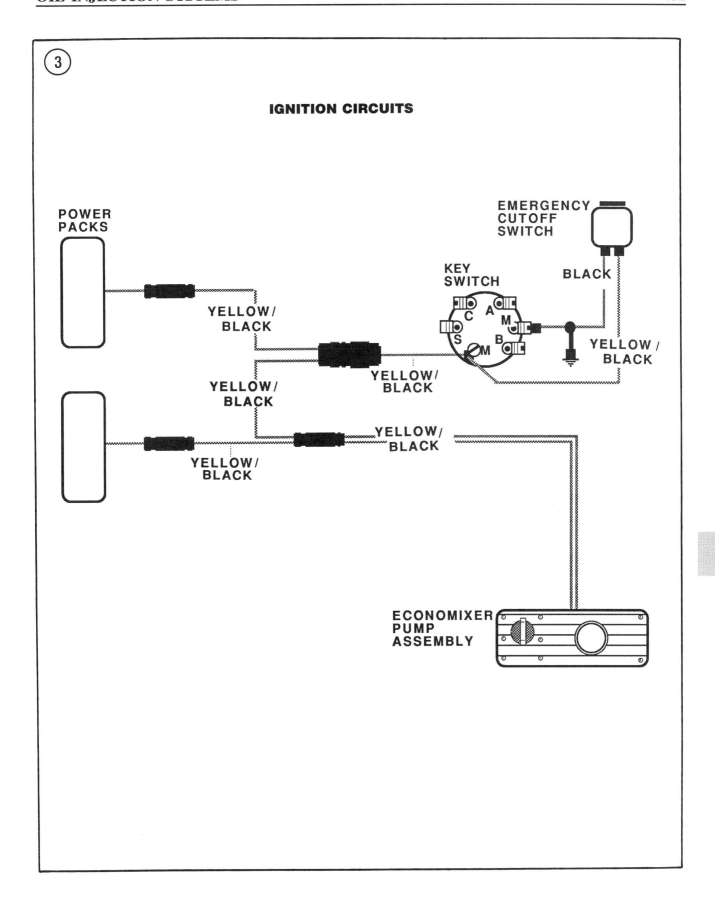

one second and a red light in the gauge flashes. This audible and visual signal sequence indicates the Economixer microprocessor is checking its system operation. When the alarm goes off and the light goes out, the pump primes itself and the system is ready to function.

A yellow light in the gauge (operated by the oil tank float) comes on whenever the oil level in the tank has dropped below the one-third full mark. Refilling the tank turns the light off.

If the operator ignores the yellow warning light and runs the engine until the reservoir is empty, the red light will flash and the horn will sound for 30 seconds, then the system will automatically shut the engine down to prevent damage. The engine can be restarted immediately and will run for another 30 seconds. At that time, the warning sequence will repeat itself and shut the engine down.

The Economixer system requires a minimum battery voltage of 11.5 volts to function. Voltage in excess of 16 volts will cause the system to shut down.

Fuel and Oil Requirements

As with all other Johnson and Evinrude outboards, engines equipped with the OMC Economixer require the use of the proper fuel (see Chapter Four) and Johnson or Evinrude Outboard Lubricant. If Johnson or Evinrude Outboard Lubricant is not available, another BIA certified TC-W oil may be used. Oil additives should *not* be used with the Economixer system.

The advantage of the Economixer system is that fuel and oil do not have to be mixed before using. Keep the fuel tank filled with the recommended gasoline and the Economixer tank filled with Johnson or Evinrude Outboard Lubricant. The system will automatically mix the according to engine requirements.

Economixer Programming

As an accessory system, the Economixer must be properly programmed for the engine with which it will be used. A plastic cap on the Economixer microprocessor covers 2 small wires (programming leads) with sockets on their ends and 4 terminal pins within a circular area of potting compound.

WARNING
Connecting the programming leads to the wrong terminal pins can result in serious engine damage.

To program or check the programming of an Economixer unit, proceed as follows:

1. Insert a small screwdriver blade under the edge of the plastic cover on the microprocessor. Lift the cover up slightly and rotate it with a pulling motion.
2. Refer to **Figure 4** and make sure that the programming leads are connected to the terminal pins according to engine designation. All V4 engines use the same connection pattern; V6 engines have different connection patterns according to engine size.
3. Check the programming lead sleeve and make sure that it is flush with the socket terminal and that the socket terminal is fully seated on its pin. A good connection at this point is essential to proper Economixer operation.
4. Reinstall the plastic cover and make sure it is fully seated to keep the potted compound area free of moisture and contamination.

Troubleshooting

An Economixer tester SE-81 is available from Johnson and Evinrude dealers. This tester will determine immediately whether the problem is in the Economixer system or in the engine.

The procedures described in this chapter require special test equipment. Johnson and Evinrude recommend the use of a Stevens or Electro-Specialties CD voltmeter and CD adapter.

When performing any of the procedures which require the engine to be cranked or started, the

OIL INJECTION SYSTEMS

engine should be on the boat in the water or in a test tank with the proper test wheel installed.

Make sure battery voltage is between 11.5-16 volts before performing any of the procedures. If it is not, charge the battery or the test results will be misleading.

Preliminary Procedure

Before performing any of the tests described in this chapter, check the following and correct as required.

1. Check the 5 amp (starboard) and 20 amp (port) fuse connectors for corroded or loose terminal connections. Make sure the fuses are good.
2. Check the spark plug leads. Some models were manufactured without resistor spark plug leads. If the leads are reddish-brown in color, replace them with part No. 174134.
3. Check the battery cable terminal connections at the battery and the starter solenoid. Clean and tighten as required.
4. Make sure the Economixer is properly programmed and the programming lead connections make good contact as described in this chapter.
5. Check the oil reservoir level. Add Johnson or Evinrude Outboard Lubricant if the reservoir is less than one-third full.

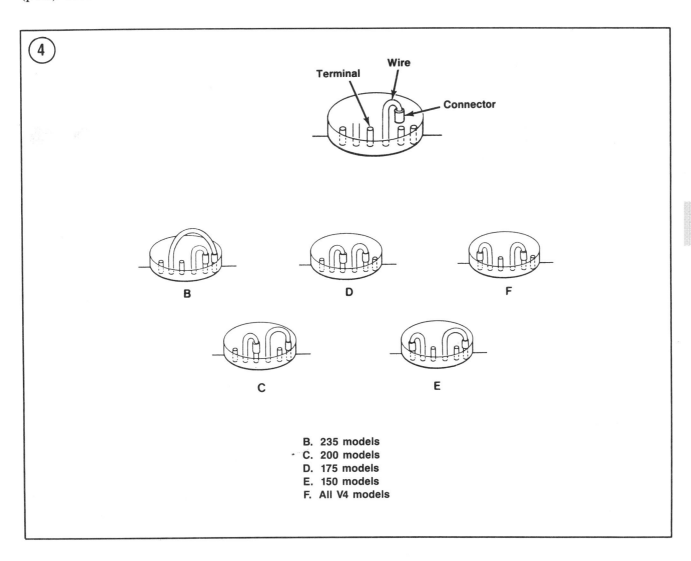

B. 235 models
C. 200 models
D. 175 models
E. 150 models
F. All V4 models

6. Check the Economixer ground connections at the terminal board (starboard side of engine). Clean and tighten as required.

7. Check the throttle sender adjustment. The sender lever should just touch the upper stop when the throttle is fully open. If it does not, loosen the linkage clamp screw and reposition the link, then tighten the screw. Close the throttle and reopen it to check adjustment. The lever should not bind against the stop.

Engine Fuel System Test

1. Disconnect the 4-wire Economixer harness connector.
2. Disconnect and cap the Economixer fuel line.
3. Connect the engine to a 6 gallon portable fuel tank filled with a 50:1 fuel-oil mixture.
4. Start the engine and run at idle. If the engine idles properly, gradually open the throttle and check running quality. If satisfactory, the problem is in the Economixer.
5. If running quality is not satisfactory in Step 4, check the carburetors for dirt or restrictions (rough running) or the fuel system for restrictions (surging). For further fuel system troubleshooting, see Chapter Three.

RPM Signal Test

1. Disconnect the 4-wire Economixer harness receptacle and plug on the port side of the engine.
2. Connect the CD adapter to the harness receptacle and turn the adapter ON.
3. With the CD voltmeter switch on POSITIVE and the meter set to the 50 volt scale, connect the black test lead to a good engine ground and the red test lead to the adapter C terminal.
4. Start the motor and note the meter reading. If it is approximately 12 volts, the rpm signal is satisfactory. A lower voltage indicates a partially discharged battery or a problem in the electrical system. See Chapter Three.

Rectifier Test

If the system passes the rpm signal test but does not operate satisfactorily, the rectifier may be at fault. Since a rectifier can pass the ohmmeter test (Chapter Three) and still not work properly with an Economixer system, Johnson and Evinrude recommend that a known-good rectifier (part No. 582399) be substituted. If the rectifer is not at fault, this substitution will minimize diagnosis time.

Key Switch Test

See Chapter Three.

Instrument Cable Connector Check

1. Disconnect the large red instrument cable connector on the port side of the engine.
2. Check the plug and receptacle sides of the connector for damaged or corroded pins or sockets.
3. Clean connectors by flushing with rubbing alcohol and blowing dry with compressed air.
4. Wipe outside of cable plug with OMC Triple-Guard grease. Align plug and receptacle arrows and press the two together firmly.

Ignition Signal Test

1. Disconnect the 2-wire Economixer receptacle and plug on the starboard side of the engine.
2. Connect the CD adapter to the harness receptacle and turn the adapter ON.
3. With the CD voltmeter switch on NEGATIVE and the meter set to the 500 volt scale, connect the black test lead to a good engine ground.
4. Crank or start the engine while alternatcly probing the CD adapter terminals A and B with the red test lead. If the meter reads approximately 300 volts, the ignition signal is satisfactory. A lower reading indicates a poor connection or problem in the ignition system. See Chapter Three.

OIL INJECTION SYSTEMS

Throttle Sender Test

The throttle sender may pass this test yet still be the cause of the problem due to an intermittent condition resulting from engine heat and vibration. When there is strong evidence that the sender is defective but it passes the static test, repeat the procedure with the engine running and connected to a portable fuel tank containing a 50:1 fuel-oil mixture.

1. Disconnect the 4-wire Economixer harness receptacle and plug on the port side of the engine.
2. Connect the CD adapter to the harness receptacle and turn the adapter ON.
3. Insert the CD jumper leads (included with CD voltmeter) in terminals A and D of the CD adapter.
4. Connect an ohmmeter between the CD jumper leads. With the throttle in the idle position, the meter should read 750-900 ohms.
5. Slowly open the throttle and note the meter reading. At wide open throttle, it should read 425-450 ohms. If the readings are unsteady as the throttle is opened, recheck the jumper lead and ohmmeter connections. If the connections are good and the reading is still unsteady (even though it is within specifications at wide-open throttle), replace the sender.

Battery Signal Test

1. Disconnect the 4-wire Economixer harness receptacle and plug on the port side of the engine.

2. Connect the CD adapter to the harness receptacle and turn the adapter ON.
3. With the CD voltmeter switch on POSITIVE and the meter set to the 50 volt scale, connect the black test lead to a good engine ground and the red test lead to the adapter B terminal.

NOTE
The CD voltmeter used in Step 4 may give a slightly low voltage reading. If the reading is 11-12 volts, recheck battery voltage with a standard voltmeter.

4. Turn the key switch ON. The meter should read 11.5-16 volts. A lower reading indicates poor connections or a partially discharged battery. A higher reading indicates an overcharged battery or an electrical system problem. See Chapter Three.

Oil Pump Test

1. Disconnect the Economixer wiring harness purple lead at terminal 5 of the terminal board.
2. Connect a low reading ammeter between the disconnected wire and terminal 5.
3. Start the engine and run at idle. The ammeter should read 3 amps discharge and return to zero each time the pump cycles. The V4 pump will require approximately one minute per pulse at idle.
4. If there is no discharge noted in Step 3, recheck all connections and repeat Step 3. If the ammeter discharges but does not return to zero, replace the Economixer oil pump.

OMC VRO SYSTEM

The variable ratio oiling (VRO) system is a factory-installed standard feature on 1984-on V4 and V6 models and 1985-on 2- and 3-cylinder models except 48 and 88 SPL models. This system uses a self-contained pump on the power head instead of the conventional fuel pump. **Figure 5** shows location for V4 models and

Figure 6 shows location for all other models. A remote mounted oil tank with pump and primer bulb and a warning horn complete the basic system.

One design uses a spark or flame arrestor installed in the pulse hose leading to the VRO pump and clamped in position to prevent it from moving and causing damage to the pump. On these models, an oil inlet filter is installed in the oil tank pickup unit and an inline filter is installed in the transparent oil inlet hose to the VRO pump to protect the pump from contamination.

The other design has the arrestor installed in the pulse hose fitting. The arrestor should not be removed from the fitting; they are serviced as an assembly. These models have the oil inlet filter installed in the oil tank pickup unit and an inline filter canister (serviced as a complete assembly) installed in the fuel line on the power head.

A 2 version of the VRO pump was introduced in 1988. The new pump is designed to be alcohol resistant and to protect against damage from other harmful additives found in some gasoline blends.

Operation

The VRO system works on crankcase pressure in a manner similar to the conventional fuel pump, drawing fuel and oil from separate tanks and mixing them in the VRO pump at a ratio varying from 50:1 to 150:1 according to engine requirements. VRO-equipped models can also be run on a 50:1 fuel-oil mixture drawn from the fuel tank if the user premixes the fuel and oil as described in Chapter Four.

Break-in Procedure

All VRO models should be run on a 50:1 fuel-oil mixture from the fuel tank (see Chapter Four) *in addition* to the lubricant supplied by the VRO system. Mark the oil level on the translucent VRO remote oil tank and periodically check to make sure the system is working (oil level drops) before switching over to plain gasoline at the end of the 10-hour break-in period.

Warning Horn

A warning horn is installed in the accessory or remote control wiring harness. The horn has 2 functions on 2-cylinder, 3-cylinder and V4 VRO models produced prior to 1986 and 3 functions on all other VRO models.

The sending unit in the remote oil tank is connected to the warning horn on all models through the key switch and grounded to the engine. If the oil level in the tank drops below the ¼ full point, the warning horn sounds for ½ second every 20 seconds to alert the user to a low oil level. A no-oil warning feature is incorporated in 1986 and later circuits. If the system runs completely out of oil or oil flow to the pump is obstructed, the warning horn sounds for ½ second every second.

A temperature sending unit is installed in each cylinder head on all models and connected to the warning horn through the key switch to warn of an overheat condition. If the power head temperature exceeds 211° F, the horn sounds continuously. Backing off on the throttle will shut the horn off as soon as power head temperature reaches 175° F, unless a restricted engine water intake is causing the overheat condition. If the

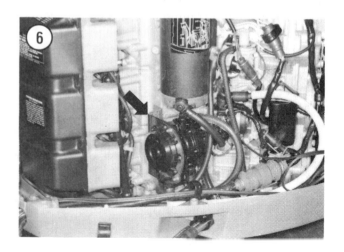

OIL INJECTION SYSTEMS

water pump indicator does not deliver a steady stream or if the horn continues sounding after 2 minutes, the engine should be shut off immediately to prevent power head damage.

CAUTION
If the engine overheats and the warning horn sounds, retorque the cylinder heads after the engine cools to minimize the possibility of power head damage from a blown head gasket.

V6 VRO models produced prior to 1986 also use the horn for a fuel warning system. A vacuum switch on the port side of the power head is connected to the warning horn through the key switch and grounded to the engine. A vacuum line to the fuel inlet on the VRO pump monitors vacuum in the fuel line. If the vacuum in the line reaches 7 in. Hg, the switch turns the warning horn on continuously to alert the user to a restricted fuel system.

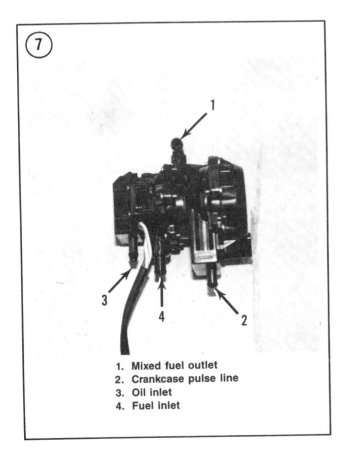

1. Mixed fuel outlet
2. Crankcase pulse line
3. Oil inlet
4. Fuel inlet

Warning Horn Test

The warning horn should be tested periodically to make sure it is functioning properly.

1. Locate the electrical wire between the warning horn and temperature switch. Move the insulating sleeve back to provide access to the disconnect point in the wire.
2. Turn the key switch ON and ground the disconnect point to the engine.
3. If the horn does not sound, check the wiring and horn.
4. Reposition the insulating sleeve over the disconnect point in the wire.

Troubleshooting

The VRO pump is sealed at the factory and is serviced by replacement if defective. Any attempt to disassemble the pump will void the factory warranty.

If the oil inlet hose is disconnected from the pump, it must be reinstalled with the same type of clamps as removed. The use of worm clamps will damage the vinyl hose while tie straps will not provide sufficient clamping pressure.

Fuel Pressure Test

Refer to **Figure 7** for inlet and outlet identification.

CAUTION
The pump nipples are plastic and can be broken if excess pressure is used in disconnecting or connecting the lines.

1. Disconnect the fuel outlet line. Install a tee in the end of the line and connect a 4 inch length of 5/16 in. ID hose to the tee.
2. Lubricate the pump outlet fitting with a drop of oil and connect the vinyl hose and tee assembly to the pump.
3. Connect a 0-15 psi pressure gauge to the tee.
4. Fasten all hose connections with tie straps or hose clamps.
5A. 1984-1986 Models—Start the engine and run at wide-open throttle in gear. The gauge should

read between 3-15 psi and drop to 1-2 psi (accompanied by a clicking sound) each time the pump discharges oil.

5B. *1987-on Models*—Start the engine and run approximately 800 rpm in gear. The gauge should read *no less than* 3 psi.

6. If no pressure is shown in Step 5:
   a. Check for fuel in the tank.
   b. Check for a pinched, kinked or restricted fuel line.
   c. Check for a pinched or leaking VRO pump pulse line (2, **Figure 7**).
7. If low pressure is shown in Step 5:
   a. Check fuel filter for restrictions.
   b. Check for a pinched or leaking VRO pump pulse line (2, **Figure 7**).
   c. Check for a pinched, kinked or restricted fuel line.
   d. Squeeze the fuel primer bulb several times to clear any possible vapor lock condition in the line.
8. If no pressure or low pressure is shown in Step 5 and the items in Step 6 or Step 7 are satisfactory, replace the VRO pump.

Oil Flow Test

1. Make sure there is sufficient oil in the VRO tank. Top up as required.
2. Connect a remote fuel tank containing the 50:1 fuel/oil mixture to the engine.
3. With the engine in a test tank or on the boat in the water, start the engine and note the flow through the transparent hose at the VRO pump.
4. If a reduced or no flow is noted in Step 3, shut the engine off. Check the oil pickup filter as described in this chapter.
5A. If filter is not clogged or obstructed, proceed to Step 6.
5B. If filter required cleaning or replacement, repeat Step 3 to see if this service restored full oil flow. If it did not, continue with Step 6.
6. Disconnect the oil hose at the inlet fitting on the lower engine cover. Have an assistant hold the hose in a suitable clean container.

7. Loosen oil hose clamp at pickup unit. Grasp pickup unit firmly to prevent it from moving on its support rods and disconnect the hose. Blow the line out with low-pressure compressed air.
8. Remove the oil pickup mounting screws with a T-25 Torx ® driver. Remove the pickup unit from the tank and let it drain into a suitable clean container.
9. Insert a suitable plug in the end of the hose attached to the pickup unit. Install a clamp to hold the plug in place.
10. Connect a Stevens gearcase vacuum tester or a hand vacuum pump to the oil hose at the lower engine cover end and install a clamp to secure the connection.
11. Draw approximately 7 in. Hg vacuum. If the system does not hold the vacuum, check the oil hose for damage. If the hose is good, apply oil at each connection while drawing a vacuum to determine the point of leakage. Correct as required.
12. If the system holds vacuum in Step 11, replace the VRO pump assembly.

Excessive Engine Smoke

Check fuel system and filters for restrictions. The engine may smoke on a cold start. This is normal for cold starting a 2-cycle engine. It will also smoke if the oil hose primer bulb is squeezed prior to starting the engine. This priming is unnecessary and loads the carburetors with an excessively rich mixture.

Low Oil Warning Sounds at 20 Second Intervals

If the oil level in the tank is satisfactory, the pickup unit disc/contacts are either out of position or dirty. Remove the pickup unit from the tank and clean the float chamber in fresh solvent. If the float does not raise the disc clear of the contacts, replace the pickup unit.

OIL INJECTION SYSTEMS

Pickup Unit and Filter Service

Other than filter replacement, the oil pickup unit is serviced as an assembly if it does not function properly.

1. Remove the oil pickup mounting screws with a T-25 Torx® driver. Remove the pickup unit from the tank and let it drain into a suitable clean container.
2. Note the position of the foam baffle (if so equipped) for reinstallation and remove from the pickup unit.
3. Pull the plastic filter assembly from the end of the pickup tube with needlenose pliers.
4. Clean filter in fresh solvent and blow dry with low-pressure compressed air, if available. Replace filter if damaged or badly clogged.
5. To reinstall filter, insert it in the plastic cap from a felt-tip marker. Marker cap should be large enough to hold filter but no larger than the outer diameter of the filter head.
6. Use the marker cap to press the filter into the pickup tube.
7. Reinstall the foam baffle (if so equipped) in the position noted in Step 2.
8. Reinstall pickup unit in oil tank and tighten the 4 retaining screws securely.

CAUTION
Failure to properly purge air from the system in Step 9 can result in serious engine damage caused by lack of proper lubrication.

9. Disconnect the oil hose at the inlet fitting on the lower engine cover. Hold the hose in a suitable clean container and squeeze the VRO primer bulb *50 (fifty)* times to purge any air from the line.
10. Reinstall the oil hose to the engine inlet fitting and tighten the clamp securely.

Table 1 ECONOMIXER TROUBLESHOOTING GUIDE

Symptom	Probable cause
Red light on and horn gives one second alarm	Normal operation.
Yellow light in gauge comes on	Low oil level in reservoir. Float mechanism is defective. Check gauge yellow light circuit. Pump assembly is defective. Short circuit in wire harness.
Yellow light, red light, 30 second horn alarm and engine shut-down	Low oil level in reservoir. Float mechanism is defective. Check gauge yellow light circuit. Short circuit in wire harness. Loose or missing pickup tube. Restricted pickup tube screen or hose. Pump assembly is defective.
Red light, 30 second horn and engine shut-down	Low oil level in reservoir. Defective key switch or wiring. Non-suppression ignition leads. Poor contact @ instrument cable connector. Poor connections @ battery or engine.
(continued)	

Table 1 ECONOMIXER TROUBLESHOOTING GUIDE (continued)

Symptom	Probable cause
Red light, 30 second horn and engine shutdown (continued)	Check gauge yellow light for defective bulb or circuit. Loose or missing pickup tube. Restricted pickup tube screen or hose. Pump assembly is defective. Battery is providing more than 16 volts or less than 11.5 volts. Defective battery.
Red light horn beeps but engine runs	Check rpm signal circuit. Defective throttle sender. Throttle sender leads open or shorted. Pump assembly is defective. Connector pins and sockets are corroded or not making good contact.
Engine runs, but not above 1,000 rpm	Check 20 amp fuse or fuse holder. Check 5 amp fuse or fuse holder. Defective key switch or wiring. Poor instrument cable connector contact. Poor connections @ battery or engine. Connector pins and sockets are corroded or not making good contact. Poor terminal board ground connections. Plugged or restricted Economixer outlet. Defective battery. Broken purple or battery leads.
Engine shut-down, no lights or alarm	Poor instrument cable connector contact. Non-suppression ignition leads. Defective pump assembly.
Engine will not start	Problem in engine fuel system. Check 20 amp fuse and fuse holder. Defective key switch or wiring. Poor instrument cable connector contact. Poor connections @ battery or engine. Defective pump assembly. Defective battery. Broken purple or battery leads. Battery is providing less than 11.5 volts.
Engine runs rough at slow speed with excess smoking	Problem in engine fuel system. Incorrect programming of Economixer. Throttle sender out of adjustment. Economixer programming lead socket not making good contact with pin. Defective pump assembly.
Fouled plugs, excess carbon, dry cylinders, piston and cylinder scoring	Problem in engine fuel system. Incorrect programming of Economixer. Economixer programming lead socket not making good contact with pin. Plugged or restricted Economixer outlet. Defective pump assembly.

Chapter Twelve

Sea Drives

OMC Sea Drives were introduced in 1982. The models available by year are as follows: 1982—2.5L and 2.6L; 1983—1.6L, 2.5L and 2.6L; 1984—1.6L (S-type), 2.5L and 2.6L (S-type) and 2.6L (ducted); 1985—1.6L (S-type), 2.5L and 2.6L (S-type) and 2.6L (ducted); 1986—1.6L and 2.6L; 1987—1.6L, 1.8L, 2.6L and 2.7L; 1988—1.6L, 1.8L, 2.6L and 2.7L; 1989—1.6L, 2.0L and 3.0L; 1990—1.6L, 2.0L and 3.0L. Service on the outboard motor assembly of the Sea Drive is the same as a standard outboard motor. The outboard motor assembly on a 1.6L Sea Drive is in relation to a V-4 cross-flow outboard motor (110-115 hp). The outboard motor assembly on a 1.8L Sea drive and a 2.0L Sea Drive is in relation to a V-4 loop-charged outboard motor (140 hp). The outboard motor assembly on a 2.5L Sea Drive is in relation to a V-6 cross-flow outboard motor (175-185 hp). The outboard motor assembly on a 2.6L Sea Drive is in relation to a V-6 cross-flow outboard motor (200 hp). The outboard motor assembly on a 2.7L Sea Drive and a 3.0L Sea Drive is in relation to a V-6 loop-charged outboard motor (225 hp). Only service that is unique to the Sea Drive will be covered in this chapter. For service that is related to the outboard motor portion of the Sea Drive assembly, refer to the respective chapter covering desired maintenance or service information. Note model year, engine design and/or related outboard motor horsepower when determining maintenance or service information applicable to your year and model of Sea Drive.

SELECTRIM/TILT™ (1983-1990 1.6L)

Components

The unit consists of a valve body assembly, oil pump, electric motor, trim cylinder and combination tilt cylinder/shock absorber mounted on an anchor bracket. A rubber hose connects the anchor bracket reservoir to the remote reservoir.

Operation

Moving the trim/tilt switch to the UP position closes the pump motor circuit. The motor drives the oil pump, forcing oil into the UP side of the trim and tilt cylinders. Because of system design, the trim cylinder functions first and moves the outboard motor upward the first 0° to 17°. At 17°, the trim cylinder is fully extended and the hydraulic fluid is diverted into the tilt cylinder, which moves the outboard motor through the final 45° of travel.

Moving the trim/tilt switch to the DOWN position also closes the pump motor circuit. The reversible motor runs in the opposite direction, forcing oil into the tilt cylinder and bringing the outboard motor back to the 17° position, where the trim cylinder lowers the outboard motor the remainder of the way.

The power tilt will temporarily maintain the engine at any angle within its range to allow shallow water operation at slow speed, launching, beaching or trailering. At about 3,000 rpm, however, the trim-up relief valve will open and automatically lower the engine to its fully trimmed out (17°) position.

A manual release valve with a slotted head permits manual raising and lowering of the outboard motor through the tilt range (17-45°).

A trim gauge sending unit is mounted on a bracket located on the inside of the outboard motor mounting brackets. The sending unit uses a calibrated lever positioned against the trim cylinder to calculate trim angle. Access to the sending unit requires the outboard motor to be fully tilted.

Hydraulic Pump Fluid Check

1. Tilt the outboard motor to its full UP position.
2. Check the fluid level in the remote reservoir (**Figure 1**, typical). The fluid should be level with the reservoir "FULL" mark.

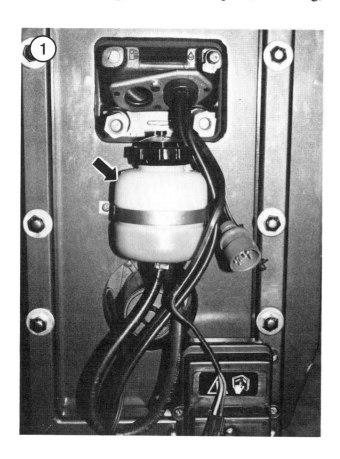

SEA DRIVES

3. If needed, remove the remote reservoir fill cap and add OMC Power Trim/Tilt fluid until proper level is obtained.

Hydraulic Pump Fluid Refill

Follow this procedure when a large amount of fluid has been lost due to an overhaul of the system or a leakage that has been corrected.

1. Remove the remote reservoir (**Figure 1**, typical) fill cap and add OMC Power Trim/Tilt fluid as needed to bring the reservoir to the "FULL" mark. Make sure the pistons are fully extended.
2. Loosen oil pump bleed screw (**Figure 2**) and allow fluid to escape around plug threads until no air bubbles are noted in fluid. Then retighten plug securely.
3. Recheck remote reservoir fluid level.
4. Run the pump motor while rechecking the oil level. Add fluid to the remote reservoir if needed. Cycle the trim/tilt assembly full up and full down at least 5 times, adding oil as required between cycles when the trim and tilt cylinders are fully extended.
5. Recharge the battery. See Chapter Seven.

Troubleshooting

Whenever a problem develops in the power trim/tilt system, the initial step is to determine whether the problem is in the electrical or hydraulic system. Electrical tests are given in this chapter. If the problem appears to be in the hydraulic system, refer it to a dealer or qualified specialist for necessary service.

To determine whether the problem is in the electric or hydraulic system, proceed as follows:

1. Make sure the plug-in connectors are properly engaged and all terminals and wires are free of corrosion. Tighten and clean as required.
2. Make sure the battery is fully charged. Charge or replace as required.
3. Check the system fuse.
4. Check the fluid level as described in this chapter. Add fluid if needed.

NOTE
The power trim/tilt mechanism does not need to be removed from the boat's transom to perform the following test.

5. Proceed as follows to install OMC Trim and Tilt Pressure Tester part No. 390010 (**Figure 3**):
   a. Remove manual release valve retaining ring (**Figure 4**).
   b. Operate trim/tilt motor to completely retract all cylinders.

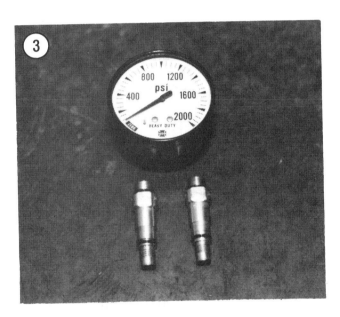

c. Rotate manual release valve 1 full turn counterclockwise.
d. Place a drain pan under the manual release valve.
e. Momentarily operate trim/tilt motor in the "UP" direction and then in the "DOWN" direction.

CAUTION
Residual hydraulic pressure may be present when manual release valve is removed. To prevent personal injury, make sure protective eye wear is worn and cover manual release valve with a cloth during removal.

f. Slowly rotate manual release valve counterclockwise to remove.
g. Securely screw pressure gauge and adapter together. Adapter "B" is used to test the UP circuit and adapter "A" is used to test the DOWN circuit.
h. Install pressure gauge and adapter into manual release valve passage and tighten to 5-10 in.-lb. Do not overtighten as damage to O-ring in end of adapter may result.
i. Operate unit UP and DOWN several cycles, then recheck and add fluid to remote reservoir as needed.

6. Perform the following operations and note the pressure gauge readings:
a. Install pressure gauge and adapter "B" into manual release valve passage.
b. Operate trim/tilt motor to extend cylinders (UP direction). Pressure gauge should read approximately 200 psi as the trim cylinder is extending and approximately 450 psi as the tilt cylinder is extending. Pressure gauge should read approximately 1250 psi as the unit stalls. Pressure gauge should not drop by more than 200 psi from the stall reading when the trim/tilt motor is stopped.
c. Install pressure gauge and adapter "A" into manual release valve passage.
d. Operate trim/tilt motor to retract cylinders (DOWN direction). Pressure gauge should read approximately 800 psi as cylinders are retracting. Pressure gauge should read approximately 800 psi as the unit stalls. Pressure gauge should not drop by more than 200 psi from the stall reading when the trim/tilt motor is stopped.

7. If the system fails any part of Step 6, the problem is hydraulic. If the hydraulic system checks out satisfactorily, the problem is in the electric system.

Trim and Tilt Switch Circuit Test

Refer to **Figure 5** for this procedure.

1. With the key switch ON, connect the voltmeter red test lead to point E1 in the trim/tilt junction box. Connect the black test lead to point G in the trim/tilt junction box. The meter should read battery voltage.
2. If battery voltage is not shown in Step 1, move the red test lead to point J1 in the battery junction box. If battery voltage is shown, check the battery junction box 50 amp fuse. If fuse is good, look for poor connections or an open circuit in the fuse holder and connecting wiring.
3. Move the red test lead to point E3 in the trim/tilt junction box. Move the trim/tilt switch to the DOWN position. If battery voltage is shown, move the red test lead to points E4 and E5. With the trim/tilt switch in the DOWN position, there should be voltage at each point. If voltage is shown and the trim motor does not run, probe connector C2 at the pump motor. If voltage is shown at the connector and the motor does not run, test the motor as described in this chapter.
4. If no voltage is shown at point E3 in Step 3, check connector C1 between the trim/tilt junction box and trim/tilt switch. If the connector is good, check for an open circuit in the following:
a. Green/orange wire between trim/tilt switch and trim/tilt junction box.

SEA DRIVES

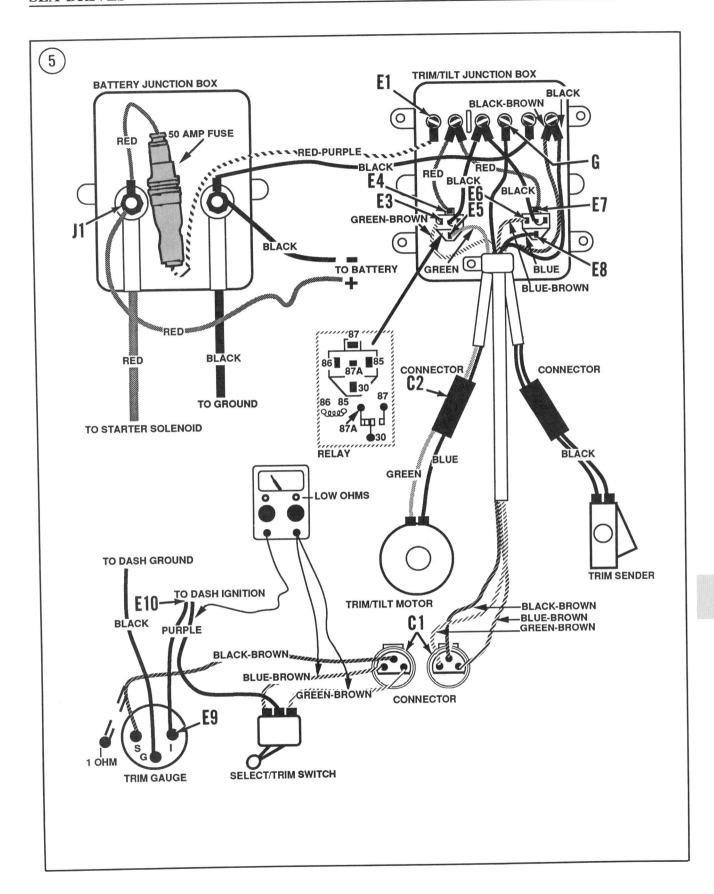

b. Trim/tilt switch between the green/orange and purple leads with the switch in the DOWN position.

5. Move the red test lead to point E6 in the trim/tilt junction box. Move the trim/tilt switch to the UP position. If battery voltage is shown, move the red test lead to points E7 and E8. With the trim/tilt switch in the UP position, there should be voltage at each point. If voltage is shown and the trim motor does not run, probe connector C2 at the pump motor. If voltage is shown at the connector and the motor does not run, test the motor as described in this chapter.
6. If no voltage is shown at point E6 in Step 5, check connector C1 between the trim/tilt junction box and trim/tilt switch. If the connector is good, check for an open circuit in the following:
   a. Blue/orange wire between trim/tilt switch and trim/tilt junction box.
   b. Trim/tilt switch between the blue/orange and purple leads with the switch in the UP position.

Trim Indicator Circuit Test

Refer to **Figure 5** for this procedure.
1. Turn the key switch ON.
2. Connect the voltmeter red test lead to point E9 on the indicator gauge. Connect the black test lead to the indicator gauge G terminal. If battery voltage is shown, proceed with Step 5.
3. If battery voltage is not shown in Step 2, move the red test lead to point E10 (accessory plug). If voltage is shown, there is an open in the circuit between the accessory plug and the trim indicator gauge.
4. If there is no voltage in Step 3, move the red test lead to accessory terminal on key switch. If no voltage is shown at accessory terminal, look for a bad fuse or open in the key switch wiring. If the fuse and wiring are good, replace the switch.
5. Turn the key switch OFF.
6. Disconnect the black/orange wire at the indicator gauge S terminal. With an ohmmeter on the low scale, connect the red test lead to the disconnected black/orange wire and the black test lead to the indicator gauge G terminal. Move the trim switch to the DOWN position and run the outboard motor to the full trim in (down) and then the full tilt (up) positions. The ohmmeter should read 82-88 ohms (full trim) and 0-10 ohms (full tilt).
7. If the resistance readings are not as specified in Step 6, check for an open circuit in the ground wire between the indicator gauge and the trim/tilt junction box or in the black/orange wire disconnected from the indicator gauge S terminal.
8. If the wiring is good, disconnect the black sending leads at the terminal strip in the trim/tilt junction box. Connect an ohmmeter between the

SEA DRIVES

disconnected sender leads and check the resistance of the sending unit. If not within specifications in Step 6, replace the sending unit (**Figure 6**).

Trim and Tilt Motor Testing

Amperage Test

This test evaluates the pump output pressure by determining the current requirements of the pump motor (**Figure 7**). Readings must be taken with the boat at rest.

1. Connect an ammeter between the positive battery terminal and the heavy red cable at the battery junction box.
2. Move the trim/tilt switch to the UP position and note the meter reading, then move the trim/tilt switch to the DOWN position and note that reading.
3. Compare the readings obtained in Step 2 to **Table 1**:
   a. If the current draw is within specifications, the motor, hydraulic pump and relief valves are working properly.
   b. A low current draw indicates a malfunctioning pump, leaking valves, weak relief valves or leaking O-rings on valve bodies.
   c. A high current draw in either the UP or DOWN position indicates a problem in the electric motor, hydraulic pump or relief valves opening at too high of pressure. With the motor separated from the hydraulic pump, perform a no-load current draw test as described in this chapter.

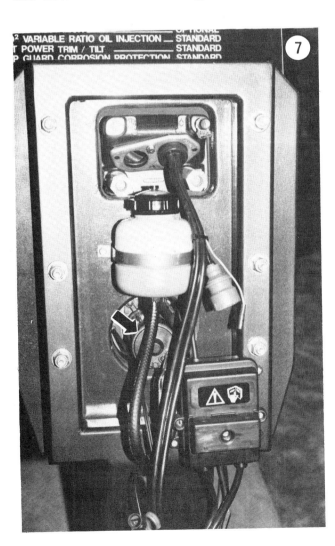

No-load Current Draw Test

1. Separate the electric motor (**Figure 7**) from the hydraulic pump.
2. Connect an ammeter in series with the motor and a fully charged battery. Connect the motor blue wire (up) to the negative battery terminal and the motor green wire (down) to the positive battery terminal. The motor shaft should rotate counterclockwise. Note the ammeter reading. Reverse the motor leads and note the ammeter reading.
3. Ammeter should read a current draw of 7 amps at a minimum of 6700 rpm.
4. If the current draw is not within specifications, repair or replace the electric motor.

System Removal/Installation

1. Disconnect the battery cables from the battery terminals.
2. Remove the remote oil reservoir from the mounting bracket and pour the oil into a container to prevent oil spillage.

3. Disconnect the trim sender and electric motor wiring harnesses at connectors located between respective component and trim/tilt junction box.

4. Rotate the manual release valve counterclockwise 2 full turns.

5. Use a lifting device or physically tilt the outboard motor outward until the motor tilt lock arms (**Figure 8**) can be engaged to secure the motor at the highest tilt position.

6. Mark sending unit bracket for correct repositioning on anchor bracket, then loosen sending unit bracket mounting screws.

7. Loosen cylinder support mounting screws, then slide support up on slotted holes and retighten screws.

8. Remove transom bracket cover to expose tilt cylinder upper mounting bracket.

9. Note four wire ties and the hoses each tie secures for correct reassembly, then use a suitable tool and remove the four wire ties.

10. Remove trailer lock rod spring by removing one end of the spring from the trailer lock rod and the other end of the spring from the anchor bracket.

11. Remove the 4 screws retaining the tilt cylinder upper mounting bracket to the port and starboard engine bracket assemblies.

12. Support the outboard motor to secure its tilt position, then release the tilt lock arms. Allow the outboard motor to lower until the tilt lock arms can be engaged in the lowest tilt lock position. Remove the pin from the tilt cylinder upper eyelet and upper mounting bracket. Remove the upper mounting bracket.

13. Remove the 4 screws securing the anchor bracket to the port and starboard transom bracket assembly. Pull the SelecTrim/Tilt assembly forward enough to allow the sending unit wiring harness and electric motor wiring harness to be pulled through lower transom cutout.

14. Withdraw the SelecTrim/Tilt assembly while slowly guiding the remote reservoir through the lower transom cutout.

15. Installation is the reverse of removal. Note the following during installation:

   a. Apply Scotch-Grip Rubber Adhesive 1300 on anchor bracket gasket.

   b. The 2 disc locks for each of the 4 anchor plate mounting screws must be installed on the screws with the interlocking (teethed) side of the discs facing each other. Tighten screws to 45-50 ft.-lb.

   c. Tighten the 4 screws retaining the tilt cylinder upper mounting bracket to 45-50 ft.-lb.

   d. Use 4 wire ties and resecure the hoses as noted during the removal procedure.

   e. Apply terminal grease on terminals in connector C2 (**Figure 5**). Apply terminal grease on back side of connector ends after assembling connector.

   f. Refill hydraulic system as outlined under Hydraulic Pump Fluid Refill in this chapter.

SEA DRIVES

SELECTRIM/TILT™
(1987-1988 1.8L, 1989-1990 2.0L, 1984-1985 2.5L [S-type], 1984-1985 2.6L [S-type], 1986-1988 2.6L, 1987-1988 2.7L and 1989-1990 3.0L)

Components

The unit consists of a valve body assembly, oil pump, electric motor and combination trim and tilt cylinder/shock absorber mounted on an anchor bracket. A rubber hose connects the anchor bracket reservoir to the remote reservoir.

Operation

Moving the trim/tilt switch to the UP position closes the pump motor circuit. The motor drives the oil pump, forcing oil into the UP side (base) of the trim and tilt cylinder. The outboard motor first moves upward through the 0° to 17° trim range. The trim angle is shown on the dash mounted trim gauge. At 17° and through the remaining 45° of tilt, the trim gauge will show full bow up.

Moving the trim/tilt switch to the DOWN position also closes the pump motor circuit. The reversible motor runs in the opposite direction, forcing oil into the DOWN side (top) of the trim and tilt cylinder to lower the outboard motor.

A manual release valve with a slotted head permits manual raising and lowering of the outboard motor through the tilt range (17-45°).

A trim gauge sending unit is mounted on a bracket located on the inside of the outboard motor mounting brackets. The sending unit uses a calibrated lever positioned against the trim cylinder to calculate trim angle. Access to the sending unit requires the outboard motor to be fully tilted.

Hydraulic Pump Fluid Check

1. Tilt the outboard motor to its full UP position.
2. Check the fluid level in the remote reservoir (**Figure 1**, typical). The fluid should be level with the reservoir "FULL" mark.
3. If needed, remove the remote reservoir fill cap and add OMC Power Trim/Tilt fluid until proper level is obtained.

Hydraulic Pump Fluid Refill

Follow this procedure when a large amount of fluid has been lost due to an overhaul of the system or a leakage that has been corrected.

1. Remove the remote reservoir (**Figure 1**, typical) fill cap and add OMC Power Trim/Tilt fluid as needed to bring the reservoir to the "FULL" mark. Make sure the pistons are fully extended.
2. Loosen oil pump bleed screw (**Figure 2**) and allow fluid to escape around plug threads until no air bubbles are noted in fluid. Then retighten plug securely.
3. Recheck remote reservoir fluid level.
4. Run the pump motor while rechecking the oil level. Add fluid to the remote reservoir if needed. Cycle the trim/tilt assembly full up and full down at least 5 times, adding oil as required between cycles when the trim and tilt cylinders are fully extended.
5. Recharge the battery. See Chapter Seven.

Troubleshooting

Whenever a problem develops in the power trim/tilt system, the initial step is to determine whether the problem is in the electrical or hydraulic system. Electrical tests are given in this chapter. If the problem appears to be in the hydraulic system, refer it to a dealer or qualified specialist for necessary service.

To determine whether the problem is in the electric or hydraulic system, proceed as follows:
1. Remove the SelecTrim/Tilt assembly from the port and starboard transom bracket assembly as outlined in this chapter.
2. Mount the SelecTrim/Tilt assembly in a suitable holding fixture.

3. Make sure the plug-in connectors are properly engaged and all terminals and wires are free of corrosion. Tighten and clean as required.
4. Make sure the battery is fully charged. Charge or replace as required.
5. Check the system fuse.
6. Check the fluid level as described in this chapter. Add fluid if needed.
7. Use OMC Trim/Tilt In-Line Pressure Tester part No. 983977 to test hydraulic system. Attach tester as follows to isolate specific components or circuits:

CAUTION
Momentarily operate trim/tilt motor in the "UP" direction and then in the "DOWN" direction prior to loosening any line fittings. Open the manual release valve one full turn to relieve any residual hydraulic pressure may be present, then close the valve. To prevent personal injury, make sure protective eye wear is worn and cover line fittings with a cloth prior to loosening.

  a. Test trim-out/tilt-up side of valve body and base of hydraulic cylinder (isolate cylinder from gauge)—Remove oil line from base of cylinder. Connect tester line between fitting at base of cylinder and port "B" on tester. Connect oil line removed from base of cylinder to port "A" on tester.
  b. Test trim-out/tilt-up side of valve body and base of hydraulic cylinder (isolate valve body from gauge)—Remove oil line from base of cylinder. Connect tester line between fitting at base of cylinder and port "A" on tester. Connect oil line removed from base of cylinder to port "B" on tester.
  c. Test trim-in/tilt-down side of valve body and top of hydraulic cylinder (isolate cylinder from gauge)—Remove oil line from top of cylinder. Connect tester line between fitting at top of cylinder and port "B" on tester. Connect oil line removed from top of cylinder to port "A" on tester.
  d. Test trim-in/tilt-down side of valve body and top of hydraulic cylinder (isolate valve body from gauge)—Remove oil line from top of cylinder. Connect tester line between fitting at top of cylinder and port "A" on tester. Connect oil line removed from top of cylinder to port "B" on tester.

8. Open shut-off valve in tester (opposite side of port "B") one complete turn counterclockwise.
9. Operate unit UP and DOWN several cycles, then recheck and add fluid to remote reservoir as needed.
10. When the trim/tilt motor is operated in the direction to extend the cylinder (UP direction) and OMC Trim/Tilt In-Line Pressure Tester part No. 983977 is attached as outlined in Step 7a or Step 7b, the following pressures should be read:

  a. Pressure gauge should read approximately 250 psi as the trim/tilt cylinder is extending and approximately 1800 psi as the unit stalls. Pressure gauge should not drop by more than 200 psi in 5 minutes from the stall reading when the trim/tilt motor is stopped. If leakage is noted, refer to Step 7a or 7b. With the cylinder operated through a complete extended (UP) circuit, close shut-off valve in tester (opposite side of port "B") and watch pressure gauge. Pressure gauge will show if leakage is present in valve body if pressure gauge is attached as outlined in Step 7a or in hydraulic cylinder if pressure gauge is attached as outlined in Step 7b.

11. When the trim/tilt motor is operated in the direction to retract the cylinder (DOWN direction) and OMC Trim/Tilt In-Line Pressure Tester part No. 983977 is attached as outlined in Step 7c or 7d, the following pressures should be read:

  a. Pressure gauge should read approximately 800 psi as the trim/tilt cylinder is retracting and as the unit stalls. Pressure gauge should not drop by more than 200 psi in 5 minutes from the stall reading when the

SEA DRIVES

trim/tilt motor is stopped. If leakage is noted, refer to Step 7c or 7d. With the cylinder operated through a complete retracted (DOWN) circuit, close shut-off valve in tester (opposite side of port "B") and watch pressure gauge. Pressure gauge will show if leakage is present in valve body if pressure gauge is attached as outlined in Step 7c or in hydraulic cylinder if pressure gauge is attached as outlined in Step 7d.

12. If the system fails any part of Step 10 or Step 11, the problem is hydraulic. If the hydraulic system checks out satisfactorily, the problem is in the electric system.

Trim and Tilt Switch Circuit Test

Refer to **Figure 5** for this procedure.

1. With the key switch ON, connect the voltmeter red test lead to point E1 in the trim/tilt junction box. Connect the black test lead to point G in the trim/tilt junction box. The meter should read battery voltage.
2. If battery voltage is not shown in Step 1, move the red test lead to point J1 in the battery junction box. If battery voltage is shown, check the battery junction box 50 amp fuse. If fuse is good, look for poor connections or an open circuit in the fuse holder and connecting wiring.
3. Move the red test lead to point E3 in the trim/tilt junction box. Move the trim/tilt switch to the DOWN position. If battery voltage is shown, move the red test lead to points E4 and E5. With the trim/tilt switch in the DOWN position, there should be voltage at each point. If voltage is shown and the trim motor does not run, probe connector C2 at the pump motor. If voltage is shown at the connector and the motor does not run, test the motor as described in this chapter.
4. If no voltage is shown at point E3 in Step 3, check connector C1 between the trim/tilt junction box and trim/tilt switch. If the connector is good, check for an open circuit in the following:
   a. Green/orange wire between trim/tilt switch and trim/tilt junction box.
   b. Trim/tilt switch between the green/orange and purple leads with the switch in the DOWN position.
5. Move the red test lead to point E6 in the trim/tilt junction box. Move the trim/tilt switch to the UP position. If battery voltage is shown, move the red test lead to points E7 and E8. With the trim/tilt switch in the UP position, there should be voltage at each point. If voltage is shown and the trim motor does not run, probe connector C2 at the pump motor. If voltage is shown at the connector and the motor does not run, test the motor as described in this chapter.
6. If no voltage is shown at point E6 in Step 5, check connector C1 between the trim/tilt junction box and trim/tilt switch. If the connector is good, check for an open circuit in the following:
   a. Blue/orange wire between trim/tilt switch and trim/tilt junction box.
   b. Trim/tilt switch between the blue/orange and purple leads with the switch in the UP position.

Trim Indicator Circuit Test

Refer to **Figure 5** for this procedure.

1. Turn the key switch ON.
2. Connect the voltmeter red test lead to point E9 on the indicator gauge. Connect the black test lead to the indicator gauge G terminal. If battery voltage is shown, proceed with Step 5.
3. If battery voltage is not shown in Step 2, move the red test lead to point E10 (accessory plug). If voltage is shown, there is an open in the circuit between the accessory plug and the trim indicator gauge.
4. If there is no voltage in Step 3, move the red test lead to accessory terminal on key switch. If no voltage is shown at accessory terminal, look for a bad fuse or open in the key switch wiring. If the fuse and wiring are good, replace the switch.
5. Turn the key switch OFF.

6. Disconnect the black/orange wire at the indicator gauge S terminal. With an ohmmeter on the low scale, connect the red test lead to the disconnected black/orange wire and the black test lead to the indicator gauge G terminal. Move the trim switch to the DOWN position and run the outboard motor to the full trim in (down) and then the full tilt (up) positions. The ohmmeter should read 82-88 ohms (full trim) and 0-10 ohms (full tilt).

7. If the resistance readings are not as specified in Step 6, check for an open circuit in the ground wire between the indicator gauge and the trim/tilt junction box or in the black/orange wire disconnected from the indicator gauge S terminal.

8. If the wiring is good, disconnect the black sending leads at the terminal strip in the trim/tilt junction box. Connect an ohmmeter between the disconnected sender leads and check the resistance of the sending unit. If not within specifications in Step 6, replace the sending unit (**Figure 6**).

Trim and Tilt Motor Testing

Amperage Test

This test evaluates the pump output pressure by determining the current requirements of the pump motor (**Figure 7**). Readings must be taken with the boat at rest.

1. Connect an ammeter between the positive battery terminal and the heavy red cable at the battery junction box.
2. Move the trim/tilt switch to the UP position and note the meter reading, then move the trim/tilt switch to the DOWN position and note that reading.
3. Compare the readings obtained in Step 2 to **Table 1**:
   a. If the current draw is within specifications, the motor, hydraulic pump and relief valves are working properly.
   b. A low current draw indicates a malfunctioning pump, leaking valves, weak relief valves or leaking O-rings on valve bodies.
   c. A high current draw in either the UP or DOWN position indicates a problem in the electric motor, hydraulic pump or relief valves opening at too high of pressure. With the motor separated from the hydraulic pump, perform a no-load current draw test as described in this chapter.

No-load Current Draw Test

1. Separate the electric motor (**Figure 7**) from the hydraulic pump.
2. Connect an ammeter in series with the motor and a fully charged battery. Connect the motor blue wire (up) to the negative battery terminal and the motor green wire (down) to the positive battery terminal. The motor shaft should rotate counterclockwise. Note the ammeter reading. Reverse the motor leads and note the ammeter reading.
3. Ammeter should read a current draw of 7 amps at a minimum of 6700 rpm.
4. If the current draw is not within specifications, repair or replace the electric motor.

System Removal/Installation

1. Disconnect the battery cables from the battery terminals.
2. Remove the remote oil reservoir from the mounting bracket and pour the oil into a container to prevent oil spillage.
3. Disconnect the trim sender and electric motor wiring harnesses at connectors located between respective component and trim/tilt junction box.
4. Rotate the manual release valve counterclockwise 2 full turns.
5. Use a lifting device or physically tilt the outboard motor outward until the motor tilt lock arms (**Figure 8**) can be engaged to secure the motor at the highest tilt position.

SEA DRIVES

6. Mark sending unit bracket for correct repositioning on anchor bracket, then loosen sending unit bracket mounting screws.

7. Loosen cylinder support mounting screws, then slide support up on slotted holes and retighten screws.

8. Place a sling around midsection of outboard motor and use a suitable lifting device to support assembly, thus relieving load on hydraulic cylinder.

9. Remove one screw from hydraulic cylinder upper pin (**Figure 9**).

10. Use a suitable punch and mallet to drive hydraulic cylinder upper pin from swivel bracket.

11. With the outboard motor supported by the sling and suitable lifting device in the full tilt position, release the tilt lock arms. Allow the outboard motor to lower until the tilt lock arms can be engaged in the lowest tilt lock position.

12. Remove the bushings from the hydraulic cylinder upper eyelet.

13. Remove the 4 screws securing the anchor bracket to the port and starboard transom bracket assembly. Pull the SelecTrim/Tilt assembly forward enough to allow the sending unit wiring harness and electric motor wiring harness to be pulled through lower transom cutout.

14. Withdraw the SelecTrim/Tilt assembly while slowly guiding the remote reservoir through the lower transom cutout.

15. Installation is the reverse of removal. Note the following during installation:

   a. Apply Scotch-Grip Rubber Adhesive 1300 on anchor bracket gasket.

   b. The 2 disc locks for each of the 4 anchor plate mounting screws must be installed on the screws with the interlocking (teethed) side of the discs facing each other. Tighten screws to 45-50 ft.-lb.

   c. Tighten the screw retaining the hydraulic cylinder upper tilt pin to 18-20 ft.-lb. (**Figure 9**).

   d. Apply terminal grease on terminals in connector C2 (**Figure 5**). Apply terminal grease on back side of connector ends after assembling connector.

   e. Refill hydraulic system as outlined under Hydraulic Pump Fluid Refill in this chapter.

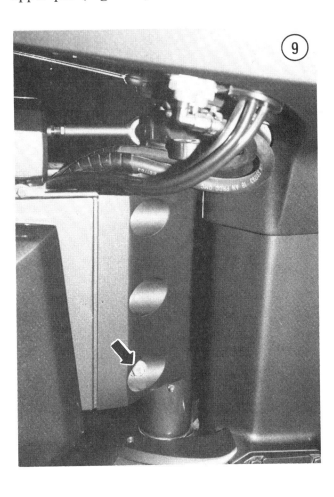

Table 1 is on the following page.

Table 1 CURRENT DRAW (AMPS) AND DURATION (UNDER LOAD CONDITIONS)

1983-1990 1.6L		
	Amps	**Time in Seconds**
Trimming up	12-14	11-14
Tilting up	15-17	12-16
Full tilt up (stall)	25-30	
Tilting down	11-13	12-16
Trimming down	13-15	11-14
Full trim down (stall)	22-26	
Trim in to full tilt up		24-29
Full tilt up to trim in		23-28

1987-1988 1.8L, 1989-1990 2.0L, 1984-1985 2.5L [S-type], 1984-1985 2.6L [S-type], 1986-1988 2.6L, 1987-1988 2.7L and 1989-1990 3.0L		
	Amps	**Time in Seconds**
Trimming up	13-15	10-12
Tilting up	13-15	27-31
Tilting down	11-13	25-29
Trimming down	11-13	9-12
Trim in to full tilt up		38-44
Full tilt up to trim in	11-13	36-41

Index

A

Anti-siphon devices 286
Armature plate 308-309

B

Battery
   care/inspection 293-295
   charging 296-299
   installation in aluminum
     boats 293
   jump starting 299
   testing 295-296
Battery charging system
   rectifier 301
   rectifier/regulator assembly
     replacement 301-302
   stator and charge coil 300
   voltage regulator 301

C

Carburetors
   cleaning/inspection 269-270
   core plugs 270-271
   high-elevation 269
   lead shot 270-271
   48-75 HP 271-275
   V4 and V6 275-284

Choke and primer solenoid
   service 284-286
Compression check 204-205
Connecting rod and crankshaft
   assembly 356-360
Crankcase and connecting rod
   bearings 346-347
Crankshaft cleaning and
   inspection 348-349
Cylinder block and crankcase
   assembly 360-364
   cleaning/inspection 343-346

E

Electrical system
   battery 292-302
   electric starting system 302
   starter motor 302-306
Engine
   flushing 203-204
   fuel filter 210
   operation 2
   timing 217
   troubleshooting 180-184
Engine synchronization 217
   engine timing 217
   equipment 217-218
   1973-1988
     50 HP 219-223
     48, 50, 55 and 60 HP
      (2 cyl. remote control) .. 223-227

     1989 48 and 50 HP 227-229
     60, 65, 70 and 75 HP
      (3 cyl.) 230-234
     65 HP (3 cyl. tiller) 234-237
   V4 engines
     1985-on 120-140 HP 246-252
     1986-on 88-115 HP 243-246
     others 238-243
   V6 engines
     1975-1985 252-257
     1986-on 150-175 HP 257-259
     1986-on 200-225 HP 259-262

F

Fasteners and torque 2-8, 319
Flywheel 319-320
Fuel mixture 195-196
Fuel pump 211-213, 265-268
Fuel selection 193-194
Fuel system
   anti-siphon devices 286
   carburetors 268-284
   choke and primer solenoid
     service 284-286
   fuel line and primer bulb 286
   fuel pump 265-268
   fuel tank 286-287
   service 210
   troubleshooting 177-180

INDEX

G

Galvanic corrosion 11-14
Gaskets and sealants 10-11, 319
Gasohol 194-195
Gearcase
   hydro-mechanical 381-397
   mechanical 397-416
   pressure and vacuum test 416-417
   propeller 375-376
   removal/installation 379-380
   water pump 376-379

H

Hydraulic pump 422, 428,
   462-463, 469

I

Ignition system
   armature plate 308-309
   charge coil, sensor coil or
     stator coil 310-312
   connector terminal 314-315
   ignition module 311-312
   operation 306-308
   power pack 313-314
   secondary ignition coil
     replacement 312-313
   stator and timer base 312
   troubleshooting 45-177

J

Jump starting 299

L

Lower drive unit 196-197
Lubricants 8-10
Lubrication
   fuel mixture 195-196
   fuel selection 193-194
   gasohol 194-195
   lower drive unit 196-197
   other lubrication points 197
   salt water corrosion of gear
     housing bearing 197-200
   sour fuel 194

M

Maintenance, anti-corrosion 203
Mechanic's techniques 31-32

O

Oil injection systems
   OMC Economixer 448-455
   OMC VRO system 455-458
OMC Economixer
   battery signal test 455
   Economixer programming 452
   engine fuel system test 454
   fuel and oil requirements 452
   ignition signal test 454
   instrument cable connector
     check 454
   key switch test 454
   oil pump test 455
   operation 448-452
   preliminary procedure 453-454
   rectifier test 454
   RPM signal test 454
   throttle sender test 455
   troubleshooting 452-453
OMC VRO system
   break-in procedure 456
   excessive engine smoke 458
   fuel pressure test 457-458
   low oil warning sounds at
     20 second intervals 458
   oil flow test 458
   operation 456
   pickup unit and filter 459
   troubleshooting 457
   warning horn 456-457

P

Performance test (on boat) 213
Piston and connecting rod
   assembly 349-356
Piston cleaning and
   inspection 347-348
Power head 317-368
   connecting rod and crankshaft
     assembly 356-360
   crankcase and connecting rod
     bearings 346-347
   crankshaft cleaning and
     inspection 348-349
   cylinder block and crankcase
     assembly 360-364
     cleaning/inspection 343-346
   disassembly 328-343
   engine serial number 318-319
   fasteners and torque 319
   flywheel 319-320
   gaskets and sealants 319
   piston and connecting rod
     assembly 349-356
   piston cleaning
     and inspection 347-348
   reed block service 364
   removal/installation 320-328
   thermostat service 365-368
Power trim/tilt system, external
   components 421
   hydraulic pump 422
   motor testing 445
   operation 421-422
   pump motor circuit
     test 426-427
   removal/installation 445
   switch circuit test 423-426
   trim gauge test 427
   trim indicator circuit
     test 426
   trim switch test 427
   troubleshooting 422-423
Power trim/tilt system, internal
   components 427-428
   hydraulic pump 428
   motor testing 445
   operation 428
   removal/installation 445-446
   switch circuit test 431-434
   trim indicator circuit
     test 433-445
   troubleshooting 429-431
Pressure and vacuum test 416-417
Propeller 14-20, 375-376

R

Reed block service 364

S

Safety 21
Salt water corrosion of gear
   housing bearing 197-200
Sea drives 461-473
SelecTrim/Tilt
   components 461, 469
   hydraulic pump 462-463, 469
   operation 462, 469
   system removal/instal-
     lation 467-468, 472-473
   trim and tilt motor
     testing 467, 472
   trim and tilt switch circuit
     test 464-466, 471
   trim indicator circuit
     test 466-467, 471-472
   troubleshooting 463-464,
     469-471
Serial number, engine 318-319

INDEX

Service hints 28-30
Sour fuel 194
Spark plugs 205-210
Starter motor 302-306
Starting system, electric 302
Storage 200-202
Submersion, complete 202-203

T

Test equipment 26-28
Thermostat service 365-368
Tools, hand 21-26
Torque specifications 2

Troubleshooting 33-184
  CD 2 ignition 46-65
  CD2USL ignition (1989-1990) . . 65-69
  CD 3 ignition 69-75, 82-100
  CD 4 ignition 100-143
  CD 6 ignition 75-87, 143-176
  charging system 39-44
  engine 182-184
  engine temperature and
    overheating 180-182
  fuel system 177-180
  ignition and neutral start
    switch 176-177
  ignition system 45-177
  operating requirements 34
  SelecTrim/Tilt 463-464, 469-471

starting system 34-39
Tune-up
  compression check 204-205
  engine fuel filter 210
  fuel pump 211-213
  fuel system service 210
  inline filter 211
  lower unit/water pump
    check 210
  performance test (on boat) 213
  spark plugs 205-210

W

Water pump 376-379

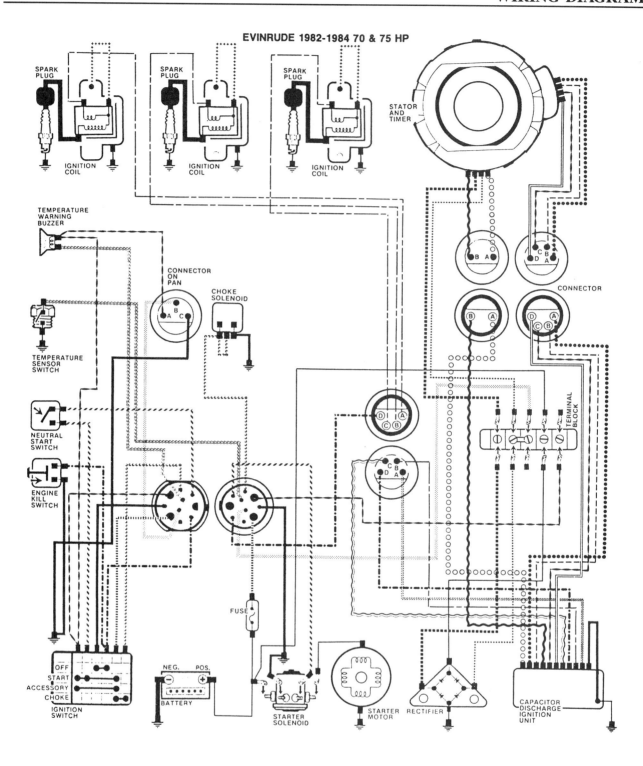

Wiring Diagrams

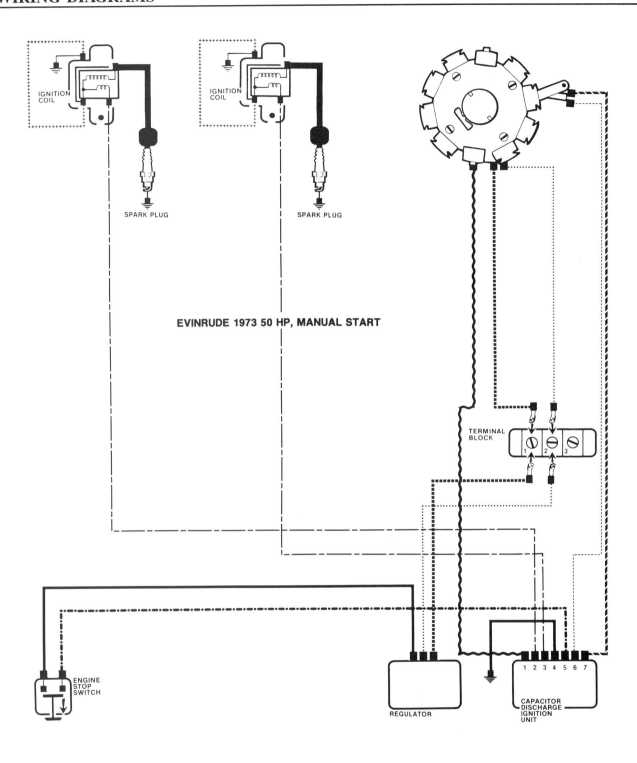

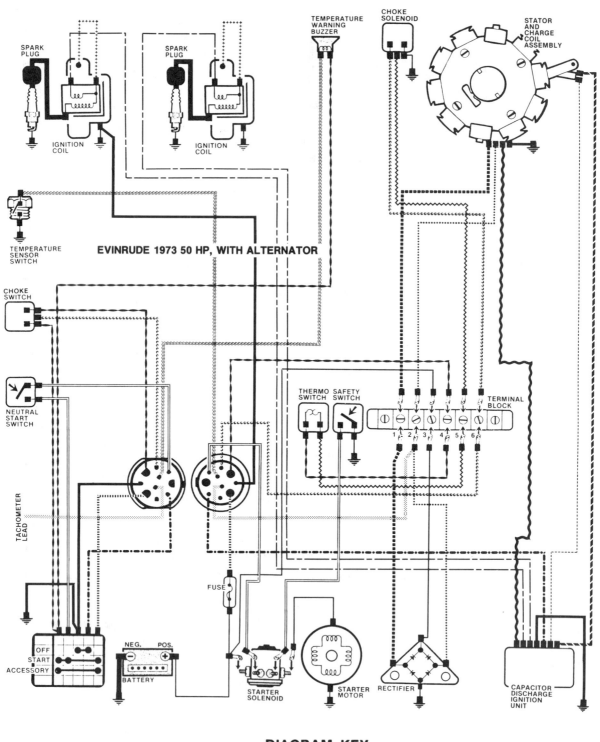

Wiring Diagrams

481

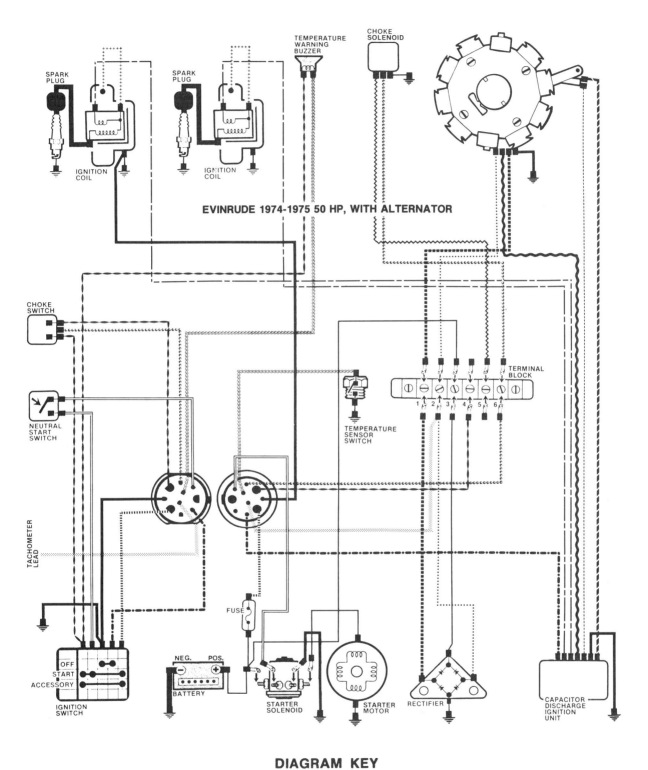

EVINRUDE 1974-1975 50 HP, WITH ALTERNATOR

DIAGRAM KEY

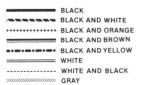
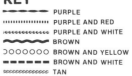

14

WIRING DIAGRAMS

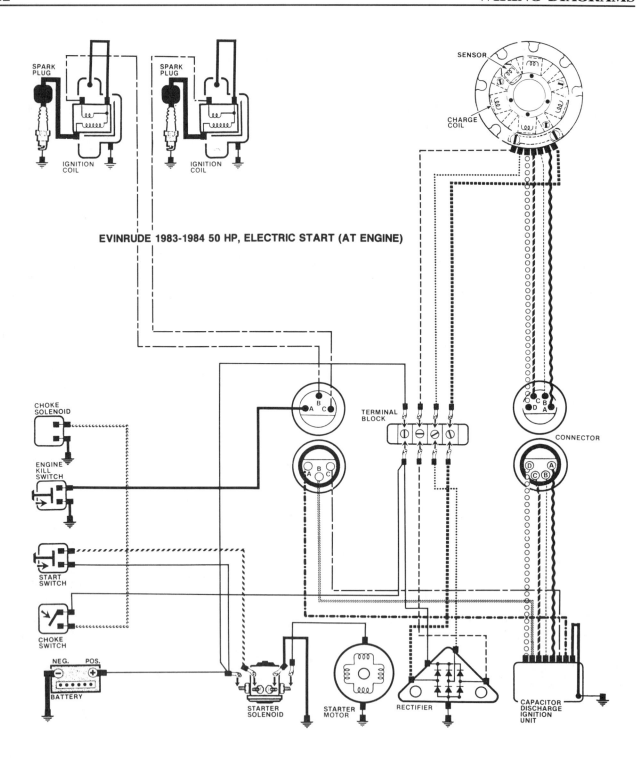

Evinrude 1983-1984 50 HP, Electric Start (At Engine)

Wiring Diagrams

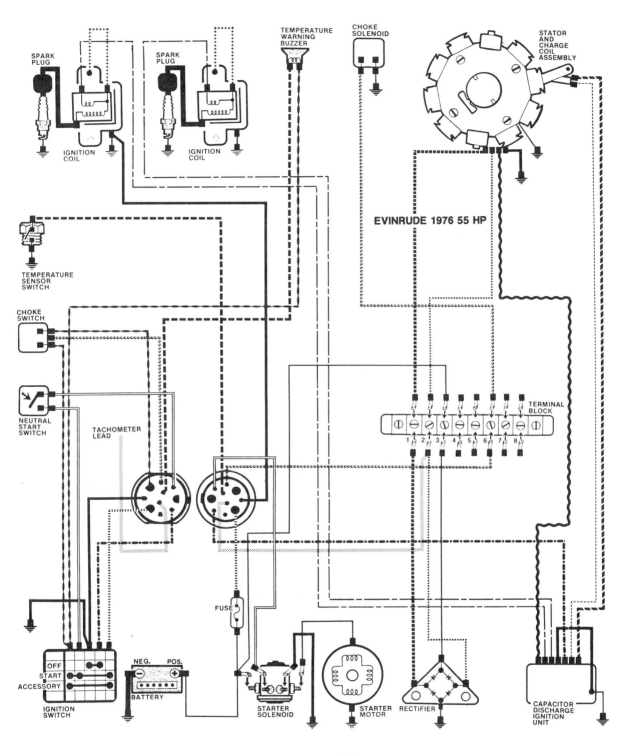

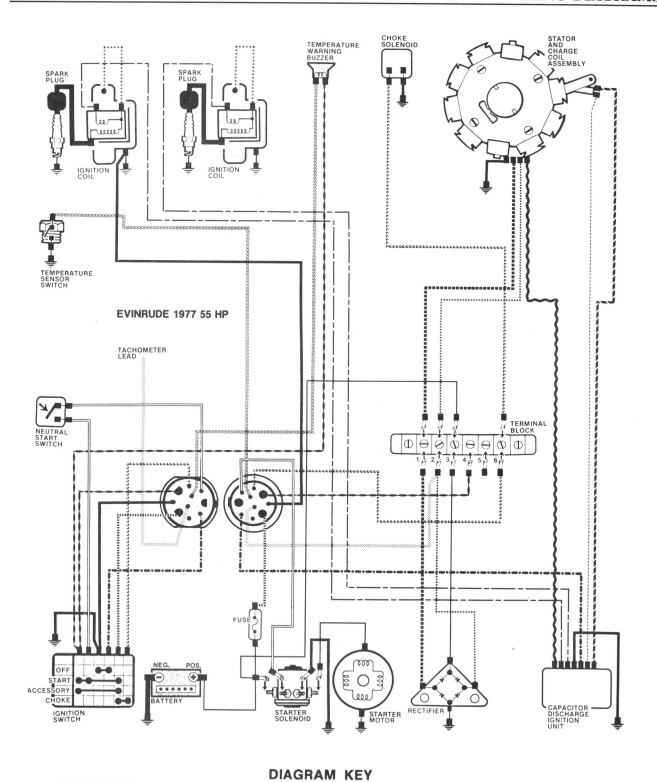

WIRING DIAGRAMS

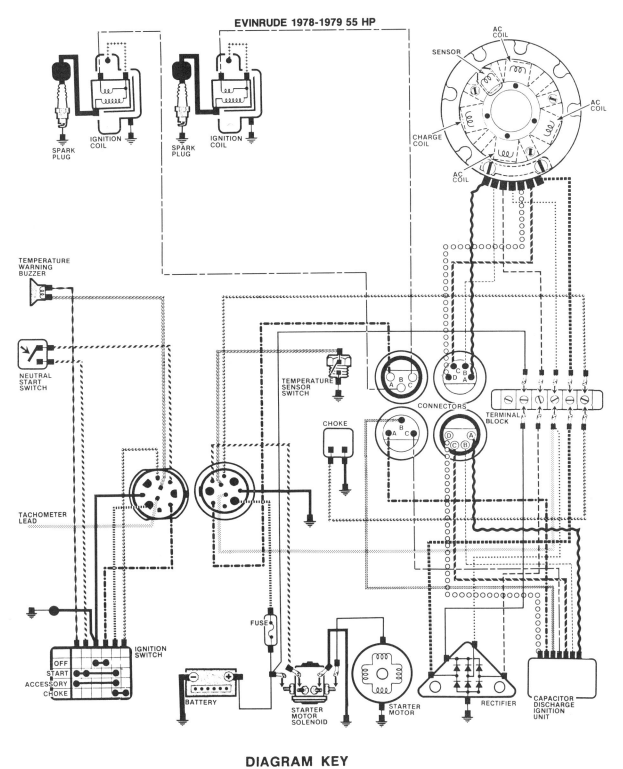

EVINRUDE 1978-1979 55 HP

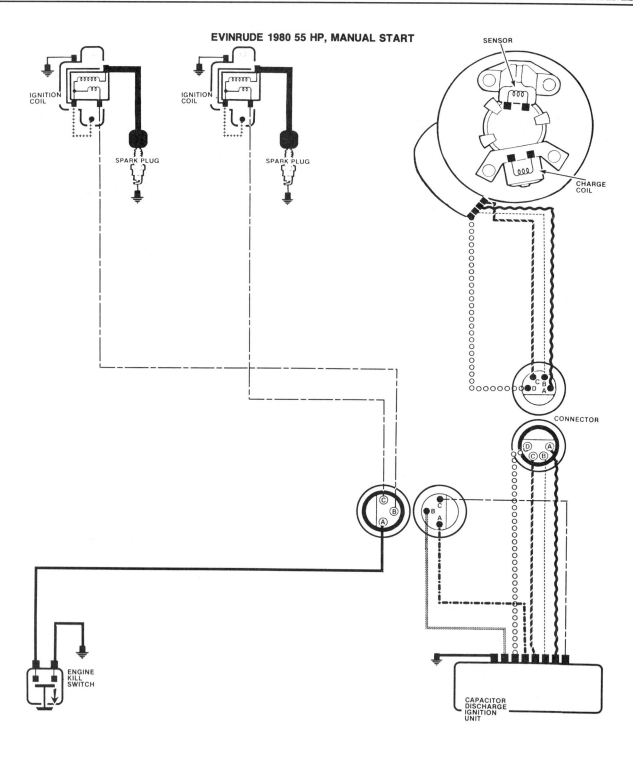

Wiring Diagrams

487

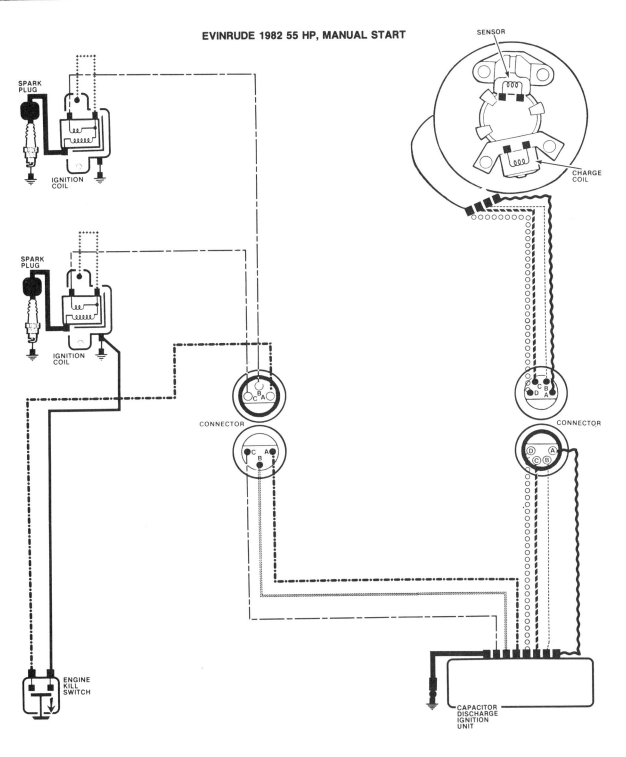

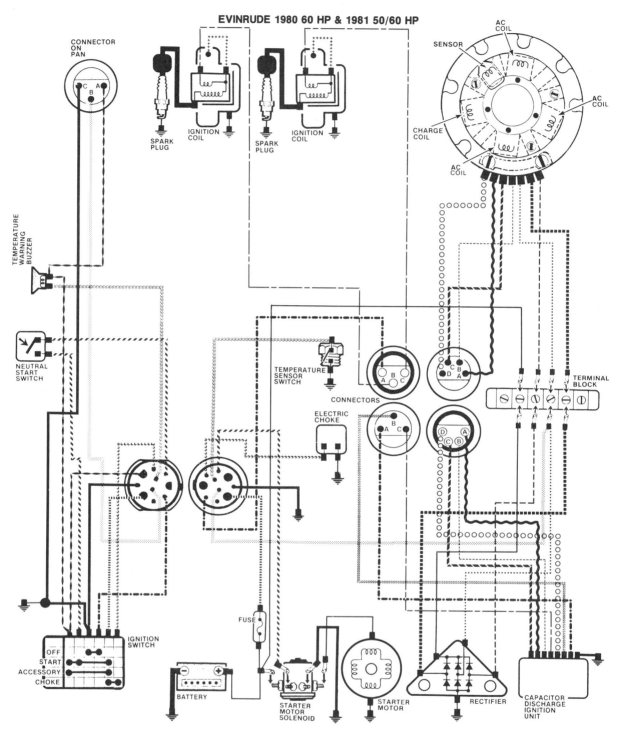

WIRING DIAGRAMS

489

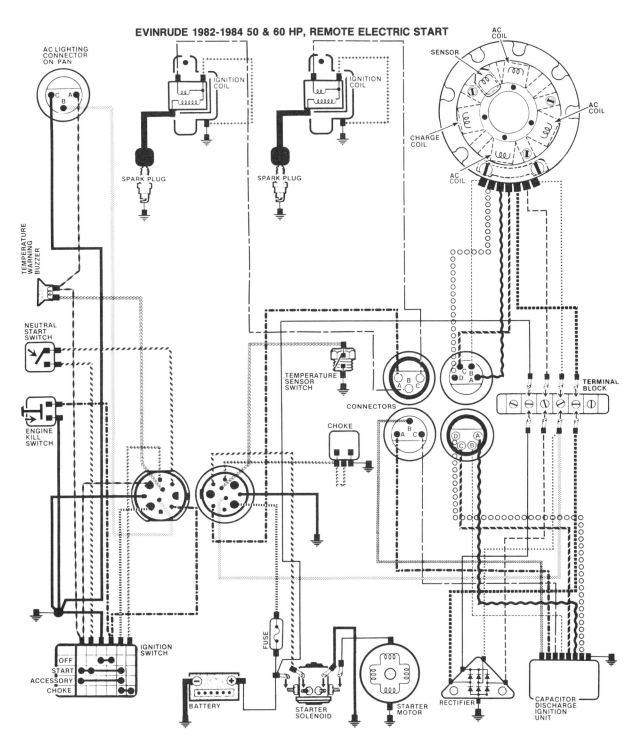

WIRING DIAGRAMS

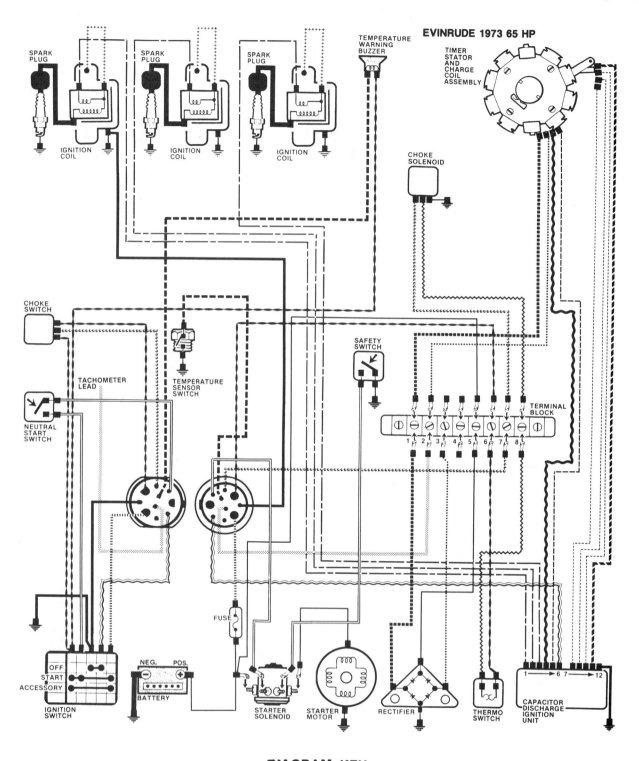

Wiring Diagrams

491

EVINRUDE 1974-1976 70 & 75 HP

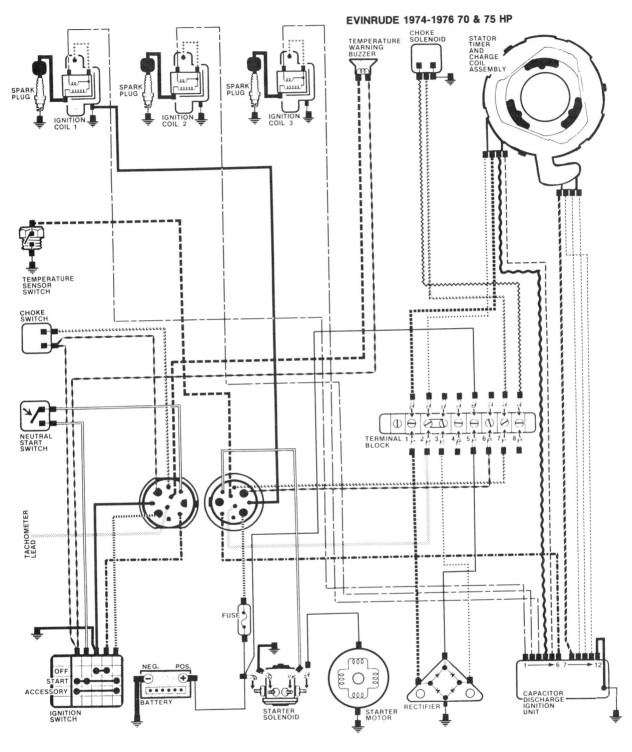

14

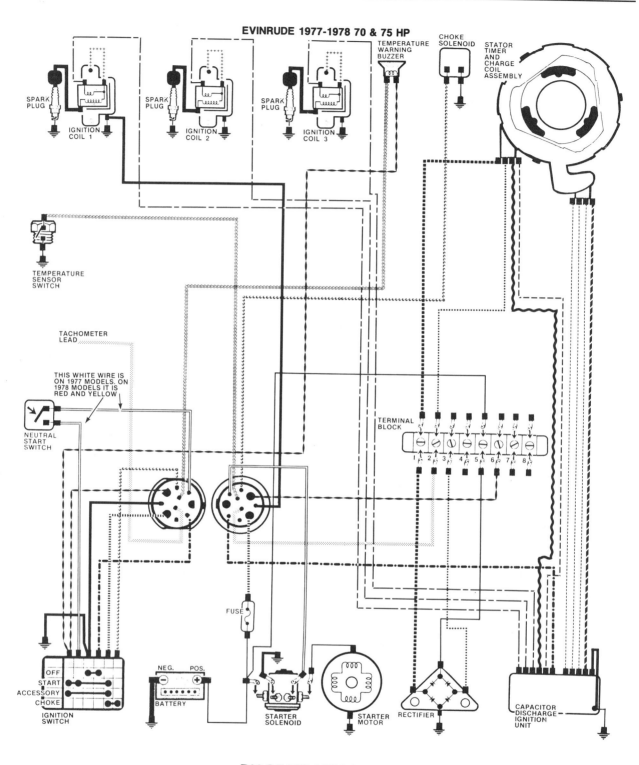

Wiring Diagrams

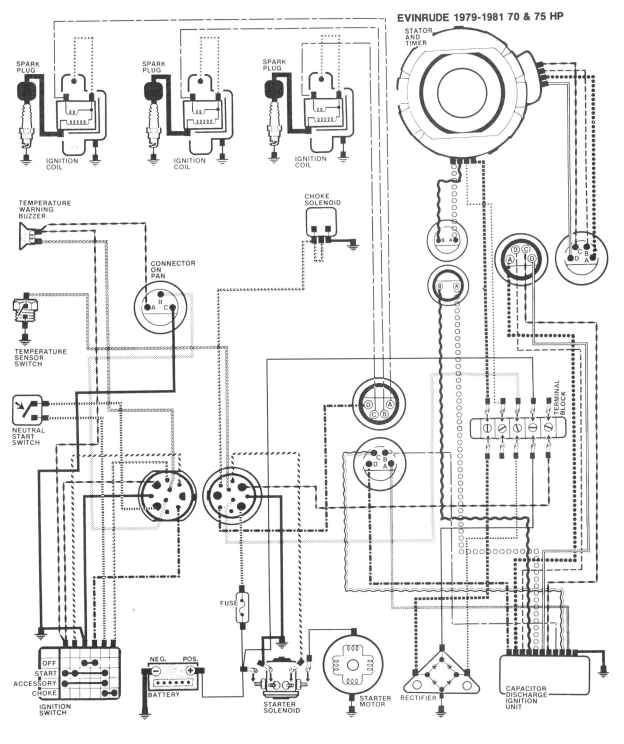

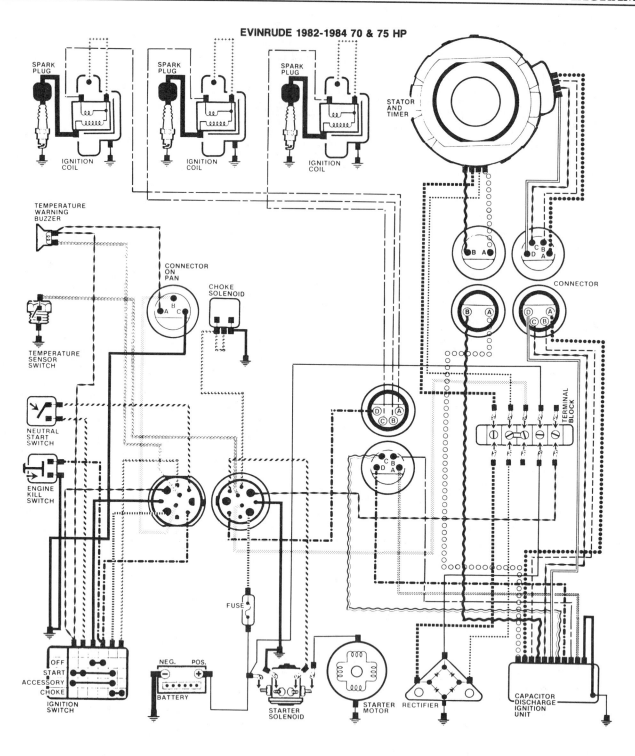

Wiring Diagrams

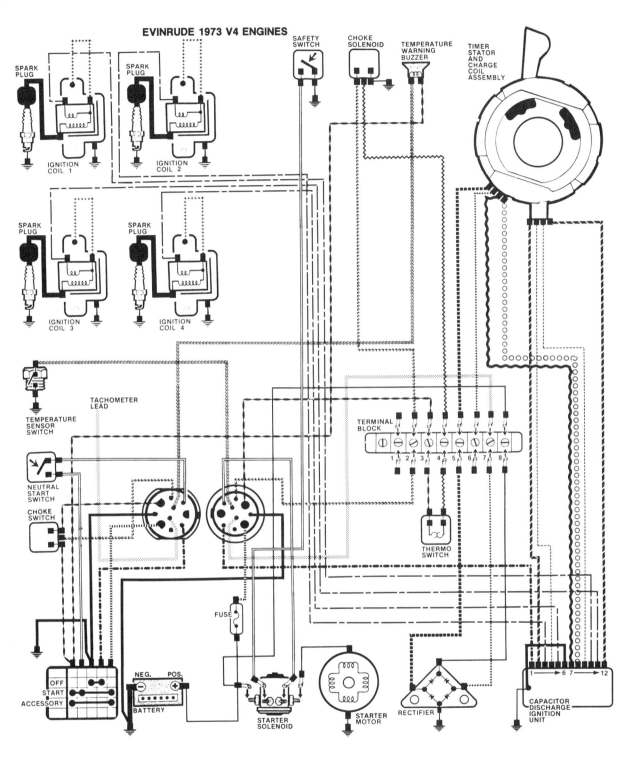

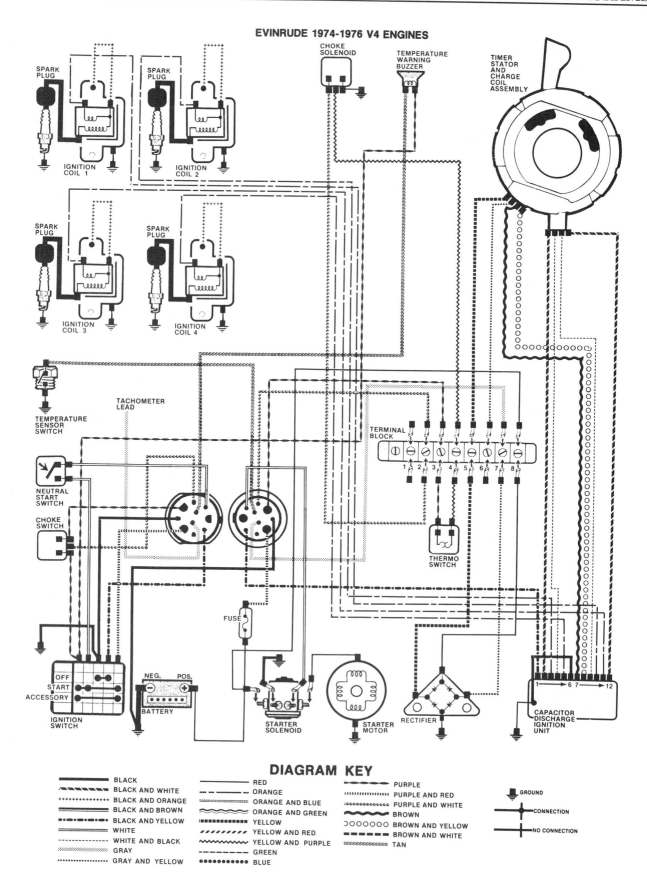

WIRING DIAGRAMS

497

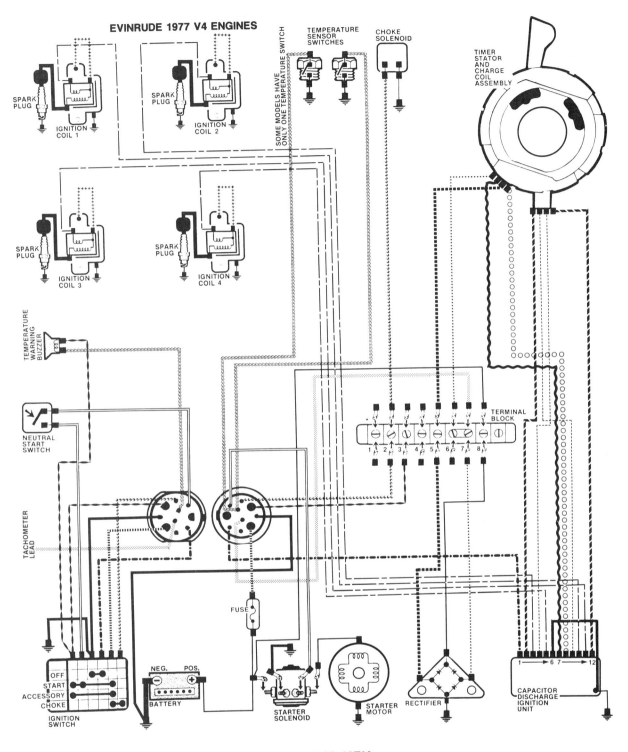

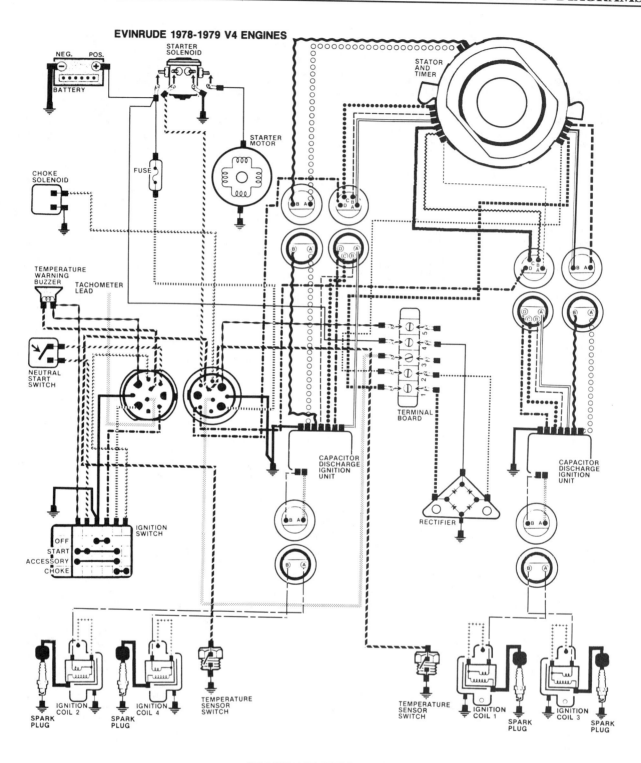

Wiring Diagrams

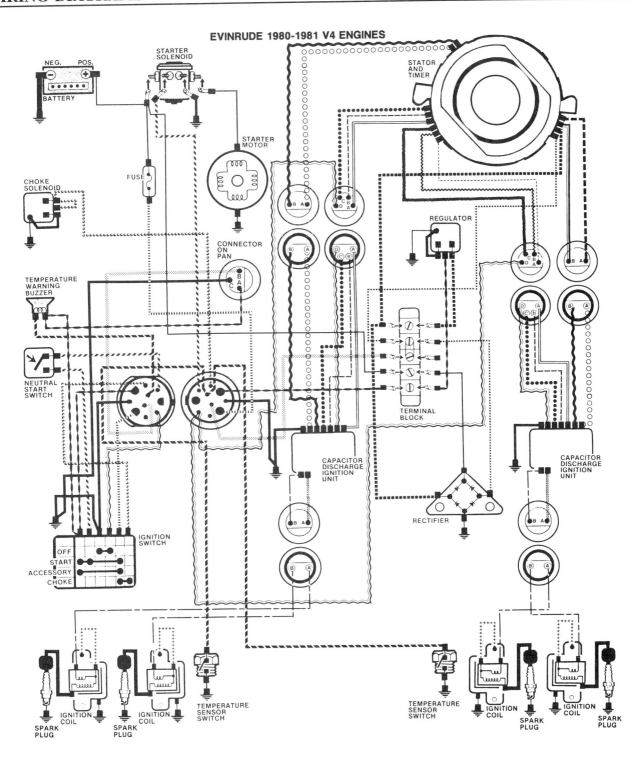

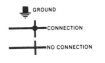

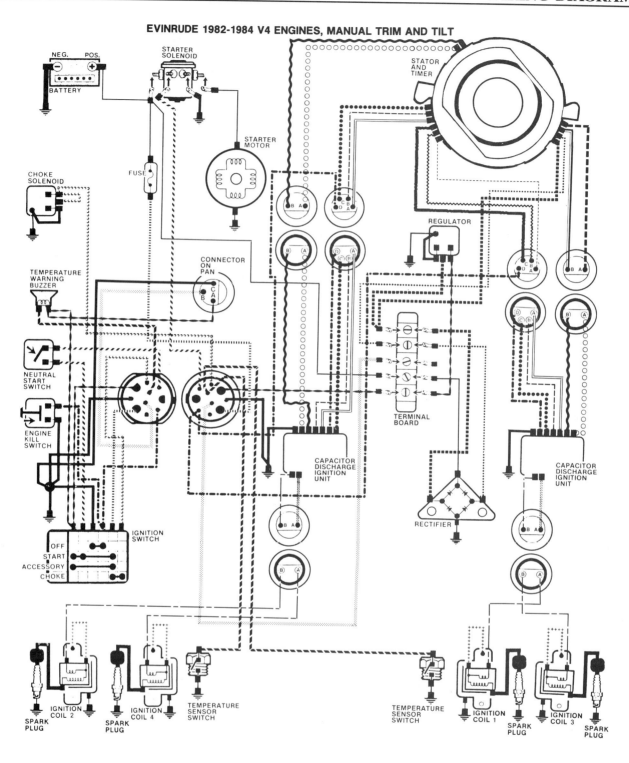

Wiring Diagrams

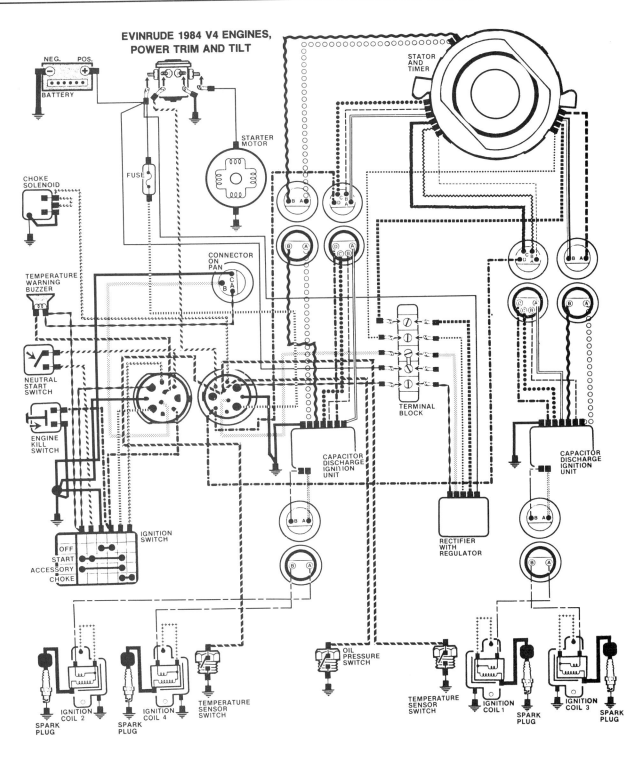

WIRING DIAGRAMS

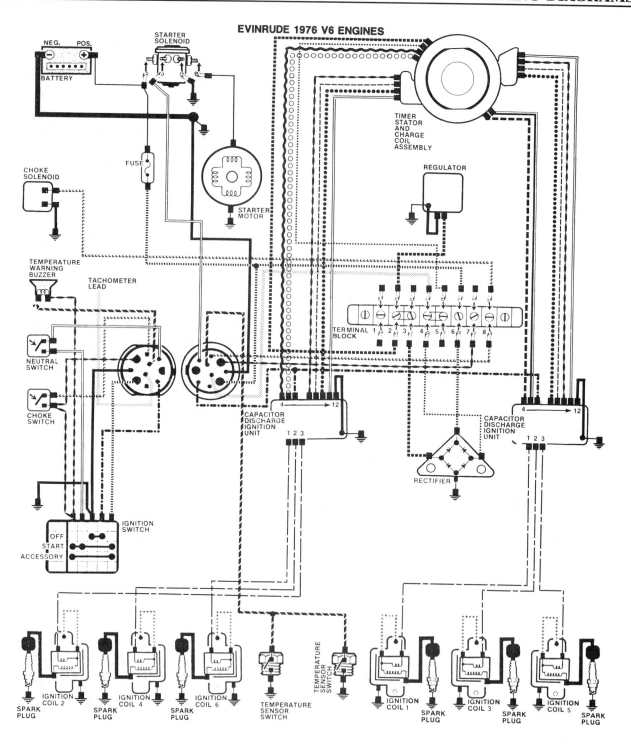

WIRING DIAGRAMS

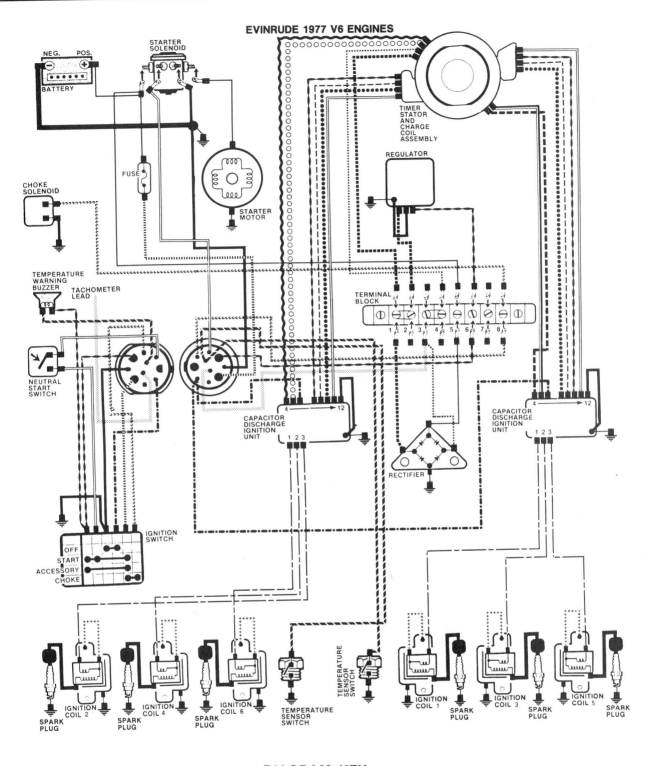

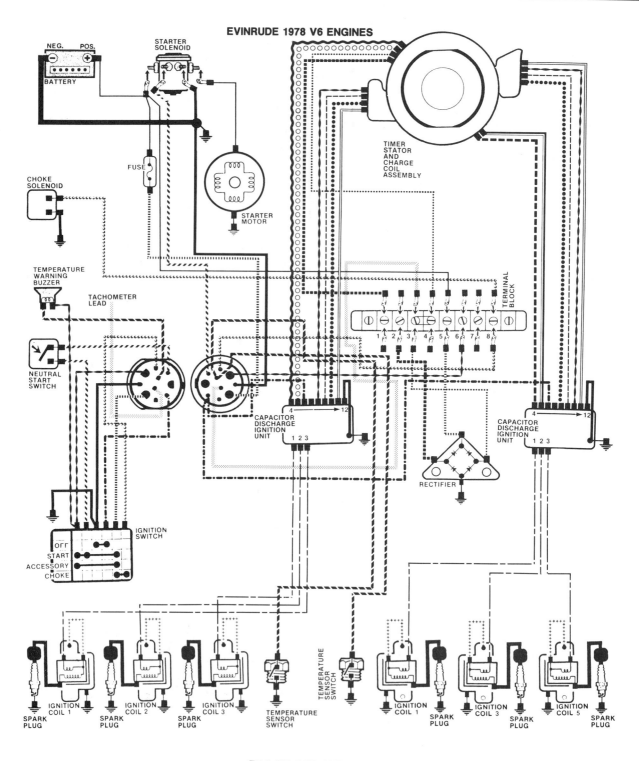

WIRING DIAGRAMS

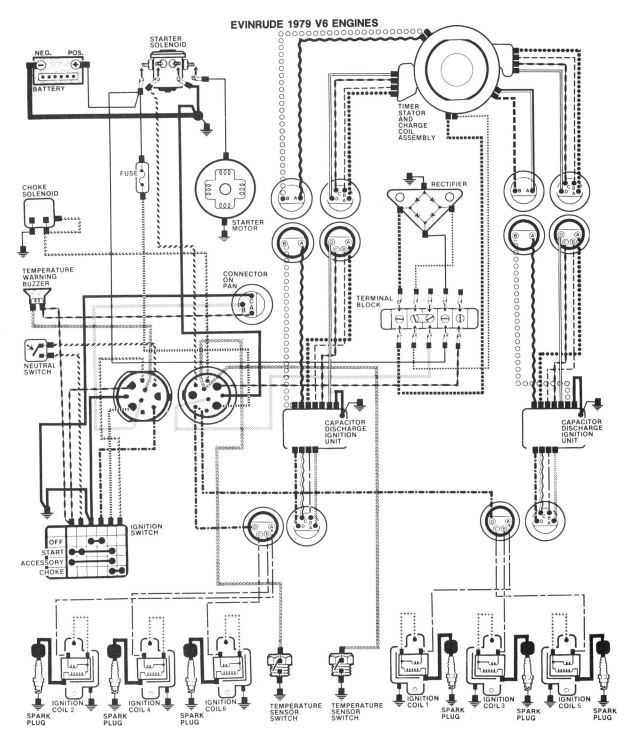

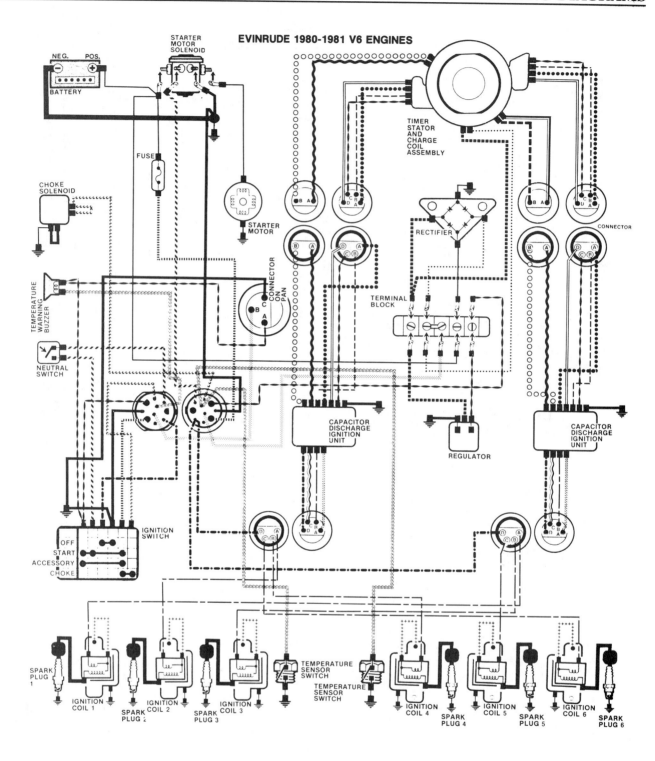

WIRING DIAGRAMS

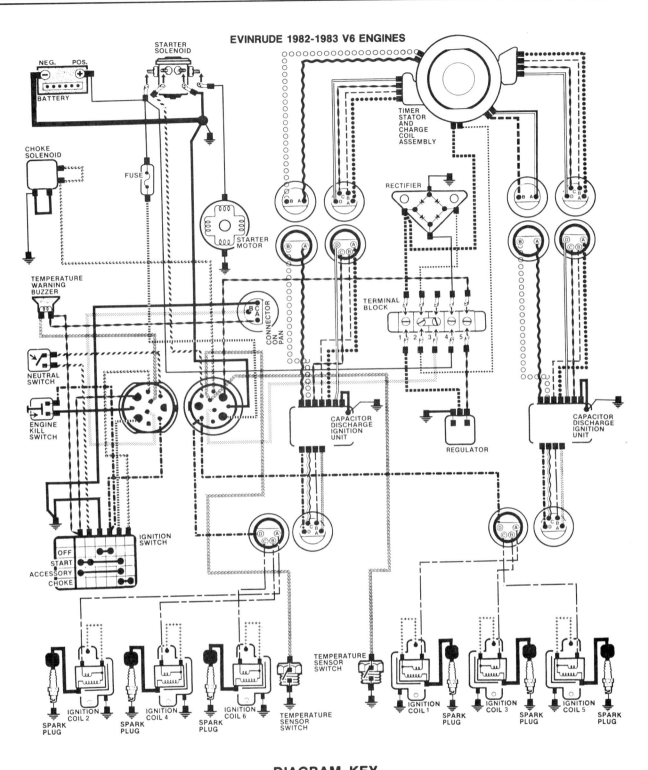

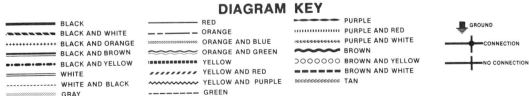

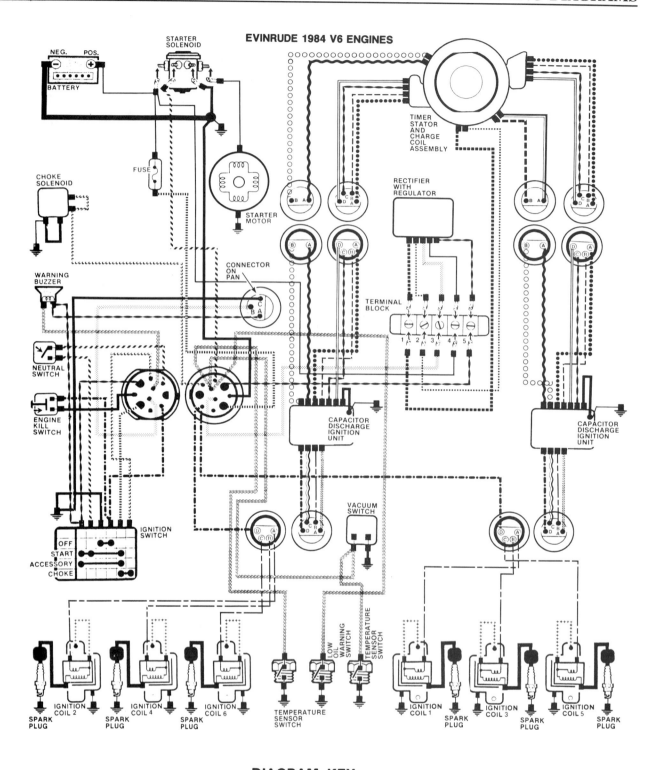

WIRING DIAGRAMS

509

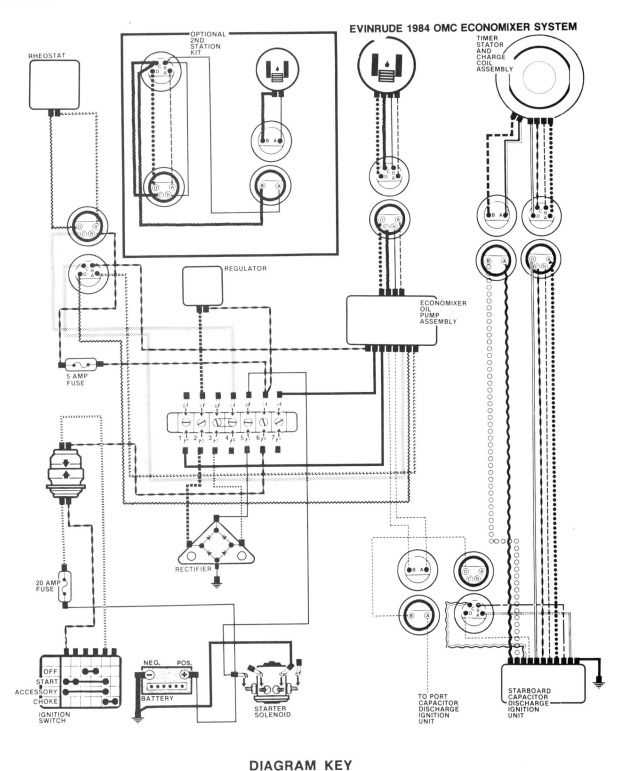

EVINRUDE 1984 OMC ECONOMIXER SYSTEM

WIRING DIAGRAMS

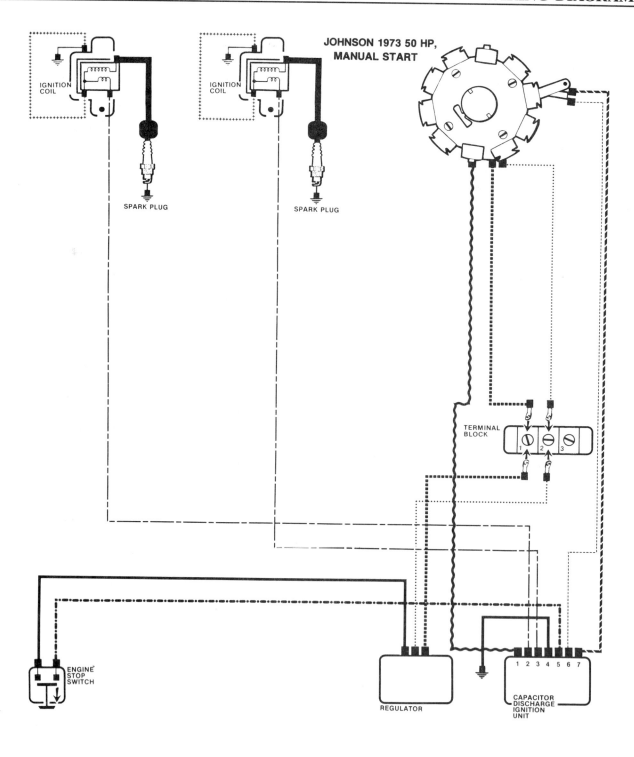

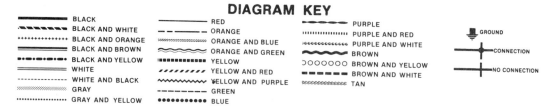

WIRING DIAGRAMS

511

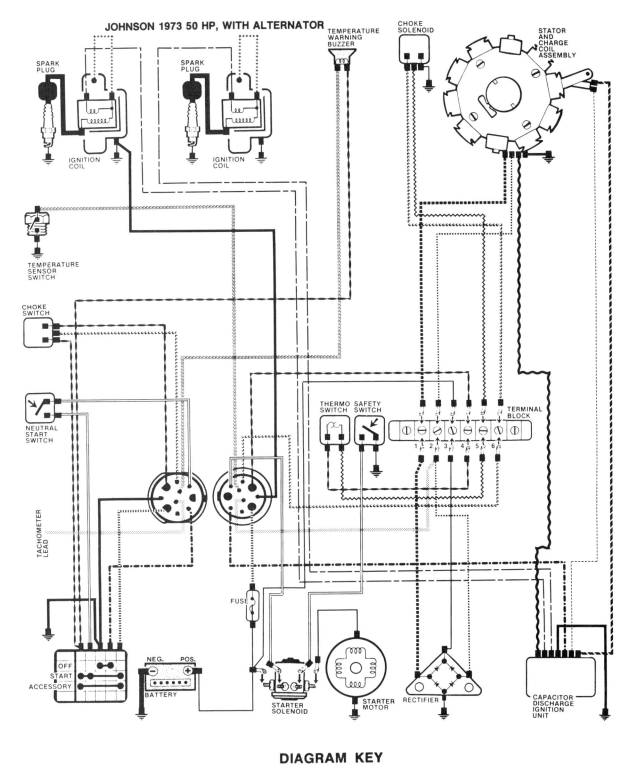

14

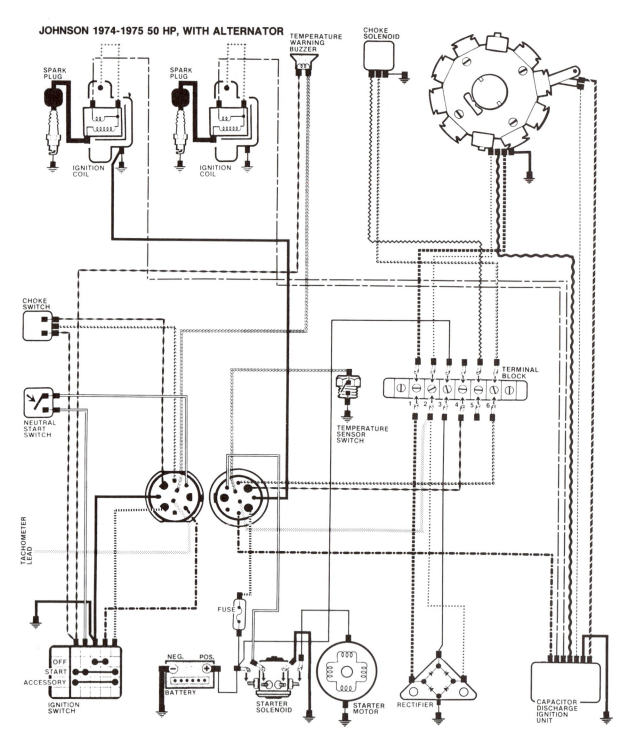

Wiring Diagrams

513

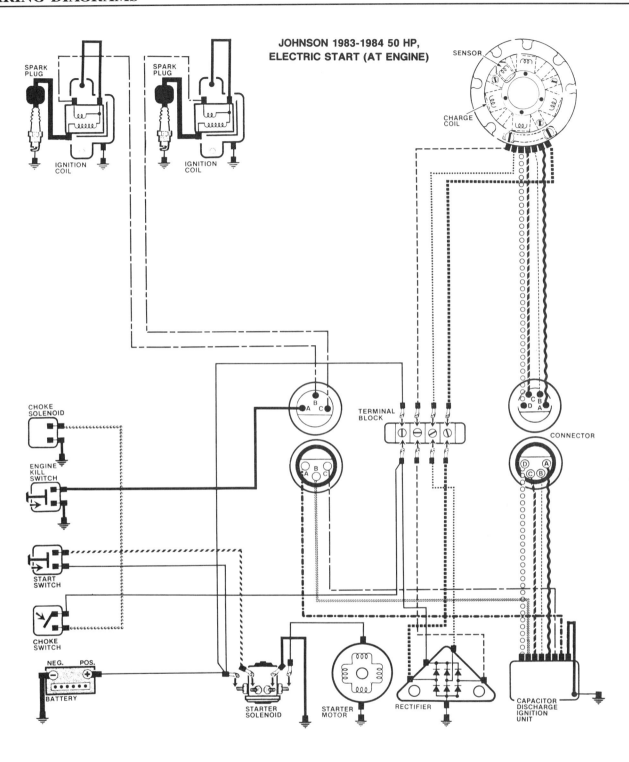

Diagram Key

14

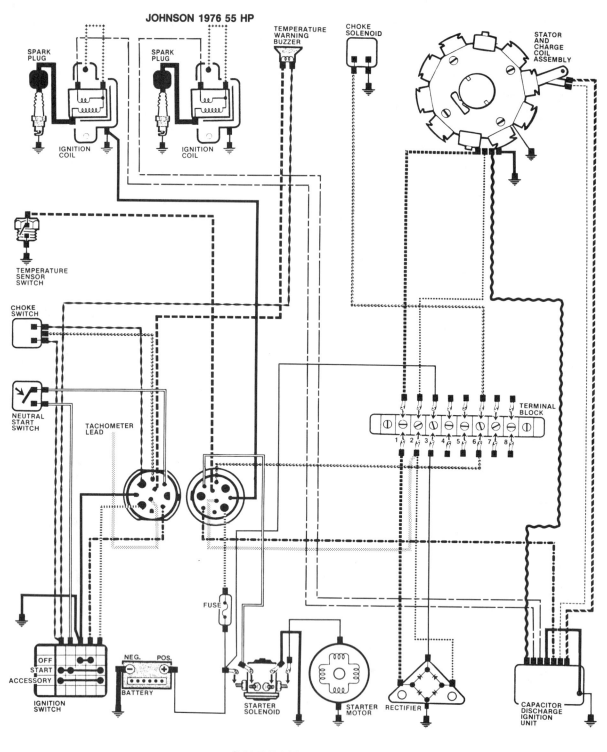

WIRING DIAGRAMS

515

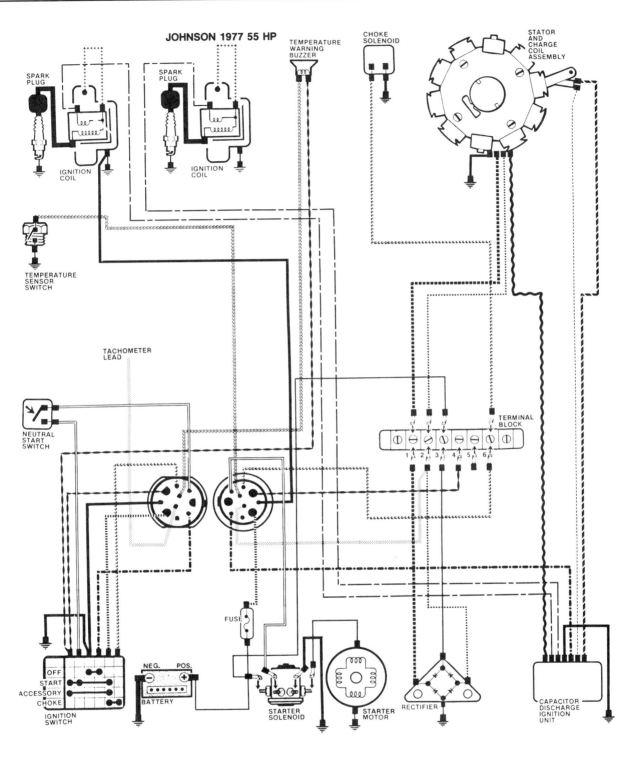

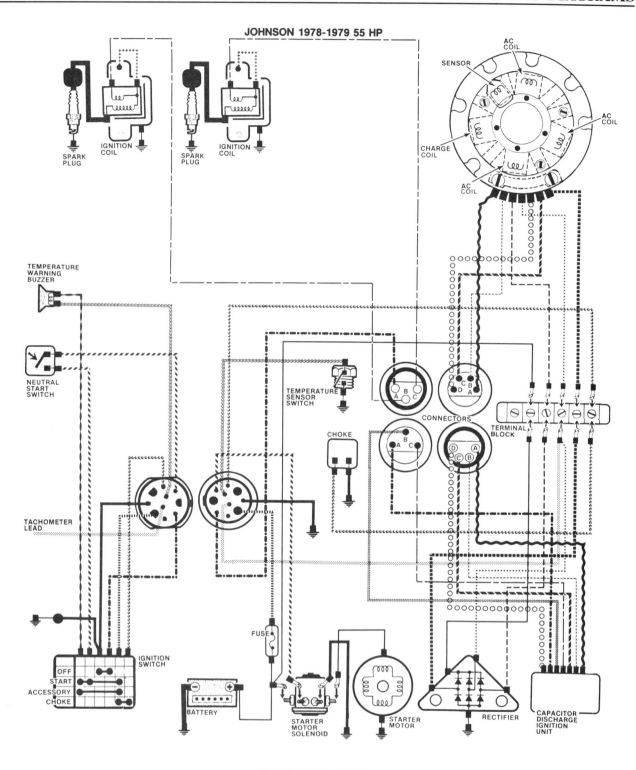

Wiring Diagrams

517

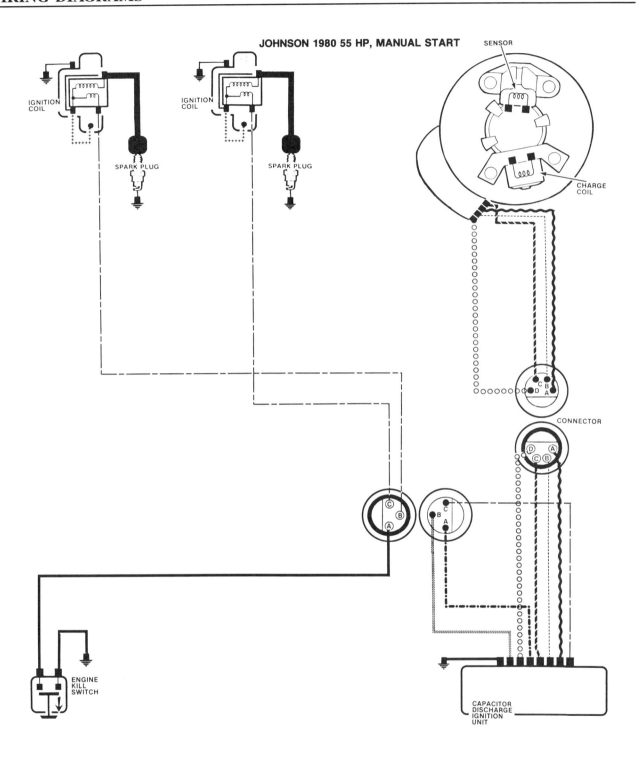

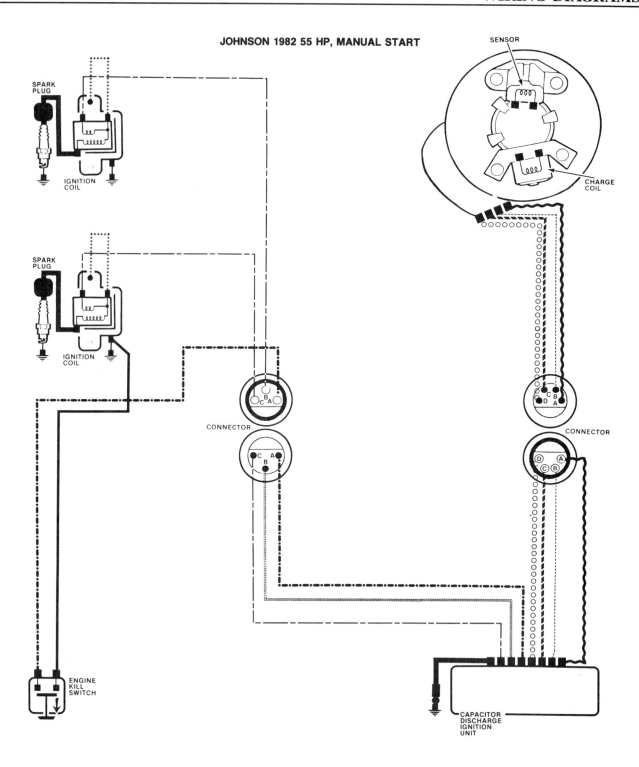

Wiring Diagrams

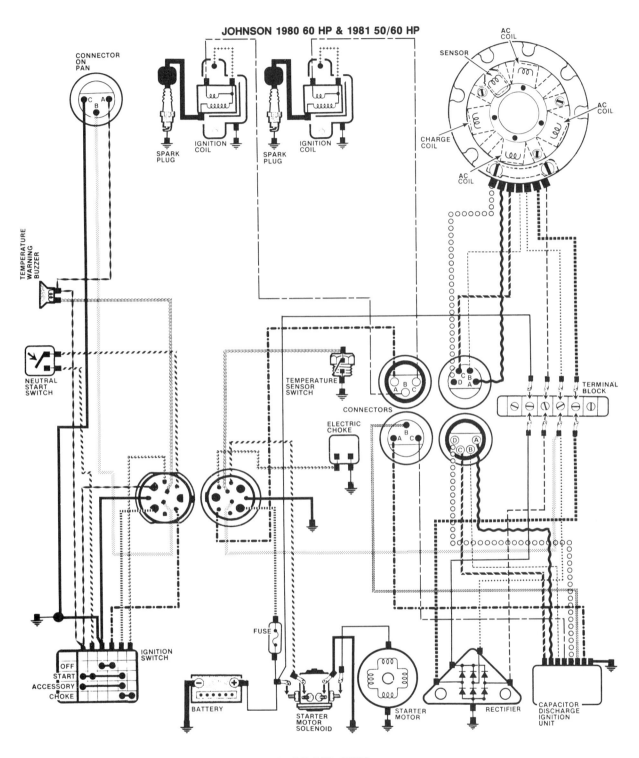

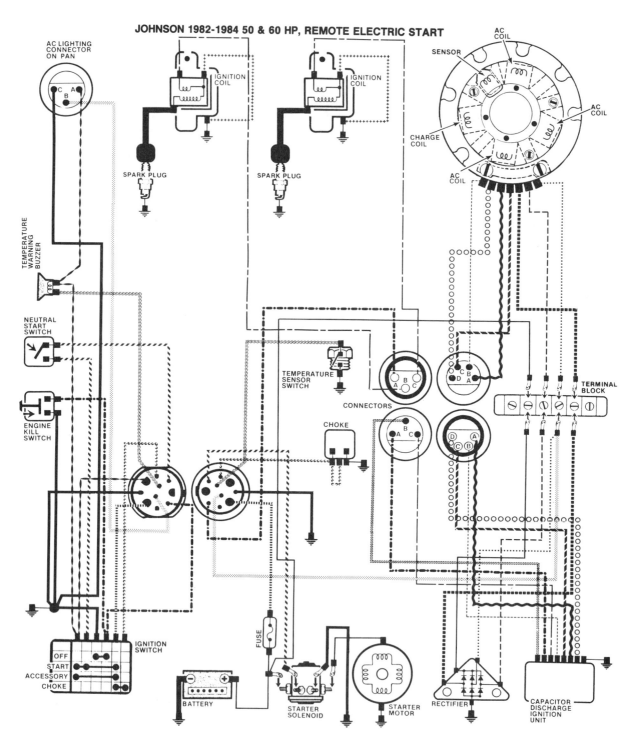

Wiring Diagrams

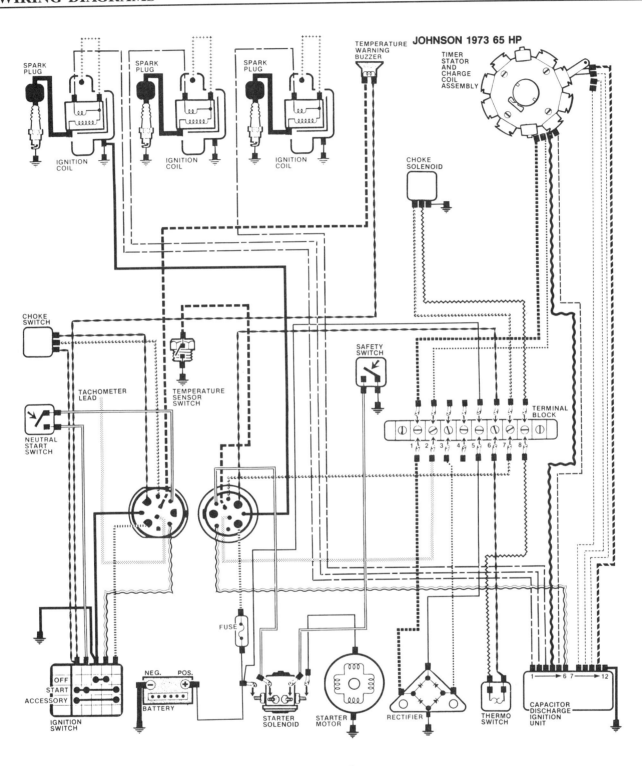

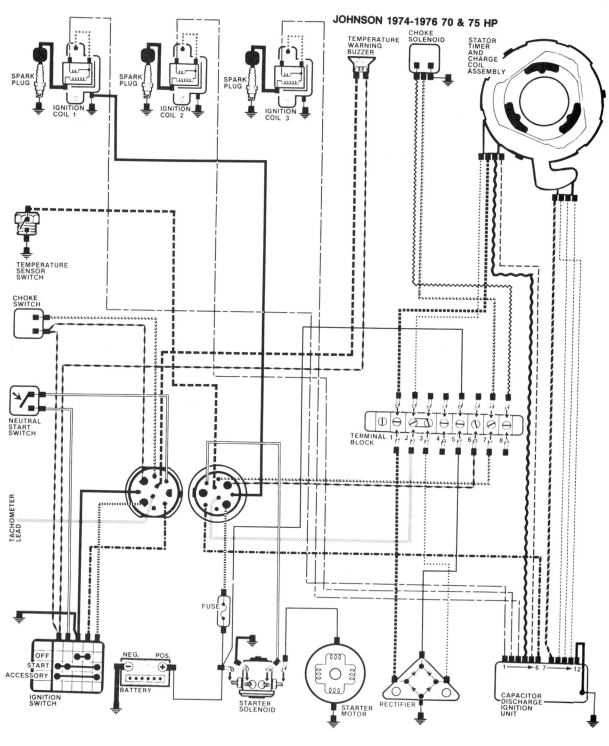

Wiring Diagrams

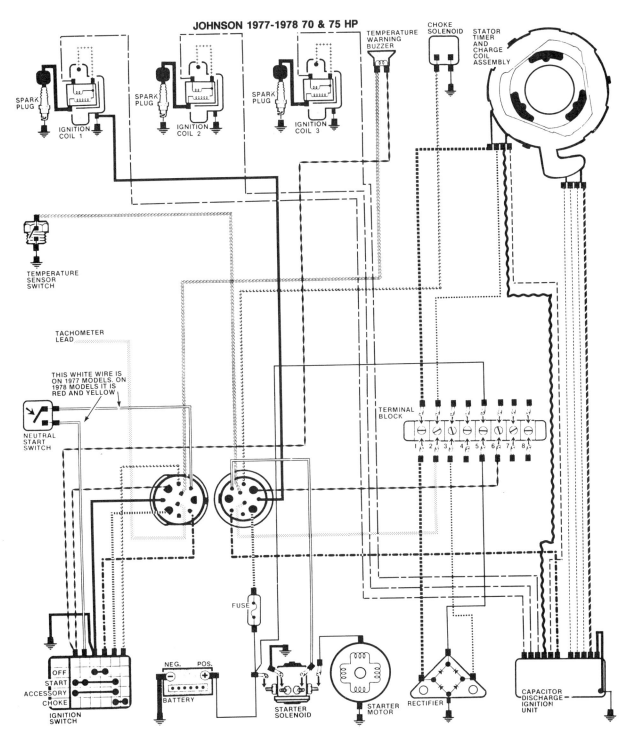

WIRING DIAGRAMS

524

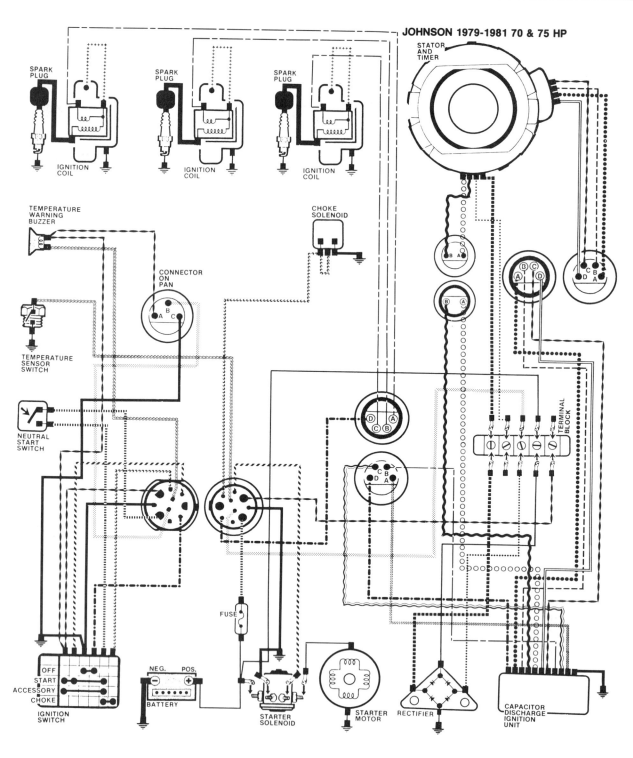

WIRING DIAGRAMS

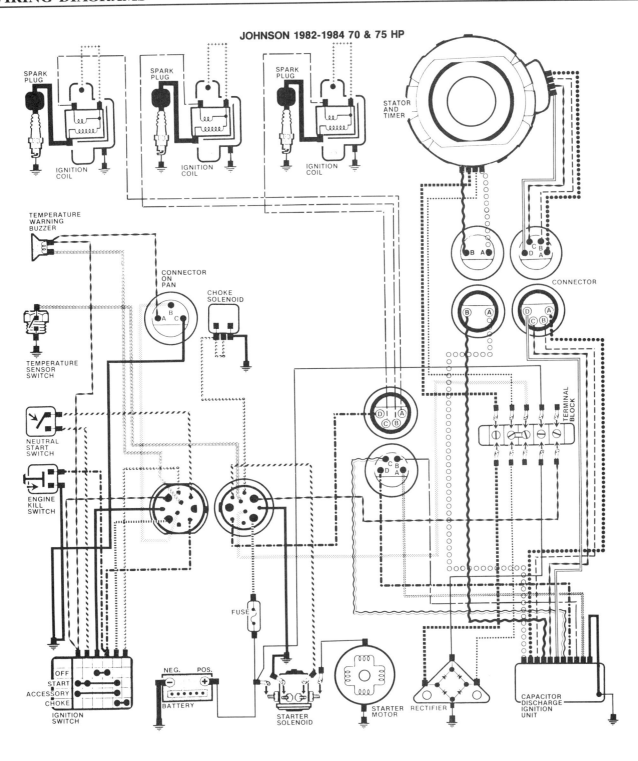

WIRING DIAGRAMS

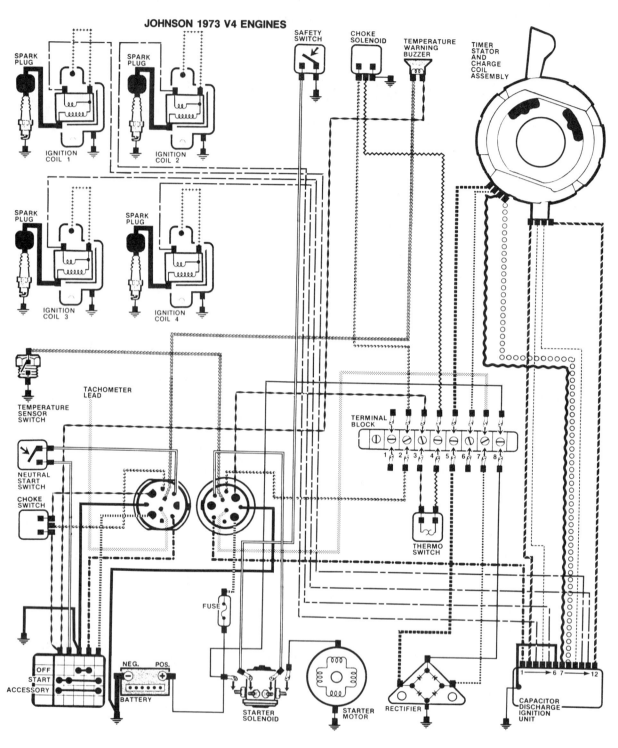

WIRING DIAGRAMS

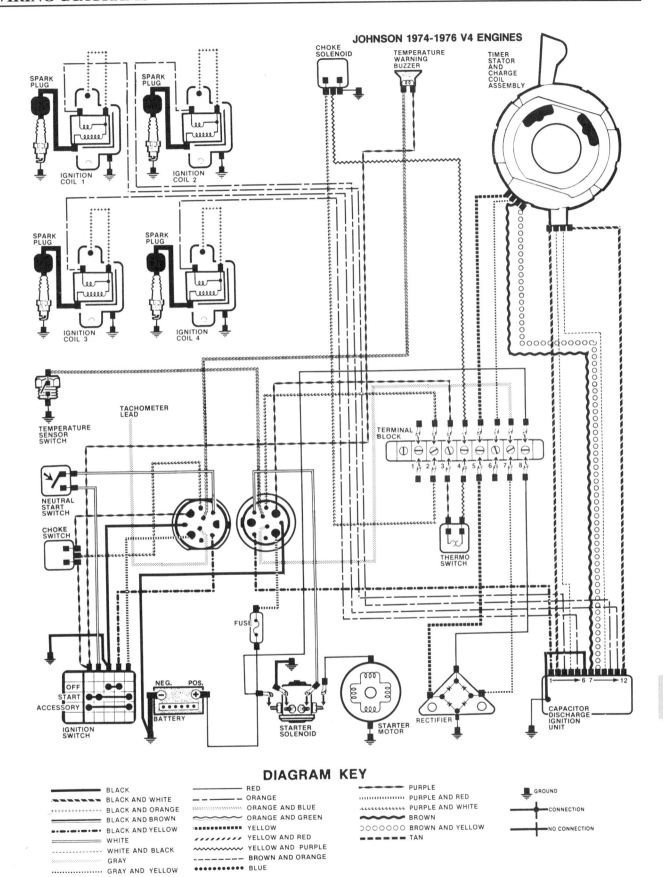

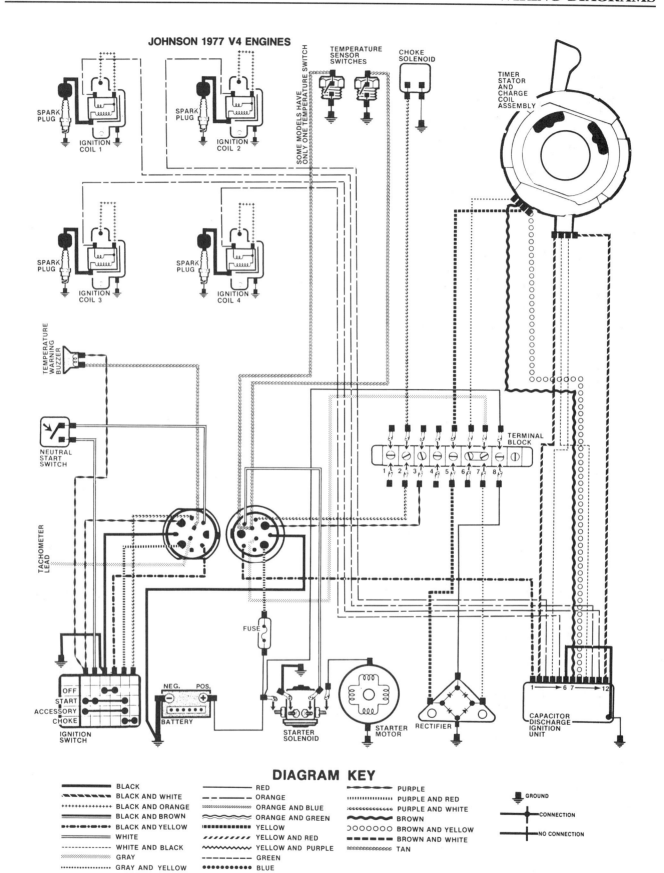

WIRING DIAGRAMS

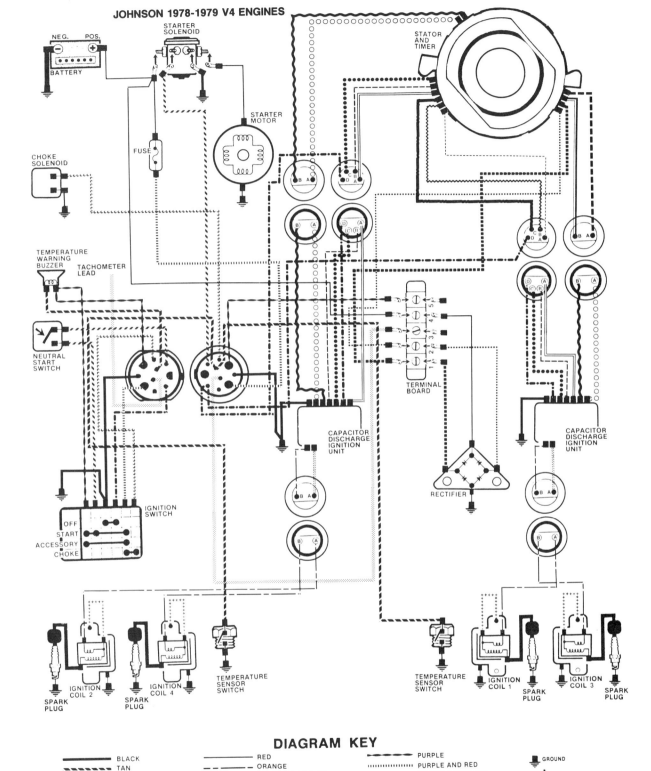

WIRING DIAGRAMS

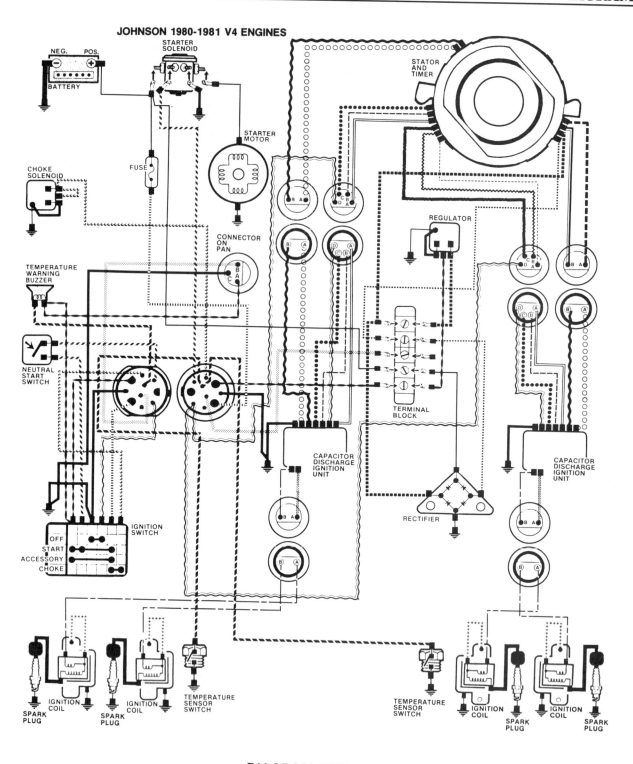

JOHNSON 1980-1981 V4 ENGINES

WIRING DIAGRAMS

531

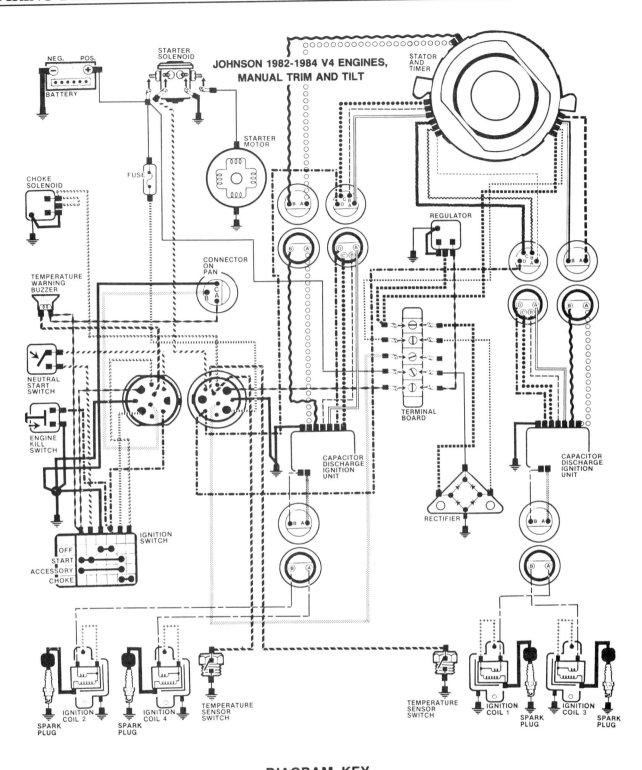

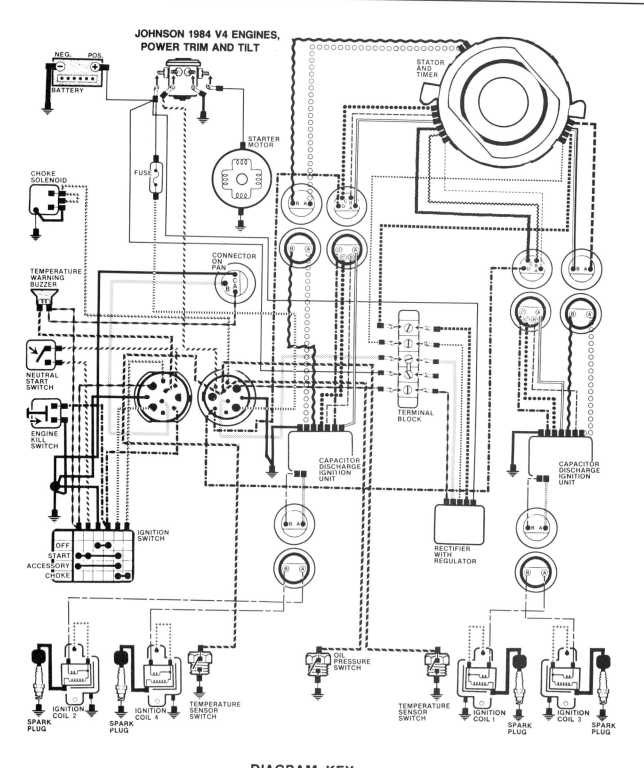

WIRING DIAGRAMS

533

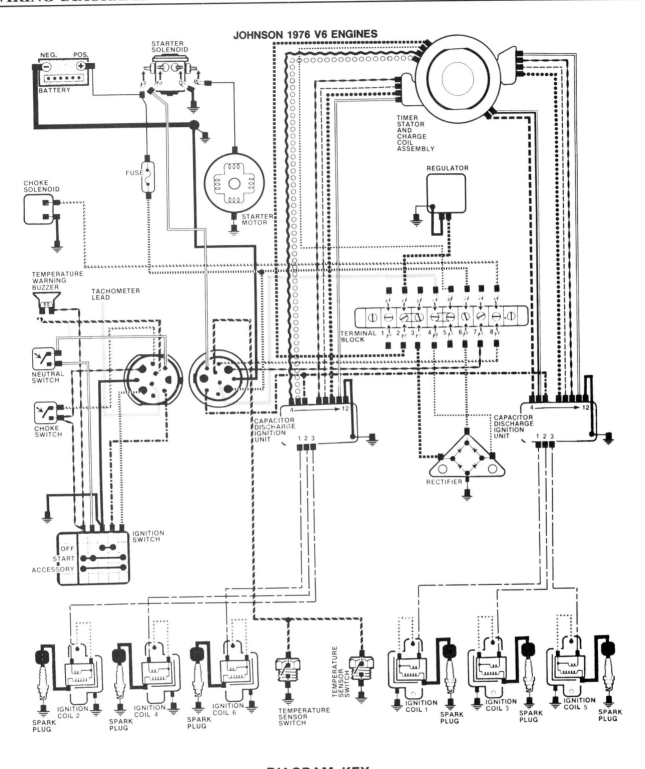

WIRING DIAGRAMS

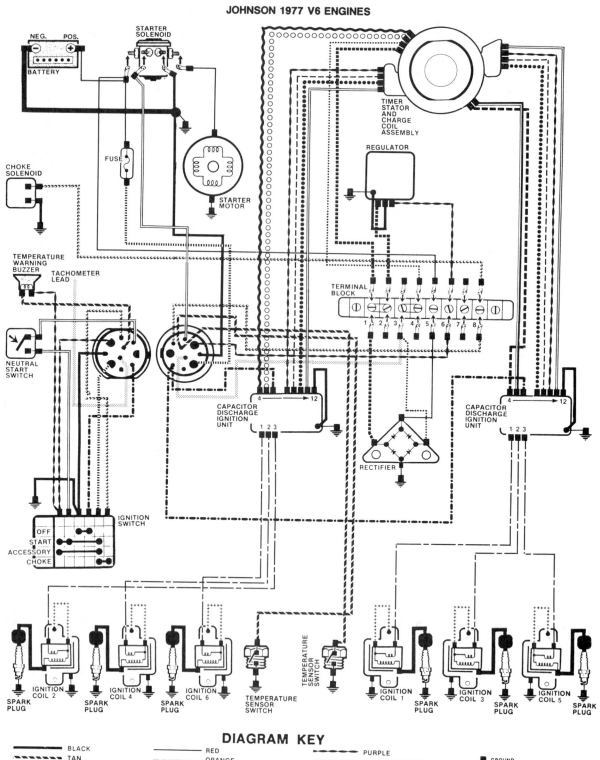

WIRING DIAGRAMS

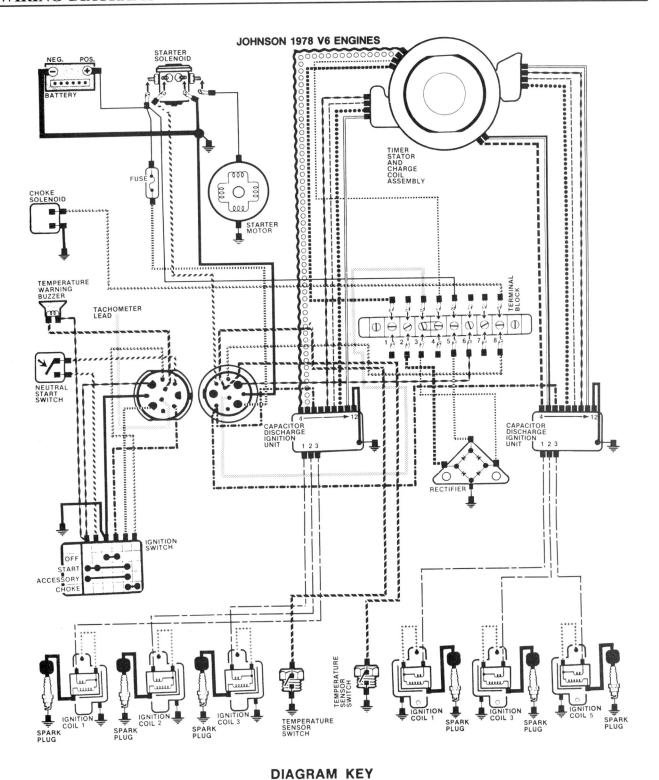

WIRING DIAGRAMS

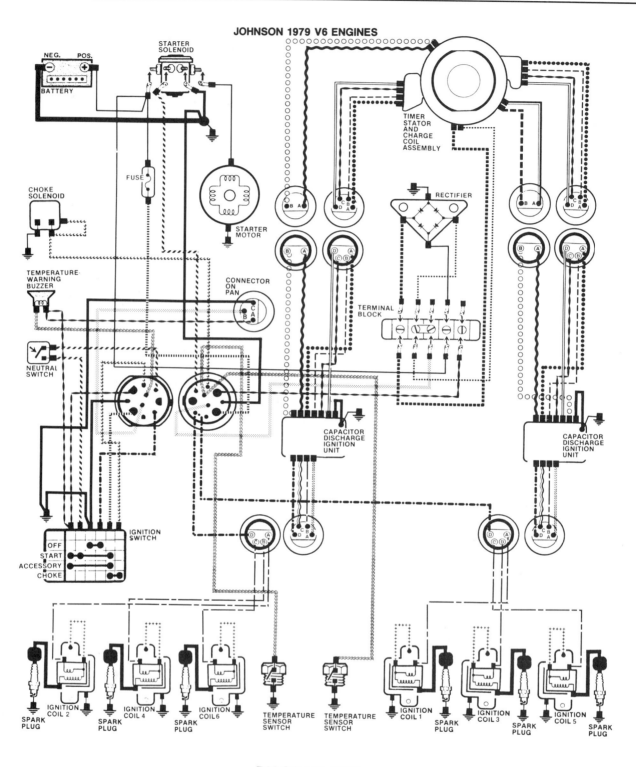

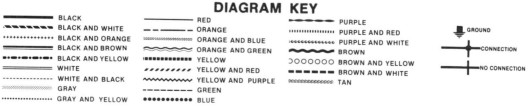

Wiring Diagrams

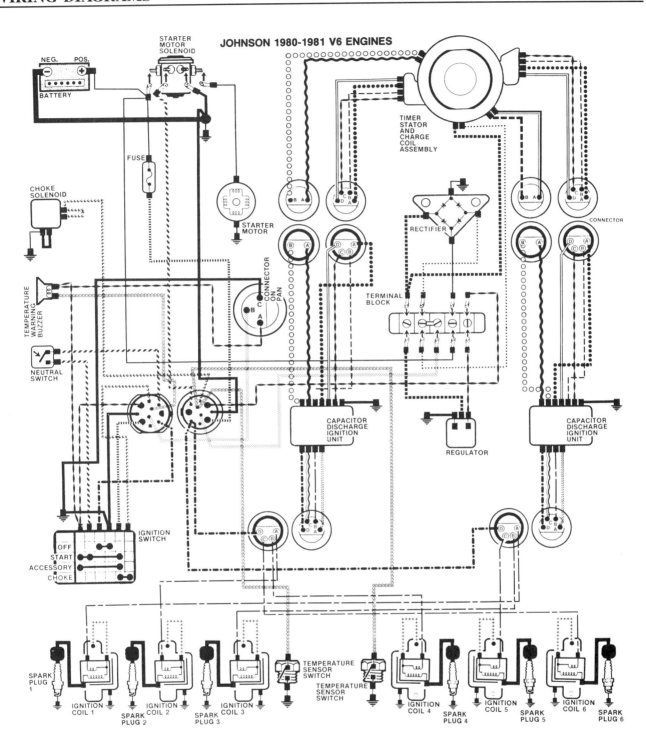

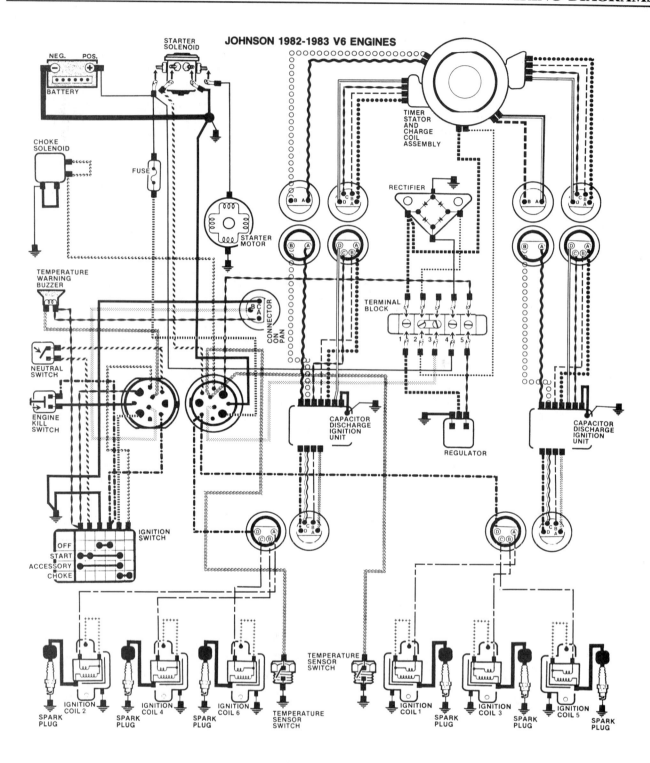
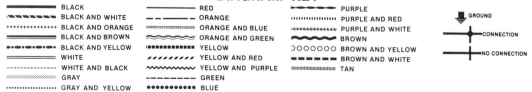

WIRING DIAGRAMS

539

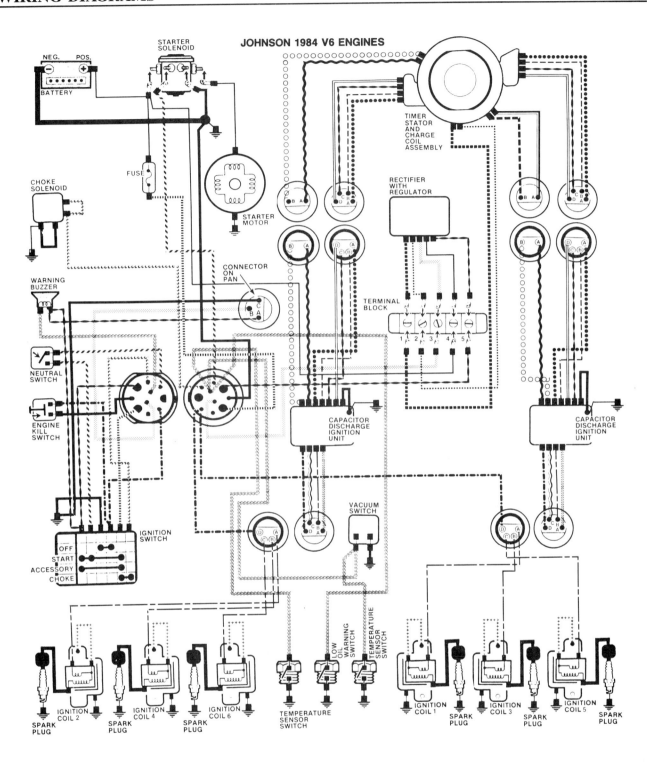

DIAGRAM KEY

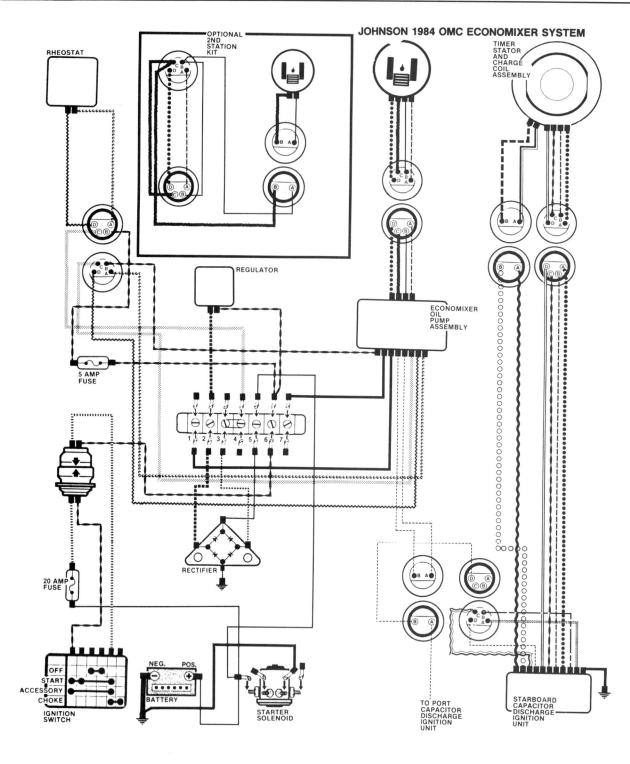

Wiring Diagrams

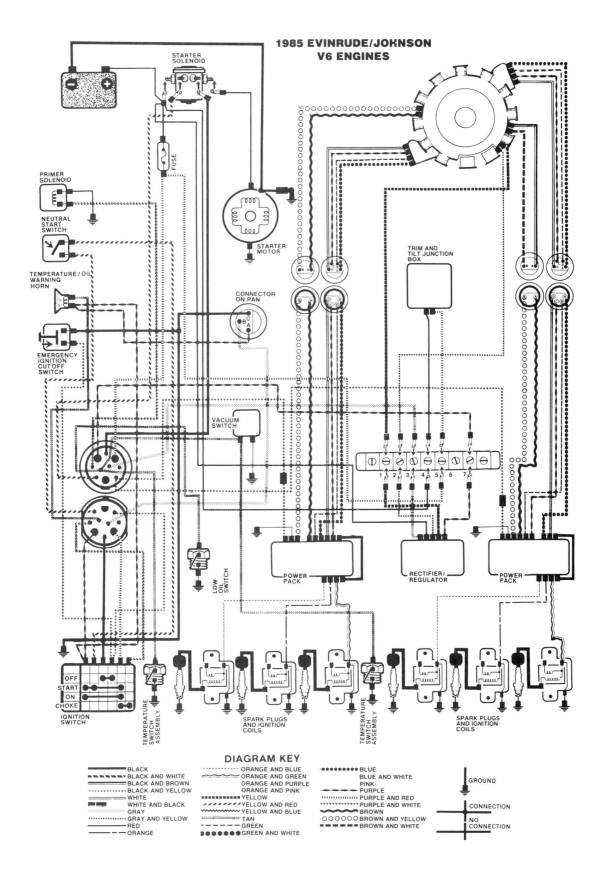

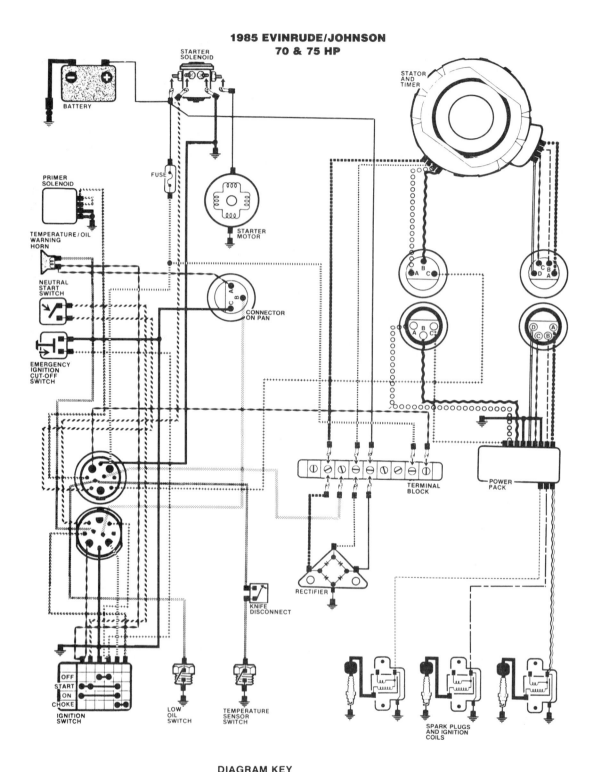

Wiring Diagrams

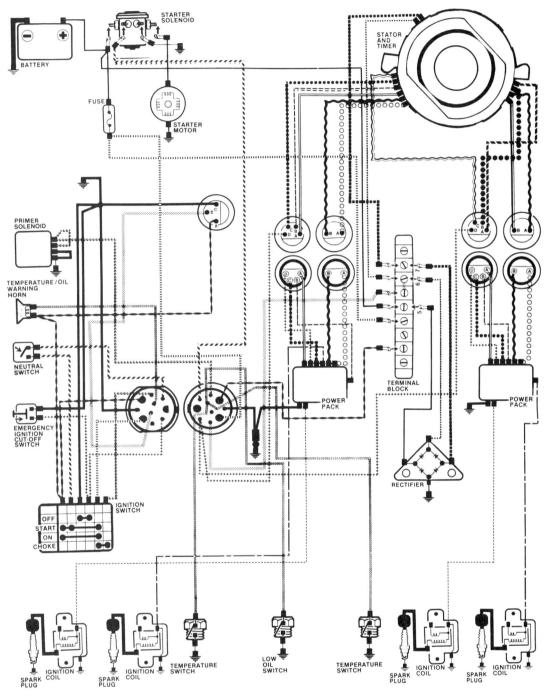

1985 EVINRUDE/JOHNSON 90 & 115 HP WITH MANUAL TILT

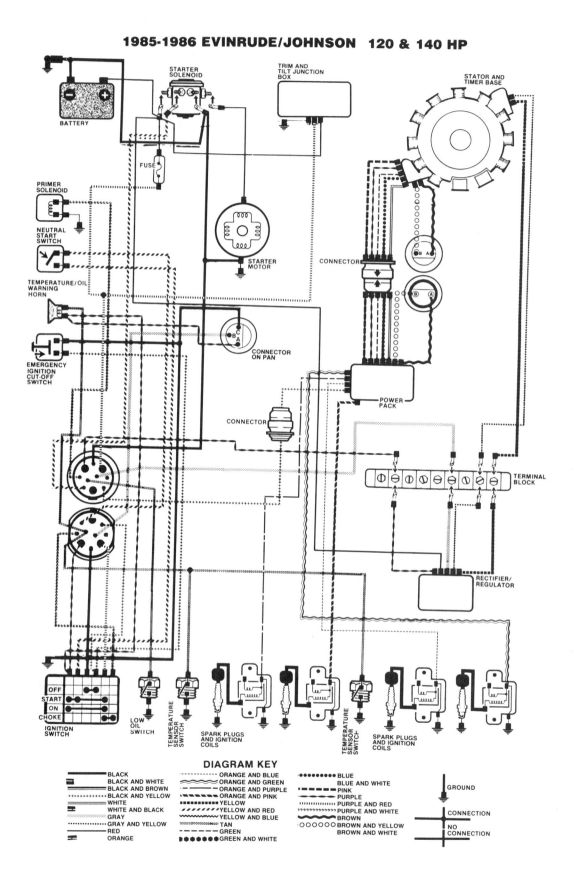

Wiring Diagrams

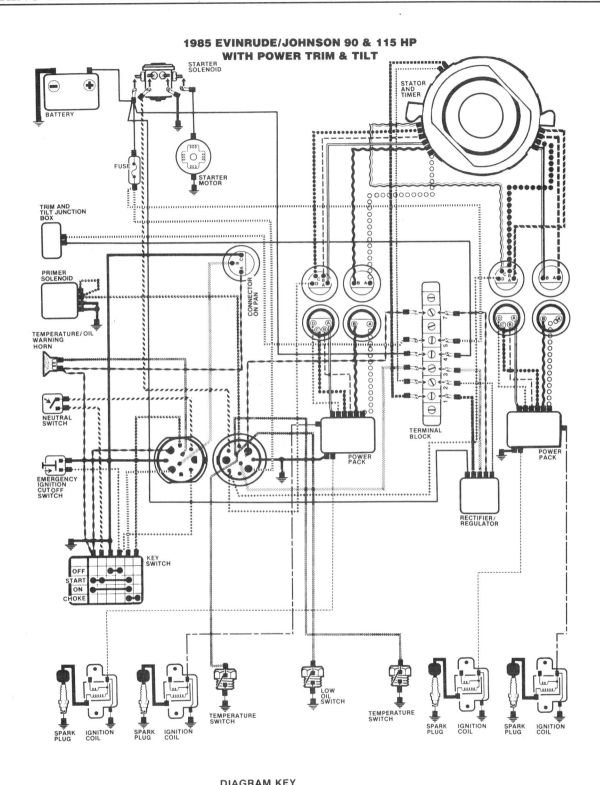

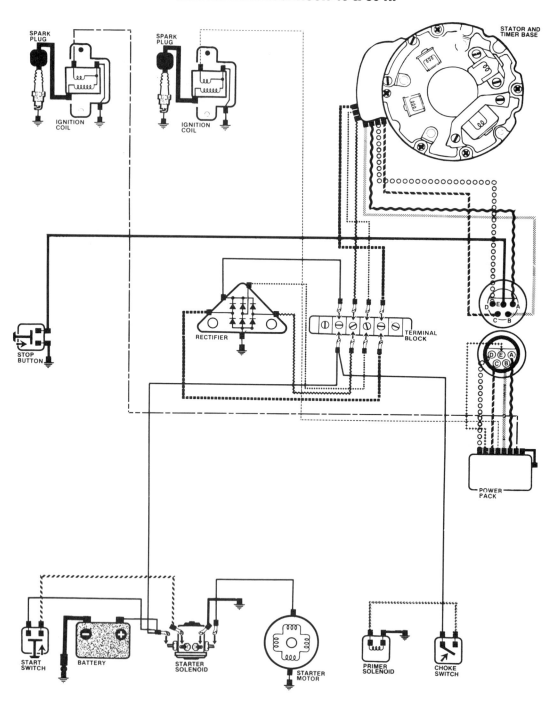

Wiring Diagrams

1986 60-75 HP WITH REMOTE ELECTRIC START

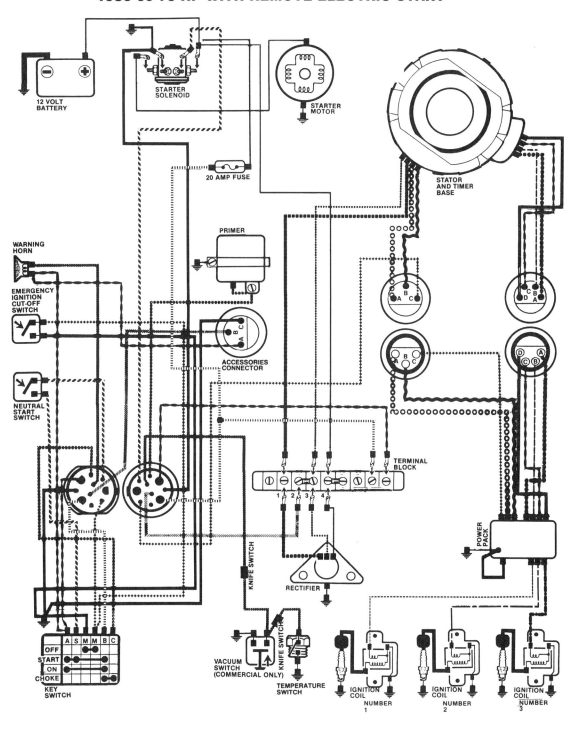

DIAGRAM KEY

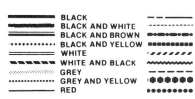
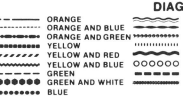

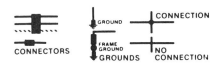

1986 90-110 WITH MANUAL TILT

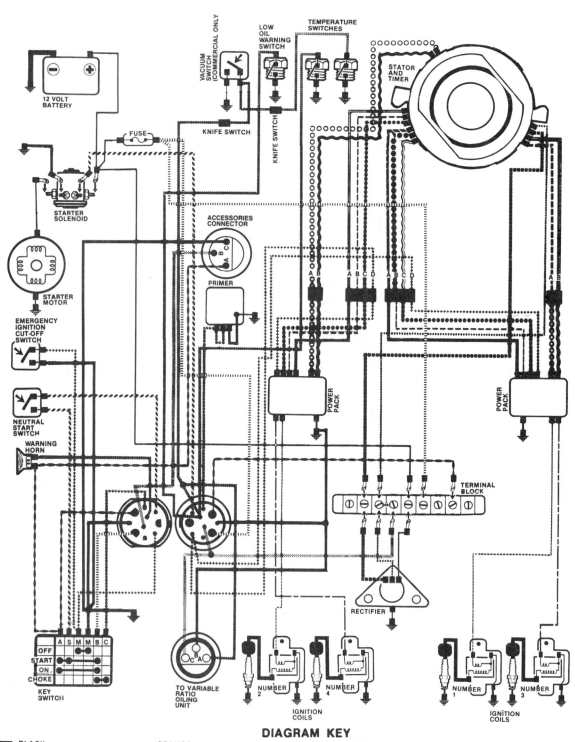

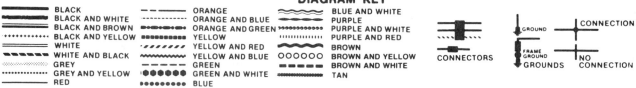

Wiring Diagrams

1986 90-110 HP WITH REMOTE ELECTRIC START

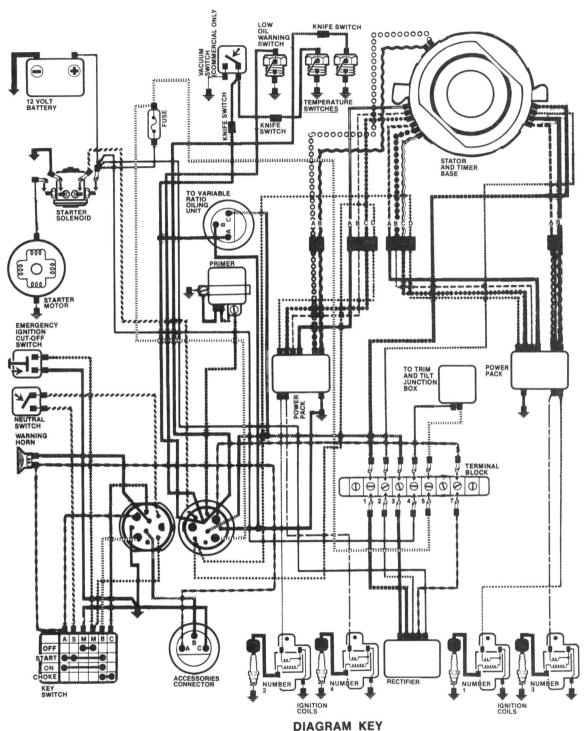

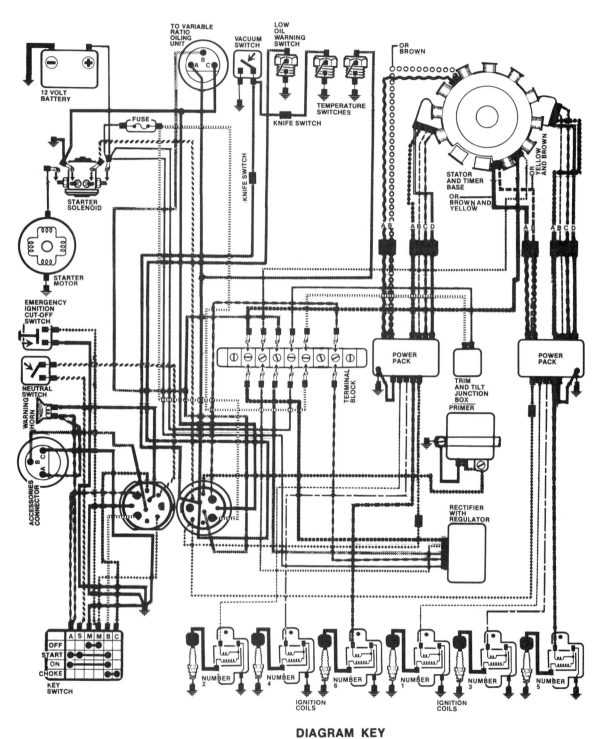

WIRING DIAGRAMS

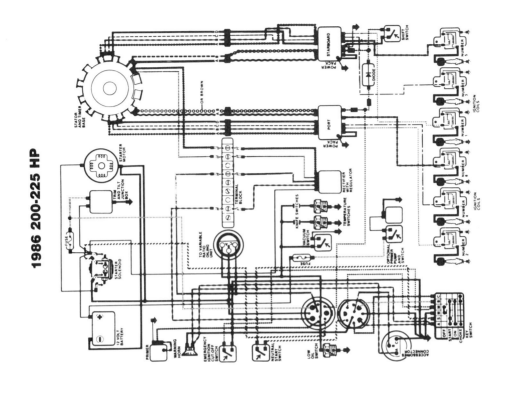

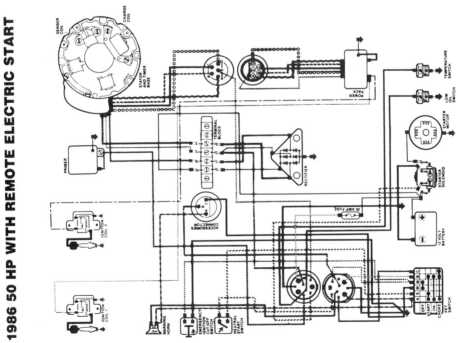

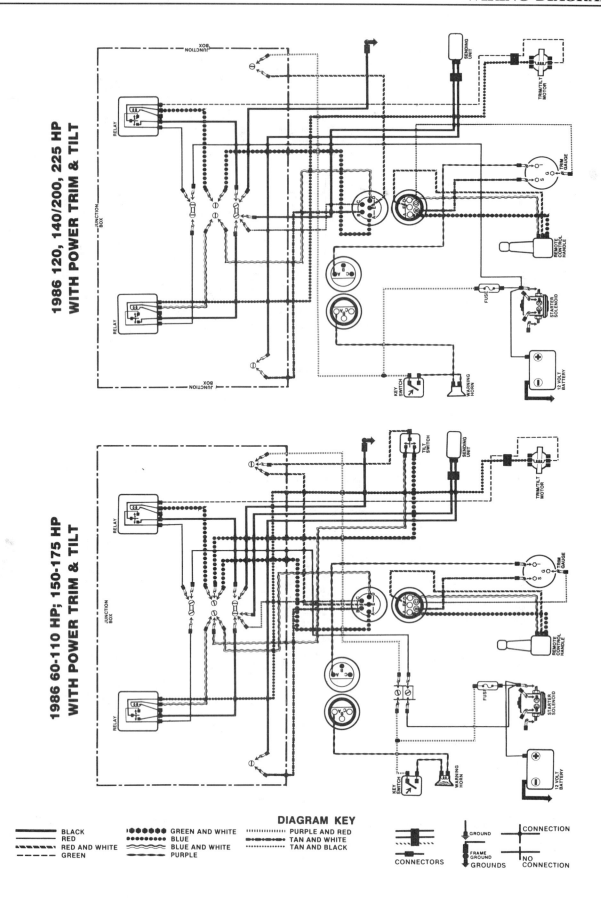

WIRING DIAGRAMS

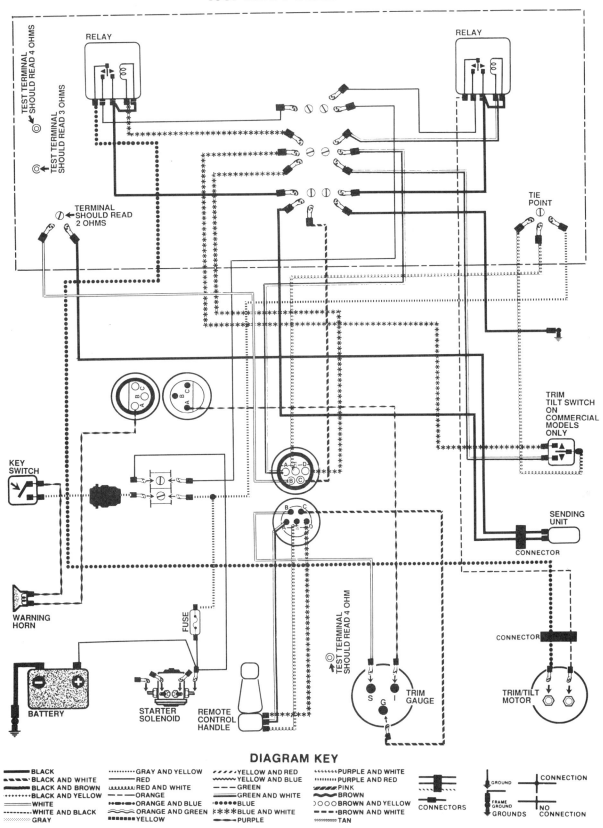

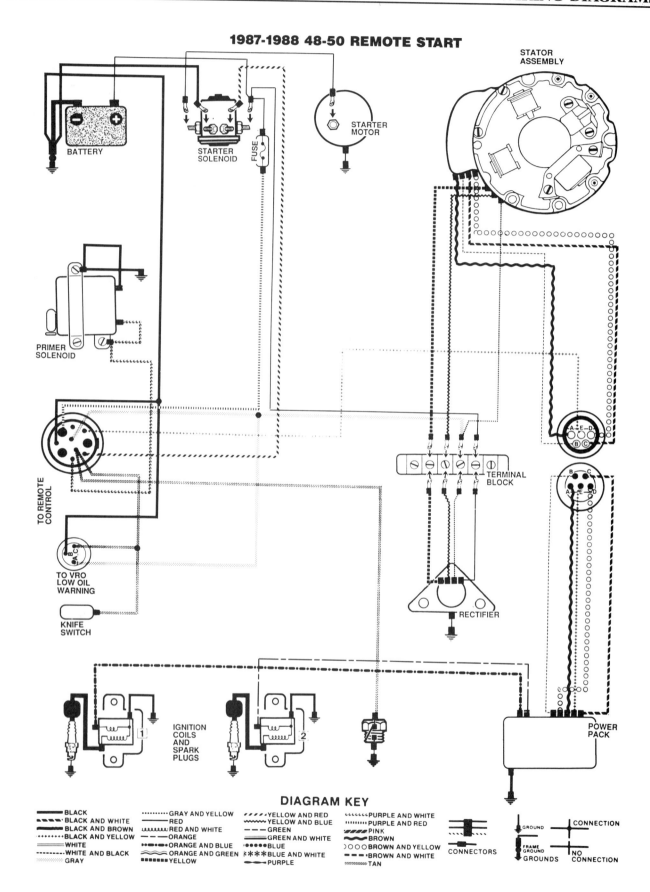

WIRING DIAGRAMS

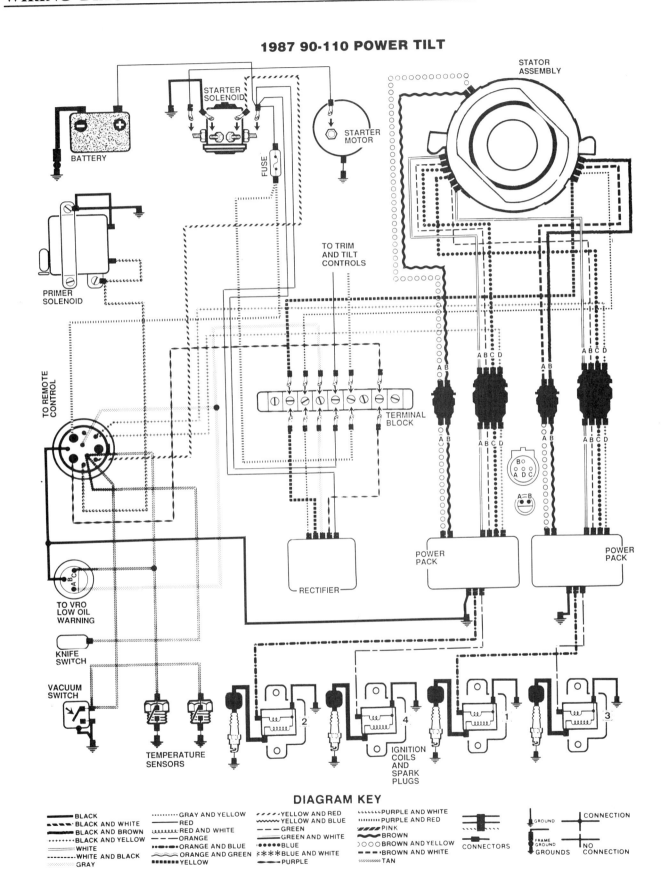

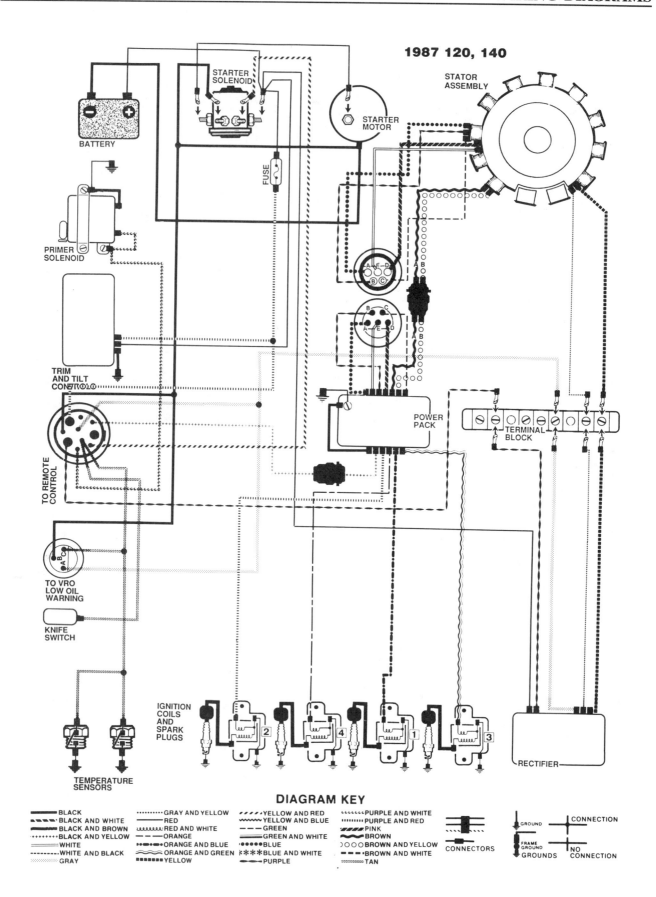

WIRING DIAGRAMS

1987-1988 150-175

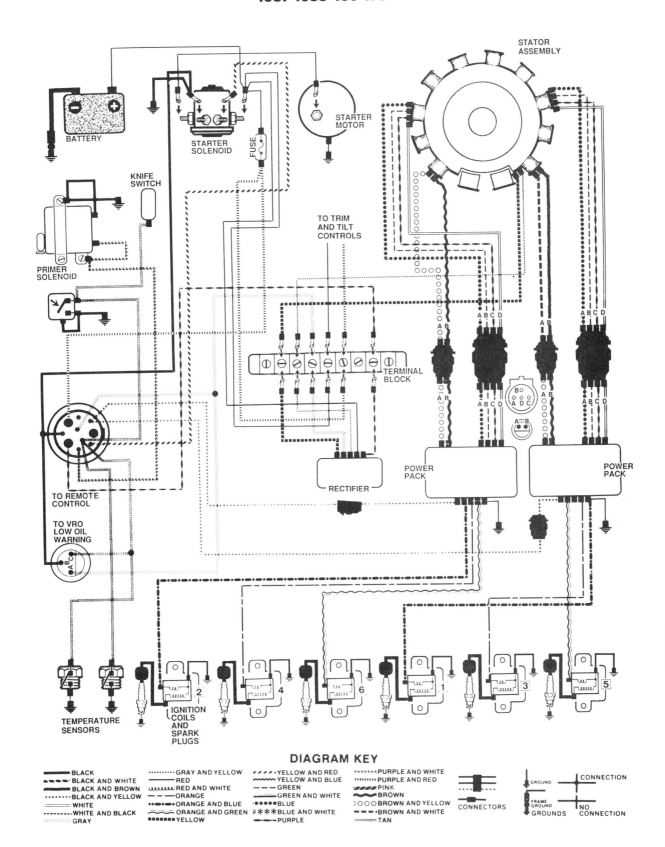

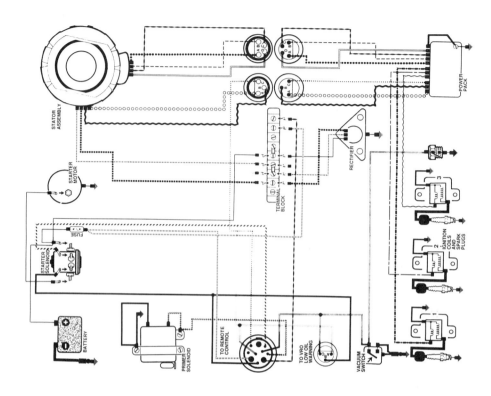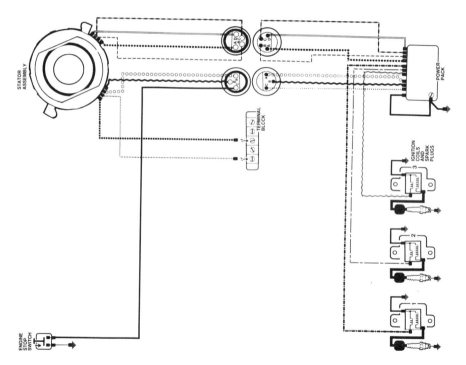

WIRING DIAGRAMS

1987 200, 225

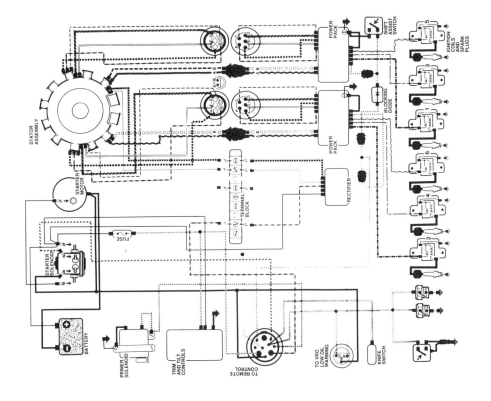

1987-1988 48-50 TILLER ELECTRIC

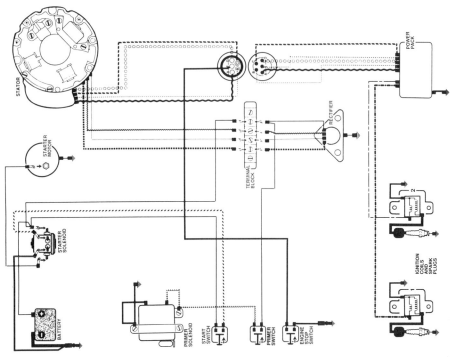

DIAGRAM KEY

BLACK	GRAY AND YELLOW	YELLOW AND RED	PURPLE AND WHITE	GROUND — CONNECTION
BLACK AND WHITE	RED	YELLOW AND BLUE	PURPLE AND RED	
BLACK AND BROWN	RED AND WHITE	GREEN	PINK	FRAME GROUND — NO CONNECTION
BLACK AND YELLOW	ORANGE	GREEN AND WHITE	BROWN	CONNECTORS
WHITE	ORANGE AND BLUE	BLUE	BROWN AND YELLOW	GROUNDS
WHITE AND BLACK	ORANGE AND GREEN	BLUE AND WHITE	BROWN AND WHITE	
GRAY	YELLOW	PURPLE	TAN	

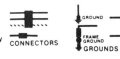

WIRING DIAGRAMS

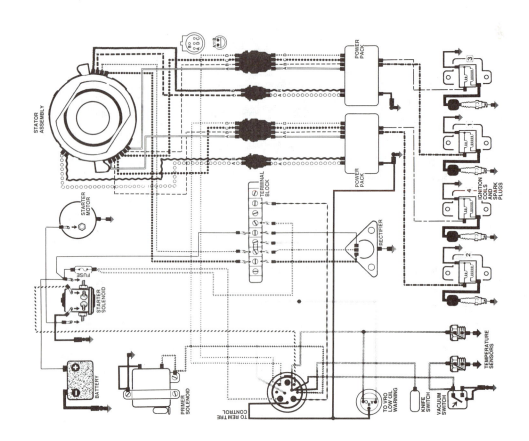

1987 88-110 MANUAL TILT

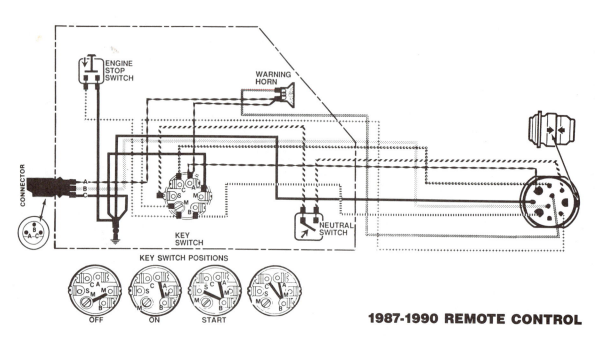

1987-1990 REMOTE CONTROL

Wiring Diagrams

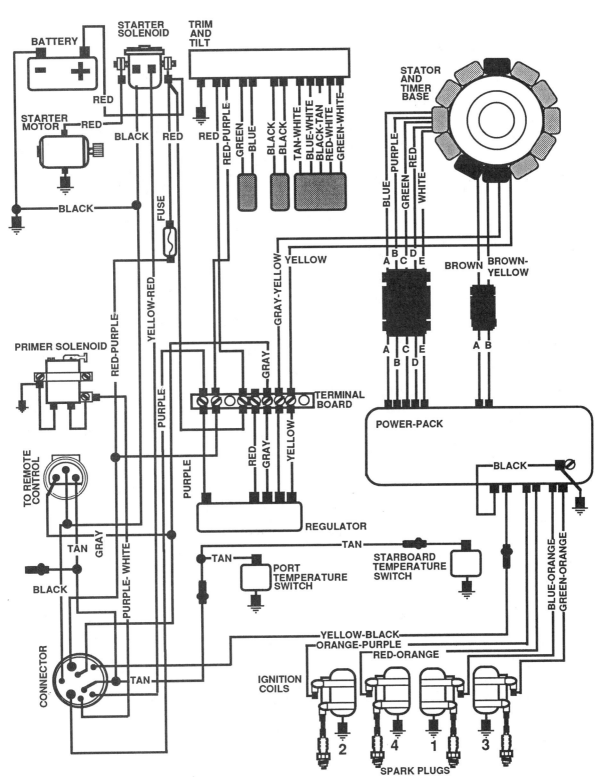

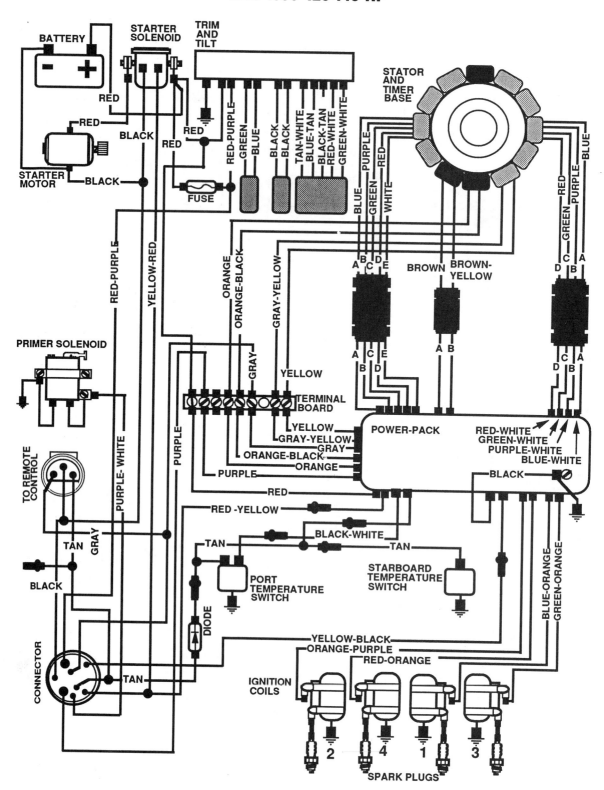

WIRING DIAGRAMS

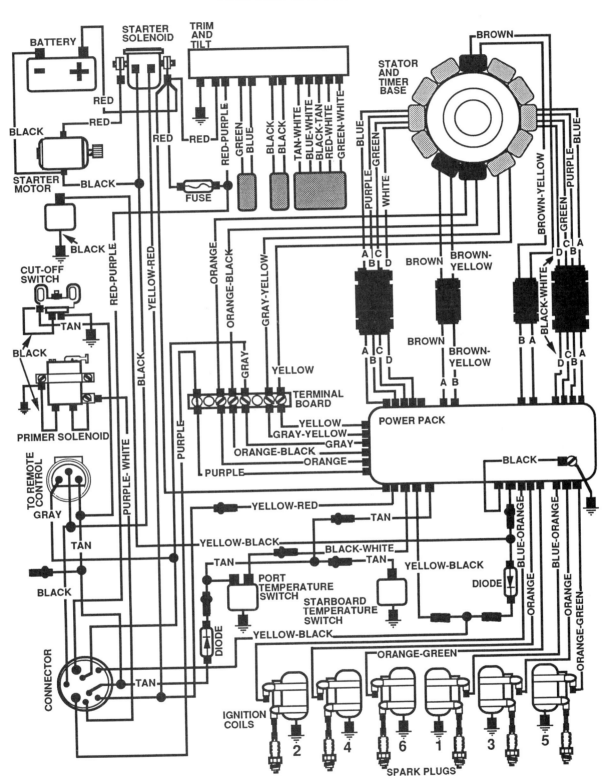

1988-1990 200-225 HP

WIRING DIAGRAMS

1989-1990 48-50 HP REMOTE START

WIRING DIAGRAMS

1989-1990 60-70 HP REMOTE START

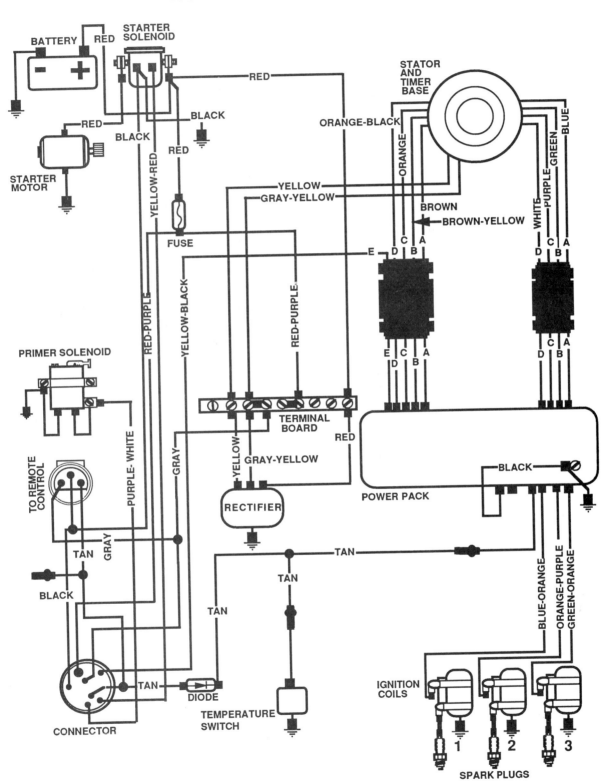

1989-1990 65 HP MANUAL START

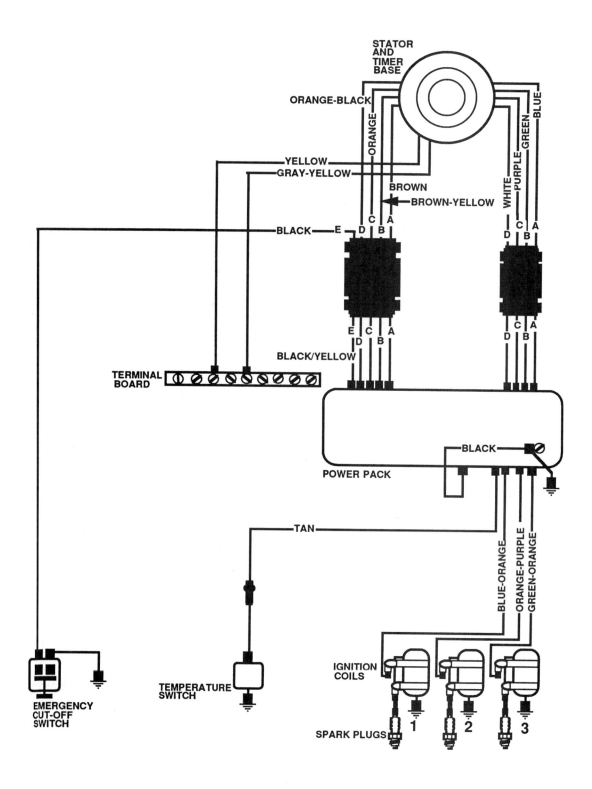

WIRING DIAGRAMS

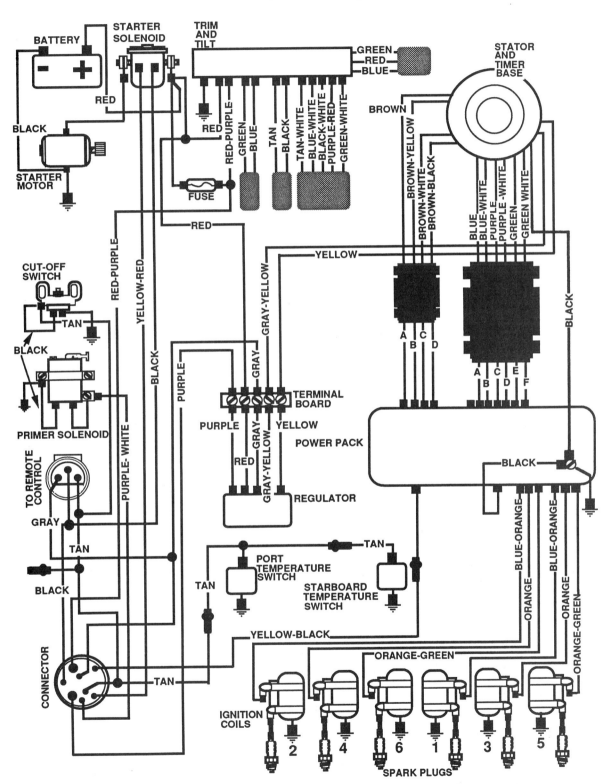

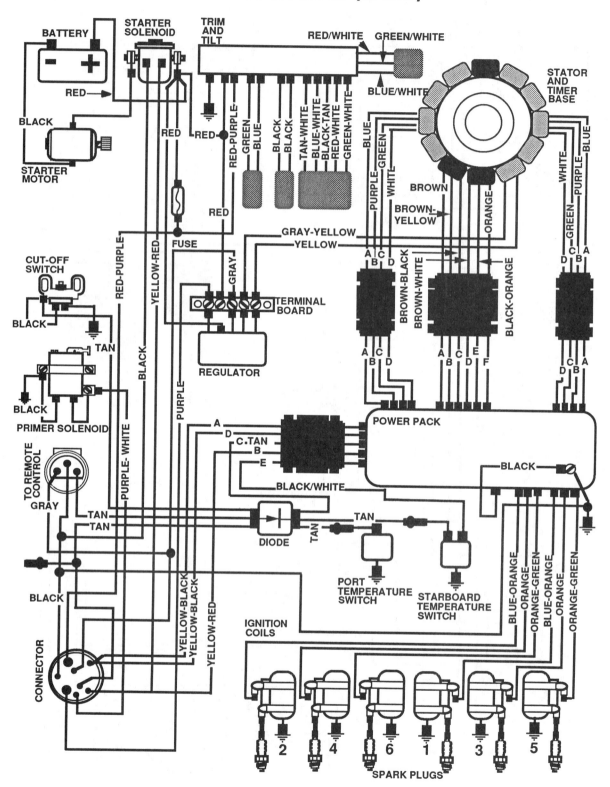

WIRING DIAGRAMS

569

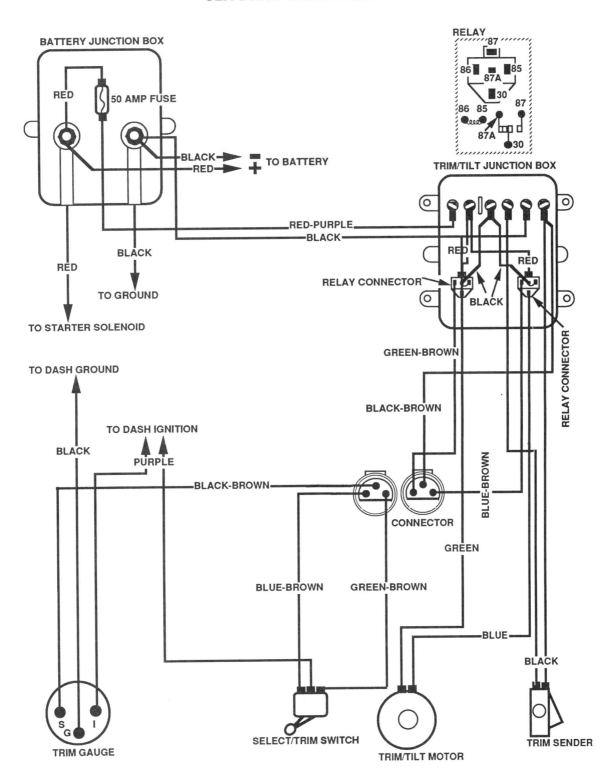

NOTES

NOTES

NOTES

NOTES

NOTES

NOTES

MAINTENANCE LOG